H. Schneider and S. Komarneni (Eds.)
Mullite

Further Titles of Interest:

Ajayan, P., Schadler, L.S., Braun, P.V.

Nanocomposite Science and Technology

2003
ISBN 3-527-30359-6

Köhler, M., Fritzsche, W.

Nanotechnology
An Introduction to Nanostructuring Techniques

2004
ISBN 3-527-30750-8

Leyens, C., Peters, M. (Eds.)

Titanium and Titanium Alloys
Fundamentals and Applications

2003
ISBN 3-527-30534-3

Lütjering, G., Albrecht, J. (Eds.)

Ti-2003 Science and Technology (5 Vols.)

2004
ISBN 3-527-30306-5

Scheffler, M., Colombo, P. (Eds.)

Cellular Ceramics
Structure, Manufacturing, Properties and Applications

2005
ISBN 3-527-31320-6

Schmidt, G. (Ed.)

Nanoparticles
From Theory to Application

2004
ISBN 3-527-30507-6

Hartmut Schneider and Sridhar Komarneni (Eds.)

Mullite

WILEY-VCH Verlag GmbH & Co. KGaA

Prof. Hartmut Schneider
Institute of Materials Research
German Aerospace Center (DLR)
Linder Höhe
51147 Köln
Germany

Prof. Sridhar Komarneni
205 Materials Research Laboratory
The Pennsylvania State University
University Park, PA 16802
USA

■ All books published by Wiley-VCH are carefully produced. Nevertheless, authors, editors, and publisher do not warrant the information contained in these books, including this book, to be free of errors. Readers are advised to keep in mind that statements, data, illustrations, procedural details or other items may inadvertently be inaccurate.

Library of Congress Card No. applied for.

British Library Cataloguing-in-Publication Data: A catalogue record for this book is available from the British Library

Die Deutsche Bibliothek –
CIP Cataloguing-in-Publication-Data:
A catalogue record for this book is available from Die Deutsche Bibliothek.

© 2005 WILEY-VCH Verlag GmbH & Co. KGaA, Weinheim

Composition: Typomedia GmbH, Ostfildern
Printing: Strauss GmbH, Mörlenbach
Bookbinding: Litges & Dopf Buchbinderei GmbH, Heppenheim
Cover Design: Gunther Schulz, Fußgönheim

Printed in the Federal Republic of Germany.

ISBN-13:   978-3-527-30974-0
ISBN-10:   3-527-30974-8

Hartmut Schneider dedicates this book to his daughters Heike and Silke and to his granddaughter Ella for their love

Sridhar Komarneni dedicates this book to his wife Sreedevi Komarneni, M.D., and his son, Jayanth S. Komarneni for their love and support.

Table of Contents

General introduction   *XV*

1   **Crystal Chemistry of Mullite and Related Phases**   *1*
1.1   The Mullite-type Family of Crystal Structures   *1*
            R.X. Fischer, H. Schneider
1.1.1   Introductory Remarks   *1*
1.1.2   The Derivation of Mullite-type Crystal Structures   *4*
1.1.3   Members of the Mullite-type Family of Crystal Structures   *12*
1.1.3.1   Introduction   *12*
1.1.3.2   MUL-II.1, $P4_2/mbc$: Schafarzikite Group   *13*
1.1.3.3   MUL-VI.11, $P4_2/mbc$: Apuanite Group   *15*
1.1.3.4   MUL-IV.12, $Pbam$: $Bi_2M_4O_9$ Group   *16*
1.1.3.5   MUL-VIII.12, $Pbam$: Versiliaite Group   *18*
1.1.3.6   MUL-VIII.2, $Pbnm$: Grandidierite Group   *19*
1.1.3.7   MUL-II.3, $Pbam$: Mullite Group   *20*
1.1.3.8   MUL-IV.31, $Pnnm$: Andalusite Group   *33*
1.1.3.9   MUL-VIII.31, $P2_1/n11$: Olivenite Group   *35*
1.1.3.10   MUL-IV.32, $Pbnm$: Sillimanite Group   *35*
1.1.3.11   MUL-VIII.33, $A2_1am$: A_9B_2 Boron Aluminates   *37*
1.1.3.12   MUL-IV.34, $P2_12_12$: $Al_5Ge_{0.972}Pb_{0.2}O_{9.71}$ Phase   *40*
1.1.3.13   MUL-VIII.34, $P2_12_12_1$: Mozartite Group   *40*
1.1.3.14   MUL-XVI.351, $A112/m$: Boralsilite Group   *42*
1.1.3.15   MUL-XXXII.352, $P\bar{1}$: Werdingite Group   *46*
1.2   The Real Structure of Mullite   *46*
            S. Rahman, S. Freimann
1.2.1   Introduction   *46*
1.2.2   High-resolution Electron Microscopy   *47*
1.2.3   X-ray Investigation   *54*
1.2.4   Real-structure Determination Using Videographic Reconstruction and
            Simulation Techniques   *56*
1.2.4.1   The Videographic Method   *57*
1.2.4.2   Structure Variants of Mullite   *60*
1.2.4.3   Two-dimensional Videographic Reconstructions   *62*

1.2.4.4   Three-dimensional Videographic Simulations
           for 2/1-and 3/2-mullite   *64*
1.2.4.5   Conclusions   *68*
1.3   Foreign Cation Incorporation in Mullite   *70*
           H. Schneider
1.3.1   Transition Metal Incorporation   *70*
1.3.1.1   Titanium Incorporation   *74*
1.3.1.2   Vanadium Incorporation   *76*
1.3.1.3   Chromium Incorporation   *77*
1.3.1.4   Manganese Incorporation   *81*
1.3.1.5   Iron Incorporation   *83*
1.3.1.6   Cobalt Incorporation   *89*
1.3.1.7   General Remarks on Transition Metal Incorporation   *89*
1.3.2   Other Foreign Cation Incorporation   *90*
1.4   Mullite-type Gels and Glasses   *93*
           M. Schmücker, H. Schneider
1.4.1   Type I (Single Phase) Mullite Precursors and Glasses   *94*
1.4.1.1   Preparation of Type I Mullite Precursors and Glasses   *94*
1.4.1.2   Temperature-induced Structural Evolution of Type I
           Mullite Precursors and Glasses   *97*
1.4.1.3   Mechanisms of Mullite formation From Type I Precursors and
           Glasses   *103*
1.4.2   Type II (Diphasic) Mullite Precursors   *105*
1.4.2.1   Synthesis of Type II Mullite Precursors   *105*
1.4.2.2   Temperature-induced Structural Evolution of Type II
           Mullite Precursors   *106*
1.4.3   Type III (Single Phase/Diphasic) Mullite Precursors   *114*
1.4.3.1   Synthesis of Type III Mullite Precursors   *115*
1.4.3.2   Temperature-induced Structural Evolution of Type III Mullite
           Precursors   *115*
1.4.3.3   Mechanisms of Mullite Formation From Type III Mullite
           Precursors   *117*
1.4.4   General Remarks on the Structure and Crystallization Behavior of
           Mullite Precursors and Glasses   *117*
1.4.4.1   Mullite Precursors: Similarities and Differences   *117*
1.4.4.2   The Coordination of Aluminum in Mullite Precursors and
           Glasses   *120*
1.4.4.3   The Origins of Mullite Crystallization   *125*
           References   *128*

2   **Basic Properties of Mullite**   *141*
2.1   Mechanical Properties of Mullite   *141*
           H. Schneider
2.1.1   Strength, Toughness and Creep   *141*
2.1.2   Elastic Moduli and Compressibility   *142*

2.1.3   Microhardness of Mullite   *146*
2.1.4   Mechanical Response to Dynamic Stress   *149*
2.2   Thermal Properties of Mullite   *149*
           H. Schneider
2.2.1   Thermochemical Data   *149*
2.2.1.1   Enthalpy, Gibbs Energy and Entropy   *149*
2.2.1.2   Heat Capacity   *152*
2.2.2   Thermal Expansion   *152*
2.2.3   Thermal Conductivity   *155*
2.2.4   Atomic Diffusion   *156*
2.2.4.1   Oxygen Diffusion   *156*
2.2.4.2   Silicon Diffusion   *158*
2.2.4.3   Aluminum Diffusion   *159*
2.2.5   Grain Growth   *160*
2.2.6   Wetting Behavior   *162*
2.3   Miscellaneous Properties   *164*
           H. Schneider
2.3.1   Optical and Infrared Properties   *164*
2.3.2   Electrical Properties   *165*
2.4   Structure-controlled Formation and Decomposition of Mullite   *167*
           H. Schneider, M. Schmücker
2.4.1   Temperature-induced Formation   *167*
2.4.1.1   Formation from Kaolinite and Related Minerals   *167*
2.4.1.2   Formation from Andalusite and Sillimanite   *172*
2.4.1.3   Formation from X-sialon   *178*
2.4.1.4   General Remarks   *178*
2.4.2   Pressure-induced Decomposition   *179*
2.4.2.1   Decomposition to Sillimanite   *179*
2.4.2.2   Decomposition to γ-Alumina plus silica   *180*
2.5   Mullite-mullite Phase Transformations   *180*
           H. Schneider, M. Schmücker
2.5.1   Compositional Transformations   *180*
2.5.2   Structural Transformations   *186*
2.5.2.1   Transformation at about 450 °C   *186*
2.5.2.2   Transformation above 1000 °C   *186*
2.6   Spectroscopy of Mullite and Compounds with Mullite-related
           Structures   *189*
           K. J. D. MacKenzie
2.6.1   Solid-state Nuclear Magnetic Resonance (NMR) Spectroscopy   *190*
2.6.1.1   Brief Principles of Solid-state NMR Spectroscopy   *190*
2.6.1.2   NMR Spectroscopic Structural Studies of Aluminosilicate
           Mullite   *191*
2.6.1.3   NMR Spectroscopic Studies of Amorphous Materials of Mullite
           Composition   *195*
2.6.1.4   NMR Spectroscopic Studies of Mullite Formation from Minerals   *198*

2.6.1.5   NMR Spectroscopic Studies of Other Compounds with Mullite
          Structure   *199*
2.6.2   Electron Paramagnetic Resonance (EPR) Spectroscopy   *204*
2.6.3   Infrared (IR), Fourier-transform Infrared (FTIR) and Raman
          Spectroscopy   *207*
2.6.3.1   The IR Spectrum of Mullite   *207*
2.6.3.2   IR Spectroscopic Studies of Mullite Formation   *208*
2.6.4   Mössbauer Spectroscopy   *210*
          References   *215*

3   **Phase Equilibria and Stability of Mullite**   *227*
3.1   The Al_2O_3–SiO_2 Phase Diagram   *227*
          J.A. Pask, H. Schneider
3.1.1   Experimental Observations   *227*
3.1.2   Processing Parameters and Reaction Mechanisms   *230*
3.1.3   Solid-solution Range of Mullite   *234*
3.1.4   Melting Behavior of Mullite   *234*
3.1.5   Simulations of the Al_2O_3–SiO_2 Phase Diagram   *234*
3.1.6   General remarks   *235*
3.2   Influence of Environmental Conditions on the Stability
          of Mullite   *235*
          H. Schneider
3.2.1   Interactions with Reducing Environments   *236*
3.2.2   Interaction with Water Vapor-rich Environments   *239*
3.2.3   Interactions with Molten Sodium Salts   *241*
3.2.4   Interactions with Fluorine Salt Environments   *242*
3.3   Ternary X–Al_2O_3–SiO_2 Phase-equilibrium Diagrams   *243*
          H. Schneider
3.3.1   Alkaline Oxide–Al_2O_3–SiO_2 *243*
3.3.2   Iron Oxide–Al_2O_3–SiO_2 *244*
3.3.3   Alkaline Earth Oxide–Al_2O_3–SiO_2 *245*
3.3.4 MnO–Al_2O_3–SiO_2 *245*
3.3.5 TiO_2–Al_2O_3–SiO_2 *245*
3.4   Multicomponent Systems   *246*
          H. Schneider
          References   *246*

4   **Mullite Synthesis and Processing**   *251*
4.1   Mullite Synthesis   *251*
          S. Komarneni, H. Schneider, K. Okada
4.1.1   Solid-state-derived Mullite   *251*
4.1.1.1   Formation from Kaolinite and Related Phases   *251*
4.1.1.2   Formation from Kyanite, Andalusite and Sillimanite   *252*
4.1.1.3   Formation from Staurolite and Topaz   *252*
4.1.1.4   Reaction Sintering of Alumina and Silica   *253*

4.1.1.5   Effects of Mineralizers, Reaction Atmosphere and
           Structural Defects   *254*
4.1.1.6   Commercial Production (Sinter-mullite)   *256*
4.1.2   Liquid-state-derived Mullite   *259*
4.1.2.1   Crystal Growth Techniques   *259*
4.1.2.2   Commercial Production (Fused-mullite)   *260*
4.1.3   Solution-sol-gel-derived Mullite   *262*
4.1.3.1   Solution-plus-solution Process   *263*
4.1.3.2   Solution-plus-sol Process   *272*
4.1.3.3   Sol-plus-sol Process   *275*
4.1.4   Spray Pyrolysis Approach   *277*
4.1.5   Hydrothermally Produced Mullite   *281*
4.1.6   Vapor-state-derived Mullite   *283*
4.1.7   Mullite Produced by Miscellaneous Methods   *283*
4.1.8   General Remarks on the Different Chemical Synthesis Methods
           of Mullite   *284*
4.2   Processing of Mullite Ceramics   *286*
           S. Komarneni, H. Schneider
4.2.1   General Sintering Characteristics   *286*
4.2.2   Sintering of Powder Compacts   *288*
4.2.3   Reaction Sintering of Alumina and Silica   *291*
4.2.4   Reaction Bonding from Different Starting Materials   *298*
4.2.5   Reaction Sintering of Chemically Produced Mullite Precursors   *299*
4.2.6   Transient Viscous Sintering of Composite Powders   *306*
4.3   Mechanical Properties of Mullite Ceramics   *307*
           K. Okada, H. Schneider
4.3.1   Mechanical Strength and Fracture Toughness   *308*
4.3.1.1   Mechanical Strength and Fracture Toughness at Room
           Temperature   *308*
4.3.1.2   Mechanical Strength and Fracture Toughness at High
           Temperatures   *310*
4.3.2   Elastic Modulus   *312*
4.3.3   Hardness   *314*
4.3.4   Thermal Shock Resistance   *315*
4.3.5   Wear Resistance   *316*
4.3.6   Fatigue Behavior   *316*
4.3.7   Creep Resistance   *317*
4.4   Thermal Properties of Mullite Ceramics   *321*
           K. Okada, H. Schneider
4.4.1   Thermal Conductivity   *322*
4.4.2   Thermal Expansion   *322*
4.5   Miscellaneous Properties of Mullite Ceramics   *323*
           K. Okada, H. Schneider
4.5.1   Electrical Properties   *323*
4.5.2   Optical Properties   *324*

4.5.3   Chemical Corrosion Behavior   *324*
4.6   Application of Mullite Ceramics   *327*
          K. Okada, H. Schneider
4.6.1   Engineering Materials   *327*
4.6.1.1   Refractory Materials   *327*
4.6.1.2   High Temperature Engineering Materials   *328*
4.6.1.3   Materials for Heat Exchangers   *329*
4.6.1.4   Structural Materials   *331*
4.6.2   Electronic Packaging Materials   *331*
4.6.3   Optical Materials   *333*
4.6.4   Tribological Materials   *334*
4.6.5   Porous Materials for Filters and Catalyst Supports   *335*
4.6.6   Materials for Miscellaneous Applications   *336*
          References   *337*

5   **Mullite Coatings**   *349*
5.1   Chemical Vapor-deposited Coatings (CVD Coatings)   *350*
          S. Basu, V. K. Sarin
5.1.1   Thermodynamics of Chemical Vapor-deposited Coatings   *350*
5.1.2   Growth Kinetics of Chemical Vapor-deposited Coatings   *352*
5.1.3   Microstructure of Chemical Vapor-deposited Coatings   *353*
5.1.4   The Structure of Chemical Vapor-deposited Al_2O_3-rich Mullite   *358*
5.1.5   High Temperature Phase Transformations in Chemically
          Chemical Vapor-deposited Mullite   *359*
5.1.6   Oxidation, Hot Corrosion and Recession Protection of
          Chemical Vapor-deposited Mullite   *364*
5.2   Plasma- and Flame-sprayed Coatings   *367*
          S. Basu, V. K. Sarin
5.2.1   Microstructural Characteristics and Stability of Coatings   *367*
5.2.2   Plasma-sprayed Environmental Barrier Coatings (EBCs)   *370*
5.2.3   Plasma-sprayed Thermal Barrier Coatings (TBCs)   *372*
5.3   Deposition of Mullite Coatings by Miscellaneous Techniques   *373*
          S. Basu, V. K. Sarin
          References   *374*

6   **Mullite Fibers**   *377*
6.1   Mullite Whiskers   *377*
          H. Schneider
6.1.1   Whisker Formation From Melts   *377*
6.1.2   Whisker Formation via Gas-transport Reactions   *378*
6.2   Sol-gel-derived Continuous Mullite Fibers   *381*
          H. Schneider
6.2.1   Laboratory Produced Fibers   *381*
6.2.2   Commercially Produced Fibers   *384*
6.2.2.1   Altex Fibers (Sumitomo Chemicals, Japan)   *384*

6.2.2.2   Nivity Fibers (Denka-Nivity, Japan)   *385*
6.2.2.3   Nextel Fibers (3M Company, U.S.A.)   *385*
6.3   Continuous Melt-grown Mullite Fibers   *391*
           H. Schneider
6.4   Application of Mullite Fibers   *393*
           H. Schneider
           References   *394*

7   Mullite Matrix Composites   *397*
7.1   Whisker-reinforced Mullite Matrix Composites   *398*
           H. Schneider
7.1.1   Fabrication Routes   *399*
7.1.2   Mechanical Properties   *399*
7.1.3   Thermal Properties   *402*
7.1.4   Miscellaneous Properties   *403*
7.2   Continuous Fiber-reinforced Mullite Matrix Composites   *403*
           H. Schneider
7.2.1   Fabrication Routes   *403*
7.2.1.1   Non-oxide Fiber-reinforced Composites   *407*
7.2.1.2   Oxide Fiber-reinforced Composites   *409*
7.2.2   Mechanical Properties   *421*
7.2.2.1   Non-oxide Fiber-reinforced Composites   *423*
7.2.2.2   Oxide Fiber-reinforced Composites   *425*
7.2.3   Thermal Properties   *434*
7.2.3.1   Thermal Expansion   *434*
7.2.3.2   Thermal Conductivity   *435*
7.2.4   Miscellaneous Properties   *435*
7.2.5   The Role of the Fiber/Matrix Interphase   *436*
7.2.5.1   Easy-cleavage Interphases   *437*
7.2.5.2   Low-toughness Interphases   *438*
7.2.5.3   Porous or Gap-producing Interphases   *438*
7.2.6   Applications   *440*
7.2.6.1   Spacecraft Applications   *440*
7.2.6.2   Aircraft Engine and Powerplant Applications   *440*
7.2.6.3   Other Industrial Applications   *442*
7.3   Platelet- and Particle-reinforced Mullite Matrix Composites   *443*
           H. Schneider, K. Okada
7.3.1   Basic Principles   *443*
7.3.1.1   Transformation Toughening   *443*
7.3.1.2   Crack-deflection Toughening   *444*
7.3.1.3   Toughening by Modulus Load Transfer   *444*
7.3.1.4   Nanoparticle Toughening   *445*
7.3.2   Zirconia Particle-reinforced Composites   *445*
7.3.2.1   Fabrication Routes   *445*
7.3.2.2   Mechanical Properties   *453*

7.3.2.3   Thermal Properties *455*
7.3.3   Silicon Carbide Platelet- or Particle-reinforced Composites *455*
7.3.3.1   Fabrication Routes *456*
7.3.3.2   Mechanical Properties *459*
7.3.4   Miscellaneous Oxide Particle-reinforced Composites *461*
7.4   Metal-reinforced Mullite Matrix Composites *463*
           H. Schneider
7.4.1   Aluminum Metal-reinforced Composites *463*
7.4.2   Molybdenum Metal-reinforced Composites *466*
           References *467*

Index *477*

General introduction

H. Schneider

Mullite is certainly one of the most prominent ceramic materials. In a rather poetic travel report to the Isle of Mull (Western Scotland) Sosman in 1956 mentioned that *"When the good Moslem looks about for an excuse to make a journey and see the world, he packs up his travel kit and heads toward the Black Stone in the Kaaba at Mekka. This object of reference is believed by some petrographers to have been a meteorite, hence endowed with the special qualities which we humans would naturally associate with a stone from the sky. To what mineral or rock should a good ceramist turn under similar circumstances? To **mullite**, of course. In manufacturing most of his products the ceramist heats clay or other materials containing alumina and silica to a temperature at which the silica and alumina combine to produce crystalline mullite. It is the ceramist's silicate, par excellence."* Though our less sentimental generation may have problems to compare a mineralogical location with the capital of a world religion, many of us certainly agree with Robert B. Sosman that mullite is the ceramic phase par excellence.

It is most interesting to have a look on the history of mullite research. Oschatz and Wächter (1847) described crystallization processes in the glass phase of porcelains. They observed an Al_2O_3-rich, acicular silicate phase, which has been designated as sillimanite, a polymorph of the Al_2SiO_5 (Al_2O_3 SiO_2) alumino silicate group. Deville and Caron (1865) identified the existence of a compound with a composition of about 70.5 wt.% Al_2O_3, which they also associated with the same mineral. Vernadsky (1890) described sillimanite-like phases with a formula of $11Al_2O_3 \cdot 8SiO_2$ occurring in porcelain. More than hundred years ago geologists of the Scottish branch of the British Geological Survey were exploring the Hebridean Island of Mull at the west coast of Scotland (Fig. I.1). They collected mineral samples from various locations of ancient lava flows from the volcano Ben More, which was active some 65 million years ago. They found Al_2O_3-rich, acicular minerals, intergrown in feldspar crystals (Fig. I.2), which initially have also been identified as sillimanite. The first phase equilibrium diagram in the Al_2O_3-SiO_2 system was published by Shepherd et al. (1909), and sillimanite still was believed to be the stable binary compound. It was in 1924 that Bowen and Greig in their basic work pointed out that the stable alumino silicate in the Al_2O_3-SiO_2 system has 3/2- ($3Al_2O_3 \cdot 2SiO_2$) instead of the 1/1- ($Al_2O_3 \cdot SiO_2$) composition. In a footnote the

Fig. I.1 Island of Mull (Western Scotland) in a view from Oban.

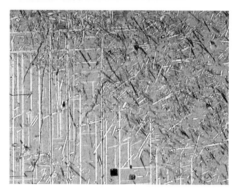

Fig. I.2 Thin section micrograph of an ancient lave from the Ben More volcano (Island of Mull, Western Scotland). Note the occurrence of mullite needles over-grown by plagioclase feldspar.

authors mentioned: *"Since this paper was written we have discovered crystals of the 3 : 2 compound in natural rocks from the island of Mull. We propose the name* **mullite** *to designate the compound in a forthcoming paper of the Journal of the Washington Academy of Science where this natural occurrence of the mineral is described."* The 3/2-phase described by Bowen and Greig originated from the contact zone of a hot magma with clay sediments.

Due to its high temperature-low pressure formation conditions, mullite occurs only rarely in nature, especially at the contact of superheated basaltic lavas with Al_2O_3-rich sediments (so-called buchites or sillimanite buchites). It sometimes can be also be found in high temperature metamorphosed rocks of the sandinite facies (Winkler, 1974). Mullite containing rocks typically consist of quartz, cristobalite and tridymite embedded in a glassy matrix. Anorthite, amphibole and α-alumina often occur as accessory phases. Mullite formation was also described in horn-felses (porcellanite), which develop at the contact of bauxite with olivine dolerite intrusions. Furthermore mullite has been found in SiO_2-rich lechatelerite glasses produced by lightning impact into quartz sands (Tröger, 1969), but also in small druses of volcanic rocks, e. g. in the Eifel mountains (Germany, Fig. I.3).

In spite of its rareness in natural rocks, mullite is perhaps the most frequent phase in conventional ceramics comprising primarily clay products. This can be explained by its occurrence as a main constituent in "conventional" ceramics like

Fig. I.3 Scanning electron micrograph of mullite needles formed hydrothermally in small druses of volcanic rocks of the Eifel mountain (Germany. Material: Courtesy B. Ternes).

pottery, porcelains, sanitary ceramics, refractories and in structural clay products like building bricks, pipes and tiles. The phase mullite thus had a big influence on the development of human civilization and culture, since these ceramic materials reach far back in the history. "Advanced" mullite ceramics are also rather old: Chinese manufactories produced so-called "white pottery" already in the Shang period (1500 to 1000 BC). First information on the fabrication of porcelain has been recognized during the Tsang dynasty (about 620 AD). This early porcelain was produced from kaolin at a firing temperature of about 1300 °C (after Litzow, 1984). In Europe the first successful production of porcelain was performed in Saxony (Germany) at the beginning of the eighteenth century. Mullite-bearing refractories and technical porcelains with a wide variety of special products and uses have become particularly important after the industrial revolution in the nineteenth century.

Besides its importance for conventional ceramics, mullite has become a strong candidate material for advanced structural and functional ceramics in recent years. The reasons for this development are the outstanding properties of mullite: Low thermal expansion, low thermal conductivity, and excellent creep resistance. Other favorable characteristics of mullite are suitable high temperature strength and outstanding stability under harsh chemical environments. The importance of mullite for both conventional and advanced ceramic applications has been documented by the enormous number of studies which have been performed in recent years (Fig. I.4, and Schneider et al. 1994): Since the appearance of the previous mullite monograph in 1994 about 1400 scientific papers directly related with mullite have been published with an up going tendency. Congresses and workshops especially held on processing, characterization, and properties of mullite took place in Tokyo, Japan, 1987, Seattle, U.S.A, 1990, Irsee, Germany, 1994, and in Oban, Scotland near to Isle of Mull in 2000.

The present mullite monograph provides an overview of the recent knowledge of mullite ceramics, and mullite matrix composites. Another focus of the book is to deepen the knowledge of crystal structure, thermodynamics and basic properties of mullite for a better prediction of the behavior of ceramics and composites. The parts of the book form a network, in which each section is correlated to the others.

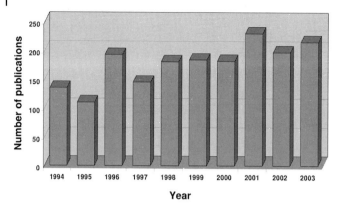

Fig. I.4 Nummber of publications on mullite-related topics, which appeared between 1994 and 2003 (Web of Science, 2005).

Some chapters of the book, although not following this general scheme are essential for understanding of mullite: The contributions of Rahman and Freimann (Section 1.2) and of MacKenzie (Section 2.6), dealing with real structure and the spectroscopy of mullite, and the part of Basu and Sarin (Part 5) compiling the knowledge on mullite coatings. As in the previous mullite book (Schneider et al. 1994) it is a major aim of the authors to combine the knowledge from crystallographers, chemists, material scientists, and engineers in order to deduce a comprehensive understanding of the properties of mullite ceramics, and to identify recent and future application fields of the material. Though it is rather difficult to bring together the different viewpoints and interests of the indicated research fields, it certainly is most worthwhile: It can contribute to a reduction of the information gaps existing between different activities, and thus may help to improve the understanding of mullite ceramics properties, and to shorten research and development cycles prior to a commercial application of materials. So, what is the future of mullite in the 21th Century? Mullite will certainly keep its present importance in traditional ceramics from now on to the future. Moreover, significant progress in the importance of advanced mullite ceramics, mullite matrix composites, metal/ mullite materials and mullite coatings for a number of technical applications is predicted.

H. Schneider in the name of all authors thanks the colleagues, who contributed to the progress of the book by contributions and discussions. The German Aerospace Center (DLR) enabled a sabbatical leave for H. Schneider to the University of Vienna, where much of the text of the book has been written. This is gratefully acknowledged. The works were supported by several institutions and agencies. Of special importance was that of the German Research Foundation (Deutsche Forschungsgemeinschaft, DFG). The authors are indebted to many colleagues, who provided information or photographs, which improved the significance of the book. Many colleagues at DLR helped to produce figures and diagrams, and to review the text. Special thanks go to Dr. M. Schmücker. Without his help it would not have been possible to complete the book in the planned time frame.

Last but not least: The authors thank their families for their patience during the preparation of the book and for their understanding and encouragement.

References

Bowen, N. L. and Greig, J. W. (1924) The system Al_2O_3-SiO_2. J. Amer. Ceram. Soc. **7**, 238, 410.

Deville, S. C. and Caron, D. (1865). Cited in: Pask, J. A. (1990). Critical review of phase equilibria in the Al_2O_3–SiO_2 system. *Ceram. Trans.* **6**, 1–13.

Litzow, K. (1984) Keramische Technik. Vom Irdengut zum Porzellan. Callwey: München.

Oschatz and Wächter (1847). Cited in: Litzow, K. (1984) Vom Irdengut zum Porzellan. Callwey: München.

Schneider, H., Okada, K. and Pask, J. A. (1994). Mullite and mullite ceramics. John Wiley and Sons, Chichester, pp. 1–251.

Shepherd, E. S., Rankin, G. A. and Wright, W.

(1909) The binary system of alumina and silica, lime and magnesia. Amer. J. Sci. **28**, 301.

Sosman, R. B. (1956) A pilgrimage to Mull. Amer. Ceram. Soc. Bull. **35**, 130–131.

Tröger, W. E. (1969) Optische Bestimmung der gesteinsbildenden Minerale. Teil 2: Textband. Schweizerbart: Stuttgart.

Vernadsky, W. I. (1890). Cited in: Pask, J. A. (1990). Critical review of phase equilibria in the Al_2O_3–SiO_2 system. *Ceram. Trans.* **6**, 1–13.

Web of Science (2005). Science Citation Index. The Thompson Corporation.

Winkler, H.G.F. (1974) Petrogenesis of metamorphic rocks. Third edition. Springer: New York – Heidelberg – Berlin.

List of Contributors

Prof. Dr. S. Basu
Manufacturing Engineering
Boston University
15 Saint Mary's Street
Boston, MA 02215
USA

Prof. Dr. R. X. Fischer
Fachbereich Geowisenschaften
Universität Bremen
Klagenfurter Strasse
28334 Bremen
Germany

Dr. S. Freimann
Minerlogisches Institut
Universität Hannover
Welfengarten 1
30167 Hannover
Germany

Prof. Dr. S. Komarneni
205 Materials Research Laboratory
The Pennsylvania State University
University Park, PA 16802
USA

Prof. Dr. K. J. D. MacKenzie
New Zealand Institute for Industrial
Research and Development
Gracefield Research Center
Lower Hutt
New Zealand

Prof. Dr. K. Okada
Department of Metallurgy and
Ceramics Science
Tokyo Institute of Technology
2-12-1 O-okayama, Meguro
Tokyo 152-8552
Japan

Prof. Dr. J. A. Pask
University of California at Berkeley
USA

Prof. Dr. S. Rahman
Mineralogisches Institut
Universität Hannover
Welfengarten 1
30167 Hannover
Germany

Prof. Dr. V. K. Sarin
Manufacturing Engineering
Boston University
15 Saint Mary's Street
Boston, MA 02215
USA

Dr. J. Schmücker
Institut für Werkstoff-Forschung
Deutsches Zentrum für Luft- und
Raumfahrt (DLR)
Linder Höhe
51170 Köln Germany

Prof. Dr. H. Schneider
Institut für Werkstoff-Forschung
Deutsches Zentrum für Luft- und
Raumfahrt (DLR)
Linder Höhe
51170 Köln
Germany

1
Crystal Chemistry of Mullite and Related Phases

1.1
The Mullite-type Family of Crystal Structures
R. X. Fischer and H. Schneider

1.1.1
Introductory Remarks

Mullite in the strict sense, with the composition $3Al_2O_3 \cdot 2SiO_2$, has been the subject of numerous crystal structure and crystal-chemical studies, not only for basic reasons but also because of its outstanding electrical, mechanical and thermal properties, and its wide use in traditional and advanced ceramics.

The fundamental building unit in mullite is the chain of edge-sharing AlO_6-octahedra. The topological arrangement of these chains is a common feature of a whole group of compounds with various chemical compositions. From its most prominent member, this group is designated as the "mullite-type structure family" or just the "mullite-type family". The individual members can have various chemical compositions and different linkages of the chains by cations and cationic groups. They are distinguished by their crystallographic subgroup relationships as described in Section 1.1.2. However, there are some specific requirements that define the mullite-type structural arrangement:

(1) The space group of a mullite-type structure must be a subgroup of the aristotype in space group $P4/mbm$ (see Section 1.1.2).

(2) The chains[1] of edge-sharing MO_6 octahedra (M = octahedrally coordinated cation, see Tables in Section 1.1.3 for examples of observed cations) must be linear representing single Einer-chains in their highest topological symmetry in space group $P4/mbm$.

(3) The axis through the terminating atoms (non-edge-sharing atoms) of the octahedra must point towards the edges (parallel to the chain direction) of adjacent octahedra ($30° \leq \omega \leq 90°$).

---

1) Note that the overall composition of the chain is MO_4 because of the edge-sharing linkage of the MO_6 octahedra where two neighboring octahedra share two O-atoms.

(4) The chain structure should resemble the orthogonal metric of the aristotype perpendicular to the chain direction as closely as possible ($\gamma' = 90 \pm 5°$).

The third criterion is strictly achieved only in those cases where the octahedral axes in the *ab*-plane are perpendicular to each other (see Fig. 1.1.1a). The typical angle ω between the octahedral axes of neighboring octahedra in the mullites listed in Table 1.1.16 is close to 60°. Essentially all chain structures with angles between the respective octahedral axes from >0° to 90° match criterion (3). At exactly 0° the structure assumes *Cmmm* symmetry which is not a translationengleiche subgroup of *P4/mbm* (Note that the hypothetical II.2 structure listed in Fig. 1.1.4 in space group *Cmmm* has a different unit cell setting and consequently does conform to the subgroup relationships). Therefore, we have set an arbitrary limit of 30° intermediate between 0° and 60° representing the limiting angle above which a structure is considered to belong to the mullite-type family. Fig. 1.1.2 shows the frequency of occurrence of the inclination angles ω of neighboring octahedra observed in the crystal structures listed in Section 1.1.3. The angles are grouped in bars rounded in steps of 5°. The highest bar represents the mullites with angles close to 60°. No compounds were found with angles less than 40° except those close to 0° resembling the *Cmmm* type of structures. Therefore, 30° seems to be a suitable value representing the gap between the two structure types. Consequently, structures like, for example, $Na_2Li_3CoO_4$ (Fig. 1.1.1c, Birx and Hoppe 1991), which crystallize in the andalusite-type space group *Pnnm* with chains of edge-sharing NaO_6 octahedra 5.8° inclined to each other, essentially resemble the *Cmmm* symmetry and are therefore not assigned to the mullite-type family. The *Cmmm* type is represented by the cadmium uranate compound $CdUO_4$ (Yamashita et al. 1981) which consists of two subsets of chains of octahedra, the UO_4 chains and the bigger CdO_4 chains perpendicularly oriented to each other as shown in Fig. 1.1.1d. The linkage of the two subsets corresponds to the rutile-type of crystal structures with edge-sharing MO_6 octahedra (TiO_6 in rutile).

Table 1.1.1 Atomic coordinates and site definitions of the chain structure in *P4/mbm* ($a = b = 8.0862$ Å, $c = 2.6659$ Å) calculated from distance least square refinement (DLS).

Atom	x	y	z	Site symmetry	Wyckoff position	No. of atoms in unit cell
Al 1	0	0	0	*m.mm*	2(d)	2
O1	0.38348	x	$^1/_2$	*m.2m*	4(h)	4
O2	0.16492	x	0	*m.2m*	4(g)	4

Nonstandard setting with origin at center (2/m) at 2/m 1 2/m, $^1/_2$ 0 0 from 4/m at 4/m 1 $2_1/g$.

Symmetry operators:

x, y, z $-y+^1/_2, x+^1/_2, z$ $-x, -y, z$ $y+^1/_2, -x+^1/_2, z$ $-x+^1/_2, y+^1/_2, z$

y, x, z $x+^1/_2, -y+^1/_2, z$ $-y, -x, z$

and equivalent positions related by center of symmetry $\bar{1}$

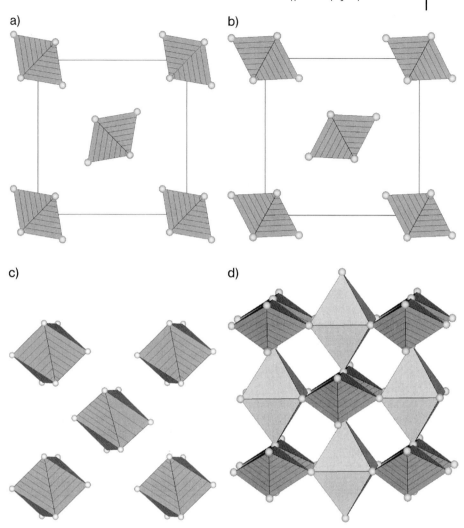

Fig. 1.1.1 Various chain structures with octahedral inclination angles ω between 90° and 0°. (a) The aristotype structure in *P4/mbm* (see chapter 1.1.2). ω = 90°. (b) The mullite structure in *Pbam*. (Angel and Prewitt, 1986). ω = 59.6°. (c) The Na$_2$Li$_3$CoO$_4$ structure in *Pnnm* (Birx and Hoppe, 1991). ω = 5.8°. (d) The CdUO$_4$ structure in *Cmmm* (Yamashita et al. 1981). ω = 0°. CdO$_6$ octahedra are light grey, UO$_6$ octahedra are dark grey and hatched.

The fourth criterion concerning the orthogonal metric is applied, for example, when the lattice symmetry is nonorthogonal. Consequently, werdingite (Section 1.1.3.15) with an angle of γ' = 92.0° between the basis vectors of the unit cell projected onto the plane perpendicular to the chain axes (the cross-section of the unit cell) is assigned to the mullite family whereas the epidote minerals

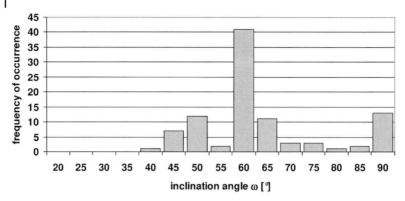

Fig. 1.1.2 Frequency of occurrence of inclination angles ω in mullite-type crystal structures listed in Section 1.1.3.

($Ca_2Al_2Fe(OH)Si_3O_{12}$) are excluded. The monoclinic chain structure of epidote (Kvick et al. 1988) in $P2_1/m$ could be derived from the aristotype as a subgroup of index 2 of *Pbnm* (sillimanite, IV.32) if the epidote unit cell is transformed according to −**a**, **a**+**c**, **b**. Even the angle ω = 41.0° between adjacent octahedra matches criterion 3. However, the angle $\gamma = \gamma' = 116°$ between the **a** and **b** axes of the unit cell after transformation is too far away from 90° and, therefore, epidote is not assigned to this family. Arbitrarily, the angle is assumed here to be within $\gamma' = 90 \pm 5°$ but it could deviate for individual compounds if, for example, their physical properties are mullite-like.

Close structural relationships with mullite have been described frequently in the literature, as, for example, for sillimanite, andalusite, grandidierite, werdingite, the boron and alkali aluminates and gallates, and X-SiAlON. Other phases exhibiting the same topology of the octahedral chain configuration, however, have not yet been assigned to this structure family. Therefore we propose a general definition of the mullite-type structure family. It has the advantage of including the above described phases and to be open for other compounds not yet identified.

It is the aim of this part of the book to define the structural features forming the mullite-type family of crystal structures, to show the group-subgroup relationships between these structures and to describe the individual crystal structures in a general context within this family. In order to achieve an optimal platform for the descriptions of the mullite-type family members, the crystal structures of the individual compounds are derived from the hypothetical tetragonal aristotype structure which represents the highest possible topological symmetry of the octahedral root structure.

1.1.2
The Derivation of Mullite-type Crystal Structures

The most commonly used and generally applicable standardization procedure for inorganic crystal structure data was introduced in 1984 by Parthé and Gelato

(1984) and further developed in the computer program STRUCTURE TIDY (Gelato and Parthé 1987) as a tool for standardizing crystal structures. However, the standardization is applied to unit cell settings and atomic coordinates just within a given space group and does not consider structural relationships between related phases crystallizing in different space groups. Therefore, the comparison and the statistical evaluation of crystal structures within a framework-type family remains difficult even when standardization routines such as STRUCTURE TIDY are applied.

The standardization method used for the mullite-type family classification is that introduced by Baur and Fischer (2000, 2002) and Fischer and Baur (2004) for zeolite-type structures. The approach can easily be applied to the mullite-type crystal structures, since the zeolites are standardized according to their framework structure and the mullite-type compounds are described based on their octahedral chains. The individual symmetry in either case is determined by the additional groups linking the framework atoms or the chains, respectively.

The following rules are applied to the standardization of the mullite-type family adopted from Baur and Fischer (2000) following the concept of Bärnighausen (1980):

(1) The hypothetical structure with the highest possible topological symmetry of the pure chain structure is called the "aristotype" (Fig. 1.1.3).

(2) Crystal structures are assigned to the mullite-type family if their chain structure can be described in a subgroup of the aristotype.

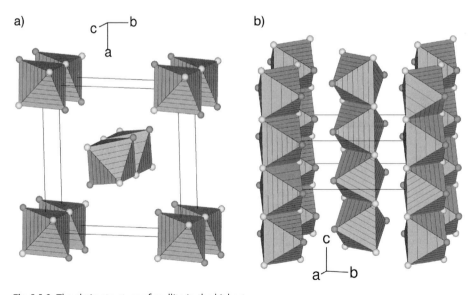

Fig. 1.1.3 The chain structure of mullite in the highest possible topological symmetry *P4/mbm*. O1 atoms are light grey, O2 atoms are dark grey. (a) View parallel **c** rotated by 20° about **a** and 20° about **b**. (b) View parallel **a** rotated by 10° about **b** and 5° about **c**. Chains extend in **c**.

Table 1.1.2 Atomic site relationships of the mullite-type crystal structures. Branch 11 in Figure 1.1.4.

MUL-I *P4/mbm*		MUL-II.1 *P4$_2$/mbc*		MUL-VI.11 *P4$_2$/mbc*	
M1 [2(d), *m.mm*]	⟶	M1 [4(d), 2.22]	⟶	M11	[4(d), 2.22]
				M12	[8(f), 2..]
O1 [4(h), *m.2m*]	⟶	O1 [8(h), *m..*]	⟶	O11	[8(h), *m..*]
				O12	[16(i), 1]
O2 [4(g), *m.2m*]	⟶	O2 [8(g), ..2]	⟶	O21	[8(g), ..2]
				O22	[16(i), 1]

Table 1.1.3 Atomic site relationships of the mullite-type crystal structures. Branch 12 in Figure 1.1.4.

MUL-I *P4/mbm*		MUL-II.1 *P4$_2$/mbc*		MUL-IV.12 *Pbam*		MUL-VIII.12 *Pbam*	
M1 [2(d), *m.mm*]	⟶	M1 [4(d), 2.22]	⟶	M1	[4(f), ..2]	⟶ M11	[4(f), ..2]
						M12	[4(f), ..2]
O1 [4(h), *m.2m*]	⟶	O1 [8(h), *m..*]	⟶	O11	[4(g), ..*m*]	⟶ O11a	[4(g), ..*m*]
						O11b	[4(h), ..*m*]
				O12	[4(h), ..*m*]	⟶ O12	[8(i), 1]
O2 [4(g), *m.2m*]	⟶	O2 [8(g), ..2]	⟶	O2	[8(i), 1]	⟶ O21	[8(i), 1]
						O22	[8(i), 1]

(3) Space group and unit cell setting of the aristotype are chosen to conform to the setting of the mullite type-material which crystallizes in *Pbam*.

(4) The O-atoms in the aristotype are sorted according to the descending order of their Wyckoff-site symbols.

(5) Atom positions listed in the subgroups follow strictly the sequence assigned in the corresponding aristotype, independently of their Wyckoff notation.

(6) The names of atoms in the subgroups carry the root name of the corresponding name in the aristotype appended by numbers and characters. That way, for example, an oxygen position designated O1 in the aristotype setting might be split into an O11 and an O12 site in the subgroup, perhaps further transformed to O11a, O11b, O12a, and O12b in another subgroup of lower index.

This standardization procedure is applied to all derivations listed in Fig. 1.1.4 and Tables 1.1.2 to 1.1.10 as a tool for an easy comparison of the structures in different space groups and different unit cell settings. However, it should be admitted that the systematic classification is not as rigorous as in the case of zeolites, where the

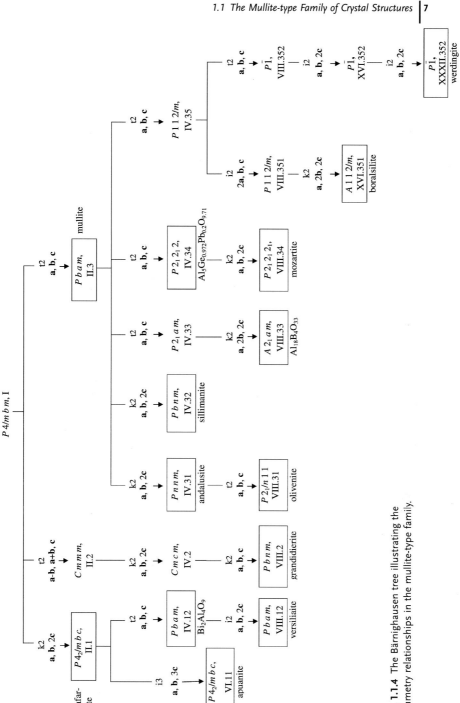

Fig. 1.1.4 The Bärnighausen tree illustrating the symmetry relationships in the mullite-type family.

Table 1.1.4 Atomic site relationships of the mullite-type crystal structures. Branch 2 in Figure 1.1.4.

MUL-I P4/mbm	MUL-II.2 Cmmm	MUL-IV.2 Cmcm	MUL-VIII.2 Pbnm
M1 [2(d), *m.mm*] →	M11 [2(a), *mmm*] →	M11 [4(a), 2/*m..*] →	M11 [4(a), 1̄]
	M12 [2(b), *mmm*] →	M12 [4(b), 2/*m..*] →	M12 [4(b), 1̄]
O1 [4(h), *m.2m*] →	O11 [4(h), 2*mm*] →	O11 [8(g), ..*m*] →	O11a [4(c), ..*m*]
			O11b [4(c), ..*m*]
	O12 [4(j), *m2m*] →	O12a [4(c), *m2m*] →	O12a [4(c), ..*m*]
		O12b [4(c), *m2m*] →	O12b [4(c), ..*m*]
O2 [4(g), *m.2m*] →	O21 [4(g), 2*mm*] →	O21 [8(e), 2..] →	O21 [8(d), 1]
	O22 [4(i), *m2m*] →	O22 [8(f), *m..*] →	O22 [8(d), 1]

Table 1.1.5 Atomic site relationships of the mullite-type crystal structures. Branch 31 in Figure 1.1.4.

MUL-I P4/mbm	MUL-II.3 Pbam	MUL-IV.31 Pnnm	MUL-VIII.31 P2₁/n11
M1 [2(d), *m.mm*] →	M1 [2(a), ..2/*m*] →	M1 [4(e), ..2] →	M1 [4(e), 1]
O1 [4(h), *m.2m*] →	O1 [4(h), ..*m*] →	O11 [4(g), ..*m*] →	O11 [4(e), 1]
		O12 [4(g), ..*m*] →	O12 [4(e), 1]
O2 [4(g), *m.2m*] →	O2 [4(g), ..*m*] →	O2 [8(h), 1] →	O21 [4(e), 1]
			O22 [4(e), 1]

Table 1.1.6 Atomic site relationships of the mullite-type crystal structures. Branch 32 in Figure 1.1.4.

MUL-I P4/mbm	MUL-II.3 Pbam	MUL-IV.32 Pbnm
M1 [2(d), *m.mm*] →	M1 [2(a), ..2/*m*] →	M1 [4(a), 1̄]
O1 [4(h), *m.2m*] →	O1 [4(h), ..*m*] →	O11 [4(c), ..*m*]
		O12 [4(c), ..*m*]
O2 [4(g), *m.2m*] →	O2 [4(g), ..*m*] →	O2 [8(d), 1]

family members in one group exhibit exactly the same topology and are commonly described in the same fashion. Here we are dealing with individual compounds of different chemical compositions, some of which have significantly different properties. Therefore, we are not aiming for a complete standardization of the mullite-type compounds and we are describing the individual type-materials in Section

Table 1.1.7 Atomic site relationships of the mullite-type crystal structures. Branch 33 in Figure 1.1.4.

MUL-I *P4/mbm*	MUL-II.3 *Pbam*	MUL-IV.33 *P2₁am*	MUL-VIII.33 *A2₁am*
M1 [2(d), *m.mm*] ⟶	M1 [2(a), *..2/m*] ⟶	M1 [2(a), *..m*]	M1 [8(b), 1]
O1 [4(h), *m.2m*] ⟶	O1 [4(h), *..m*] ⟶	O11 [2(b), *..m*] ⟶	O11a [4(a), *..m*] O11b [4(a), *..m*]
		O12 [2(b), *..m*] ⟶	O12a [4(a), *..m*] O12b [4(a), *..m*]
O2 [4(g), *m.2m*] ⟶	O2 [4(g), *..m*] ⟶	O21 [2(a), *..m*] ⟶	O21 [8(b), 1]
		O22 [2(a), *..m*] ⟶	O22 [8(b), 1]

Table 1.1.8 Atomic site relationships of the mullite-type crystal structures. Branch 34 in Figure 1.1.4.

MUL-I *P4/mbm*	MUL-II.3 *Pbam*	MUL-IV.34 *P2₁2₁2*	MUL-VIII.34 *P2₁2₁2₁*
M1 [2(d), *m.mm*] ⟶	M1 [2(a), *..2/m*] ⟶	M1 [2(a), *..2*] ⟶	M1 [4(a), 1]
O1 [4(h), *m.2m*] ⟶	O1 [4(h), *..m*] ⟶	O1 [4(c), 1] ⟶	O11 [4(a), 1] O12 [4(a), 1]
O2 [4(g), *m.2m*] ⟶	O2 [4(g), *..m*] ⟶	O2 [4(c), 1] ⟶	O21 [4(a), 1] O22 [4(a), 1]

1.1.3 usually in their original settings (see also discussion in last paragraph of Section 1.1.3.10 for the sillimanite group compounds).

The symmetry derivations with complete group-subgroup relationships are illustrated in Fig. 1.1.4 by Bärnighausen trees (Bärnighausen, 1980). This is a generally accepted and widely used concept to represent structural relationships (see, for example, the symmetry relationships between various ABW type structures (Kahlenberg et al. 2001), the complex relationships among the derivative structures of perovskite (Bock and Müller, 2002) or the rutile-type structures (Baur, 1994)). All space groups are given as maximal subgroups derived from the space group representing the highest possible topological symmetry (i.e. *P4/mbm*, Fig. 1.1.4, Table 1.1.1). Branches contain information on the type of the subgroup relationship (isomorphic (i), translationengleich (t), or klassengleich (k)) with the respective subgroup index representing the factor of the symmetry reduction or the number of cosets derived from the supergroup.

The hierarchical order of the space groups is represented by roman numerals assigned to the space group symbols. The aristotype has the number I, while the subgroups have numerals corresponding to their index of symmetry reduction

Table 1.1.9 Atomic site relationships of the mullite-type crystal structures. Branch 351 in Figure 1.1.4.

MUL-I P4/mbm	MUL-II.3 Pbam	MUL-IV.35 P112/m	MUL-VIII.351 P112/m	MUL-XVI.351 A112/m
M1 [2(d), m.mm]	M1 [2(a), .. 2/m]	M11 [1(a), 2/m]	M11a [1(a), 2/m]	M11a [4(e), 1̄]
			M11b [1(c), 2/m]	M11b [4(f), 1̄]
		M12 [1(g), 2/m]	M12 [2(m), m]	M12 [8(j), 1]
O1 [4(h), m.2m]	O1 [4(h), .. m]	O11 [2(n), m]	O11a [2(n), m]	O11a1 [4(i), m]
				O11a2 [4(i), m]
			O11b [2(n), m]	O11b1 [4(i), m]
				O11b2 [4(i), m]
		O12 [2(n), m]	O12a [2(n), m]	O12a1 [4(i), m]
				O12a2 [4(i), m]
			O12b [2(n), m]	O12b1 [4(i), m]
				O12b2 [4(i), m]
O2 [4(g), m.2m]	O2 [4(g), .. m]	O21 [2(m), m]	O21a [2(m), m]	O21a [8(j), 1]
			O21b [2(m), m]	O21b [8(j), 1]
		O22 [2(m), m]	O22a [2(m), m]	O22a [8(j), 1]
			O22b [2(m), m]	O22b [8(j), 1]

Table 1.1.10 Atomic site relationships of the mullite-type crystal structures. Branch 352 in Figure 1.1.4.

MUL-II.3 Pbam	MUL-IV.35 P112/m	MUL-VIII.352 P1̄	MUL-XVI.352 P1̄	MUL-XXXII.352 P1̄
M1 [2(a), ..2/m]	M11 [1(a), 2/m]	M11 [1(a), 1̄]	M11a [1(a), 1̄]	M11a1 [1(a), 1̄]
				M11a2 [1(b), 1̄]
			M11b [1(b), 1̄]	M11b [2(i), 1]
	M12[1(g), 2/m]	M12 [1(e), 1̄]	M12a[1(e), 1̄]	M12a1 [1(e), 1̄]
				M12a2 [1(h), 1̄]
			M12b [1(h), 1̄]	M12b [2(i), 1]
O1 [4(h), ..m]	O11 [2(n), m]	O11 [2(i), 1]	O11a [2(i), 1]	O11a1 [2(i), 1]
				O11a2 [2(i), 1]
			O11b [2(i), 1]	O11b1 [2(i), 1]
				O11b2 [2(i), 1]
	O12 2(n), m]	O12 [2(i), 1]	O12a [2(i), 1]	O12a1 [2(i), 1]
				O12a2 [2(i), 1]
			O12b [2(i), 1]	O12b1 [2(i), 1]
				O12b2 [2(i), 1]
O2 [4(g), ..m]	O21 [2(m), m]	O21 [2(i), 1]	O21a [2(i), 1]	O21a1 [2(i), 1]
				O21a2 [2(i), 1]
			O21b [2(i), 1]	O21b1 [2(i), 1]
				O21b2 [2(i), 1]
	O22 [2(m), m]	O22 [2(i), 1]	O22a [2(i), 1]	O22a1 [2(i), 1]
				O22a2 [2(i), 1]
			O22b [2(i), 1]	O22b1 [2(i), 1]
				O22b2 [2(i), 1]

relative to the aristotype. Entries with the same index are drawn on the same height, thus representing members of the same hierarchical order. Members on one level occurring in different branches are distinguished by arabic numerals. Furthermore, the set of basis vectors is given which describes the transformation of a unit cell to its setting in the subgroup. Space groups representing observed crystal structures are put in frames. Space groups presented without frames are needed to indicate intermediate steps in describing symmetry transformations. The atomic site relationships are tabulated based on the derivations in the Bärnighausen tree. Arrows indicate the splitting of atom sites upon symmetry reduction. Wyckoff positions and site symmetries are given in square brackets according to the notation in the International Tables for Crystallography (Hahn, 2002). Atom names are derived from the corresponding names in the aristotype, keeping the numeral assigned to the atom in its highest symmetry and adding a second numeral expressing the second level derivation. If an atom site is affected by the symmetry reduction, it keeps the name of its supergroup root name. Upon further reduction of symmetry, lower case letters are added to the atom names. If further subdivisions are needed, numerals and lower case letters follow each other consecutively.

The highest possible topological symmetry for the pure chain structure of edge-sharing MO_6 octahedra of mullite-type compounds can be described in the tetragonal space group *P4/mbm*. Atomic parameters are calculated by distance least squares (DLS) methods using the program DLS76 (Baerlocher et al. 1976). Distances are prescribed by Al–O = 1.885 Å corresponding to the sum of the ionic radii of Al and O (Shannon, 1976), and O–O = 2.666 Å calculated from the octahedral geometry. The distance between O atoms of neighboring octahedra (the distance between O1 and O2 atoms with the coordinates given in Table 1.1.1) is arbitrarily set to 2.833 Å corresponding to the O–O distance in an AlO_4 tetrahedron assumed to bridge adjacent octahedra. Lattice constants refined to $a = b = 8.0862$ Å and $c = 2.6659$ Å and final atomic parameters are listed in Table 1.1.1. The distance least squares residual is $R = 0.0003$ expressing the ideal geometry of this refined chain structure shown in Fig. 1.1.3.

1.1.3
Members of the Mullite-type Family of Crystal Structures

1.1.3.1 Introduction
In the following chapters compounds are described which are proven by crystal-structure refinements to belong to the mullite-type family. Individual members are sorted into groups of the same subgroup symmetry. The groups are named after their most prominent member. However, neither the number of groups nor the number of entries within these groups is aiming at completeness. Unit cell settings and cell parameters are standardized to the most abundant compound within this group preferably matching the setting of the aristotype as closely as possible. However, many of the type-materials are established in the literature with their own history and their own individual unit cell settings independent of the mullite-

type standardization. Therefore, we describe the type-materials in their original setting in the text and the figures. Chemical compositions are given in two notations. In the first column of the tables, the chemical composition is listed as given in the original literature. In the second column, the chemical composition is given as the unit cell contents with the octahedrally coordinated M cation listed as the first element. The atomic positions of oxygen atoms are according to strict geometrical rule. Their designations (e.g. 01, 02 ...) allow easy description of subgroups (e.g. 01 → 011, 012 ..., 02 → 021, 022 ...). However, an other setting (e.g. 0(A,B), 0(D) ...) has also frequently been in literature. Both settings correlate in the way given in Table 1.1.17.

Crystal structure drawings are performed with the program STRUPLO (Fischer and Messner, 2004) as part of the BRASS program package (Birkenstock et al. 2004). MO_6 coordinations forming the MO_4 chains are drawn in polyhedral representation with the O ligands marked as small circles. Extra chain cations are drawn as spheres with a radius of about half their ionic radius (Shannon, 1976).

1.1.3.2 MUL-II.1, $P4_2/mbc$: Schafarzikite Group

This group consists of the crystals of the highest symmetry (except for orthorhombic mullite, which has the same index of symmetry reduction) observed so far in the mullite-type family of crystal structures, crystallizing in a tetragonal space group. Schafarzikite ($FeSb_2O_4$) is a prominent member of this group. The FeO_4 chains of edge-sharing FeO_6 octahedra are the mullite-type backbones of this structure. A characteristic feature of these compounds is the presence of ions with lone-

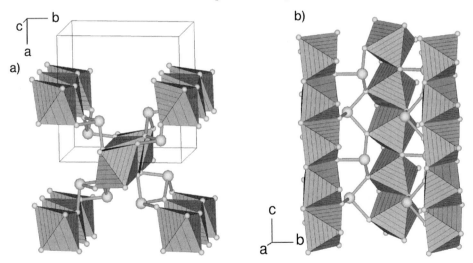

Fig. 1.1.5 Projections of the crystal structure of schafarzikite (Fischer and Pertlik, 1975). Sb atoms are shown as spheres bonded to three O atoms of the AlO_6 octahedra. (a) View parallel **c** rotated by 10° about **a** and **b**. (b) View parallel **a** rotated by 12° about **b** and 10° about **c**.

Table 1.1.11 MUL-II.1 compounds: Schafarzikite group.

Chemical composition	Unit cell contents	Mineral name	a [Å]	c [Å]	V [Å³]	ω [°]	Reference
$MnSb_2O_4$	$Mn_4Sb_8O_{16}$		8.7145(2)	6.0011(2)	455.74	90	Chater and Gavarri (1985)
$FeSb_2O_4$	$Fe_4Sb_8O_{16}$	Schafarzikite	8.590(5)	5.913(5)	436.31	90	Fischer and Pertlik (1975)
		synthetic	8.6181(2)	5.9225(2)	439.87	90	Chater et al. (1985)
$NiSb_2O_4$	$Ni_4Sb_8O_{16}$		8.3719(3)	5.9079(3)	414.08	90	Gavarri (1981)
$CuAs_2O_4$	$Cu_4As_8O_{16}$	Trippkeite	8.592(4)	5.573(4)	411.41	90	Pertlik (1975)
$ZnSb_2O_4$	$Zn_4Sb_8O_{16}$		8.527(2)	5.942(2)	432.04	90	Puebla et al. (1982)
Pb_3O_4	$Pb_4Pb_8O_{16}$		8.811(5)	6.563(3)	509.51	90	Gavarri and Weigel (1975)

pair electrons. These cations form three short bonds to neighboring oxygen atoms completed by the lone-pair electrons to an approximately tetrahedral coordination. In Fig. 1.1.5 the chains of corner-sharing SbO_3 groups are shown crosslinking the octahedral pillars. Other compounds of this group with NiO_4, CuO_4, PbO_4, ZnO_4 and MnO_4 chains are given in Table 1.1.11. There, the chains are crosslinked by SbO_3, AsO_3, and PbO_4 groups with cations above the center of the planes formed by the O atoms.

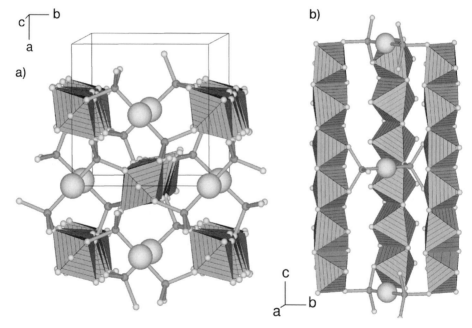

Fig. 1.1.6 Projections of the crystal structure of apuanite with a subset of Fe cations and S anions (Mellini and Merlino, 1979). Big spheres are S atoms, small spheres are Fe atoms.
(a) View parallel **c** rotated by 2° about **b** and 4° about **a**.
(b) View parallel **a** rotated by 5° about **b** and 8° about **c**.

1.1.3.3 MUL-VI.11, $P4_2/mbc$: Apuanite Group

Apuanite can be regarded as a derivative structure of schafarzikite (MUL-II.1) by substituting one-quarter of the antimony (Sb) atoms by iron (Fe) and adding sulphide ions bridging adjacent iron cations. The insertion of sulfur atoms yields four-coordinated iron, while the remaining antimony atoms remain bonded to three oxygen atoms of the octahedral chain. Owing to the ordered distribution of iron and sulfur, lattice constant c is tripled relative to the schafarzikite cell (Figs. 1.1.6 and 1.1.7, Table 1.1.12). Apuanite is found as massive black aggregates in veinlets at the contact between dolomite and phyllite as a product of a metasomatic process (Mellini and Merlino, 1979).

Table 1.1.12 MUL-VI.11 compound: Apuanite group.

Chemical composition	Unit cell contents	Mineral name	a [Å]	c [Å]	V [Å³]	ω [°]	Reference
$Fe_5Sb_4O_{12}S$	$(Fe_{4.15}Zn_{0.32}Fe_{7.40})$ $(Fe_{6.87}Sb_{15.64}As_{1.49})$ $O_{48}S_{3.57}$	Apuanite	8.372(5)	17.974(10)	1259.80	90	Mellini and Merlino (1979)

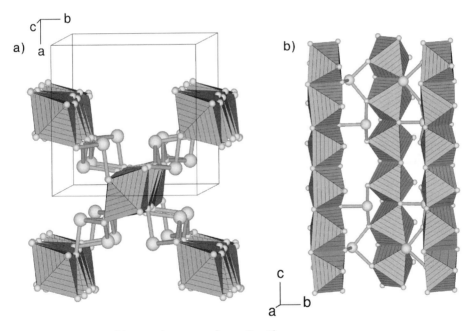

Fig. 1.1.7 Projections of the crystal structure of apuanite with a subset of Sb cations shown as medium sized spheres (Mellini and Merlino, 1979). (a) View parallel **c** rotated by 2° about **b** and 4° about **a**. (b) View parallel **a** rotated by 5° about **b** and 8° about **c**.

1.1.3.4 MUL-IV.12, *Pbam*: $Bi_2M_4O_9$ Group

In $Bi_2M_4O_9$ (M = Al^{3+}, Ga^{3+}, Fe^{3+}) and $Bi_2Mn_4O_{10}$ the mullite-type columns of edge-shared MO_6 octahedra are linked by dimers of MO_4 tetrahedra ($Bi_2M_4O_9$) or MO_5 square-pyramidal polyhedra ($Bi_2Mn_4O_{10}$), respectively. Perpendicular to the octahedral chain axes the polyhedra form five-membered rings of three tetrahedra or pyramids and two octahedra. The Bi^{3+} atoms are located in the structural channels running parallel to the octahedral chains (Fig. 1.1.8). $Bi_2Al_4O_9$, $Bi_2Ga_4O_9$ and $Bi_2Fe_4O_9$ show complete solid solution (Niizeki and Wachi, 1968).

On the basis of infrared studies, Voll et al. (2005) provide some interesting ideas on the distribution of the tetrahedrally bound Al^{3+} and Fe^{3+} cations in $Bi_2(Al,Fe)_4O_9$ mixed crystals. This has been derived from the portion of the infrared spectrum which is exclusively associated with stretching vibrations of tetrahedrally coordinated Al^{3+} (near 915 cm^{-1}) and Fe^{3+} (near 810 cm^{-1}). However, while the end members of the solid solution $Bi_2Al_4O_9$ and $Bi_2Fe_4O_9$ display single and sharp signals at the expected wave numbers, the mixed crystals with $Bi_2(Al_2Fe_2)O_9$ composition show an infrared band being splitted into three components, centering near to those of the aluminum and iron end members and to an average position near 860 cm^{-1} (Fig. 1.1.9). The sub-bands occurring in the $Bi_2(Al_2Fe_2)O_9$ phase near to the band positions of the end members $Bi_2Al_4O_9$ and $Bi_2Fe_4O_9$ are ascribed to tetrahedral Al-Al and Fe-Fe clusters whereas the infrared peak in between those of the end members is ascribed to the contacts between tetrahedral Al-Al and Fe-Fe clusters.

The crystal structure of $Bi_2Mn_4O_{10}$ deviates slightly from that of $Bi_2M_4O_9$ because manganese occurs in two oxidation states (Mn^{3+} and Mn^{4+}, Nguyen et al.

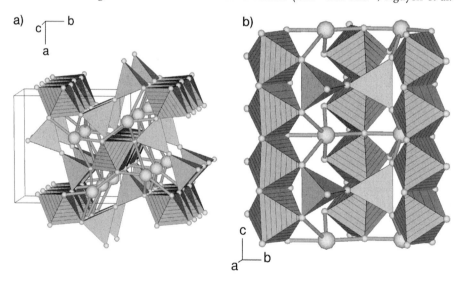

a)

b)

Fig. 1.1.8 Projections of the crystal structure of $Bi_2Al_4O_9$ (Abrahams et al. 1999). (a) View parallel **c** rotated by 5° about **b** and 8° about **a**. (b) View parallel **a** rotated by 5° about **b** and 8° about **c**.

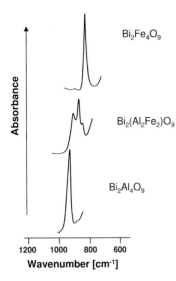

Fig. 1.1.9 Infrared absorption spectra of $Bi_2M_4O_9$ phases ($M = Al^{3+}$, Fe^{3+}), showing the region of stretching vibrations of TO_4 tetrahedra. Note the splitting of the signal of the $Bi_2(Al_2Fe_2)O_9$ compound, which is associated with the occurrence of tetrahedrally coordinated aluminum and iron clusters (after Voll et al. 2005).

1999). The end members of phases form mixed crystals (e.g. Giaquinta et al. 1995), or can be doped by other cations (e.g. Sr^{2+}, Ti^{4+}, Zha et al. 2003).

$Bi_2M_4O_9$ phases exhibit good oxygen conductivity. This can be explained structurally by the sequence of tetrahedral dimers and of large voids in the octahedral chain direction. A schematic view of the one-dimensional oxygen conductivity of $Bi_2M_4O_9$ after Abrahams et al. (1999) is given in Fig. 1.1.10. It involves hopping of oxygen atoms bridging MO_4 dimers (O_{bri} = bridging oxygen atom) to a vacancy in between two adjacent dimers. In a second step these dislocated O_{bri} atoms migrate to an available regular O_{bri} site. No Bi^{3+} conductivity has been taken into account, possibly due to the asymmetric polarization of the Bi^{3+} electronic shell (see Abrahams et al. 1999). According to Zha et al. (2003), the electrical conductivity achieves a maximum if 20 % of Bi^{3+} is replaced by Sr^{2+}. $Bi_2M_4O_9$ ceramic conductors are suitable materials for membranes, electrodes and sensors. Of special interest is its potential use for electrodes in solid oxide fuel cells (SOFCs) for intermediate temperature operation.

The oxygen conduction mechanism in $Bi_2M_4O_9$ is also interesting with respect to a potential high-temperature oxygen conductivity, especially in Al_2O_3-rich aluminum-silicon mullite, with its high number of oxygen vacancies. Al_2O_3-rich aluminum-silicon mullites display a sequence of tetrahedral dimers and of oxygen vacancies in the octahedral chain direction quite similar to that in $Bi_2M_4O_9$. Although this sequence is disordered and incomplete in the former case, but ordered and complete in the latter, it may indicate similar oxygen conduction mechanisms in $Bi_2M_4O_9$ and mullite in the strict sense.

Besides the above described compounds, a number of phases with PbO_6 and especially MnO_6 octahedra, connected by SnO_4 or various rare earth oxygen polyhedra, has been mentioned in the literature. Prominent representatives are summarized in Table 1.1.13.

Table 1.1.13 MUL-IV.12 compounds: $Bi_2M_4O_9$ group.

Chemical composition	Unit cell contents	*a* [Å]	*b* [Å]	*c* [Å]	*V* [Å³]	ω [°]	Reference
$Bi_2Al_4O_9$	$Al_4Al_4Bi_4O_{18}$	7.7134(1)	8.1139(2)	5.6914(1)	356.20	61.1	Abrahams et al. (1999)
$Bi_2Fe_2Al_2O_9$	$(Al_2Fe_2)Bi_4$ $Fe_2Al_2O_{18}$	7.8611(3)	8.2753(3)	5.8535(3)	380.80	59.7	Giaquinta et al. (1995)
$Bi_2Ga_4O_9$	$Ga_4Ga_4Bi_4O_{18}$	7.934	8.301	5.903	388.77	61.7	Müller-Buschbaum and Chales de Beaulieu (1978)
$BiMn_2O_5$	$Mn_4Mn_4Bi_4O_{20}$	7.540(5)	8.534(5)	5.766(5)	371.02	59.6	Niizeki and Wachi (1968)
$LaMn_2O_5$	$Mn_4Mn_4La_4O_{20}$	7.6891(7)	8.7142(7)	5.7274(5)	383.76	43.8	Alonso et al. (1997a)
$PrMn_2O_5$	$Mn_4Mn_4Bi_4O_{20}$	7.5583(9)	8.6481(9)	5.7119(6)	373.36	42.3	Alonso et al. (1997a)
$NdMn_2O_5$	$Mn_4Mn_4Nd_4O_{20}$	7.5116(8)	8.6270(8)	5.7060(5)	369.76	46.7	Alonso et al. (1997a)
$SmMn_2O_5$	$Mn_4Mn_4Sm_4O_{20}$	7.4332(7)	8.5872(7)	5.6956(5)	363.55	44.3	Alonso et al. (1997a)
$EuMn_2O_5$	$Mn_4Mn_4Eu_4O_{20}$	7.3986(8)	8.5666(9)	5.6925(6)	360.80	47.3	Alonso et al. (1997a)
$TbMn_2O_5$	$Mn_4Mn_4Tb_4O_{20}$	7.3251(2)	8.5168(2)	5.6750(2)	354.04	46.5	Alonso et al. (1997b)
$HoMn_2O_5$	$Mn_4Mn_4Ho_4O_{20}$	7.2643(3)	8.4768(3)	5.6700(2)	349.15	47.2	Alonso et al. (1997b)
$ErMn_2O_5$	$Mn_4Mn_4Er_4O_{20}$	7.2360(2)	8.4583(2)	5.6655(1)	346.75	47.5	Alonso et al. (1997b)
$Bi_2Fe_4O_9$	$Fe_4Fe_4Bi_4O_{18}$	7.950(5)	8.428(5)	6.005(5)	402.35	62.4	Niizeki and Wachi (1968)
$Bi_2Fe_2Ga_2O_9$	$(Fe_{2.4}Ga_{1.6})Bi_4$ $Fe_{1.6}Ga_{2.4}O_{18}$	7.946(1)	8.335(1)	5.929(1)	393.62	64.0	Giaquinta et al. (1992)
$SnPb_2O_4$	$Sn_4Pb_8O_{16}$	8.7215(3)	8.7090(3)	6.2919(3)	477.90	84.9	Gavarri et al. (1981)
Pb_3O_4	$Pb_4Pb_8O_{16}$	8.8189(3)	8.8068(2)	6.5636(1)	509.77	89.3	Gavarri et al. (1978)

1.1.3.5 MUL-VIII.12, *Pbam*: Versiliaite Group

Versiliaite is orthorhombic with space group *Pbam* (see Table 1.1.14). With respect to its physical, chemical and crystallographical properties the phase is closely related to schafarzikite (MUL-II.1) and apuanite (MUL-VI.11). Octahedral chains are connected by $Fe^{3+}O_4$ tetrahedral chains, in which, however, every forth Fe^{3+} ion is replaced by Sb^{3+}. Adjacent tetrahedral chains are in turn linked by $Fe_2^{3+}O_7$ groups, with S^{2-} ions bridging both tetrahedra. Charge balance is achieved by partial Fe^{3+} for Fe^{2+} substitution at octahedral sites (Fig. 1.1.11). Versiliaite belongs to the same metasomatic paragenesis as apuanite and occurs, for example, at the contact between dolomite and phyllite (see Mellini and Merlino, 1979).

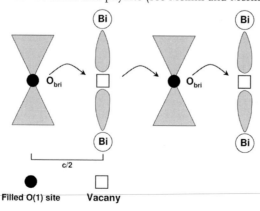

Fig. 1.1.10 Schematic view of the conduction mechanism of bridging oxygen atoms (O_{bri}) of the $Bi_2M_4O_9$ phases. Filled triangles represent oxygen tetrahedra and pestles signify electron orbitals of bismuth (after Abrahams et al. 1999).

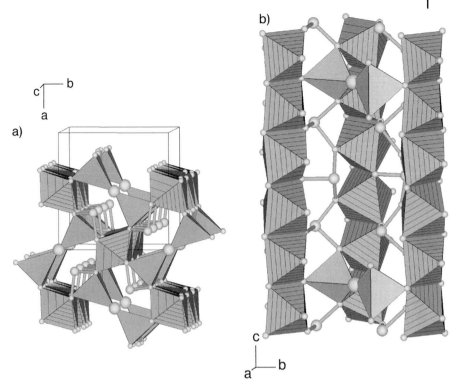

Fig. 1.1.11 Projections of the crystal structure of versiliaite (Mellini and Merlino, 1979). (a) View parallel **c** rotated by 2° about **b** and 4° about **a**. (b) View parallel **a** rotated by 10° about **b** and 8° about **c**.

Table 1.1.14 MUL-VIII.12 compound: Versiliaite group.

Chemical composition	Unit cell contents	Mineral name	a [Å]	b [Å]	c [Å]	V [Å^3]	ω [°]	Reference
$Fe_6Sb_6SO_{16}$	$(Fe_7Zn)Fe_3$ $Sb_{12}AsS_2O_{32}$	Versiliaite	8.492(5)	8.326(5)	11.938(7)	844.07	89.5 79.4	Mellini and Merlino (1979)

1.1.3.6   **MUL-VIII.2, *Pbnm*: Grandidierite Group**

Grandidierite occurs in nature as large elongated bluish-green crystals. It has or-thorhombic symmetry like most mullite-type phases (Table 1.1.15). Grandidierite, with the ideal composition $MgAl_3BSiO_9$, can schematically be derived from anda-lusite $Al_4Si_2O_{10}$ by a combined substitution of $B^{3+} = Si^{4+} + 0.5\ O^{2-}$ and $Mg^{2+} = Al^{3+} + 0.5\ O^{2-}$ in the ratio 1:1 (Stephenson and Moore, 1968). The mullite-type edge-shared AlO_6 octahedral chains are linked by a complex array of polyhedra: AlO_5

Table 1.1.15 MUL-VIII.2 compounds: Grandidierite group.

Chemical composition	Unit cell contents	Mineral name	a [Å]	b [Å]	c [Å]	V [Å^3]	ω [°]	Reference
$(Mg,Fe)Al_3$ $SiBO_9$	$Al_8Al_4Mg_{3.6}$ $Fe_{0.4}Si_4B_4O_{36}$	Grandidier-ite	10.335(2)	10.978(2)	5.760(2)	653.52	86.8 88.1	Stephenson and Moore (1968)
$(Fe,Mg)Al_3$ $SiBO_9$	$Al_8Al_4Fe_{3.4}$ $Mg_{0.4}Mn_{0.1}Si_4$ B_4O_{36} with traces of P	Ominelite	10.343(2)	11.095(1)	5.7601(8)	661.01	87.5 88.1	Hiroi et al. (2002)

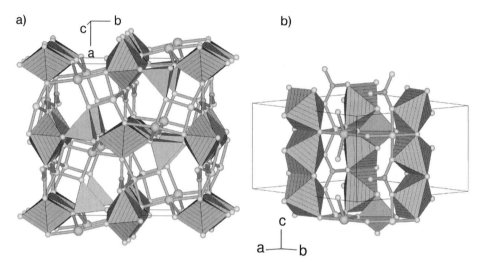

a)

b)

Fig. 1.1.12 Projections of the crystal structure of grandidierite (Stephenson and Moore, 1968). (a) View parallel [001] rotated by 4° about [010] and 6° about [100]. (b) View parallel [110] rotated by 2° about [001] and 6° about [$\bar{1}$10].

and MgO$_5$ bipyramids, SiO$_4$ tetrahedra and BO$_3$ triangles (Fig. 1.1.12). Ominelite is the iron-rich analogue of grandidierite with a Fe/Mg ratio of about 8.5, while grandidierite displays values near 0.1.

1.1.3.7 MUL-II.3, *Pbam*: Mullite Group

Mullite and mullite solid solution Mullite was first mentioned by Bowen and Greig (1924) as a natural mineral on the island of Mull occurring at the contact zone of a hot magma with Al$_2$O$_3$-rich clay sediments. It is worth noting that the phase has been described much earlier (Oschatz and Wächter, 1847, Deville and Caron, 1865,

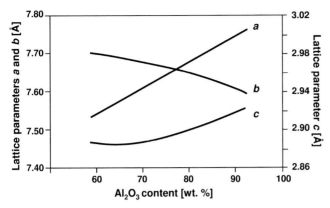

Fig. 1.1.13 Development of the lattice constants *a*, *b* and *c* of mullite with the Al_2O_3 content (after Fischer et al. 1994).

Vernadsky, 1890, Thomas, 1922), although it has been mistaken for sillimanite. Since the first citation in 1924, an enormous number of papers has been published on the synthesis, properties, phase equilibria and crystal chemistry of mullite, on mullite ceramics and mullite matrix composites (see also the General Introduction). Most of this work has been focused on synthetic materials.

Mullite is an aluminosilicate with a variable aluminum to silicon ratio represented by the solid solution series $Al_2[Al_{2+2x}Si_{2-2x}]O_{10-x}$. The composition of mullite observed so far ranges between $0.18 \leq x \leq 0.88$, corresponding to 57 to 92 mol% Al_2O_3 (Fischer et al. 1996). The chemical composition of mullite can be derived from its lattice constant *a* according to the linear relationship $m = 144.5 \ a - 1029.5$ (Fischer et al. 1996), where *m* denotes the molar content of Al_2O_3 in mol%. Fig. 1.1.13 shows the relationship between the lattice parameters of mullite and the molar Al_2O_3 content.

Table 1.1.16 gives an overview of the crystal structure refinements performed on mullite and other MUL-II.3 compounds. Mullites have been described in two different unit cell settings. Burnham (1963a) introduced a unit cell setting shifted by 1/2, 1/2, 0 relative to the origin in the initial structure determination of Ďurovič (1962a) and Sadanaga et al. (1962). Most of the subsequent structure refinements are based on the Burnham setting which is also used here. It has the advantage to conform to the setting and the origin of the closely related sillimanite. The atomic parameters of the two different settings can be considered enantiomorphous, related to each other by an inversion center $\bar{1}$ in 1/4, 1/4, 0 which is equivalent with an origin shift of 1/2, 1/2, 0. All entries in Table 1.1.16 referring to the initial setting by Ďurovič (1962a), Sadanaga et al. (1962) and by Brunauer et al. (2001) are marked by an asterisk.

The average crystal structure of mullite can be best described by comparison with the closely related but structurally simpler sillimanite (composition Al_2SiO_5, i.e. $Al_2O_3 \cdot SiO_2$, see Section 1.1.3.10 for further details). In sillimanite, the aluminum oxygen (AlO_6) octahedral chains are crosslinked by alternating AlO_4 and SiO_4

Table 1.1.16 MUL-II.3 compounds: Mullite group.

Chemical composition	Unit cell contents	Mineral / compound name	a [Å]	b [Å]	c [Å]	V [Å³]	ω [°]	Reference
$Al_{4.6}Si_{1.4}O_{9.7}$	$Al_2Al_{2.6}Si_{1.4}O_{9.7}$	Mullite-ss*	7.549(8)	7.681(8)	2.884(3)	167.23	58.5	Ďurovič (1962a)
$Al_{4.8}Si_{1.2}O_{9.6}$	$Al_2Al_{2.8}Si_{1.2}O_{9.6}$	Mullite-ss*	7.583(2)	7.681(2)	2.8854(5)	168.06	59.6	Sadanaga et al. (1962)
$Al_{5.38}Si_{2.44}O_{9.70}$	$Al_2Al_{3.38}Si_{2.44}O_{9.70}$	Mullite-ss	7.584(3)	7.693(3)	2.890(1)	168.61	59.5	Burnham (1963a)
$Al_{5.66}Si_{2.16}O_{9.70}$	$Al_2Al_{3.66}Si_{2.16}O_{9.70}$	Mullite-ss	7.584(3)	7.693(3)	2.890(1)	168.61	59.5	Burnham (1963a)
$Al_{4.72}Si_{1.32}O_{9.66}$	$Al_2Al_{2.72}Si_{1.32}O_{9.66}$	Mullite-ss*	7.566(5)	7.682(5)	2.884(2)	167.62	59.4	Ďurovič (1969)
$Al_{4.5}Si_{1.5}O_{9.75}$	$Al_2Al_{2.5}Si_{1.5}O_{9.75}$	3/2 mullite*	7.553(1)	7.686(1)	2.8864(7)	167.56	59.0	Saalfeld and Guse (1981)
$Al_{4.72}Si_{1.32}O_{9.66}$	$Al_2Al_{2.72}Si_{1.32}O_{9.66}$	Mullite-ss	7.5785(6)	7.6817(7)	2.8864(3)	168.04	59.8	Angel and Prewitt (1986)
$Al_{4.8}Si_{1.2}O_{9.6}$	$Al_2Al_{2.8}Si_{1.2}O_{9.6}$	2/1 mullite	7.588(2)	7.688(2)	2.8895(6)	168.56	59.8	Angel et al. (1991)
$Al_{4.95}Si_{1.05}O_{9.53}$	$Al_2Al_{2.95}Si_{1.05}O_{9.53}$	Mullite-ss	7.6110(3)	7.6803(4)	2.8872(1)	168.77	60.8	Ban and Okada (1992)
$Al_{4.98}Si_{1.02}O_{9.51}$	$Al_2Al_{2.98}Si_{1.02}O_{9.51}$	Mullite-ss	7.6156(6)	7.6780(6)	2.8859(2)	168.75	61.7	Ban and Okada (1992)
$Al_{4.87}Si_{1.13}O_{9.57}$	$Al_2Al_{2.87}Si_{1.13}O_{9.57}$	Mullite-ss	7.5964(5)	7.6803(5)	2.8824(2)	168.17	60.5	Ban and Okada (1992)
$Al_{4.64}Si_{1.36}O_{9.68}$	$Al_2Al_{2.64}Si_{1.36}O_{9.68}$	Mullite-ss	7.5640(4)	7.6923(4)	2.8806(1)	167.61	61.1	Ban and Okada (1992)
$Al_{4.75}Si_{1.25}O_{9.63}$	$Al_2Al_{2.75}Si_{1.25}O_{9.63}$	Mullite-ss	7.5811(3)	7.6865(3)	2.8821(1)	167.95	60.1	Ban and Okada (1992)
$Al_{4.59}Si_{1.41}O_{9.70}$	$Al_2Al_{2.59}Si_{1.41}O_{9.70}$	Mullite-ss	7.5539(2)	7.6909(2)	2.88391(6)	167.54	59.2	Ban and Okada (1992)
$Al_{4.54}Si_{1.46}O_{9.73}$	$Al_2Al_{2.54}Si_{1.46}O_{9.73}$	Mullite-ss	7.5421(1)	7.6957(1)	2.88362(5)	167.37	59.1	Ban and Okada (1992)
$Al_{4.52}Si_{1.48}O_{9.74}$	$Al_2Al_{2.52}Si_{1.48}O_{9.74}$	Mullite-ss	7.5459(2)	7.6937(2)	2.88346(7)	167.40	59.0	Ban and Okada (1992)
$Al_{4.56}Si_{1.44}O_{9.72}$	$Al_2Al_{2.56}Si_{1.44}O_{9.72}$	Mullite-ss	7.5473(1)	7.6928(1)	2.88391(4)	167.44	58.9	Ban and Okada (1992)
$Al_{4.5}Si_{1.5}O_{9.75}$	$Al_2Al_{2.5}Si_{1.5}O_{9.75}$	3/2 mullite	7.54336(6)	7.69176(6)	2.88402(2)	167.34	58.9	Balzar and Ledbetter (1993)
$Al_{5.65}Si_{0.35}O_{9.175}$	$Al_2Al_{3.65}Si_{0.35}O_{9.175}$	Mullite-ss	7.7391(6)	7.6108(5)	2.9180(1)	171.87	64.5	Fischer et al. (1994)
$Al_{4.82}Si_{1.18}O_{9.59}$	$Al_2Al_{2.82}Si_{1.18}O_{9.59}$	Mullite-ss	7.5817(8)	7.6813(9)	2.8865(5)	168.10	60.0	Voll et al. (2001)
$Al_{4.68}Si_{1.32}O_{9.66}$	$Al_2Al_{2.68}Si_{1.32}O_{9.66}$	Mullite-ss	7.5655(4)	7.6883(4)	2.8851(2)	167.81	59.1	Voll et al. (2001)
$Al_{4.98}Si_{1.02}O_{9.51}$	$Al_2Al_{2.98}Si_{1.02}O_{9.51}$	Mullite-ss	7.616(2)	7.678(2)	2.8891(4)	168.97	63.0	Johnson et al. (2001)
$Al_{5.16}Si_{0.84}O_{9.42}$	$Al_2Al_{3.16}Si_{0.84}O_{9.42}$	Mullite-ss	7.606(1)	7.682(1)	2.8871(4)	168.70	61.4	Johnson et al. (2001)
$Al_{4.68}Si_{1.32}O_{9.66}$	$Al_2Al_{2.68}Si_{1.32}O_{9.66}$	Mullite-ss	7.5454(2)	7.6956(2)	2.88398(6)	167.46	58.6	Johnson et al. (2001)
$Al_{4.66}Si_{1.34}O_{9.67}$	$Al_2Al_{2.66}Si_{1.34}O_{9.67}$	Mullite-ss	7.5499(3)	7.6883(3)	2.88379(9)	167.39	58.5	Johnson et al. (2001)

Table 1.1.16 (cont.)

Chemical composition	Unit cell contents	Mineral / compound name	a [Å]	b [Å]	c [Å]	V [Å³]	ω [°]	Reference
$Al_{4.67}Si_{1.17}P_{0.17}O_{9.76}$	$Al_2Al_{2.67}Si_{1.17}P_{0.17}O_{9.76}$	P-doped mullite	7.5722(4)	7.6960(4)	2.8856(1)	168.16	60.0	Ronchetti et al. (2001)
$Al_{4.20}Fe_{0.30}Si_{1.50}O_{9.74}$	$Al_2Al_{2.20}Fe_{0.30}Si_{1.50}O_{9.74}$	Fe-doped mullite	7.5693(2)	7.7159(2)	2.89470(5)	169.06	58.5	Ronchetti et al. (2001)
$Al_{4.43}Fe_{0.1.50}Si_{1.35}P_{0.08}O_{9.76}$	$Al_2Al_{2.43}Fe_{0.1.50}Si_{1.35}P_{0.08}O_{9.76}$	Fe,P-doped mullite	7.5673(2)	7.7016(2)	2.88895(6)	168.37	59.1	Ronchetti et al. (2001)
$Al_{4.52}Ge_{1.48}O_{9.74}$	$Al_2Al_{2.52}Ge_{1.48}O_{9.74}$	Ge-aluminate	7.650(2)	7.779(2)	2.925(2)	174.06	61.5	Ďurovič and Fejdi (1976)
$Al_{4.72}Ge_{1.28}O_{9.64}$	$Al_2Al_{2.72}Ge_{1.28}O_{9.64}$	Ge-aluminate	7.6530(4)	7.7779(4)	2.9252(2)	174.12	61.5	Voll et al. (2001)
$Al_{4.5}Ge_{1.5}O_{9.75}$	$Al_2Al_{2.5}Ge_{1.5}O_{9.75}$	Ge-aluminate	7.644	7.779	2.925	173.93	61.5	Saalfeld and Gerlach (1991)
$Al_{4.76}Ge_{1.24}O_{9.62}$	$Al_2Al_{2.76}Ge_{1.24}O_{9.62}$	Ge-aluminate	7.644(1)	7.760(1)	2.923(1)	173.38	61.8	Saalfeld and Gerlach (1991)
$Al_{3.92}Cr_{0.5}Si_{1.58}O_{9.79}$	$Al_2Al_{1.92}Cr_{0.5}Si_{1.58}O_{9.79}$	Cr-doped mullite	7.56712(6)	7.70909(6)	2.90211(2)	169.30	58.2	Fischer and Schmeider (2000)
$Ga_{4.62}Ge_{1.38}O_{9.69}$	$Ga_2Ga_{2.62}Ge_{1.38}O_{9.69}$	Ge-gallate	7.8674(4)	8.0305(4)	3.0148(2)	190.47	61.0	Voll et al. (2001)
Al_5BO_9	$Al_2Al_3BO_9$	B-aluminate	7.621	7.621	2.833	164.54	77.3	Mazza et al. (1992)
$Al_4B_2O_9$	$Al_2Al_2B_2O_9$	B-aluminate	7.617	7.617	2.827	164.02	71.1	Mazza et al. (1992)
$Al_6NaO_{9.5}$	$Al_2Al_4NaO_{9.5}$	Na-aluminate	7.640	7.640	2.937	171.43	59.5	Mazza et al. (1992)
$Al_6Na_{0.67}O_{9.33}$	$Al_2Al_4Na_{0.67}O_{9.33}$	Na-aluminate	7.6819(4)	7.6810(4)	2.91842(8)	172.20	63.8	Fischer et al. (2001)
$Al_6KO_{9.5}$	$Al_2Al_4KO_{9.5}$	K-aluminate	7.708	7.708	2.906	172.65	73.7	Mazza et al. (1992)
$Al_6K_{0.67}O_{9.33}$	$Al_2Al_4K_{0.67}O_{9.33}$	K-aluminate	7.6934(3)	7.6727(3)	2.93231(7)	173.09	66.1	Fischer et al. (2001)
$Ga_6Rb_{0.67}O_{9.33}$	$Ga_2Ga_4Rb_{0.67}O_{9.33}$	Rb-gallate	8.0087(3)	8.0177(3)	3.04158(3)	195.30	65.5	this work

An asterisk in the mineral name column indicates that the unit cell setting corresponds to the setting introduced by Ďurovič (1962a).

tetrahedra forming double chains of tetrahedra running parallel to **c**. The double chains of ordered AlO_4 and SiO_4 tetrahedra in sillimanite are replaced in mullite by a disordered arrangement of $(Al,Si)O_4$ tetrahedra.

Formally mullite can be derived from sillimanite by the coupled substitution

$$2Si^{4+} + O^{2-} \rightarrow 2Al^{3+} + \text{vacancy } (\square) \tag{1}$$

The reaction involves removal of oxygen atoms from the structure and the formation of oxygen vacancies. Structure refinements have shown that the oxygen atoms bridging two polyhedra in the tetrahedral double chain (so called O(C) oxygen sites, corresponding to O3 in this work) are removed. As a consequence the two tetrahedral (T) sites are displaced to a position designated as T* (large arrows in Fig. 1.1.14a,b) and O3 becomes threefold coordinated by cations forming T_3O groups (so called triclusters).

The cation occupation at the T* sites of mullite is a point of controversy. According to the basic work of Angel and Prewitt (1986) the occupancy of silicon in tetrahedral T* sites is very small or even zero. In order to maintain charge balance, the removal of O3 oxygen atoms is accompanied by replacement of tetrahedrally coordinated silicon by aluminum. Thus, the introduction of an oxygen vacancy causes the two tetrahedral cation (T) atom sites to migrate towards adjacent oxygen O3 atoms which are pulled towards these cations. These oxygen atoms, slightly displaced from the special position (Wyckoff site 2(d) in space group *Pbam*) have been designated O(C*) corresponding to O4 in the new configuration. According to Angel and Prewitt (1986) the number of oxygen vacancies corresponds to the x value of the mullite mixed crystal series $Al_{4+2x}Si_{2-2x}O_{10-x}$. The c lattice constant of mullite is halved compared with the corresponding parameter in sillimanite because silicon and aluminum atoms are randomly distributed over the tetrahedral cation sites. The structural data obtained for 3/2-mullite ($x = 0.25$) and 2/1-mullite ($x = 0.40$), respectively, are compiled in Tables 1.1.17 and 1.1.18.

The crystal structure analysis from X-ray or neutron diffraction data is based on the deconvolution of electron densities of a large number of atoms projected into one unit cell. The local structural order of oxygen vacancies and of the aluminum to silicon distribution, however, may deviate from the average situation. Information on the structural short-range-order can be obtained from the additional diffuse scattering of mullite ("superstructure reflections"). This additional scattering has been attributed to partial ordering of O3 oxygen vacancies and to aluminum and silicon distribution over tetrahedral (T) sites. According to Paulmann (1996) the ordering scheme of the oxygen vacancies in mullite is very stable persisting up to the melting point. A hypothetically ordered mullite, with all tetrahedrally bound cations in T_3O triclusters, is shown in Fig. 1.1.14. Order-disorder phenomena in mullite are associated with its real structure which is considered in Section 1.2. Information on the degree of the tetrahedral aluminum to silicon ordering in mullite has also been derived from ^{29}Si nuclear magnetic resonance (NMR) spectroscopy by Schmücker et al. (2005). The silicon spectrum of mullite displays three well resolved maxima near −88, −92 and −95 ppm. Schmücker et al. assigned the

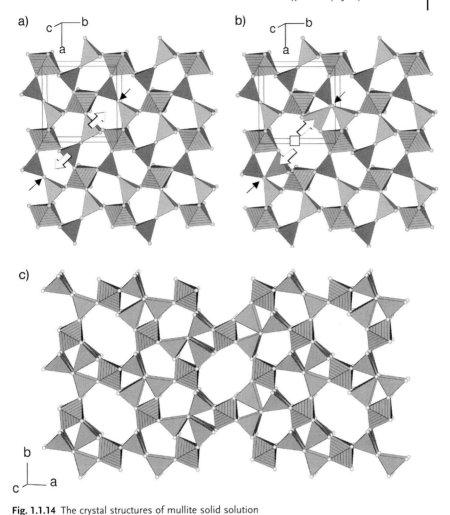

Fig. 1.1.14 The crystal structures of mullite solid solution compounds derived from sillimanite. All projections viewed parallel to **c** rotated by 4° about **a** and **b**. (a) Linkage of octahedral chains in sillimanite by T_2O_7 groups. Big arrows indicate the migrations directions of the T-atoms, thin arrows point to the O atoms which are dislocated off the special position. (b) Modeling of an oxygen vacancy in the layer shown in (a). Big arrows indicate the migration directions corresponding to (a), thin arrows indicate the dislocation directions of the O3 atoms migrating to the O4 position. The square indicates the oxygen vacancy. (c) Hypothetically ordered crystal structure of a 4/1-mullite ($x = 0.67$) corresponding to a $3 \times a$ orthorhombic supercell (after Fischer et al. 1994). All octahedra are linked by T_3O-groups.

Table 1.1.17 Atomic coordinates and occupancy factors for the average structure of 3/2- (x = 0.25) and 2/1-mullite (x = 0.40).

Atom name §	Alternative designations commonly used §	x		y		z		Occupancy	
		3/2-mullite	2/1-mullite	3/2-mullite	2/1-mullite	3/2-mullite	2/1-mullite	3/2-mullite	2/1-mullite
Al1	Al	0.0	0.0	0.0	0.0	0.0	0.0	1.0	1.0
(Al,Si)	T	0.1485	0.14901(2)	0.3411	0.34026(2)	0.5	0.5	0.863	0.56 Al 0.25 Si
(Al,Si)*	T*	0.2621	0.26247(9)	0.2057	0.20529(9)	0.5	0.5	0.137	0.13 Al 0.06 Si
O1	O(A,B)	0.3579	0.3590(1)	0.4221	0.4218(1)	0.5	0.5	1.0	1.0
O2	O(D)	0.1265	0.1273(1)	0.2201	0.2186(1)	0.0	0.0	1.0	1.0
O3	O(C)	0.0	0.0	0.5	0.5	0.5	0.5	0.590	0.39
O4	O(C*)	0.4507	0.4498(4)	0.0518	0.0505(4)	0.5	0.5	0.137	0.19

References:

3/2-mullite data: Saalfeld and Guse (1981), unit cell transformed to the setting used by Angel and Prewitt (1986), Al and Si are not distinguished

2/1-mullite data: Angel and Prewitt (1986).

§ Both designations are used in this book depending on literature reference.

Table 1.1.18 Interatomic distances and bond angles for the average structures of 3/2- (x = 0.25) and 2/1-mullite (x = 0.40).

	Distances [Å]			Structural site	Angles [°]		
	3/2-mullite	2/1-mullite	Multiplicity		3/2-mullite	2/1-mullite	Multiplicity
M1-site				**M1**			
Al 1–O1	1.896	1.8936(5)	4×	O1–Al 1–O1	180	180	2×
Al 1–O2	1.943	1.9366(9)	2×	O2–Al 1–O2	180	180	1×
				O1–Al 1–O1	99.17	99.31(2)	2×
mean	1.912	1.908		O1–Al 1–O1	80.83	80.69(2)	2×
				O1–Al 1–O2	89.80	89.65(3)	4×
				O1–Al 1–O2	90.20	90.35(3)	4×
T-site				**T**			
(Al,Si)–O1	1.700	1.7102(8)	1×	O1–(Al,Si)–O3	111.08	111.15(3)	1×
(Al,Si)–O2	1.725	1.7273(5)	2×	O1–(Al,Si)–O2	106.69	106.66(3)	2×
(Al,Si)–O3	1.658	1.6676(2)	1×	O2–(Al,Si)–O2	113.56	113.34(5)	1×
mean	1.702	1.708		O2–(Al,Si)–O3	109.39	109.50(3)	2×
T*-site				**T***			
Al*–O1	1.814	1.8169(10)	1×	O1–(Al,Si)*–O4	106.19	106.21(7)	1×
Al*–O2	1.773	1.7728(6)	2×	O1–(Al,Si)*–O2	99.97	100.37(5)	2×
Al*–O4	1.852	1.8518(31)	1×	O2–(Al,Si)*–O4	118.96	118.68(5)	2×
mean	1.803	1.804		O2–(Al,Si)*–O2	108.96	109.01(6)	1×

References:

3/2-mullite data: Saalfeld and Guse (1981)
2/1-mullite data: Angel and Prewitt (1986).

signal near −88 ppm with the highest intensity to three AlO_4 tetrahedra in the next nearest environment of SiO_4, similar to the situation in sillimanite. The nuclear magnetic resonance peaks near −92 and −95 ppm have been assigned respectively to two AlO_4 and one SiO_4 tetrahedra, and to one AlO_4 and two SiO_4 tetrahedra being next nearest neighbors of SiO_4. Comparison of the measured data with simulated nuclear magnetic resonance data yields a moderate degree of aluminum to silicon odering in the tetrahedra of mullite.

Burnham (1964a) remarked that the mullite structure theoretically fits any composition between (disordered) sillimanite ($x = 0.00$) and aluminum oxide (ι-alumina, $x = 1.00$), and that there is no obvious reason why the composition should be restricted to 3/2-mullite ($x = 0.25$) and 2/1-mullite ($x = 0.40$), or to mixed crystals between them. In reality, mullite mixed crystal formation is limited towards SiO_2-rich compositions ($x \approx 0.20$). The existence of the miscibility gap between sillimanite and mullite has been explained by the different ordering schemes of sillimanite and mullite. The extent of the miscibility gap has remained a point of dispute. Mullites with a composition midway between sillimanite and 3/2-mullite (termed Al_2O_3-rich sillimanite) were described in a contact metamorphosed rock (lithomarge from Rathlin Island, Northern Ireland, Cameron, 1976a) and in iron-rich and titanium-poor compounds (see Cameron, 1977b). A complete solubility between sillimanite and mullite, which has been described as occuring at higher pressures (Hariya et al. 1969, Grigor'ev, 1976), seems to be less probable.

If mullites are simple solid solutions with little structural variation, their cell parameters should depend linearly on the Al_2O_3 content. The plot of the a constant versus the Al_2O_3 content has frequently been used for discussion of the dependence (Fig. 1.1.13). Many authors (Trömel et al. 1957, Gelsdorf et al. 1961, Ďurovič 1962b, Hariya et al. 1969, Cameron 1977c, Schneider and Wohlleben, 1981, Schneider, 1990, Ban and Okada, 1992, Fischer et al. 1994, Rehak et al. 1998) have shown that the a lattice constants of mullites in the composition range 60–74 mol% Al_2O_3 actually increases linearly with the Al_2O_3 content, while the b constant decreases slightly and non-linearly. Extrapolation of the curves towards lower Al_2O_3 content results in the a and b constants of sillimanite (50 mol% Al_2O_3). Continuing the a and b curves of the mullite solution series against higher Al_2O_3 contents shows that the lines cross at about 80 mol% Al_2O_3 ($x \approx 0.67$). Actually a compound with $a = b$ was first described by Ossaka (1961), and was designated as tetragonal mullite phase. In spite of the coincidental identity of a and b lattice constants this mullite is orthorhombic. Therefore it could be designated as "pseudo-tetragonal" mullite (see Schneider and Rymon-Lipinski, 1988). To be precise, it should read mullite with "pseudo-tetragonal metric", since the symmetry of the crystal structure is clearly orthorhombic.

Only a few results are available on the phase relations of Al_2O_3-rich mullites beyond the "pseudo-tetragonal" point. Alumina phases with mullite-like structures (ι-alumina) have been discussed in the literature and were believed to be either tetragonal (Cameron, 1977a, Foster, 1959, Perrotta and Young, 1974), or they have been assigned to orthorhombic symmetry (Saalfeld, 1962, Duvigneaud, 1974). Careful own microchemical studies have shown that such phases belong to the

Table 1.1.19 Site occupancies in mullite phases.

	$x \leq 0.67$	$x \geq 0.67$
No. of T atoms (Si + Al)	$4 - 2x$	$4 - 2x$
No. of T* atoms (all Al)	$2x$	$2 - x$
No. of T** atoms (all Al)	0	$-2 + 3x$
No. of Al atoms in T site	2	2
No. of Si atoms in T site	$2 - 2x$	$2 - 2x$
No. of O3 [O(C)] atoms	$2 - 3x$	0
No. of O4 [O(C*)] atoms	$2x$	$2 - x$

After Fischer et al. (1994).

mullite-type alkali aluminates rather than to mullites. Investigations on Al_2O_3-rich mullite precursors synthesized via a specific chemical route from aluminum-*sec*-butylate and silicon chloride yielded mullites that have unusual lattice constants with $a > b$ ($a = 7.760$ Å, $b = 7.595$ Å, $c = 2.9192$ Å, data from a sample annealed at 1000 °C, Schneider et al. 1993a), corresponding to Al_2O_3 up to about 89 mole% Al_2O_3 ($x = 0.83$ corresponding to a 9/1-mullite composition of $Al_{5.65}Si_{0.35}O_{9.18}$) which lies far beyond the crossover of the a and b curves ($a > b$ at about 80 mol% Al_2O_3, Fig. 1.1.13). A low temperature orthorhombic mullite, probably with $a > b$, has also been described by Huling and Messing (1992).

Understanding the crystal structure of the $a > b$ phase turned out to be difficult, since the conventional structure model of mullite is restricted to $x \leq 0.67$ (≤ 80 mol% Al_2O_3, Angel and Prewitt 1986), where all possible O3 vacancies have formed. Angel and Prewitt mentioned that at $x > 0.67$ some additional Al^{3+} ions are incorporated at interstitial sites. Another approach to accommodating the additional Al involves the formation of T_4O groups (so-called tetraclusters), where four(!) tetrahedra are connected via a common oxygen atom. Fischer et al. (1994) performed a structure refinement of the mullite with a very high Al_2O_3 content ($x = 0.83$ i.e. 9/1-mullite see Tables 1.1.19 and 1.1.20), and proved that the average

Table 1.1.20 Chemical compositions and site assignments for selected compounds in the mullite solid solution series.

Compound	x	Composition of octahedral chain	Residue	No. of groups		
				T_2O	T_3O	T_4O
Sillimanite	0	Al_2O_8	$Al_2Si_2O_2$	2	0	0
3/2-mullite	0.25	Al_2O_8	$Al_{2.5}Si_{1.5}O_{1.75}$	1.25	0.5	0
2/1-mullite	0.4	Al_2O_8	$Al_{2.8}Si_{1.2}O_{1.6}$	0.8	0.8	0
4/1-mullite	0.67	Al_2O_8	$Al_{3.33}Si_{0.67}O_{1.33}$	0	1.33	0
9/1-mullite	0.825	Al_2O_8	$Al_{3.65}Si_{0.35}O_{1.175}$	0	0.7	0.475
ι-Al_2O_3	1	Al_2O_8	Al_4O	0	0	1

After Fischer et al. (1994).

structure remained orthorhombic. Although tetrahedral T_4O arrangements are not favored on the basis of energy calculations (Padlewski et al. 1992a), Fischer and coworkers preferred this model. They suggested that these extremely Al_2O_3-rich mullites contain sillimanite-like T_2O dimers of tetrahedra with additional pairs of AlO_4 tetrahedra (T* and T**, i.e. T_3O and T_4O groups, respectively). T_3O and T_4O groups are linked to the bridging oxygen atom, which is loosely bound with long distances to the octahedral chains. T* and T** sites can alternatively be interpreted as fivefold coordinated. In order to get further structural information on the $a > b$-mullite ($x = 0.83$) a plane image reconstruction has been carried out in comparison to that of 2/1-mullite. The theoretical background of this procedure is based on the so-called videographic simulation and reconstruction of the real structure of mullite (e.g. Rahman, 1993a, see Section 1.2). The reconstructed image of the $a > b$-mullite ($x = 0.83$) derived from its diffraction patterns (Fig. 1.1.15) is given in Fig. 1.1.16 and compared with 2/1-mullite ($x = 0.40$). The structures of both mullites are believed to consist of domains with high O3 oxygen vacancy concentrations (bright areas containing light dots) distributed in a matrix with relatively low O3 oxygen vacancy concentrations (dark areas). The domains as well as the cells in individual domains partly exhibit a quasi-antiphase relationship. The antiphase relationships are referred to the O3 sublattice in the mullite structure, whereby the "antiphase boundaries" run parallel to [001] for 2/1-mullite ($x = 0.40$), and parallel [100] for the $a > b$-mullite. The domains are distributed quasi randomly in the case of 2/1-mullite whereas for $a > b$-mullite a clear correlation distance along [101] and [$\bar{1}$01] (inter-domain vectors) exists, forming an irregular twin boundary along the main directions.

The near-tetragonal arrangement of structural units perpendicular to the **c** axis of mullite suggests that the a and b cell dimensions display comparable developments with increasing Al_2O_3 content. However, as shown in Fig. 1.1.13, a increases linearly while b decreases slightly and non-linearly (see also above). Clockwise and counter-clockwise rotations of polyhedra are unlikely to explain these phenomena since they would cause a strong deformation of the structure. Also the theory of rigid unit modes (RUMs, Hammonds et al. 1998), which says that the stiff tetrahedral units of a silicate structure show little tendency to rotate if they are

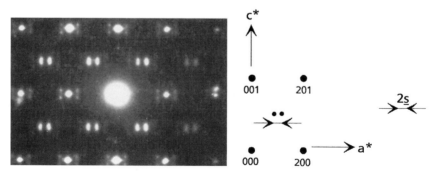

Fig. 1.1.15 **a*c*** electron diffraction pattern of Al_2O_3-rich mullite ($a > b$, $x = 0.83$, from Schneider et al. 1994a).

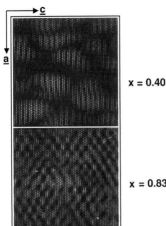

c

a

x = 0.40

x = 0.83   **Fig. 1.1.16** Reconstructed image for the *a* > *b*-mullite (*x* = 0.83) in comparison with 2/1-mullite (*x* = 0.40). Information on the procedure, its theoretical background and its significance are given in the text (see also Section 1.2, from Schneider, 1994a).

stabilized by octahedra, contradicts the rotational approach. Random Al^{3+} and Si^{4+} distribution over the tetrahedral sites with the amount of tetrahedral Al^{3+} increasing with the *x*-value of mullite can also not produce the observed development, because it has a similarly increasing effect on both lattice constants. As a last influencing factor the distribution of O3 oxygen vacancies may be considered. The number and distribution of O3 vacancies are important, since they increase the compressibility in the specific lattice direction, which in turn works against incorporation-induced expansion. Again, randomly distributed vacancies would have similar expansion-reducing effects in both **a** and **b** directions. However, things change if structurally anisotropic vacancy distributions are taken into account in a way that the number of O3 oxygen vacancies along **b** increases more strongly than along **a** in going from 3/2- to 2/1-mullite. Since the vacancies work against the expansion induced by aluminum for silicon substitution, this would have the result of a less strong expansion parallel to **b** than parallel to **a**.

Various types of mullites depending on their synthesis routes have been described as relevant for ceramics and technical processes:

Sinter-mullites, which have Al_2O_3 contents with a frequency maximum at about 60 mol% Al_2O_3 (72 wt.% Al_2O_3). These mullites have been designated as *3/2-mullite* ($3Al_2O_3 \cdot 2SiO_2$, *x* = 0.25) or stoichiometric mullite. The term sinter-mullite describes a mullite that has been produced from the starting materials essentially by solid-state reactions. The Al_2O_3 content of sinter-mullites is influenced by sintering temperatures, the duration of heat treatment, the initial bulk composition, the nature, grain size and efficiency of mixing of the starting materials, and whether α-alumina nucleated (see also Section 3.1).

Fused-mullites[2] which have Al_2O_3 contents with a frequency maximum at about

---

2) The term *fused-mullite* has frequently been used in the literature (e. g. Cameron, 1977c), although it is misleading and can be mistaken for molten mullite, i. e. a glass phase with mullite composition. A hyphen should therefore be put between the terms "fused" and "mullite".

64 mole% Al_2O_3 (76 wt.% Al_2O_3) and the ideal composition 66.7 mol% Al_2O_3 (78 wt.% Al_2O_3). These mullites have been designated as *2/1-mullites* ($2Al_2O_3 \cdot SiO_2$, $x = 0.40$). The term fused-mullite describes mullites that are produced either by melting the raw materials in an electric arc furnace above 2000 °C with subsequent crystallization of mullite during cooling of the bath, or by laboratory scale crystal growth techniques (see Section 4.1.2). Higher Al_2O_3 contents can be achieved by rapid quenching, or alternatively by a very slow cooling process. The composition of mullites crystallized from liquids is primarily a function of temperature and to a lesser degree of the initial composition (see Cameron, 1977c): Guse (1974) and Guse and Mateika (1974) investigated the solid solution range of Czochralski-grown fused-mullites by varying the starting chemical composition of the melt between 77.3 and 71.8 wt.% Al_2O_3 (22.7 to 28.2 wt.% SiO_2). In spite of this rather large variation, the Al_2O_3 content of corresponding mullites changed only from about 77.5 to 75.5 wt.%, both compositions lying close to that of 2/1-mullite. A conclusion of Guse's observation is that the formation of 2/1- or, alternatively, 3/2-type mullite is controlled by the synthesis process, while the bulk chemical composition of the system plays a less important role.

Chemical-mullites, or solution-sol-gel-derived mullites of which the chemical composition greatly depends on the starting materials and on the formation temperature. Al_2O_3 contents over 83 mol% have been described (see above and Section 4.1.3). Chemical mullites are formed from organic and inorganic precursors by polymerization and ceramization. Mullitization takes place at low temperatures between about 900 and 1300 °C, with the "low temperature" mullites normally being Al_2O_3-rich.

Germanium mullite   The chemical similarity of Si^{4+} and Ge^{4+}, especially the comparable ionic radii (Si^{4+}: 0.26 Å, Ge^{4+}: 0.39 Å) imply that germanium mullite, isotypic with silicon mullite, does exist. Actually, Gelsdorf et al. (1958) were the first to synthesize a 3/2-type germanium mullite ($3Al_2O_3 \cdot 2GeO_2$). They also described a complete solid solution between 3/2-silicon ($3Al_2O_3 \cdot 2SiO_2$) and 3/2-germanium mullite ($3Al_2O_3 \cdot 2GeO_2$). Detailed data on germanates with mullite structure were published by Perez Y Jorba (1968, 1969), and Schneider and Werner (1981, 1982). A structural refinement of a germanate ($3.08Al_2O_3 \cdot 2GeO_2$) isostructural with 2/1-mullite has been performed by Ďurovič and Fejdi (1976).

Complete gallium (Ga^{3+}) and partial iron (Fe^{3+}) substitution for aluminum (Al^{3+}) in the germanates was reported by Schneider and Werner (1981, 1982). The incorporations cause cell edge expansion, especially along **b**. This was taken as evidence for a preferred octahedral incorporation of gallium and iron, producing strong expansion along the elastic M1-O2 (M(1)-O(D)) bond lying about 30° to either side of the **b** axis but to about 60° of **a** (Fig. 1.1.14). Tetrahedral incorporation also occurs in the case of gallium. The expansion anisotropy, being much stronger in iron-substituted mullites than in gallium-substituted mullites, however, indicates that tetrahedral incorporation is much less probable for iron. There is a linear dependence between the mean radius of trivalent cations and cell volumes of the mullites. Infrared spectroscopic studies carried out by Schneider (1981) and Voll et al. (2001) on different silicon and germanium mullites showed notable band broadenings in the $3Al_2O_3 \cdot 2GeO_2$ absorption diagram, while $3Al_2O_3 \cdot 2SiO_2$ and $3Ga_2O_3 \cdot 2GeO_2$ exhibit bands with smaller half-widths. This was taken as evidence for a higher degree of cation disorder in Al,Ge-mullite. Only phases with near-stoichiometric 3/2-compositions of the germanium mullites (i.e. $3M_2O_3 \cdot 2GeO_2$, M = Al,Ga) have so far been identified. This is striking, since the radii of fourfold-coordinated cations are very similar (Al^{3+}: 0.39 Å, Ga^{3+}: 0.47 Å, Ge^{4+}: 0.39 Å). Further research is required to solve this point.

Alkali aluminates and gallates   Angerer (2001) studied the synthesis of mullite-type alkali aluminates $Al_6Na_{0.67}O_{9.33}$, $Al_6K_{0.67}O_{9.33}$ and $Al_6Rb_{0.67}O_{9.33}$. He showed that complete binary and ternary miscibility exists between these end members. Angerer also synthesized the alkali gallates $Ga_6Na_{0.67}O_{9.33}$, $Ga_6K_{0.67}O_{9.33}$ and $Ga_6Rb_{0.67}O_{9.33}$ which are isotypic with alkali aluminates. No alkaline earth aluminates and gallates, or aluminates and gallates with threefold-charged cations (except B^{3+}) with mullite-type structures have been described yet.

Fischer et al. (2001) performed structure refinements of the end members and of the mixed crystal of the series $Al_6Na_{0.67}O_{9.33}$–$Al_6K_{0.67}O_{9.33}$. The structure consists exclusively of AlO and GaO networks, respectively. Because of the charge deficiency of the networks, sodium and potassium atoms reside in the vacant O3 oxygen sites. Na^+ occurs at a split site slightly off the special position, while K^+ enters the ideal non split site. This was explained by Fischer et al. (2001) by the fact that Na^+ is too small to completely fill the large oxygen vacancies, while the larger K^+ cations fit well. The alkali ions are restricted to 2 atoms per 3 unit cells, and as a consequence a fixed number of 0.66 vacancies per unit cell occurs.

1.1.3.8   MUL-IV.31, *Pnnm*: Andalusite Group

Andalusite is an orthorhombic natural mineral occurring in high-grade metamorphic Al_2O_3-rich rocks with the composition Al_2SiO_5 ($Al_2O_3 \cdot SiO_2$). The mullite-specific chains of edge-sharing AlO_6 octahedra in andalusite are linked by edge-sharing AlO_5 bipyramids which alternate with SiO_4 tetrahedra (Burnham and Buerger, 1961, see Fig. 1.1.17). Fivefold-coordinated cations in MO_5 polyhedra have frequently been observed in mullite-type phases, e.g. grandidierite (Section 1.1.3.6), werdingite (Section 1.1.3.15), boralsilite (Section 1.1.3.14), boron aluminates (Section 1.1.3.11). The arrangement of AlO_5 bipyramids and of SiO_4 tetrahedra formally resembles the sequence of mullite-type "Zweier double chains" (M-chains according to Saalfeld, 1979), if the AlO_5 bipyramids are replaced by $(Al,Si)_2O_5$ tetrahedral groups.

Schneider and Werner (1982), Fischer and Schneider (1992) and Voll et al. (2001) described germanates of the composition Ga_2GeO_5 isotypic with andalusite. Schneider and Werner stated that about 25% of the Ga^{3+} can be replaced by Al^{3+}. There is a group of OH-bearing phosphates and arsenates with the andalusite structure and the general composition M_2TO_4OH (M = Cu^{2+}, Mn^{2+}, Zn^{2+}, Co^{2+};

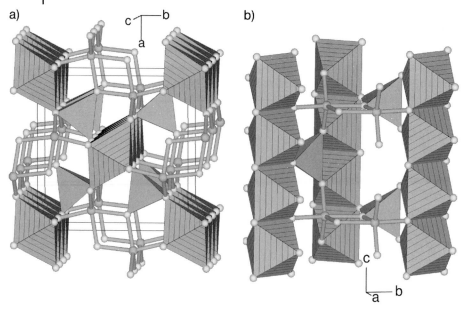

Fig. 1.1.17 Projections of the crystal structure of andalusite
(Pilati et al. 1997). (a) View parallel **c** rotated by 4° about **b** and
6° about **a**. (b) View parallel **a** rotated by 10° about **b** and **c**.

Table 1.1.21 MUL-IV.31 compounds: Andalusite group.

Chemical composition	Unit cell contents	Mineral name	a [Å]	b [Å]	c [Å]	V [Å³]	ω [°]	Reference
Al_2SiO_5	$Al_4Al_4Si_4O_{20}$	Andalusite	7.7992(6)	7.9050(6)	5.5591(5)	342.73	60.9	Pilati et al. (1997)
$Al_5Fe_3Ge_4O_{20}$	$(Al_{2.2}Fe_{1.8})$ $Al_{2.8}Fe_{1.2}$ Ge_4O_{20}		8.0376(1)	8.1673(2)	5.7726(1)	378.95	59.7	Fischer and Schneider (1992)
$Cu_2(OH)PO_4$	Cu_4Cu_4 $(OH)_4P_4O_{16}$	Libethenite	8.062(5)	8.384(4)	5.881(2)	397.51	65.2	Cordsen (1978)
$Zn_2(OH)AsO_4$	Zn_4Zn_4 $(OH)_4As_4O_{16}$	Adamite	8.306(4)	8.524(6)	6.043(3)	427.85	63.0	Hill (1976)
$Mn_2(OH)AsO_4$	Mn_4Mn_4 $(OH)_4As_4O_{16}$	Eveite	8.57(1)	8.77(1)	6.27(1)	471.25	68.4	Moore and Smyth (1968)
$(Cu_{0.42}Zn_{0.58})_2$ $(OH)AsO_4$	$(Cu_{3.36}Zn_{0.64})$ $Zn_4(OH)_4$ As_4O_{16}		8.50(2)	8.52(2)	5.99(1)	433.80	62.7	Toman (1978)
$(Mn_{0.86}Al_{1.14})$ SiO_5	$(Mn_{2.96}Al_{1.04})$ $Mn_{0.48}$ $Al_{3.52}Si_4O_{20}$	Kanonaite	7.959(2)	8.047(2)	5.616(1)	359.68	60.8	Weiss et al. (1981)

$T = P^{5+}, As^{5+}$). Prominent members of this group are libethenite, eveite and adamite (Table 1.1.21). M atoms occur in both sixfold and fivefold coordination, corresponding to Al^{3+} in andalusite, while T cations are fourfold coordinated as is Si^{4+} in andalusite. $MO_4(OH)_2$ octahedra form the edge-shared mullite-type chains, which are connected by $MO_4(OH)$ bipyramids and TO_4 tetrahedra as in andalusite. It is not known whether the phosphates and arsenates can transform to mullite-type phases at elevated temperatures. However, their relatively low thermal stability, at least at atmospheric pressure, makes this rather unlikely.

1.1.3.9 MUL-VIII.31, $P2_1/n11$: Olivenite Group

Olivenite is a hydroxyl copper arsenate (Table 1.1.22) which has been found in the oxidation zone of copper-bearing ore bodies. Olivenite is structurally related to adamite and libethenite, and thus in a wider sense belongs to the andalusite group. The structure is composed of edge-shared CuO_6 octahedra and CuO_5 bipyramids similar to the relationships in andalusite (see Figs. 1.1.17 and 1.1.18).

Table 1.1.22 MUL-VIII.31 compound: Olivenite group.

Chemical composition	Unit cell contents	Mineral name	a [Å]	b [Å]	c [Å]	α [°]	V [Å^3]	ω [°]	Reference
Cu_2OHAsO_4	$Cu_4Cu_4(OH)_4$ As_4O_{16}	Olivenite	8.5894(2)	8.2073(2)	5.9285(1)	90.088(3)	417.93	63.1 2×	Burns and Hawthorne (1995)

1.1.3.10 MUL-IV.32, *Pbnm*: Sillimanite Group

In nature sillimanite tends to form from pelitic rocks during low- or medium-pressure metamorphism. It is one of the three polymorphs with composition $Al_2O_3 \cdot SiO_2$ (i.e. Al_2SiO_5), the other two polymorphs being andalusite (Section 1.1.3.8) and kyanite, the latter not belonging to the mullite structure family. The orthorhombic phase contains the mullite-type AlO_6 octahedral chains running parallel to the crystallographic **c** axis. Octahedral chains are connected by double chains of tetrahedra with aluminum and silicon alternating in an ordered sequence (Fig. 1.1.19). The similarity between sillimanite and mullite is remarkable (see, for example, Burnham, 1963b, 1964a), and many properties are quite similar. However, there is no complete solubility between sillimanite and mullite, at least at atmospheric pressure (see also Section 1.1.3.7).

The synthetic $PbMBO_4$ phases also have an orthorhombic sillimanite-type structure with chains of edge-sharing MO_6 octahedra. These octahedra have two highly asymmetric O-M-O bridging angles but a constant M-M distance of 2.97 Å. The explanation of the deformation of the octahedra is the rigid nature of the BO_3 groups connecting octahedral chains (Fig. 1.1.20). Octahedral cations are Al^{3+}, Ga^{3+}, Fe^{3+}, Cr^{3+} and Mn^{3+} (Park and Barbier, 2001, Park et al. 2003b). The large

a)

b)

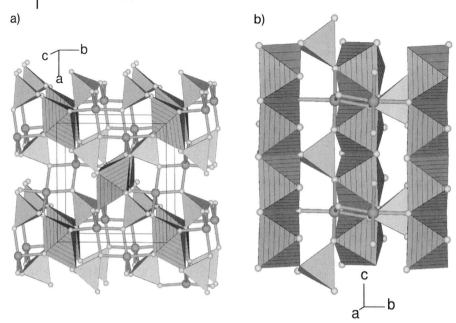

Fig. 1.1.18 Projections of the crystal structure of olivenite (Burns and Hawthorne, 1995). (a) View parallel **c** rotated by 6° about **a** and **a** × **c**. (b) View parallel **a** rotated by 2° about **a** × **c** and 4° about **c**.

a)

b)

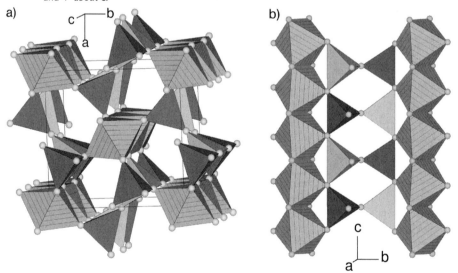

Fig. 1.1.19 Crystal structure projections of sillimanite (Yang et al. 1997). AlO_4 tetrahedra are light gray, SiO_4 tetrahedra are dark gray. (a) View parallel **c** rotated by 10° about **a** and 5° about **b**. (b) View parallel **a** rotated by 2° about **b** and **c**. Chains extended in **c**.

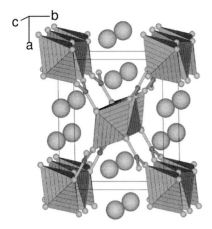

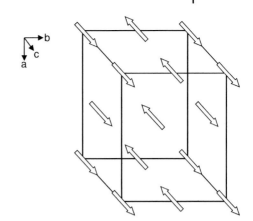

Fig. 1.1.20 Crystal structure projections of PbGaBO$_4$ (Park and Barbier, 2001). B and Pb atoms are represented by small and large spheres, respectively. View parallel **c** rotated by 10° about **a** and 5° about **b**.

Fig. 1.1.21 The magnetic structure of PbFeBO$_4$ showing antiferromagnetism. Arrows indicate areas of equal magnetization (after Park et al. 2003a).

lead atoms enter the structural channels for charge compensation and geometrical stabilization reasons. As a result of the transition metal incorporation the octahedral chains of the PbMBO$_4$ phases display one-dimensional magnetic behavior. The chromium and iron compounds are characterized by intrachain antiferromagnetism (Fig. 1.1.21), whereas the manganese shows ferromagnetic coupling. Interchain correlations are always ferromagnetic, which means that these phases belong to the class of insulating ferromagnets.

The sillimanite-group compounds are listed in Table 1.1.23. While in sillimanite the octahedral axes in the **ab**-plane are closer to the **b** axis, and the *a* lattice constant is slightly smaller than *b*, the situation is reversed in the other compounds of this group. This could be adjusted by transforming the unit cell interchanging **a** and **b** yielding a strictly standardized setting. However, the space group of the resulting structure in the new setting would be *Pnam* and therefore would represent a new branch in the Bärnighausen tree in Fig. 1.1.4. For simplicity and convenience such a strict standardization is neither applied here nor in other groups.

1.1.3.11 MUL-VIII.33, *A2₁am*: A$_9$B$_2$ Boron Aluminates

The boron aluminate usually designated as A$_9$B$_2$ (i.e. 9Al$_2$O$_3 \cdot$2B$_2$O$_3$) is a synthetic compound not yet identified in nature (Table 1.1.24). It directly forms from the oxides or by transformation of Al$_{6-x}$B$_x$O$_9$ above about 1050 °C (see Section 1.1.3.14). Without any atomic deficiency or atomic replacement, its calculated composition is 10Al$_2$O$_3 \cdot$2B$_2$O$_3$. However, if less than 2% of the aluminum in the structure are replaced by boron atoms this may result in a composition near 9Al$_2$O$_3 \cdot$2B$_2$O$_3$. Since it is extremely difficult to determine the chemical composi-

Table 1.1.23 MUL-IV.32 compounds: Sillimanite group.

Chemical composition	Unit cell contents	Mineral name	a [Å]	b [Å]	c [Å]	V [Å³]	ω [°]	Reference
Al_2SiO_5	$Al_4Al_4Si_4O_{20}$	Sillimanite	7.4857(8)	7.6750(9)	5.7751(7)	331.80	57.3	Yang et al. (1997)
$Pb(Fe,Mn)(VO_4)OH$	$(Fe_{0.8}Mn_{0.2})4Pb_4V_4O_{16}(OH)_4$	Čechite	9.435	7.605	6.099	437.62	52.1	Pertlik (1989)
$PbZn(VO_4)OH$	$Zn_4Pb_4V_4O_{16}(OH)_4$	Descloizite	9.416(2)	7.593(2)	6.057(1)	433.05	52.4	Hawthorne and Faggiani (1979)
$(Pb_{2.8}Fe_{1.2})Cu_4O_{1.6}(VO_4)_4(OH)_2$	$(Cu_4Pb_{2.8}Fe_{1.2}V_4O_{17.6}(OH)_2$	Mottramite	9.640(5)	7.525(7)	5.900(3)	427.99	48.1	Permer et al. (1993)
$Pb(Cu_{0.81}Zn_{0.12})VO_4(OH)$	$(Cu_{0.85}Zn_{0.15})4Pb_4V_4O_{16}(OH)_4$	Mottramite	9.316(4)	7.667(4)	6.053(2)	432.34	49.6	Cooper and Hawthorne (1995)
$PbMnVO_4(OH)$	$Mn_4Pb_4V_4O_{16}(OH)_4$	Pyrobelonite	9.507(2)	7.646(2)	6.179(1)	449.15	51.9	Kolitsch (2001)
$PbGaBO_4$	$Ga_4Pb_4B_4O_{16}$		8.2495(11)	6.9944(10)	5.8925(8)	340.00	74.4	Park and Barbier (2001)
$PbCrBO_4$	$Cr_4Pb_4B_4O_{16}$		8.1386(11)	6.9501(10)	5.9410(8)	336.05	75.1	Park et al. (2003b)
$PbMnBO_4$	$Mn_4Pb_4B_4O_{16}$		8.6418(11)	6.7062(9)	5.9429(8)	344.41	69.8	Park et al. (2003b)
$PbFeBO_4$	$Fe_4Pb_4B_4O_{16}$		8.3339(17)	7.0089(14)	5.9412(12)	347.03	72.5	Park et al. (2003b)
$PbAlBO_4$	$Al_4Pb_4B_4O_{16}$		8.0215(6)	6.9209(5)	5.7134(4)	317.19	77.8	Park et al. (2003a)

Table 1.1.24 MUL-VIII.33 compounds: A_9B_2 boron aluminates group.

Chemical composition	Unit cell contents	a [Å]	b [Å]	c [Å]	V [Å³]	ω [°]	Reference
$Al_{18}B_4O_{33}$	$Al_8Al_{11.6}B_{4.4}O_{36}$	7.6942(1)	15.0110(2)	5.6689(1)	654.74	67.5	Garsche et al. (1991)
$Al_{16.6}Cr_{1.4}B_4O_{33}$	$Al_8Al_{10.1}Cr_{1.5}B_{4.4}O_{36}$	7.7051(1)	15.0637(2)	5.7001(1)	661.60	66.9	Garsche et al. (1991)

tion of these nano-sized compounds the actual composition of these phases remains a point of dispute and may vary to a certain extent from sample to sample. Nevertheless it is reasonable to retain the general designation A_9B_2. Owing to the stability of A_9B_2 boron aluminate up to about 1200 °C, its low thermal expansion coefficient and the low thermal conductivity, and because of the corrosion resistance against molten B_2O_3-rich glasses, A_9B_2 boron aluminates have been widely used as refractory linings and for thermal insulation (see Wada et al. 1993, 1994, Rymon-Lipinski et al. 1985). A_9B_2 boron aluminate forms whiskers via a gas-transport reaction, thus making it available for, for example, reinforcement of metal matrices for automotive engine parts.

Ihara et al. (1980) and Garsche et al. (1991) found a A_9B_2 boron aluminate ($x = 1$) with a doubled b parameter. The mullite-type AlO_6 octahedral chains in A_9B_2 display a repeat distance of about 5.7 Å, indicating a sequence of two octahedra to identity. The octahedral chains are linked by edge-sharing AlO_5 bipyramids alternating with isolated AlO_4 tetrahedra and BO_3 triangles (Ihara et al. 1980, Garsche et al. 1991, Fig. 1.1.22). The occurrence of AlO_6, AlO_5 and AlO_4 polyhedra has been confirmed by ^{27}Al NMR satellite transition spectroscopy (SATRAS, Kunath et al. 1992). Garsche et al. found that A_9B_2 is able to incorporate up to about 10 wt.% Cr_2O_3, with Cr^{3+} substituting Al^{3+} in the octahedra.

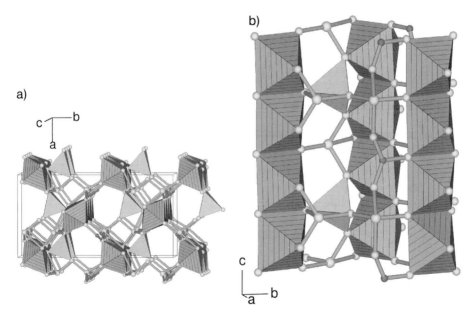

Fig. 1.1.22 Projections of the crystal structure of $Al_{18}B_4O_{33}$ (A_9B_2) (Garsche et al. 1991). (a) View parallel **c** rotated by 2° about **b** and 4° about **a**. (b) View parallel **a** rotated by 10° about **b** and **c**.

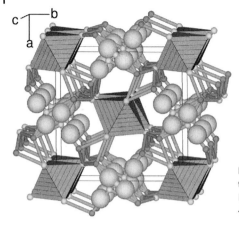

Fig. 1.1.23 Projections of the crystal structure of $Al_5Ge_{0.972}Pb_{0.2}O_{9.71}$ (Saalfeld and Klaska, 1985). View parallel to **c** rotated by 4° about **a** and **b**.

1.1.3.12 MUL-IV.34, $P2_12_12$: $Al_5Ge_{0.972}Pb_{0.2}O_{9.71}$ Phase

Saalfeld and Klaska (1985) described an Al_2O_3-rich germanium mullite with 8.2 wt.% PbO and 2.4 wt.% Nd_2O_3 (Table 1.1.25). It has been shown that the large Pb^{2+} and Nd^{3+} ions are incorporated at O3 oxygen vacancies in the mullite structure (Figs. 1.1.14b and 1.1.23), thus representing a partially filled structure. The investigations provide further evidence for the mullite structure's ability to incorporate foreign cations at various structural sites. Depending on the size and charge of the cations, they may enter octahedral or tetrahedral sites substituting Al^{3+}, or are incorporated interstitially in the structural channels running parallel to the crystallographic **c** axis or in structural voids produced by O3 oxygen vacancies.

1.1.3.13 MUL-VIII.34, $P2_12_12_1$: Mozartite Group

A large variety of mullite-type phases of the composition $CaM(OH)TO_4$ has been described (octahedrally coordinated cations M = Al^{3+}, Mn^{3+}, Cu^{2+}, Zn^{2+}, tetrahedrally coordinated cations T = Si^{4+}, V^{5+}, As^{5+}, see Table 1.1.26). Prominent members are mozartite ($CaMn(OH)SiO_4$), vuagnatite ($CaAl(OH)SiO_4$) and adelite (Ca-$(Cu,Ni,Zn)(OH)(V,As)O_4$). The silicates mozartite and vuagnatite differ from the vanadates and arsenates of the adelite group (Table 1.1.26) by their charge distribution. In mozartite and vuagnatite the edge-sharing octahedral chains are formed by the trivalent cations Al^{3+} and Mn^{3+} (Fig. 1.1.24), whereas in adelite the divalent Cu^{2+}, Ni^{2+} and Zn^{2+} ions enter the octahedral sites. In the latter case the positive charge deficit is balanced by the tetrahedral cations As^{5+} and V^{5+}, while Si^{4+} occurs in mozartite and vuagnatite. The higher valence state of tetrahedral cations in adelite leads to electrostatically more saturated oxgygen atoms of the tetrahedra. The divalent octahedral cations of these compounds, on the other hand, produce a slight undersaturation of the octahedral oxygen atoms and causes an affinity for protonization. This favors the formation of octahedrally bound hydroxyl (OH) groups.

Of special crystallochemical interest are compounds showing Jahn-Teller distor-

Table 1.1.25 MUL-IV.34 compound: $Al_5Ge_{0.972}Pb_{0.2}O_{9.71}$ phase.

Chemical composition	Unit cell contents	a [Å]	b [Å]	c [Å]	V [Å³]	ω [°]	Reference
$Al_{5.03}Ge_{0.97}Pb_{0.15}Nd_{0.06}O_{9.71}$	$Al_2Al_3Ge_{0.972}Pb_{0.2}O_{9.71}$	7.6648(4)	7.7914(3)	2.9213(1)	174.46	61.4	Saalfeld and Klaska (1985)

Table 1.1.26 MUL-VIII.34 compounds: Mozartite group.

Chemical composition	Unit cell contents	Mineral name	a [Å]	b [Å]	c [Å]	V [Å³]	ω [°]	Reference
$CaAlOHSiO_4$	$Al_4Ca_4(OH)_4Si_4O_{16}$	Vuagnatite	7.055(6)	8.542(7)	5.683(5)	342.48	52.3	McNear et al. (1976)
$CaMnOHSiO_4$	$Mn_4Ca_4(OH)_4Si_4O_{16}$	Mozartite	7.228(1)	8.704(1)	5.842(1)	367.53	46.3	Nyfeler et al. (1997)
$PbZnOHAsO_4$	$Zn_4Pb_4(OH)_4As_4O_{16}$	Arsen-descloizite	7.646(2)	9.363(2)	6.077(1)	435.05	52.6	Keller et al. (2003)
$CaZnOHAsO_4$	$Zn_4Ca_4(OH)_4As_4O_{16}$	Austinite	7.5092(8)	9.0438(9)	5.9343(8)	403.01	49.8	Clark et al. (1997)
$CaNiOHAsO_4$	$Ni_4Ca_4(OH)_4As_4O_{16}$	Nickel-austinite	7.455(3)	8.955(3)	5.916(2)	394.95	52.1	Cesbron et al. (1987)
$CaMgOH(V,As)O_4$	$Mg_4Ca_4(OH)_4V_3AsO_{16}$	Gottlobite	7.510(4)	9.004(5)	5.948(3)	402.20	51.2	Witzke et al. (2000)
$PbCuOHAsO_4$	$Cu_4Pb_4(OH)_4As_4O_{16}$	Duftite	7.768(1)	9.211(1)	5.999(1)	429.23	51.0	Kharisun et al. (1998)
$CaCuOHAsO_4$	$Cu_4Ca_4(OH)_4As_4O_{16}$	Conichalcite	7.4	5.842	9.21	398.16	47.6	Qurashi and Barnes (1963)
$CaCuOH(V,As)O_4$	$Cu_4Ca_4(OH)_4(V_{2.36}As_4)_{16}$	Calciovolborthite[a]	5.836(1)	7.430(2)	9.347(1)	405.30	47.9	Basso et al. (1989)

a) The name tangeite was proposed by Basso et al. (1989)

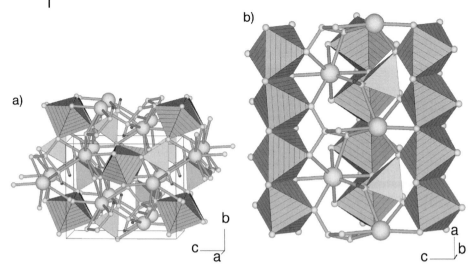

Fig. 1.1.24 Projections of the crystal structure of mozartite
(Nyfeler et al. 1997). (a) View parallel **a** rotated by 6° about **b**
and **c**. (b) View parallel **b** rotated by 5° about **c** and 10°
about **a**.

tion of the octahedra. The $Mn^{3+}O_6$ octahedra of mozartite display approximately
orthorhombic geometry, which is explained by the combination of a tetragonally
compressed Jahn-Teller effect and lattice-induced stress. Comparison with the isostructural vuagnatite, with no electronically induced distortions, indicates that the
distorted octahedral geometry in mozartite causes shifts of the valence sums of
oxygen atoms being hydrogen bonded. As a consequence the OH groups in mozartite are located at the isolated SiO_4 apex and not bonded to the octahedra as in
some of the isostructural phases. Owing to the specific structural arrangement of
mozartite the OH–O distance in mozartite is very short (<2.5 Å). The compounds
of the adelite group with tetrahedral As^{5+} and V^{5+} show a different Jahn-Teller
distortion with tetragonally elongated geometry for $Cu^{2+}O_6$.

1.1.3.14 MUL-XVI.351, A112/*m*: Boralsilite Group

Boralsilite The natural mineral boralsilite, with the ideal composition Al_{16}
$B_6Si_2O_{37}$, occurs as a high temperature phase in pegmatites of the granulite facies
in metapelitic rocks. The basic mullite-type AlO_6 octahedra chains determine a
lattice constant of about 5.6 Å (in the boralsilite setting this is the *b* constant),
reflecting twice the periodicity of the AlO_6 octahedron (twice the O–O edge length
in the octahedron of about 2.8 Å). A number of inter-chain units connect the
octahedral chains in boralsilite: dimers of corner-sharing SiO_4 and AlO_4 tetrahedra, and dimers of edge-sharing AlO_5 bipyramids joined above by BO_3 triangles
and laterally by BO_4 tetrahedra. Thus Al occurs in tetrahedral, bipyramidal and

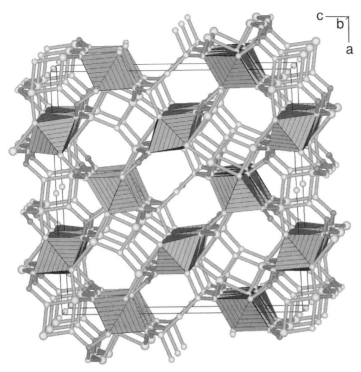

Fig. 1.1.25 Projection of the crystal structure of boralsilite. View parallel **b** rotated by 4° about **a** × **c** and **c** (Peacor et al. 1999).

octahedral coordination (Peacor et al. 1999, Fig. 1.1.25). Substitutions of Mg^{2+}, Fe^{2+} and Fe^{3+} for Al^{3+} and of Al^{3+} for Si^{4+} allows a wide range of mixed crystal formation. Furthermore, partial site occupancies in boralsilite may give rise to a solid solution with werdingite, the boron aluminates $Al_{6-x}B_xO_9$ and A_9B_2, and with aluminum-silicon mullite (mullite in the strict sense).

$Al_{6-x}B_xO_9$ boron aluminates Scholze (1956) was among the first to describe $Al_{6-x}B_xO_9$ with $x = 2$ (i.e. $Al_4B_2O_9$). Mazza et al. (1992) described members of the solid-solution series $Al_{6-x}B_xO_9$ with $1 < x < 3$ crystallizing in space group *Pbam* with lattice constants similar to mullite. According to Mazza et al. these boron aluminates display a pseudo-tetragonal metric with $a = b$. Fischer et al. (2005), however, found that the compound with $x \approx 2$ is monoclinic with space group A $2/m$. Fischer et al. (2005) also suggested that the structure of this boron aluminate is strongly related to boralsilite, with Si^{4+} being replaced by Al^{3+}, and with the lattice constants being doubled ($a = 15.05$ Å, $b = 14.81$ Å, $c = 5.54$ Å, monoclinic angle $\gamma = 90.9°$) with respect to Mazza's unit cell. The mullite-type backbone of the structure, the edge-sharing AlO_6 octahedral chains, are crosslinked by AlO_4 and BO_4 tetrahedra, AlO_5 bipyramids and BO_3 triangles. One oxygen atom is disordered on an interstitial site. Voll et al. (2005) showed that a complete solid solution

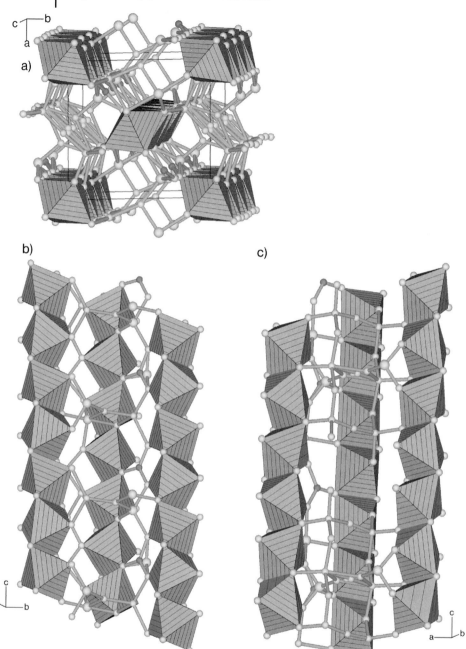

Fig. 1.1.26 Projections of the crystal structure of werdingite (Niven et al. 1991). (a) View approximately parallel **c**. (b) View approximately parallel **a**. (c) View approximately parallel **b**.

Table 1.1.27 MUL-XVI.351 compound: Boralsilite group.

Chemical composition	Mineral name	Unit cell contents	a [Å]	b [Å]	c [Å]	γ [°]	V [Å³]	Reference
$Al_{16}B_6Si_2O_{37}$	boralsilite	$Al_{12}Al_{20}B_{12}Si_4O_{74}$	15.079(1)	14.767(1)	5.574(1)	91.96(1)	1240.4	Peacor et al. (1999)

Table 1.1.28 MUL-XXXII.352 compound: Werdingite group.

Chemical composition	Mineral name	Unit cell contents	a [Å]	b [Å]	c [Å]	α [°]	β [°]	γ [°]	V [Å³]	Reference
$(Mg_{0.84}Fe_{0.16})2Al_{12}$ $(Al_{0.79}Fe_{0.21})2Si_4$ $B_2(B_{0.77}Al_{0.23})2O_{37}$	Werdingite	$Al_8(Mg_{0.84}Fe_{0.16})2Al_4$ $(Al_{0.79}Fe_{0.21})2Si_4B_2$ $(B_{0.77}Al_{0.23})2O_{37}$	7.995(2)	8.152(1)	11.406(4)	110.45(2)	110.85(2)	84.66(2)	650.5	Niven et al. (1991)

exists between Al_5BO_9 ($x = 1$, Al_2O_3-rich) and $Al_4B_2O_9$ ($x = 2$, B_2O_3-rich). Nuclear magnetic resonance inspection clearly shows that the Al_2O_3-rich compound Al_5BO_9 has boron in triangular coordination only, while the B_2O_3-rich phase $Al_4B_2O_9$ contains both triangular and tetrahedrally coordinated boron. The $Al_{6-x}B_xO_9$ phases are stable in a small temperature region (950 to 1050 °C). At higher temperatures they transform to A_9B_2 boron-aluminate ($Al_{18}B_4O_{33}$, see Section 1.1.3.11). Since both compounds belong to the mullite structure family a topotactical transformation with preservation of the octahedral chains is predicted.

1.1.3.15 MUL-XXXII.352, *P* $\bar{1}$: Werdingite Group

Werdingite (ideal formula: $Mg_2Al_{14}Si_4B_4O_{37}$, Moore et al. 1990) occurs in nature in high grade Al_2O_3-rich metamorphic rocks of the granulite facies. In the triclinic phase (space group: $P\bar{1}$) part of Mg^{2+} can be substituted by Fe^{2+}, although the pure magnesium end member of werdingite has been synthesized at elevated temperatures and pressures as well (Werding and Schreyer, 1992). In werdingite the mullite-type octahedral AlO_6 chains are crosslinked by Si_2O_7 and $(Al,Fe)_2O_7$ dimers with typical Si-Si, Al-Al, and (Al,Fe)-(Al,Fe) pairs, $(Al,Mg)O_5$ bipyramids and by BO_3 triangles (Fig. 1.1.26). Formally, the werdingite crystal structure can be derived from that of mullite by the following substitutions:

$$Al^{3+} \to Si^{4+} + 0.5\, O^{2-}; \qquad B^{3+} \to Si^{4+} + 0.5\, O^{2-}; \qquad Mg^{2+} \to Al^{3+} + 0.5\, O^{2-}$$

Cation substitutions go along with the formation of oxygen vacancies (Niven et al. 1991). Because of the various cation substitutions a wide range of mixed crystals can form with coupled cation replacements of Fe^{2+} for Mg^{2+}, Fe^{3+} for Al^{3+} and Al^{3+} for B^{3+} and with various cation ordering schemes. In spite of similar local arrangements no long-range structural correspondence exists between werdingite and grandidierite, andalusite and the boron-aluminates $Al_{6-x}B_xO_9$ and A_9B_2.

Acknowledgments

We thank Werner H. Baur (Evanston) for his comments on the manuscript and for his valuable suggestions, and Elke Eggers (Bremen) for typesetting and formatting the tables.

1.2

The Real Structure of Mullite

S. Rahman and S. Freimann

1.2.1

Introduction

X-ray and electron diffraction patterns of 3/2- and 2/1-mullite show besides the Bragg reflections diverse diffuse scattering phenomena in nearly the whole reciprocal space. The reason for such diffuse scattering is structure disorder. The diffuse scattering in mullite is caused by a certain distribution of oxygen vacancies coupled with cation shifts, resulting in a complicated real-structure configuration. Generally, a real structure cannot be described by a single structure unit (unit cell), which repeats periodically in all three dimensions. In fact, the real structure contains many structure variants, which are distributed by certain rules or functions.

The physical (electrical, mechanical, thermal, optical) properties of crystalline solids depend on the atomic structure. According to the disorder type (point defects, stacking faults, modulations, short-range order, domain formation) a deviation from the ideal structure is present and hence the physical properties are influenced. In order to describe the physical behavior of a crystalline solid a real-structure determination is needed. In the case of mullite, the complex diffuse scattering phenomena cannot be interpreted by classical disorder theories (Daniel and Lipson, 1944, Jagodzinski, 1949, 1964a, 1964b, Kunze, 1959, Korekawa, 1967, Korekawa et al. 1970, De Wolff, 1974, Böhm, 1977, Boysen et al. 1984). For evaluation of the mullite real structure, X-ray single-crystal diffraction and high-resolution electron microscopy (HREM) were applied. The results obtained from the X-ray and HREM investigations were used as input data for a simulation and reconstruction technique, called the videographic method (Rahman, 1991), to determine and describe the real structure of 3/2- and 2/1-mullite.

1.2.2
High-resolution Electron Microscopy

High-resolution electron microscopy (HREM) is one of the most important experimental methods for determining the real structure of disordered single crystals. Deviations from the ideal periodic arrangement (point defects, modulations, short-range order, domain boundaries, etc.) are typical examples that can be examined applying HREM. A proper interpretation of the experimental real-structure images is only possible via contrast simulations using predefined superstructure models.

HREM investigations of mullite were first performed for Al-rich mullite ($0.25 < x < 0.4$, see Section 1.1.3.7) by Nakajima et al. (1975), who directly correlated varying spot intensities with different vacancy concentrations along [010]. The authors ascertained a periodic arrangement of increased vacancy concentrations every third or fourth layer parallel to [100]. Based upon image simulations, Ylä-Jääski and Nissen (1983) observed in mullites with $x = 0.48$ and 0.54 antiphase domain boundaries running parallel to [100] and slightly inclined against [601], respectively. In an investigation of 1.71/1-mullite, which was obtained from sintering Al_2O_3 and $ZrSiO_4$, Schryvers et al. (1988) reported a random distribution of vacancies without any direct correlation between contrast pattern and vacancy distribution. Comparable results to Schryvers et al. (1988) were published for 3/2-mullite by Epicier et al. (1990) and Epicier (1991), who stated that less than 50 % vacancies cannot be detected along [001] and the vacancies are randomly

Fig. 1.2.1 Electron diffraction pattern (top) *h0l* plane, (bottom) *0kl* plane of 2/1-mullite.

distributed. However, a random distribution of oxygen vacancies in the mullite structure would cause a monotone diffuse background in contrast to the observed diffuse maxima with different shapes in the diffraction patterns of mullite (Fig. 1.2.1).

Using systematically developed structure models, an attempt was made by Rahman and Weichert (1990) and Rahman (1993b, 1994a) to determine the arrangement of oxygen vacancies semi-quantitatively (always coupled to an occupation of an Al* site) by analyzing observed differences of the image contrast in HREM images of the *ab* plane. In further investigations (Paulmann et al. 1994, Rahman et al. 1996) 200-kV and 300-kV HREM images of mullite with beam direction parallel to [010] and [100] have been compared with extensive multi-slice calculations of structure models with different oxygen-vacancy arrangements.

This section presents the experimental results of HREM and their interpretation regarding the oxygen-vacancy distribution in the three main planes via contrast simulation. Hitachi H-800 and H-9000 electron microscopes operating at 200 and 300 kV, respectively, were used for the high-resolution observations. Both microscopes were equipped with LaB$_6$ cathodes and top-entry high-resolution specimen stages with a ±15° tilting angle. The images were recorded either on photographic film or, in special cases (image reconstruction), by using a video unit connected to

Table 1.2.1 Microscopic constants (H-800 and H-9000) used for image simulations.

	H-800	H-9000
Accelerating voltage [kV]	200	300
Wavelength λ [nm]	0.0025079	0.0019688
Spherical aberration C_s	1.0	0.9
Chromatic aberration C_c	1.1	1.5
Focus spread Δ [nm]	5	5
Divergence α [rad]	5×10^{-4}	5×10^{-4}

an image-processing system. All of the HREM images were taken from very thin mullite crystals (t = 2-4 nm, x = 0.25 and 0.4) with the electron beam parallel to the [001], [010] and [100] zone axes. Preparation of the crystallites for the investigations was carried out by crushing them with propanol in an agate mortar and transferring the suspension onto copper grids covered with a perforated carbon film. The microscope constants adopted for the image simulations are listed in Table 1.2.1. The image contrast simulations were done using the program system SIM40 (Rahman and Rodewald, 1992) based upon the multi-slice approximation (Cowley and Moodie, 1957). The program system was specially developed for the contrast simulation of real structures.

Diffuse scattering with different intensity distributions is observed when $h0l$ and $0kl$ electron-diffraction patterns are examined (Fig. 1.2.1). The diffuse maxima of the $h0l$ plane at the positions $2h\pm2/3$, 0, $l\pm1/2$ were interpreted in the past by several authors (McConnell and Heine, 1985, Angel and Prewitt, 1987, Angel et al. 1991, Padlewski et al. 1992b) as satellite reflections, indicating a twofold modulation of the c axis and an incommensurate modulation along the a axis. Such an interpretation, assuming maximal ordering, ignored the diffuse streaks running approximately parallel to the c^* axis. However, the $0kl$ diffraction pattern shows diffuse maxima at positions 0, $2k\pm2/3$, $l\pm1/5$, which does not agree with a doubling of the c lattice constant within this plane. Furthermore, many other diffuse figures are present, for example cross-shaped streaks in the $0kl$ plane, and many complicated diffuse scattering phenomena in reciprocal planes perpendicular to c^* (Freimann and Rahman, 2001), which will be presented and discussed in Section 1.2.3.

In order to interpret the experimental HREM images and to determine the oxygen vacancy distribution, contrast simulations with predefined structure models were calculated for the three main directions. To investigate the influence of contrast variation with different concentrations of vacancies along the beam direction, calculations with a $5a \times 5b \times 6c$ super cell were performed with individual slice sequences as shown schematically in Fig. 1.2.2a (only the vacancy positions are shown). To avoid cut-off effects in calculations of the (010) and (100) plane, the first and last slice were chosen to be vacancy-free. The projected vacancy distributions and concentrations are presented schematically in Fig. 1.2.2b–d, while the corre-

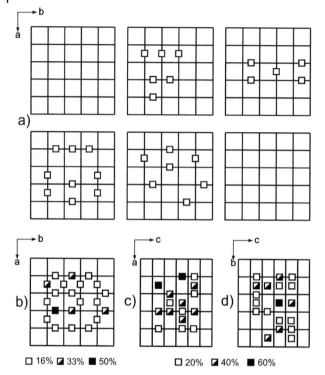

Fig. 1.2.2 (a) Slice sequences for HREM contrast simulation of vacancy distributions; (b) Vacancy concentrations projected on (001); (c) Vacancy concentrations projected on (010); (d) Vacancy concentration projected on (100).

sponding calculated images along [010], [100] and [001] are shown in Figs. 1.2.3 and 1.2.4, respectively. Depending on the actual defocus value (–35 nm, –75 nm), an accurate detection of about 20% vacancy concentration is possible in [010] and [100] (Fig. 1.2.3).

In the case of (010) simulations (Fig. 1.2.3) the contrast maxima directly coincide with the projected O_c (i.e. O3) positions and higher vacancy concentrations cause an easily detectable intensity enhancement of the corresponding dot in the calculated image due to the associated cation shifts (Al*) near the oxygen vacancy. Similarly in (100) simulations (Fig. 1.2.3) oxygen vacancies are responsible for the appearance of an additional dot between the main maxima. A simultaneous shifting of the neighbouring main maxima results in a broad dot of higher intensity with enhanced dark regions along [010] in the immediate vicinity of a vacancy position.

In comparison, simulations of the (001) plane (Fig. 1.2.4) are characterized by four more-or-less elongated maxima within a mullite unit cell, best resolved at defocus values of –50 nm and –55nm. As discussed earlier in more detail (Rahman and Weichert, 1990, Rahman, 1993b, Rahman, 1994a), the maxima can be

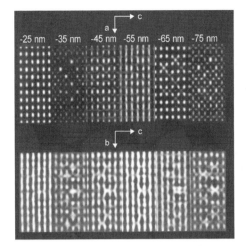

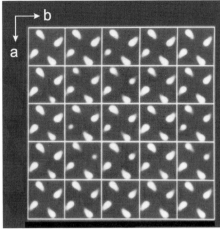

Fig. 1.2.3 HREM contrast simulations along [010] (top) and [100] (bottom) of the 3D structure model shown in Fig. 1.2.2a (thickness = 3.84 nm and 3.78 nm).

Fig. 1.2.4 HREM contrast simulation along [001] of the 3D structure models shown in Fig. 1.2.2a (thickness = 3.46 nm, $\Delta F = -50$ nm). The simulated contrast becomes reduced in the vicinity of an oxygen vacancy.

directly correlated with channels beside the O_c (O3) positions. The channels are further enhanced by introducing an oxygen vacancy and cause the contrast located clockwise next to an oxygen vacancy to become less intense. A correlation with occupied Al* positions is obvious, so that contrast changes are mainly affected by cation shifts. This simple relationship and the correspondence of the simulated contrast pattern (Fig. 1.2.4) with the experimental structure image (Fig. 1.2.5) permit a semi-quantitative determination of enhanced vacancy concentrations along [001] for a thin mullite crystal. Since comparable results with negligible differences were obtained for multi-slice simulations along the main crystallographic direc-

Fig. 1.2.5 HREM image (200 kV) of the *ab* plane of 3/2-mullite. One unit cell is outlined.

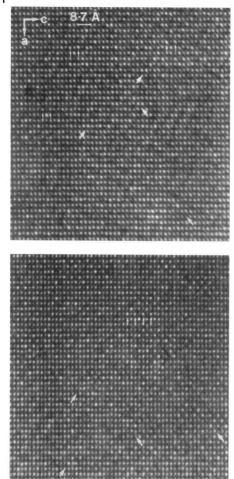

Fig. 1.2.6 200-kV HREM images of 3/2-mullite (top) and 2/1-mullite (bottom) along [010]. The arrows indicate the preferred orientations of oxygen vacancies along <102>. Small arrows indicate the twofold periodicity along [001] in the projection.

tions with an accelerating voltage of 300 kV, a separate presentation is not required.

HREM images of 3/2- and 2/1-mullite with a beam direction parallel to [010] and an approximate defocus value of -25 nm, are presented in Fig. 1.2.6. Although both images exhibit intensity variations of the contrast pattern, these variations are more pronounced in the image of Al-rich 2/1-mullite, whereas the contrast pattern of 3/2-mullite shows minor intensity deviations. Bearing the above mentioned features of 200-kV HREM contrast simulations in mind, it is obvious that an enhancement of certain dots must be attributed to higher vacancy concentrations along [010]. The most striking property of HREM images of the (010) plane are linear arrangements of enhanced dots with preferred orientations parallel to <102>. With an extension between 0.40 nm to 2.42 nm in 2/1-mullite and 0.40 nm to 1.20 nm in 3/2-mullite, the length of the rows are found to be closely related to

Fig. 1.2.7 HREM images of 2/1-mullite along [100] (top 200 kV, bottom 300 kV). The arrows indicate the preferred orientations along <012> (top) and <013> (bottom). Small arrows indicate the period of 1.5 b.

the chemical composition of the investigated mullite. More conspicuous in HREM images of 2/1-mullite, the ordering scheme of columns with enhanced vacancy concentrations along [010] furthermore often reveals a period of 1.5a and 2c.

In contrast to the appearance of distinct white spots in HREM images of the (010) plane, 200-kV images along [100] (Fig. 1.2.7) do not show a definite period parallel to [010], because the 020 reflections (d_{020} = 0.384 nm) are relatively weak, whereas the stronger 040 reflections (d_{040} = 0.192 nm) do not contribute to image formation owing to a point-to-point resolution of 0.23 nm. However, brighter dots in HREM images at a defocus value of -0.25 nm indicate higher vacancy concentrations along the illumination direction. Preferred orientations parallel to <012> and <001> are marked by arrows in Fig. 1.2.7. By examining a 300-kV image (point-to-point resolution of about 0.17 nm), arrangements of enhanced vacancy concentrations exhibit an undulatory form with an average direction along <013> (Fig. 1.2.7). Consistent with the positions of diffuse maxima in 0kl diffraction patterns, the average distances between these arrangements reveal an interval of 1.5b and a five- to sixfold periodicity along [001].

Summarizing the results of the HREM investigation and analysis of the electron diffraction pattern, it is evident that mullite does not exhibit a perfectly ordered arrangement of oxygen vacancies. Weak diffuse regions as well as the diffuse max-

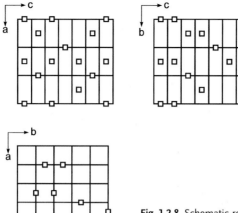

Fig. 1.2.8 Schematic representation of the ordering schemes of the oxygen vacancies in the three main crystallographic planes.

ima in $h0l$ and $0kl$ diffraction patterns imply the formation of short-range-order regions. The real structure is characterized by partly ordered regions ("domains") with enhanced vacancy concentrations showing a specific correlation along preferred directions, as indicated in experimental HREM images. A schematic representation of the vacancy arrangements in the main crystallographic planes is presented in Fig. 1.2.8.

Although HREM investigations yield valuable information about preferred vacancy arrangements, one has to deal with some restrictions regarding the 3D interpretation of the contrast pattern, since they are caused by a projection of the structure along the incident beam. However, the HREM results in the three main crystallographic directions give important parameters as input data for two-dimensional and three-dimensional videographic real-structure simulations, as discussed in Section 1.2.4.

1.2.3
X-ray Investigation

In order to solve the real structure of mullite the diffuse scattering in the whole reciprocal space was determined. To get an overview of the position of the diffuse scattering, an X-ray rotation photograph (Mo Kα) about the c axis for mullite with $x = 0.4$ is presented in Fig. 1.2.9. The rotation photograph shows diffuse layers with $l = 1/2, 1/3, 1/4, 1/6$. To give an impression of the complicated diffuse intensity distribution in these reciprocal planes a precession photograph with quartz monochromator for the $hk1/4$ plane is given in Fig. 1.2.10. In this precession photograph there are diffuse circular arcs around the positions 1,0,0.25 and 0,1,0.25 and diffuse streaks parallel to the directions <110>*. Without going into details such a complicated diffuse scattering distribution cannot be explained by the known disorder theories (Daniel and Lipson, 1944, Jagodzinski, 1949, 1964a,b,

hk1

hk0

hk1/2
hk1/3
hk1/4
hk1/6

Fig. 1.2.9 X-ray rotation photograph about the *c*-axis for
2/1-mullite.

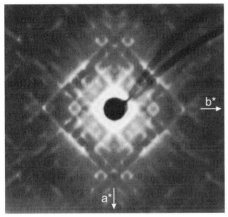

b*

a*

Fig. 1.2.10 X-ray precession photograph
(Mo Kα) of the *hk*1/4 reciprocal plane of
2/1-mullite (*h*, *k* ≤ 8).

Kunze, 1959, Korekawa, 1967, Korekawa et al. 1970, De Wolff, 1974, Böhm, 1977,
Boysen et al. 1984).

The lack of 3/2-mullite single crystals for X-ray investigations made it impossible to obtain diffraction patterns (four-circle diffractometer) for all reciprocal
planes as in the case of 2/1-mullite. In order to overcome these difficulties Rahman et al. (2001) prepared a 3/2-mullite single crystal from a sillimanite single
crystal by slow thermal heating and annealing at 1600 °C for 24 h. For a complete
characterization of the diffuse scattering in 3/2- and 2/1-mullite, a four-circle diffractometer was used to measure the intensity distribution in *hk*1/2, *hk*1/3, *hk*1/4
and *hk*1/6. The experimental results of the four-circle measurements for *hk*1/4 are
presented in Fig. 1.2.11. The inner part (close to the incident beam spot) of the
four-circle diffraction pattern is better resolved than in the precession photograph
(Fig. 1.2.10). Comparing the diffraction patterns of 3/2 and 2/1-mullite it can be
seen that both compositions in principle show the same scattering phenomena;
however, the diffuse figures of 3/2-mullite are broadened and have weaker relative

b*

a*

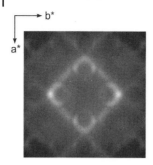 

Fig. 1.2.11 X-ray diffraction pattern of 2/1-mullite (left) and 3/2-mullite (right) for $hk1/4$ (h, $k \leq$ 2.5) showing complicated diffuse scattering.

intensities. Furthermore the exact positions of the maxima of intensity are slightly different in 2/1- and 3/2-mullite.

The information about the real-structure configuration in mullite is implicitly contained in the shape and intensity distribution of the diffuse scattering patterns. A direct method of solving the real structure of mullite with such complicated diffuse scattering phenomena is not available. In less complicated cases these difficulties can be overcome by applying optical Fourier transforms or Monte-Carlo simulations using predefined structural models. In the case of mullite, representative structure models cannot be easily derived because of the great number of varieties of the model. Rahman (1991) suggested, that the vacancy distribution in mullite can be described by inter-vacancy correlation vectors. Butler and Welberry (1994) calculated SRO parameters of mullite from the intensities of the diffuse scattering. The diffuse scattering was measured in the area hkl with $h,k < 4.5$ and $0.5 < l < 1.0$ and analyzed using least-squares techniques by deriving an equation for the diffuse scattering that only involves the local order of the oxygen vacancies. The calculated diffraction patterns of a Monte-Carlo simulation, in which 12 interaction energies were adjusted to fit 12 of the 18 presented SRO parameters, show some similarities to the experimental patterns but still show some differences. However, a complete description of the mullite real structure is not given. In the next section a simulation and reconstruction method will be applied to mullite in order to solve and describe the real structure.

1.2.4
Real-structure Determination Using Videographic Reconstruction and Simulation Techniques

The videographic method is a simulation and reconstruction procedure that uses a statistical mathematical approach and computer graphics to aid the interpretation of diffuse scattering (X-ray, electron, neutron) from a disordered crystal and to solve the real structure. A detailed description of the videographic method was given by Rahman (1991, 1992, 1993a, 1994b) and Rahman and Rodewald (1995), so the method will only be briefly outlined here.

1.2.4.1 The Videographic Method

The basic principle of the videographic method is the representation of the real structure as a computer graphic, where different atoms are replaced by picture elements (pixels) of different grey levels according to their scattering power. As the videographic method of 1991 was developed, a special graphic adapter was needed to display the image with 256 grey levels. Moreover, an array processor was used for the time-consuming calculations (Fourier transforms). Nowadays these components are no longer necessary, because most personal computers are able to meet the above-mentioned requirements. The videographic method contains two procedures complementing each other, the videographic simulation of the real structure and the videographic reconstruction, which enables to draw conclusions from the diffraction pattern about the atomic arrangements.

Videographic simulation is performed by the distribution of structure variants, which can be derived from the average structure according to the approach that an average structure $<\varphi>$ can be described as a superposition of n possible structure variants φ_j:

$$<\varphi(x,y,z)> = \frac{1}{N}\sum_{j=1}^{n}\varphi_j(x,y,z) \times N_j \qquad (2)$$

where N = total number of structure variants and N_j = number of the structure variants j.

These structure variants are distributed using a random variable J taking the values j ($j = 1, ..., n$) with certain probabilities determined by the distribution function of J. The distribution function is defined in terms of the combination probabilities W_{ij} (Rahman, 1991) for the direct combination of two structure variants along the three main simulation directions a, b and c. For the distribution behind the first shell an input of influence factors for any correlation vectors is used (Rahman and Rodewald, 1995). With the aid of the influence factors f_v a structure variant A (or φ_i) can directly influence the probability of the occurrence of the structure variant B (or φ_j) at the position of a correlation vector lmn:

$$P_B(lmn) = P_B^{Tab} \times f_v \qquad (3)$$

where P_B^{Tab} = probability of structure variant B according to the tables containing the combination probabilities and lmn = components of the correlation vector between A and B.

In this way preferences beyond the first or second shell can be considered. A three-dimensional simulation field $S(L,M,N)$ with L rows, M columns and N layers can be expressed as

$$S(L,M,N) = \sum_{l}^{L}\sum_{m}^{M}\sum_{n}^{N}\varphi_{lmn}(J_{lmn}) \qquad (4)$$

where $\varphi_{lmn}(J_{lmn})$ = structure variant of type J at an lmn position, J_{lmn} = random variable for an lmn position and l, m, n = integers.

The videographic real-structure image $S(x,y)$ can be obtained by replacing every structure variant by a videographic pixel pattern. To check the result of a simulation, the Fourier transformation of the real-structure image $F[S(x,y)]$ is calculated:

$$Q(u, v) = F\left[S(x, y)\right] = \frac{1}{LR} \sum_{x=0}^{L-1} \sum_{y=0}^{R-1} S(x, y) \exp\left[-2\pi i\left(x\frac{u}{L} + y\frac{v}{R}\right)\right]$$ (5)

where L, R = number of rows and columns of the videographic image, x, y = coordinates of the videographic image and u, v = coordinates in Fourier space.

The Fourier transforms $Q(u,v)$ are displayed for comparison with the experimental diffraction patterns. The input parameters (combination probabilities and influence factors) are systematically varied until a best fit between simulation and experiment is reached. It must be pointed out that the simulation field must have an appropriate size [(L, M, N, Eq. (4)] in order to obtain representative results. In a further step the 3D simulation field is analyzed to obtain the frequencies of correlation vectors lmn and control the chemical composition (frequency of each structure variant).

To enable the calculation of diffraction patterns for any reciprocal layers (in the case of mullite for example $hk1/6$, $hk1/4$, $hk1/3$, $hk1/2$) the 3D simulation field can be converted into a floating point data file containing the coordinates (x, y, z) of the atoms and the temperature factors. The positions of the atoms (x, y, z coordinates) contained in the files (of the variants) are converted according to the position of the variant in the simulation field into atomic positions related to the supercell of the 3D simulation field. Using the atom coordinates of the real structure (supercell) the structure factors F_{hkl} can be calculated for every reciprocal plane conventionally.

In order to get structural information for the input parameters of the simulation, the videographic reconstruction can be applied, which enables recovery of real-structure configurations by filter operations in reciprocal space. The principal idea in a reconstruction of an unknown real structure is that the diffuse scattering in the diffraction pattern of a partly ordered structure is a subset of the monotone diffuse background for a random disorder (Rahman, 1992). For this reason, a randomly disordered structure is usually assumed as a starting model. In addition to the Bragg peaks $Q(u,v)_B$, its Fourier transform shows a monotone diffuse background $\Delta Q(u,v)_d$. Image reconstruction of the real structure $S(x,y)$ can be achieved by selective filtering of certain frequencies by means of the transfer function. In this case $Q(u,v)_B$ or $\Delta Q(u,v)_d$ can be multiplied by the transfer functions $G_1(u,v)$ and $G_2(u,v)$, respectively:

$$F\left[\varphi(x,y)\right] = Q(u,v)_B \times G_1(u,v) + \Delta Q(u,v)_d \times G_2(u,v)$$ (6)

The values of the transfer functions can be chosen to select either the diffuse regions, parts of the diffuse regions or the Bragg reflections, which are accounted for by a backward Fourier transform. In order to obtain information about the real-structure arrangements causing the diffuse scattering in mullite, specific frequencies of the diffuse background $\Delta Q(u,v)_d$ are selected by the transfer function

$G_2(u,v)$. The positions of the selected areas are chosen according to the positions of diffuse scattering in the experimental diffraction pattern in mullite; the other frequencies of $\Delta Q(u,v)_d$ are excluded except the Bragg reflections. The transfer function can be expressed as a convolution between the reciprocal lattice function $F(u,v)$ and a window function $W(u,v)$, which determines the coordinates of the selected areas and corresponds to a filter mask.

$$G(u,v) = F(u,v) * W(u,v) \tag{7}$$

The backward Fourier transform of an image with L rows and R columns shows a reconstructed real-structure image $S'(x,y)$ with a different contrast distribution compared to the starting model.

$$S'(x,y) = \int_u \int_v \left(Q(u,v)_B \cdot G_1(u,v) + \Delta Q(u,v)_d \cdot G_2(u,v) \right) \cdot \exp\Omega \cdot du\,dv$$

$$\text{where } W = 2\pi i \left(x\frac{u}{L} + y\frac{y}{R} \right) \tag{8}$$

By visual inspection of the reconstructed real-structure images, preferred atomic arrangements in the real structure can be determined, but no quantitative results regarding the frequencies of correlation vectors are available. For this reason a correlation function $P(X,Y)$ can be calculated from the Fourier transforms of the reconstructed images:

$$P(X,Y) = \sum_{u=0}^{L-1} \sum_{v=0}^{R-1} |Q(u,v)|^2 \quad \cos 2\pi \left(X\frac{u}{L} + Y\frac{v}{R} \right) \tag{9}$$

where X, Y = coordinates of the correlation function.

Analysing the intensities of the correlation (Patterson) function the frequencies of correlation vectors can be estimated and used as input parameters for a new videographic simulation.

To attain a complete characterization of the ordering scheme in the mullite real-structure, 2D reconstructions for the main planes were first performed, and these results were tested by 2D videographic simulations. In order to describe the oxygen vacancy correlations in the real structure, the lmn correlation vectors between oxygen vacancies are defined as:

$$l = 1/2a, \; m = 1/2b, \; n = c \; (a, b \text{ and } c \text{ are the lattice constants}). \tag{10}$$

Then a 3D ordering scheme was derived containing the most important correlation vectors. Via 3D simulations of the mullite real structure this ordering scheme was refined and confirmed. The results of the videographic simulation and reconstruction method will be presented in the following sections.

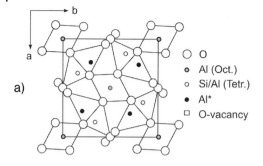

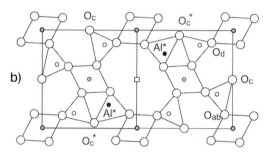

Fig. 1.2.12 Structure of mullite (a) Average structure (Angel and Prewitt 1986); the origin of the unit cell is displaced by $a+1/2$ and $b+1/2$ against Angel and Prewitt 1986); (b) Atomic displacements around an oxygen vacancy (following the nomenclature of Angel and Prewitt 1986 for the atomic positions).

1.2.4.2 Structure Variants of Mullite

The structure variants are derived from the average structure [Fig. 1.2.12a, Eq. (2)] considering crystal-chemical rules for bond lengths, atomic distances and coordination number. To derive the structure variants the following restrictions were applied: The occupation of the tetrahedral Si/Al and Al* sites depends on the position of oxygen atoms or vacancies. If the O_c site is occupied, the two adjacent Si/Al sites are occupied too. However, next to a vacancy the adjacent Al* sites are occupied instead of the Si/Al sites, and the coordinating oxygen atoms shift from the O_c to the O_c^* position (Fig. 1.2.12b). Structure variants with two vacancies building the correlation vector <110> are not allowed, because an Al or Si atom would only be threefold coordinated. Si/Al-O_c^* bond lengths of 0.173 and 0.178 nm lead to a tetrahedral occupation by Al, whereas 0.167 nm for the Si/Al-O_c bond gives Si occupation of the T position.

Considering these rules and different O_c and O_c^* occupations, 34 structure variants (Fig. 1.2.13) result from the decomposition of the mullite average structure. Each four variants (1-4, 5-8, 9-12, 13-16) exhibit an oxygen vacancy on the same cell edge and different O_c and O_c^* occupations on the remaining three edges. Structure variants 17 to 32 are vacancy-free, but with a variation on the O_c/O_c^* sites. Variants 33 and 34 represent the silica free ι-Al_2O_3 modification proposed by Saalfeld (1962).

Since it is difficult to handle simulations with such a great number of structure variants (large combination tables with a great number of combination probabilities and complicated conditional probabilities leading to great deviations between

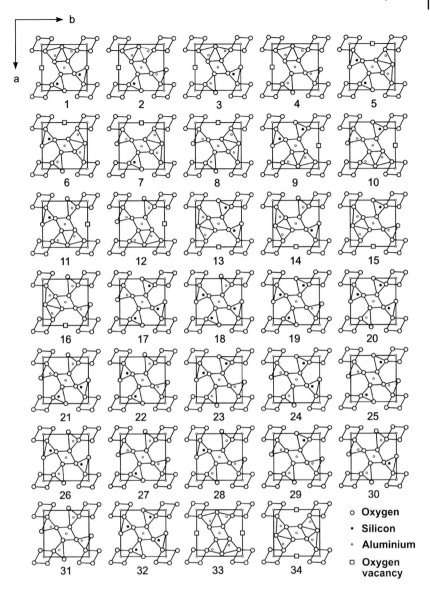

Fig. 1.2.13 34 structure variants of mullite.

the input and output parameters), the number of structure variants can be reduced to 7 by neglecting the O_c* positions. Fig. 1.2.14 shows seven structure variants for mullite, which differ in the position of oxygen vacancies. Variants 1 to 4 have one vacancy on one edge of the unit cell, in variant 5 all O_c positions are occupied and variants 6 and 7 have two vacancies on opposite edges. Since the occupation of the O_c* site is coupled with the occurrence of vacancies, it is possible to generate a

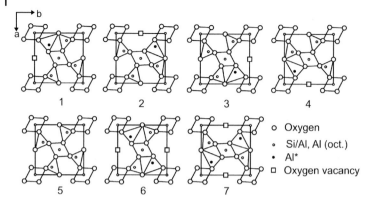

Fig. 1.2.14 7 structure variants of mullite neglecting the O_c* position.

simulation field with the seven structure variants and subsequent convert the seven into the 34 different variants (Freimann, 2001). In this way the Fourier transforms of the videographic simulations are calculated for the correct representation of the structure.

1.2.4.3 Two-dimensional Videographic Reconstructions

For the videographic reconstruction of the *ac* and *bc* planes a simplified starting model, consisting only of oxygen atoms and vacancies [both on the O_c site], was used. The oxygen atoms and vacancies were distributed randomly and the Fourier transform showed a diffuse background alongside the Bragg reflections. The selected regions in the filter masks coincide with the diffuse regions of the experimental diffraction patterns for *h0l* and *0kl*. According to Eq. (7) the filter masks are convoluted with the Fourier transform of the starting model. The filter masks contain the Bragg reflections, the diffuse maxima and the diffuse streaks. All other frequencies that are not included in the filter mask (not present in the experimental diffraction pattern) are set to zero and only the selected areas are considered for backward Fourier transformation (Fig. 1.2.15a). The resulting reconstructed real-structure images (Fig. 1.2.15b) reveal an intensity distribution typical for the applied filter masks. In the reconstructed image of the *ac* plane, specific arrangements of pixels with relatively high intensities building the correlation vectors <3*m*0>, <0*m*2> and <1*m*1> can be observed. The reconstructed image of the *bc* plane shows preferred arrangements along the approximate direction <*l*23> often in a distance of 3*b* and 5*c*. In order to determine the frequencies of correlation vectors (*lmn*) the correlation function [Eq. (9)] is calculated from the reconstructed image. Parts of the correlation (Patterson) functions for the *ac* and *bc* planes are presented in Fig. 1.2.15c. The intensities in the correlation functions are proportional to the frequencies of correlation vectors. Analysing the intensity distribution of the correlation functions the following most frequent 2D correlation vectors were estimated for the *ac* and *bc* planes:

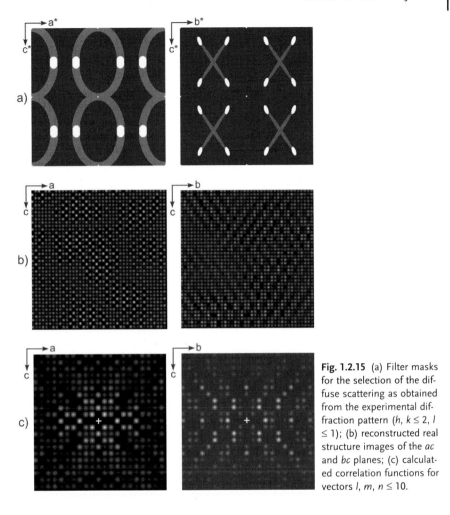

Fig. 1.2.15 (a) Filter masks for the selection of the diffuse scattering as obtained from the experimental diffraction pattern (h, $k \leq 2$, $l \leq 1$); (b) reconstructed real structure images of the ac and bc planes; (c) calculated correlation functions for vectors l, m, $n \leq 10$.

ac plane: $<3m0>$, $<1m1>$, $<0m2>$
bc plane: $<l30>$, $<l22>$, $<l23>$, $<l05>$.

The 2D correlation vectors obtained from the videographic reconstruction were used as input parameters for 2D videographic simulations. The Fourier transforms of the 2D simulations are in good agreement with the experimental diffraction patterns shown in Fig. 1.2.1, which indicates the correctness of the reconstruction procedure. Videographic reconstructions were also performed for the $hk1/2$, $hk1/3$, $hk1/4$ and $hk1/6$ reciprocal planes of 2/1-mullite (Freimann et al. 1996, Freimann, 2001). Without going into details the most frequent correlation vectors were determined for each layer as shown in Fig. 1.2.16. It must be noted, that the 2D reconstruction of $hk1/2$, $hk1/3$, $hk1/4$ and $hk1/6$ only have a two-dimensional character and cannot be directly used for 3D simulation.

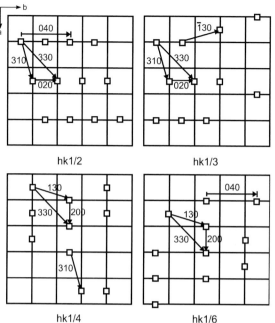

Fig. 1.2.16 Schematic representation of the oxygen vacancy correlation vectors obtained from the reconstruction for *hk*1/2, *hk*1/3, *hk*1/4 and *hk*1/6.

1.2.4.4 Three-dimensional Videographic Simulations for 2/1-and 3/2-mullite

In order to obtain representative results for the distribution function, a simulation field containing 48 × 48 × 96 structure variants, was chosen for videographic 3D simulations. From the results presented in the previous sections (HREM, X-ray, videographic 2D reconstructions and simulations) 2D correlation vectors for the three main planes were obtained, which allowed the derivation of the 3D correlation vectors presented in Table 1.2.2. The 1.5-fold period along [100] and [010] in *h0l* and *0kl* diffraction pattern (Fig. 1.2.1) are caused by a preference of the <310>, <130> and <330> correlation vectors. Furthermore, it can be assumed that the twofold period along [001] in *h0l* patterns originates from more frequent <022> correlation vectors, which already appear in (100) HREM images as preferred arrangements along the [012] and [0-12] crystallographic directions (Fig. 1.2.7). Thus, a projection of the <022> vector onto (010) results in a twofold period along [001].

The preferred vacancy arrangements along the crystallographic directions [102] and [-102] observed in HREM images of the *ac* plane (Fig. 1.2.6) can be achieved by a preference of the <111> correlation vector in agreement with the 2D reconstruction and simulation of the *ac* plane with the high frequency of the 2D vector <1*m*1>. The preference of the 3D correlation vector <201> corresponds to a high frequency of <2*m*1> in the reconstruction of the *ac* plane and - in projection on (100) - to a vacancy arrangement along [001], which appears in (100) HREM images. In addition, the correlation vectors <131> are preferred in the 3D simulations.

Table 1.2.2 Derivation of the preferred 3D correlation vectors (*lmn*) from the most frequent 2D vectors.

ab	ac	bc	3D
<31*n*>	<3*m*0>		<310>
<33*n*>	<3*m*0>	<*l*30>	<330>
<13*n*>		<*l*30>	<130>
	<1*m*1>	<*l*11>	<111>
	<2*m*1>	<*l*01>	<201>
	<1*m*1>	<*l*31>	<131>
<02*n*>	<0*m*2>	<*l*22>	<022>
<31*n*>	<3*m*2>	<*l*12>	<312>

The 3D videographic simulations [Eq. (4)] were carried out first for 2/1-mullite using the 3D vectors presented in Table 1.2.2. The correctness of the simulated structure was tested by comparing its Fourier transforms with the experimental diffraction patterns (precession photographs and four-circle diffractometer). The simulations were successively refined until the best fit was found (Rahman et al. 1996, Freimann and Rahman, 2001). The Fourier transforms calculated after projecting the 3D videographic simulation field onto the *ac* and *bc* plane are presented in Fig. 1.2.17 together with the corresponding X-ray precession photographs. In both the X-ray precession pattern and the calculated Fourier transform of the *h0l* plane there are the rounded streaks with the maximum intensities at the approximate positions 1.3 *a** and 0.5 *c**. In the calculated *0kl* plane the cross-shaped streaks and the diffuse maxima agree in position (1.4 *b** and 0.19 *c**) and relative intensity distribution with the precession pattern. The differences at greater scattering angles may result from the fact that the calculated Fourier transforms do not include the dependence of intensity on the scattering angle.

In order to compare the agreement in the reciprocal planes perpendicular to *c**, Fig. 1.2.18 presents the experimental diffraction patterns of 2/1-mullite and the Fourier transforms for the 3D simulation. In all these planes a very good agreement is observed. The calculated *hk*1/2 plane shows the diffuse maxima at the correct position on the *a** axis at 1.3*a**, corresponding to the diffuse maxima in the *h0l* plane. In the *hk*1/3 plane the maximum intensities are at the same position but additional diffuse circular arcs around the reciprocal coordinates 1, 0, 1/3 and 0, 1, 1/3 appear. In the *hk*1/4 plane the diffuse arcs change into angular shapes and there are streaks parallel to the direction <110>*, which are also present in the *hk*1/6 plane. In both the *hk*1/4 and *hk*1/6 planes the maxima of diffuse intensity are on the *b** axis at approximately 1.4*b**.

Because of the good agreement between the Fourier transforms and the experimental patterns, this 3D simulation gives a good representation of the vacancy distribution in 2/1-mullite. The simulation field was analyzed for all correlation

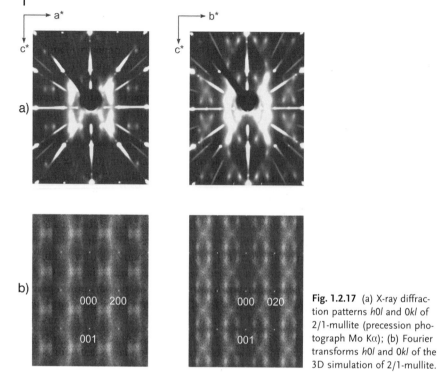

Fig. 1.2.17 (a) X-ray diffraction patterns *h0l* and *0kl* of 2/1-mullite (precession photograph Mo Kα); (b) Fourier transforms *h0l* and *0kl* of the 3D simulation of 2/1-mullite.

vectors $l,m,n \leq 6$ (Freimann, 2001). Hence the simulation and the real structure can be completely described. The frequencies of correlation vectors (up to 060) are given in Fig. 1.2.20. It is remarkable, that the preferred vectors indeed have the highest frequencies (<310>, <111>, <022>, <201>, <330>). For some of the vectors it is possible to describe their influence on specific details of the diffuse scattering: For example the diffuse maxima in the *h0l* and *hk*1/2 plane can be explained with the vectors <310> and <022>, in which <310> is responsible for the location along the *a** axis at 1.3*a** and <022> for the location along the *c** axis at 1/2*c**.

The single-crystal diffraction patterns for the planes *hk*1/2, *hk*1/3, *hk*1/4 and *hk*1/6 of 3/2-mullite are shown on the left in Fig. 1.2.19. Comparing this with the diffraction pattern of 2/1-mullite (Fig. 1.2.18) it can be seen that both compositions (2/1- and 3/2-mullite) in principle show the same scattering phenomena, although the diffuse figures of 3/2-mullite are broadened and have weaker relative intensities. Furthermore the exact positions of the maxima of intensity are slightly different in 2/1- and 3/2-mullite. Because of the similar diffraction patterns of 3/2-mullite we supposed that the results for 2/1-mullite were also valid for 3/2-mullite. Thus 3D simulations for 3/2-mullite were performed preferring the same correlation vectors but with reduced frequencies because of the lower concentration of vacancies. Again the simulation was refined by varying the frequencies of the preferred vectors until the Fourier transforms of the simulation showed the best agreement with the experimental diffraction patterns. In Fig. 1.2.19, the

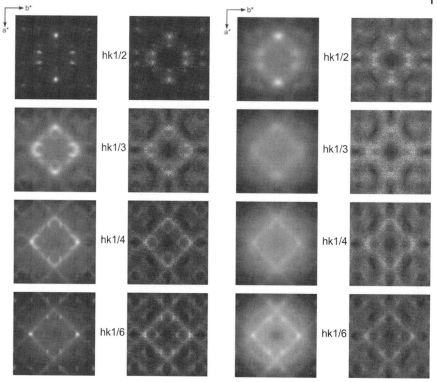

Fig. 1.2.18 (Left) X-ray diffraction pattern of 2/1-mullite; (right) Fourier transforms of the 3D simulation of 2/1-mullite ($h, k \leq 2.5$).

Fig. 1.2.19 (Left) X-ray diffraction pattern of 3/2-mullite; (right) Fourier transforms of the 3D simulation of 3/2-mullite ($h, k \leq 2.5$).

experimental diffraction patterns of 3/2-mullite are presented on the left and the calculated patterns from the 3D simulation are shown on the right. The shapes and the maximum intensities of the diffuse scattering are very similar in the experiment and the simulation. The simulation field was analyzed and the vector frequencies are presented in Fig. 1.2.20. Comparing the frequencies of inter-vacancy correlation vectors in this simulation with those for 2/1-mullite, it is clear that both graphs in Fig. 1.2.20 have a similar shape, indicating that the vacancy distribution in both compositions generally obeys the same rules, except that the frequencies of the correlation vectors are lower in 3/2-mullite. There are only small differences in the sequence of correlation vectors (the sequence of 2/1-mullite: <310>, <111>, <022>, <201>, <330>; the sequence of 3/2-mullite: <022>, <201>, <111>, <310>, <130>). Hence it is evident that 2/1- and 3/2-mullite have very similar ordering schemes of oxygen vacancies and the real structures can be described via inter-vacancy correlation vectors (with slightly different frequencies). A schematic representation of the 3D ordering scheme, which is constructed with the most important correlation vectors, is given in Fig. 1.2.21.

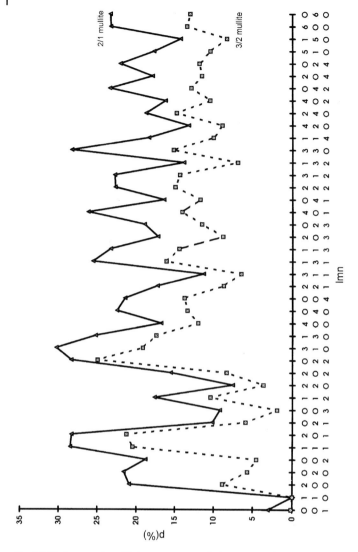

Fig. 1.2.20 Frequencies of inter-vacancy correlation vectors determined from the analysis of the 3D simulation fields for 3/2
and 2/1-mullite [p = probability (%) of finding another oxygen
vacancy at the end of an *lmn* correlation vector].

1.2.4.5 Conclusions

Videographic 3D simulations were performed for 2/1- and 3/2-mullite, whose
Fourier transforms reproduce the diffuse scattering in all examined reciprocal
planes as observed in the experimental patterns (Fig. 1.2.17, 1.2.18 and 1.2.19).
The 3D simulation fields are built up by $48 \times 48 \times 96$ structure variants and include
221 184 unit cells with about 3.5 million atoms. By analysing the 3D simulation

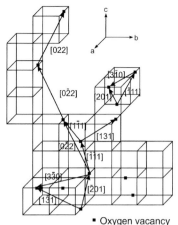

Fig. 1.2.21 3D ordering scheme of the oxygen
• Oxygen vacancy   vacancies.

fields the vacancy distribution in 2/1- and 3/2-mullite could be described completely in terms of inter-vacancy correlation vectors. For this reason mullite is the first non-metallic mineral in which the diffuse scattering can be completely explained by short-range ordering. Short-range-order parameters (correlation vectors) are usually used to describe the real structure of intermetallic alloys above the critical temperature, T_c. In most cases the first three short-range-order parameters are adequate to describe the short-range-order state. In mullite, with T_c above the melting point, the three shortest correlation vectors are not sufficient to describe the oxygen vacancy ordering scheme. In this case, higher order inter-vacancy correlation vectors (Fig. 1.2.20) are important for the complete interpretation of the mullite real structure. Unlike alloys, the short-range ordering process in mullite is related to vacancies coupled with cation displacements.

Knowledge of the real structure may help to explain some physical properties of mullite. In a part of the simulation field for 2/1-mullite (Fig. 1.2.22) differences in the vacancy arrangements along the *a* and *b* axes can be observed. In one *ab* layer (Fig. 1.2.22 top) the differences in probabilities of correlation vectors (p_{lmn}) are small: There are linear arrangements along the *a* axis with the vectors <200> and <130> as well as linear arrangements along the *b* axis with the vectors <020> and <310>. The frequencies of <200> and <020> are very similar, the vector <310> is more frequent than <130>, but both are preferred. However, for a proper description of the physical behavior of a bulk crystal it is more realistic to consider not only one layer with correlation vectors *lm*0 but at least two subsequent layers. The superposition of two subsequent layers (Fig. 1.2.22 bottom) with the correlation vectors *lm*0 and *lm*1 clearly shows differences in the vacancy arrangements (concentrations) in projection on the *a* and *b* directions. The vector <201> is very frequent (marked with black lines in Fig. 1.2.22), but the vector <021> is rare, while <131> is frequent, <311> rare. Therefore the dependence of several physical properties on the direction may be explained with the given distribution of oxygen vacancies, which can be well described with the help of inter-vacancy correlation vectors.

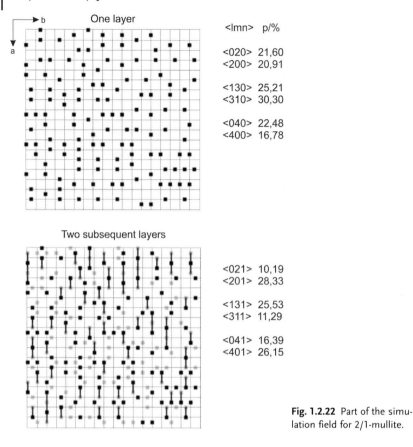

One layer

$\langle lmn \rangle$	p/%
$\langle 020 \rangle$	21,60
$\langle 200 \rangle$	20,91
$\langle 130 \rangle$	25,21
$\langle 310 \rangle$	30,30
$\langle 040 \rangle$	22,48
$\langle 400 \rangle$	16,78

Two subsequent layers

$\langle lmn \rangle$	p/%
$\langle 021 \rangle$	10,19
$\langle 201 \rangle$	28,33
$\langle 131 \rangle$	25,53
$\langle 311 \rangle$	11,29
$\langle 041 \rangle$	16,39
$\langle 401 \rangle$	26,15

Fig. 1.2.22 Part of the simulation field for 2/1-mullite.

The results presented above reflect the importance of real-structure determination in material science in order to characterize the behavior of physical properties in the case of disorder.

1.3
Foreign Cation Incorporation in Mullite
H. Schneider

1.3.1
Transition Metal Incorporation

Natural mullites contain iron, titanium, and occasionally chromium as foreign components. The foreign cation content is usually low, though Fe_2O_3- and TiO_2-rich mullites have also been described (Agrell and Smith, 1960, Table 1.3.1).

Synthesis experiments carried out at varying temperatures and atmospheres have shown that a large variety of transition metals does enter the mullite struc-

Table 1.3.1 Chemical composition of naturally occurring mullite (wt. %)[a].

	1	2	3	4	5	6
SiO_2	30	32	29	29	31	29
TiO_2	1.29	2.27	0.55	0.79	0.70	0.80
Al_2O_3	68	65	64	70	68	70
Fe_2O_3	0.62	0.94	5.90	0.50	0.50	0.30
Cr_2O_3	n.d.	n.d.	n.d.	n.d.	–	0.30

a) Total iron expressed as Fe_2O_3; n.d. = not determined.
1 = Rudh'a' Chromain, Carsaig Bay, Mull, Scotland (Agrell and Smith, 1960).
2 = Carrickmore, west of Ballycastle, Northern Ireland (Agrell and Smith, 1960).
3 = Tievebulliagh, County Antrim, Northern Ireland (Agrell and Smith, 1960).
4 = Seabank Villa, Mull, Scotland (Bowen and Greig, 1924).
5 = Rudh'a' Chromain, Loch Scridain, Mull, Scotland (Cameron, 1976b).
6 = Bushfeld Complex, Thorncliffe and Maandagshoek, South Africa (Cameron, 1976b).

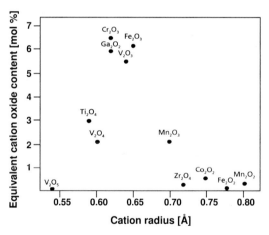

Fig. 1.3.1 Transition metal incorporation in mullite, given as maximum equivalent cation oxide contents and plotted versus the radii of substituting transition metal cations. Gallium incorporation (given as Ga_2O_3) is presented for comparison (from Schneider et al. 1994).

ture. Depending on synthesis conditions mullite may incorporate Ti^{3+}, Ti^{4+}, V^{3+}, V^{4+}, Cr^{3+}, Mn^{2+}, Mn^{3+}, Fe^{2+}, Fe^{3+} and Co^{2+}, though in strongly differing amounts (see Schneider, 1990). The upper solubility limit is controlled by the radii and oxidation states of transition metal ions: The highest degrees of incorporation are observed for V^{3+}, Cr^{3+} and Fe^{3+} followed by Ti^{4+}, while only very low amounts of Mn^{2+}, Fe^{2+} and Co^{2+} ions enter the mullite structure (Fig. 1.3.1). No systematic studies on the incorporation of scandium (Sc^{3+}), cobalt (Co^{3+}, Co^{2+}) and nickel (Ni^{3+}, Ni^{2+}) have been published so far.

Knowledge of equilibrium data for oxygen-transition metal systems with several states of oxidation are helpful for understanding the cation incorporation proc-

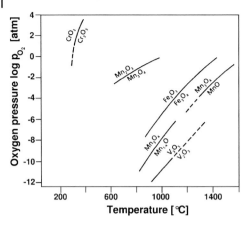

Fig. 1.3.2 Stability relations of transition metal oxides with different oxidation stages of chromium (Wilhelmi, 1975), manganese (Schwerdtfeger and Muan, 1975, Hübner and Sato, 1970), iron (Darken and Gurry, 1945, 1946) and vanadium (Anderson and Kahn, 1975) as function of temperature and oxygen pressure (from Schneider et al. 1994).

esses. A graph comparing the stabilities of Mn_2O_3/Mn_3O_4 and Mn_3O_4/MnO reaction pairs with those of Fe_2O_3/Fe_3O_4, CrO_2/Cr_2O_3 and V_3O_5/V_2O_3 as a function of temperature and oxygen partial pressure (p_{O_2}) is shown in Fig. 1.3.2. This shows that Cr^{3+} is the only stable oxidation state of chromium at higher temperatures regardless of the oxygen partial pressure, while the stability field of Fe^{3+} extends up to about 1400 °C at $p_{O_2} = 0.2$ atm. V^{3+} occurs above about 1600 °C at $p_{O2} = 0.2$ atm, but is stable under a reducing atmosphere at lower temperatures. The experimental results on vanadium, chromium and iron incorporation into mullite obtained under varying reaction temperatures and atmospheres are in good accordance with the thermodynamic data (e. g.: Schneider, 1990, Rager et al. 1990, Schneider and Rager, 1986). Experimental data (e. g.: Schneider, 1990, Rager et al. 1993), moreover, suggest that the TiO_2/Ti_2O_3 reaction pair displays a similar temperature and oxygen partial pressure-dependent redox behavior to V_3O_5/V_2O_3. The equilibrium data of the manganese oxides show that Mn^{3+} is stable only below about 1000 °C. This may explain why Mn^{3+} incorporation into mullite by oxide reaction sintering is not possible, since it would require temperatures above 1200 °C, which is beyond the stability field of Mn^{3+}. However, a possible way to incorporate Mn^{3+} in mullite is the sol-gel route synthesis at low temperatures (≤ 900 °C, Schneider and Vasudevan, 1989). The same synthesis technique may allow incorporation of Co^{3+} and Ni^{3+} into mullite. Another method of producing Mn^{3+}-, Co^{3+}- and Ni^{3+}-doped mullites may be synthesis by the hydrothermal technique, using oxidizing buffer systems.

Nucleation and growth, and associated shapes and sizes of the transition metal-doped mullite crystals, are controlled by the synthesis procedure, the synthesis temperature and atmosphere, and by the type and amount of transition metal incorporated. The size of the mullite crystals prepared from oxide powder mixtures with excess SiO_2 by reaction sintering increases within the sequence vanadium, chromium, titanium and iron. Within the same series the shapes of the mullite crystals change from small equiaxed to larger tabular grains (Fig. 1.3.3a–c). The influences controlling nucleation and growth of transition metal-doped mullites are probably complex. Growth is probably enhanced by the decreasing viscosities

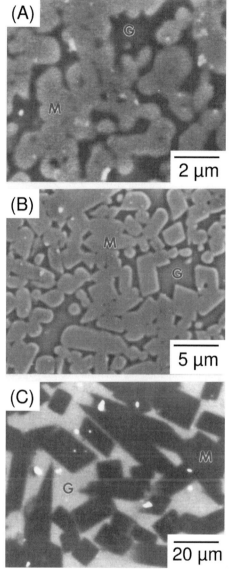

Fig. 1.3.3 Electron microprobe photographs of ceramics consisting of mullite (M) and glass phase (G). (a) Vanadium-rich ceramic with mullite containing 8.7 wt.% V_2O_3. The mullite crystals are very small with reniform contours; (b) Chromium-rich ceramic with mullite containing 11.5 wt.% Cr_2O_3. The mullite crystals are small with rounded-off contours; (c) Iron-rich ceramic with mullite containing 10.3 wt.% Fe_2O_3 (from Schneider et al. 1994).

of the partial melts coexisting with mullite in going from vanadium- to iron-rich oxide mixtures, and the associated higher velocities of diffusion species. Solid-state sintering studies show that low reaction rates and the microstructural development of samples are not only controlled by the presence of a low-viscosity liquid phase but at least in the case of the Cr_2O_3–Al_2O_3–SiO_2 system is a property inherent to transition metal-doped mullite itself (Saruhan and Schneider, 1993).

1.3.1.1 Titanium Incorporation

The extent of titanium solubility in mullite ranges between about 2 and 6 wt.% TiO_2, depending on the synthesis conditions (Gelsdorf et al. 1958, Murthy and Hummel, 1960, Green and White, 1974, Bohn, 1979, Baudin et al. 1983). Schneider and Rager (1984), Schneider (1986a, 1990), and Rager et al. (1993) found that both, Ti^{3+} and Ti^{4+} cations enter the structure.

Schneider (1986a, 1990) provided data on the dependence between the amounts of Ti_2O_3 and TiO_2 incorporated in mullite and the Al_2O_3 and SiO_2 concentrations of the phase (Fig. 1.3.4). It has been shown that a reciprocal and equimolar dependence exists between Ti_2O_3 and Al_2O_3, but not between Ti_2O_3 and SiO_2. The chemical interdependences were interpreted as implying that the entry of Ti^{3+} into mullite goes along with the removal of the same number of Al^{3+} ions from the structure. According to EPR studies, and because of the similar cation radii of octahedrally bound Al^{3+} and Ti^{3+} ions ($^{[6]}Al^{3+}$: 0.53 Å, $^{[6]}Ti^{3+}$: 0.64 Å; all cation radii are from Shannon 1976) $Ti^{3+} \rightarrow Al^{3+}$ substitution at octahedral sites is taken into account. The dependences between TiO_2 and Al_2O_3 and SiO_2, respectively, are contrary to those described above, with a reciprocal and equimolar dependence between TiO_2 and SiO_2, but not between TiO_2 and Al_2O_3 (Fig. 1.3.4). Probable Ti^{4+} incorporation mechanisms are $Ti^{4+} \rightarrow Al^{3+}$ or $Ti^{4+} \rightarrow Si^{4+}$ substitution. The cation radii of $^{[4]}Si^{4+}$ (0.26 Å), $^{[6]}Al^{3+}$ (0.53 Å), and $^{[6]}Ti^{4+}$ (0.605 Å) strongly suggest an entry of Ti^{4+} into the oxygen octahedra. An $^{[4]}Al^{3+} \rightarrow ^{[4]}Si^{4+}$ substitution necessary for charge compensation, if Ti^{4+} enters octahedral sites may explain the higher incorporation-induced expansion of the *a* rather than the *b* lattice constant, which is characteristic for Ti^{4+}-doped mullite (Table 1.3.2).

Electron paramagnetic resonance (EPR) studies carried out on Ti^{3+}-doped mullite yielded interesting details (Rager et al. 1993): Two components of the EPR peak

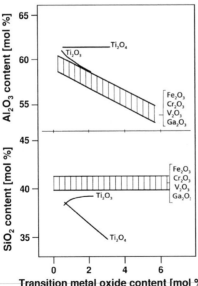

Fig. 1.3.4 Titanium oxide (Ti_2O_3, TiO_2), vanadium oxide (V_2O_3), chromium oxide (Cr_2O_3), manganese oxide (Mn_2O_3), and iron oxide (Fe_2O_3) plotted versus Al_2O_3 and SiO_2 contents, respectively. Relationships for gallium oxide (Ga_2O_3) are given for comparison (from Schneider et al. 1994).

Table 1.3.2 Chemical composition and lattice parameters of mullites containing high amounts of transition metal cations.

Chemical composition (wt. %)			Method	Lattice parameters				Reference
				a [Å]	b [Å]	c [Å]	V [Å³]	
Mullites containing transition metal cations:								
Al_2O_3 : 72.0	SiO_2 : 24.5	TiO_2 : 4.2	EMA	7.5637(5)	7.7009(6)	2.8931(3)	168.51(2)	Schneider (1990)
Al_2O_3 : 63.0	SiO_2 : 28.2	V_2O_3 : 8.7	XFA*	7.5550(2)	7.711(3)	2.8995(9)	168.92(2)	Schneider (1990)
Al_2O_3 : 72.5	SiO_2 : 24.0	V_2O_4 : 3.5	XFA†	7.5510(1)	7.698(1)	2.8936(5)	168.19(7)	Schneider (1990)
Al_2O_3 : 60.0	SiO_2 : 28.4	Cr_2O_3 : 11.5	EMA	7.5697(5)	7.7117(8)	2.9025(2)	169.43(2)	Rager et al. (1990)
Al_2O_3 : 68.4	SiO_2 : 25.9	Mn_3O_4 : 5.7	XFA‡	7.5630(2)	7.721(2)	2.8828(7)	168.33(6)	Schneider and Vasudevan (1989)
Al_2O_3 : 62.1	SiO_2 : 27.4	Fe_2O_3 : 10.3	EMA	7.5740(1)	7.726(1)	2.9004(5)	169.73(4)	Schneider (1987)
Reference mullite:								
Al_2O_3 : 71.2	SiO_2 : 28.6	–	EMA	7.5461(8)	7.6918(9)	2.8829(4)	167.33(3)	Schneider (1990)

Spectroscopic data and microchemical analyses suggest that samples marked *, †, and ‡ contain essentially V^{3+}, V^{4+} and Mn^{3+}, respectively. Therefore, chemical compositions are given as V_2O_3, V_2O_4, and Mn_3O_4, respectively.

EMA = Electron microprobe analysis; XFA = X-ray fluorescence analysis.

have been resolved at Q-band but not at X-band frequency. These two components were assumed to be due to at least two non-equivalent but structurally very similar Ti^{3+} centers in mullite. A structural model of mullite with two ordering patterns with different local atomic configurations was used to explain the EPR patterns. Rager et al. (1993) believed that the splitting of the Ti^{3+}-EPR is due to slightly different distortions of octahedra. Two octahedral positions with slightly different distortions have also been observed in iron-doped mullite by means of Mössbauer spectroscopy (see Section 1.3.1.5).

Interfacial reactions between titanium metal and mullite have been studied in the temperature range between 200 and 650 °C by Yue et al. space (1998) using secondary ion mass spectrometry (SIMS), Auger electron spectroscopy (AES) and X-ray diffractometry (XRD). At temperatures below 450 °C interfacial reactions are very slow, producing very thin interfacial layers only. The first deposited titanium atoms form Ti-O bonds with the oxygen on the mullite surface. However, intense interfacial reactions occur above about 650 °C and Ti-O-, Ti-Al- and Ti-Si-bonds can be identified.

1.3.1.2 Vanadium Incorporation

Only few data are available on vanadium-doped mullites. Schneider (1990) mentioned that maximally about 9 wt.% V_2O_3 and 3.5 wt.% V_2O_4 (Fig. 1.3.1) can be incorporated into mullite (Table 1.3.2).

Comparison of the lattice constants of V^{3+}-doped mullites prepared from reaction-sintered oxide mixtures with undoped mullites yields relatively high **c**, lower **b** and very low **a** expansion (Table 1.3.2, Schneider, 1990). The intense **b** cell edge expansion was correlated with the entry of V^{3+} ions at octahedral lattice sites. It was explained in a similar way to the thermal expansion of mullite, with a considerable lengthening of the long and "elastic" octahedral $M(1)-O(D)$[3] bond lying to about 30° to either side of **b** in mullite but with a much weaker $M(1)-O(A)$ lengthening (Fig. 1.1.17, Schneider and Eberhard, 1990).

Plots of mol SiO_2 versus V_2O_3 concentrations of V^{3+}-substituted mullites yielded no dependence, whereas reciprocal correlations occur between Al_2O_3 and V_2O_3 (Fig. 1.3.4). This indicates that the entry of V^{3+} into mullite is associated with removal of the same number of Al^{3+} from the structure. Relationships are less clear in the case of V^{4+} substitution, probably because of low foreign cation concentrations and the occurrence of different oxidation states of vanadium.

The above-described observation of V^{3+} and V^{4+} incorporation into mullite has been supported by EPR studies. Vanadium-rich mullites synthesized under a strong reducing atmosphere yield EPR spectra with a weak and broad signal near

---

3) Different designations of the oxygen positions of mullite have been described in literature: 01, 02, 03, 04 which corresponds to O(A,B), O(D), O(C) and O(C*), (Table 1.1.7). The first setting follows strict geometrical rules and allows an easy description of members of the mullite family being in group-subgroup relationship to the aristotype with the highest symmetry (Tables 1.1.2 to 1.1.10, 1.1.17 and Fig. 1.1.4). Since the second setting has frequently been cited in the crystal chemical literature of mullite it has been used in the following.

$g_{eff} = 4.1$ and a stronger peak near $g_{eff} = 1.9$. The EPR signal group near $g_{eff} = 4.1$ is attributed to V^{3+} in an octahedral environment. The band near $g_{eff} = 1.9$ with little fine structure has been correlated to the occurrence of a small amount of V^{4+} (Schneider, 1990). Vanadium-rich mullites synthesized in a moderately reducing atmosphere exhibit EPR signals near $g_{eff} = 2.0$ with characteristic hyperfine splitting. The EPR bands are attributed to isolated V^{4+} ions, probably localized in octahedral coordination.

1.3.1.3 Chromium Incorporation

Chromium-doped mullites were synthesized by Gelsdorf et al. (1958), Murthy and Hummel (1960) and Rager et al. (1990). It has been stated that up to 12 wt.% Cr_2O_3 is incorporated in mullite.

A detailed study of the crystal chemistry of chromium-doped mullites, prepared by reaction sintering of Al_2O_3, SiO_2 and Cr_2O_3 (0.5 to 11 wt.%) powders, was performed by Rager et al. (1990). Rager and coworkers found a reciprocal and equimolar dependence between Cr_2O_3 and Al_2O_3 in mullite, but not between Cr_2O_3 and SiO_2 content. They concluded that Cr^{3+} is incorporated by replacement of Al^{3+}. The structural formula of chromium-doped mullite, which corresponds to that of 3/2-type mullite indicates that the variation of chromium incorporation is not correlated with a change of the amount of O(C) oxygen vacancies (structural state x) in mullite. Chromium entry into mullite causes the largest linear lattice expansion along the crystallographic **c** axis, followed by smaller linear expansions parallel to **a** and **b** (Table 1.3.2). The incorporation-induced structural expansion of mullite, with $\Delta a > \Delta b$, does not fit with a simple $Cr^{3+} \rightarrow Al^{3+}$ substitution at octahedral sites, which should give $\Delta b > \Delta a$.

Electron paramagnetic resonance studies carried out by Rager et al. (1990) provided further information on the structural distribution of chromium in mullite. Chromium-doped mullites exhibit two rather sharp EPR signals near $g_{eff} = 5$, and a broad signal near $g_{eff} = 2.2$ (Fig.1.3.5). The peaks near $g_{eff} = 5$ were assigned to Cr^{3+} in slightly distorted octahedral M(1) positions in mullite, whereas the broad slightly asymmetric signal near $g_{eff} = 2.2$ may indicate coupling between localized magnetic moments. Rager et al. (1990) explained the signal at $g_{eff} \approx 2.2$ in terms of interstitial Cr^{3+} incorporation in mullite. According to the EPR peak intensities the entry of Cr^{3+} into the regular M(1)O_6 octahedra is favored at low bulk-Cr_2O_3 contents of mullite, whereas interstitial incorporation with formation of chromium clusters becomes more important at higher Cr_2O_3 contents (Fig. 1.3.5).

The strong preference of Cr^{3+} for octahedral coordination (Wells, 1984) suggests that interstitial Cr^{3+} is located in distorted, octahedral environments. Such sites in mullite are (1) the structural vacancies formed by removal of O(C) oxygen atoms that bridge adjacent tetrahedra near (0.1, 0.25, 0) (substitution model Cr I, Fig. 1.3.6b), and (2) the structural sites in the relatively wide structural channels running along near (0.2, 0.5, 0) (substitution model Cr II, Fig. 1.3.6c). Possible chromium octahedral sites in the O(C) vacancies are strongly distorted but become

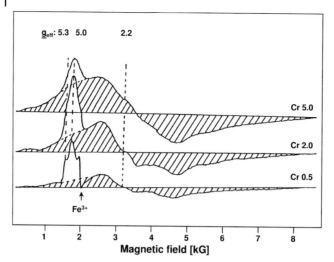

Fig. 1.3.5 EPR spectra of chromium-substituted mullites (Cr_2O_3 contents Cr 0.5 = 0.5 wt.%, Cr 2.0 = 2 wt.%, Cr. 5.0 = 5 wt.%). The hatched areas correspond to the EPR signal denoted by g_{eff} = 2.2. The arrow labeled Fe^{3+} indicates that traces of "free" Fe_2O_3 occur in the samples (from Schneider et al. 1994).

more regular if the O(C*) oxygen position is occupied. This is the case near the characteristic oxygen vacancies of the mullite structure. Such sites become almost completely regular by a small additional O(C*) oxygen shift toward the center of the octahedron. Incorporation of Cr^{3+} on equipoint near (0.1, 0.25, 0) implies that the nearest [4]Al* position cannot be occupied, which is equivalent to the substitution $^{[6]}Cr^{3+} \rightarrow ^{[4]}Al^{(*)3+}$. In addition, adjacent $Al(2)O_4$ tetrahedra, which would share an edge with the new octahedron, must be vacant, or the Al(2)-ion must occupy an additional Al* site. Chromium incorporation into the structural channels in a distorted octahedral environment near (0.2, 0.5, 0) does not require any severe changes in the configuration of adjacent polyhedra. It can be described by the substitution scheme $^{[6]}Cr^{3+} \rightarrow ^{[4]}Al(2)^{3+}$. Both types of interstitial octahedra form pairs with the $Al(1)O_6$ octahedra by sharing common faces, as in the α-alumina structure. Furthermore, the proposed substitution schemes do not need any charge compensation mechanism. Therefore, the broad EPR signal at $g_{eff} \approx 2.2$ has been interpreted by Rager et al. (1990) as being mainly due to chromium pairs, where the formation of pairs may occur via occupation of neighboring regular octahedral sites, or through occupation of adjacent regular and interstitial sites, or both.

Unpolarized crystal field spectra measured by Ikeda et al. (1992) in the wavelength range of 340 to 1540 nm by reflection from polycrystalline mullite doped with 8 wt.% Cr_2O_3 yielded further evidence for the structural distribution model of Cr^{3+} developed by Rager et al. (1990). Crystal-field spectra show two pairs of absorption peaks. One of them was attributed to Cr^{3+} ions replacing Al^{3+} at octahe-

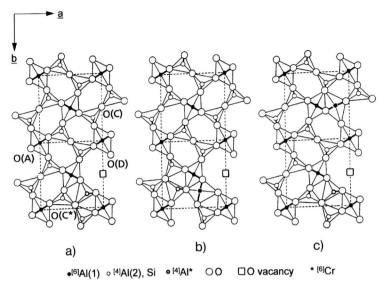

a)   b)   c)

•[6]Al(1)   ○ [4]Al(2), Si   ○ [4]Al*   ○O   □O vacancy   * [6]Cr

Fig. 1.3.6 Structural models for chromium incorporation in mullite shown in projections down the **c** axis; (a) Transition metal-free mullite, or Cr^{3+} substitution for octahedrally bound Al^{3+} in M position; (b) Substitution of Cr^{3+} for tetrahedrally bound Al^{3+} in T* position near the oxygen vacancy (substitution model Cr I); (c) Substitution of Cr^{3+} for tetrahedrally bound Al^{3+} in T position in the structural channels running parallel to to the **c** axis (substitution model II, see Rager et al. 1990, from Schneider et al. 1994).

dral M(1) sites (Fig. 1.1.17). The other pair of absorption peaks was assigned to Cr^{3+} ions occurring at interstitial octahedral lattice sites. Crystal field spectroscopic data revealed the presence of two kinds of interstitial CrO_6 octahedra corresponding to the EPR-derived substitution model Cr II (Fig. 1.3.6). Integrated absorption intensities yielded estimated Cr^{3+} site occupancies of $\approx 40\%$ in M(1) position, and of $\approx 30\%$ in both of the interstitial channel positions (Ikeda et al. 1992). Becker and Schneider (2005) re-investigated the Cr^{3+} incorporation in mullite single crystals. Although they obtained similar results to those of Ikeda et al. they suggested that the spectra can also be interpreted by a single Cr^{3+} position in mullite (Fig. 1.3.7).

Data on time-resolved fluorescence spectroscopy of chromium-doped mullites (about 2 to 11 wt.%) have been provided by Piriou et al. (1996). They showed that depending on the chromium concentration and on the excitation frequency a low-field site (LFS) transition ($^4T_2 \rightarrow {}^4A_2$), and a high-field site (HFS) transition ($^2E \rightarrow {}^4A_2$) can be distinguished. Mullites with low chromium contents (≤ 2 wt.% Cr_2O_3) exhibit high- and low-field sites, and an intermediate site with ^2E character with a long decay time. Mullites with high chromium contents (≥ 5 wt.% Cr_2O_3) produce emission spectra with one low-field site only, having the lowest $^4T_2$ level. Accord-

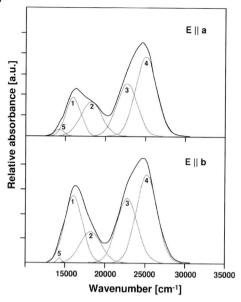

Fig. 1.3.7 Crystal field spectra of a chromium-substituted mullite single crystal plate (0.6 wt.% Cr_2O_3), cut parallel to (001), with the electrical vectors oscillating parallel to to the **a** (E//**a**) and **b** (E//**b**) axes. Full lines: Measured data, dotted lines: Fitted curves. Ikeda et al. (1992) assigned peaks 1 and 3 to interstitial Cr^{3+}, referring to substitution model Cr I and Cr II (see Fig. 1.3.6 and Rager et al. 1990), while peaks 2 and 4 were assigned to substitution of Cr^{3+} for octahedrally bound Al^{3+}. According to Becker and Schneider (2005) a single Cr^{3+} position should also be considered (according to Becker and Schneider, 2005).

ing to Piriou et al. (1996) the low-field site can be attributed to a distorted interstitial octahedral environment of chromium. The HFS peaks occurring additionally in low-chromium mullite spectra lie close to lines in chromium-doped α-alumina (ruby). They are therefore ascribed to Cr^{3+} in undistorted $M(1)O_6$ octahedra in mullite. Obviously, fluorescence spectroscopy supports the earlier suggestion of a bimodal Cr^{3+} distribution in mullite, including undistorted $M(1)O_6$ octahedra and distorted interstitial octahedral sites. According to these results, at low Cr_2O_3 concentrations mullite preferentially incorporates chromium in the $M(1)O_6$ octahedra, whereas the second substitution mode becomes more important at higher chromium contents.

Measurements of the extended X-ray absorption fine structure (EXAFS) of the Cr_K edge were performed by Bauchspiess et al. (1996) for mullites doped with Cr_2O_3 ranging between 5 and 11 wt.%. For all spectra the Fourier transform (FT), which is directly related to the radial distribution function (rdf), is characterized by two pronounced peaks. The first peak near $R = 1.65$ Å was ascribed to oxygen atoms nearest to chromium, and the second peak to those nearest to aluminum and/or silicon. The chromium-oxygen peak yielded a satisfactory fit of measured and calculated values. Although the fit of the chromium-aluminum (silicon) peak is less perfect no asymmetry of the coordination shell has been detected. However, the EXAFS data are not contradictory to the previously reported spectroscopic results, which yielded evidence for a bimodal structural distribution of chromium in mullite. Obviously, the resolution of the EXAFS spectrum is not high enough to distinguish the structurally very similar octahedral chromium sites in mullite. Bauchspiess et al. (1996) on the basis of their studies mentioned that the chromium-aluminum (silicon) distances in mullite increase with the degree of chromium incorporation by 0.007 Å from the 5 wt.% Cr_2O_3 sample to the 11 wt.% Cr_2O_3 mullite.

The Rietveld refinement of chromium-doped mullite (10 wt.% Cr_2O_3) revealed that Cr^{3+} resides preferentially in the octahedral M(1) site (Fischer and Schneider, 2000, see also Parmentier et al. 1999), with a mean octahedral M(1)-O distance close to that of the calculated alumina to chromia (Al_2O_3/Cr_2O_3) mole fraction. The predominance of the incorporation of the relatively large Cr^{3+} ions in the place of Al^{3+} should produce the strongest expansion along the long and elastic M(1)-O(D) bond with associated strong **b** axis lengthening. Since chromium incorporation produces $a > b$ expansion (see Rager et al. 1990) this obviously is not the case. According to Fischer and Schneider a possible explanation of the problem is that the strong M(1)-O(D) expansion is partially compensated by a simultaneous shortening of tetrahedral T-O bonds.

There remains, however, the discrepancy between X-ray diffraction and spectroscopic data. The latter yield evidence for two different structural Cr^{3+} positions in mullite, one at "normal" octahedral M(1) sites and the other at interstitial octahedral sites, as shown in Fig. 1.3.6. Fischer and Schneider (2000) presented a new model including both diffraction and spectroscopic data. It starts from the presence of . . . Cr^{3+}-Cr^{3+}-Cr^{3+} . . . clusters in the octahedral chains running along the **c** axis. CrO_6 octahedra then can have either CrO_6 units as next-nearest neighbors ("cluster CrO_6") or, alternatively, CrO_6 and AlO_6 polyhedra ("non-cluster CrO_6"). Since the two types of CrO_6 octahedra display slightly different distortions they may produce split signals in the respective spectra. Although this incorporation model is consistant with diffraction and spectroscopic results it is in fundamental disagreement with the chromium incorporation model published by Rossouw and Miller (1999). These authors, starting from the "Atom location by channelling enhanced microanalysis (ALCHEMI)" technique, stated that in Cr_2O_3-rich mullites (10 wt.% Cr_2O_3) most of the chromium enters an interstitial site at 0, 0.25, 0. This, however, is unlikely, since the next-nearest distance between two adjacent octahedral sites then becomes extremely short (1.93 Å). This must lead to a very high (Al^{3+}, Cr^{3+})-Cr^{3+} repulsion, which in turn should produce a very high temperature factor for these cation sites. However, this obviously is not the case.

In spite of the many detailed diffraction and spectroscopic studies, the manner of structural incorporation of Cr^{3+} in mullite remains an open question.

1.3.1.4 Manganese Incorporation

Schneider and Vasudevan (1989) provided data on manganese-doped mullites, which were synthesized from metal organic starting materials by a modified sol-gel technique at low temperature (≥ 700 °C). Schneider and Vasudevan suggested that up to 6 wt.% Mn_2O_3 can enter the mullite structure. Manganese was believed to be incorporated into mullite as Mn^{3+}, which is reasonable considering the charge and size of the cation and because of the "low" temperature stability of Mn^{3+} (see Fig. 1.3.2).

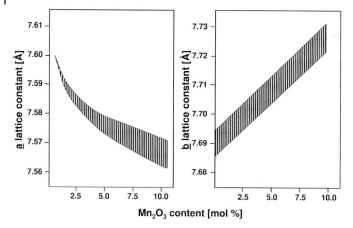

Fig. 1.3.8 Schematic dependence of lattice constants *a* (left)
and *b* (right) of mullite from the degree of manganese incorporation (given as Mn_2O_3 content). The hatched areas correspond to the scatter of experimentally determined values.
Note that this scatter depends on the Mn_2O_3 content in the
case of *a* but not of *b* constants (from Schneider et al. 1994).

With respect to the lattice constants of sinter-mullite the **a** cell edge of the undoped sol-gel mullite is considerably expanded, while **b** and **c** edge lengths are
similar to those of the reference mullite (see Table 1.3.2). The strong **a** edge expansion is caused by the high Al_2O_3 content of the low-temperature-produced sol-gel
mullite (see Sections 1.1.3.7 and 2.5.1). As manganese is incorporated, the *a* constant first displays strong contraction, which later becomes weaker. The **a** cell edge
contraction correlates with a strong and linear **b** expansion, while the **c** cell edge
length is less dependent on manganese incorporation (Fig. 1.3.8).

X-ray diffraction line profile analyses carried out by Schneider and Vasudevan
(1989) indicate crystallographically controlled strain in mullite: With manganese
incorporation this first decreases and later increases again along the [110] direction, while it gradually decreases parallel to [001]. The incorporation-induced strain
distribution is explained by the overlapping of a strain relaxation mode, and a
mode inducing strain increase. Schneider and Vasudevan (1989) attributed the
strain relaxation mode to partial rearrangement of the distorted crystal structure of
the undoped sol-gel mullite, due to manganese incorporation. The mode that increases strain was correlated with the substitution of octahedrally bound Al^{3+} by
Mn^{3+}: The $3d^4$ electrons of Mn^{3+} in an octahedral crystal field are split from the
energetic ground state to the high-spin state $(t_{2g})^3(e_g)^1$. Considering that the e_g
electron occupies the d_{z2} orbital and not the $d_{(x2-y2)}$ orbital, this has the effect of
lengthening the octahedron's **z** axis. This may be explained by repulsion of the d_{z2}
electron from the electrons of the respective oxygen ligands (Jahn-Teller distortion). Taking into account that the octahedron's **z** axis is parallel to the M(1)-O(D)
bond, and that **z** lies about 30° to either side of the crystallographic **b** axis
(Fig. 1.1.14), this may also explain the strong Mn^{3+} incorporation-induced **b** expan-

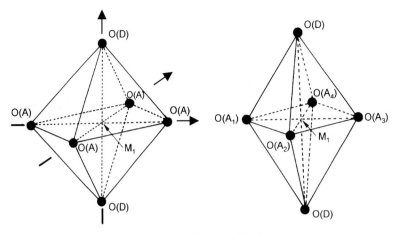

Fig. 1.3.9 Deformation of the oxygen octahedra in mullite by substitution of Al^{3+} by Mn^{3+} (right). Note that the symmetrically equivalent O(A) sites in the non-substituted less deformed octahedron (left) degenerate to non-equivalent sites in the distorted state (labeled as O(A$_1$), O(A$_2$), O(A$_3$) and O(A$_4$); from Schneider et al. 1994).

sion and the high standard deviations of the *b* spacings. The deformation of the oxygen octahedra by Mn^{3+} incorporation with associated M-O(D) lengthening can also be responsible for the **a** axis shortening in sol-gel mullites. Substitution of Al^{3+} by Mn^{3+} obviously degenerates the symmetrically equivalent O(A) positions of the octahedron to four non-equivalent states O(A$_1$), O(A$_2$), O(A$_3$) and O(A$_4$), by shortening of the O(A$_1$)-O(A$_4$) and O(A$_2$)-O(A$_3$) bonds lying close to **a**. Since no incorporation-induced change of the *c* parameter was observed, the lengths of the O(A$_1$)-O(A$_2$) and O(A$_3$)-O(A$_4$) bonds lying parallel to **c** are less affected by the entry of Mn^{3+} into the mullite structure (Fig. 1.3.9).

Undoped sol-gel mullites form very small crystallites. The incorporation of small amounts of Mn^{3+} in the sol-gel mullites induces tabular crystal growth parallel to the crystallographic **c** axis, while higher amounts of Mn^{3+} incorporated in mullite cause the formation of more equiaxed but smaller crystallites. Crystallite sizes of sol-gel mullites along [110] are similarly small over the whole manganese solid-solution range.

1.3.1.5 Iron Incorporation

A survey of literature data shows that mullite incorporates up to about 12 wt.% Fe_2O_3 (Muan, 1957, Brownell, 1958, Gelsdorf et al. 1958, Murthy and Hummel, 1960, Razumowski et al. 1977, Bohn, 1979, Schneider and Rager, 1986). The crystal chemistry of iron-doped synthetic sinter-mullites and of commercially produced fused-mullites was studied by Schneider and Rager (1986) with chemical, X-ray diffraction (XRD) and electron paramagnetic resonance (EPR) techniques. Iron incorporation into mullite causes relatively low a, but stronger b and c lattice constant expansions (Table 1.3.2). The observation was interpreted in terms of an octahedral substitution of Al^{3+} by Fe^{3+} which causes the most "elastic" bond of the octahedron, i.e. M(1)-O(D), lying about 30° to either side of **b**, but about 60° to either side of **a** (Fig. 1.1.17) to expand most strongly. The reciprocal dependence between Fe_2O_3 and Al_2O_3 but not between Fe_2O_3 and SiO_2 (Fig. 1.3.4) also indicates preferred Fe^{3+} substitution for Al^{3+}.

Schneider et al. (1994) published a Mössbauer spectrum from an iron-rich mullite (about 11 wt.% Fe_2O_3). It consists of two relatively broad symmetric lines, similar to the spectrum published by Cameron (1977c). A one-doublet computer fit yields a quadrupole splitting (QS) of about 1.10 mm sec^{-1}, an isomer shift (IS) of about 0.40 mm sec^{-1} versus α-iron, and a line width of about 0.60 mm sec^{-1} (liquid N_2 conditions: 80 K). The isomer shift is appropriate for Fe^{3+} in octahedral coordination. According to these studies there is no clear evidence for tetrahedral Fe^{3+} incorporation in mullite, although Mössbauer spectra of natural sillimanites suggest that about 80% of the incorporated iron can be in octahedral, and about 20% in tetrahedral, coordination (Rossman et al. 1982; see also the results on the electron paramagnetic resonance studies on iron-doped mullite). Since the quadrupole splitting of the Fe^{3+} doublet is moderately large in mullite, a moderately distorted octahedral iron site has been be expected, in comparison to the low distorted octahedral Fe^{3+} sites in pyroxenes, garnets, amphiboles and micas (QS: about 0.4 to 0.6 mm/sec) and the more distorted Fe^{3+} sites in epidote (QS: about 2.0 mm/sec or higher, Rossman et al. 1982). The revision of Mössbauer measurements of iron-doped mullite provides a new detailed view of the Fe^{3+}-incorporation mode in mullite: Attempts to fit a single Mössbauer doublet with a Lorentzian line shape fail (Cardile et al. 1987, Parmentier et. al. 1999, Mack et al. 2005). Mack et al. (2005) on the basis of temperature-dependent Mössbauer measurements propose a three-doublet Mössbauer curve deconvolution (Fig. 1.3.10): Two structurally very similar sites (labelled A and B) with an isomer shift of about 0.32 mm sec^{-1}, and a third somewhat deviating site (labelled C) with an isomer shift of about 0.10 mm sec^{-1} versus α-iron (relationships refer to room temperature). While site A can readily be assigned to octahedral Fe^{3+}, the temperature-dependent evolution of site B suggests that it should equally be attributed to Fe^{3+} in octahedral coordination. The third site, C, is assigned to a minor amount of Fe^{3+} in tetrahedral sites. Making use of this approach, the octahedral site A (occupancy about 65%) is slightly more distorted than octahedral site B (occupancy about 30%), while the tetrahedral site C is of minor importance. It is not clear whether both octahedral sites (A and B) reflect inherently existing differences in the structural arrangement of the octahedral chains in mullite, which, however, are so weak that they cannot be resolved with structure refinements, or if these differences are produced by the iron incorporation itself. The Mössbauer studies of Mack et al. give neither evidence for a temperature-induced increase of the amount of tetrahedrally coordinated iron, nor for significant changes of the distortion of iron oxygen polyhedra. This documents the very high thermal stability of the Fe^{3+} distribution in mullite. The finding of a favored incorporation of iron at octahedral sites has also been stressed in studies

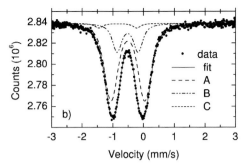

Fig. 1.3.10 Mössbauer transmission spectrum of iron-substituted mullite (10.3 wt.% Fe_2O_3) at 100 °C. Sites A and B have been assigned to octahedrally bound Fe^{3+}, while site C is assigned to tetrahedrally coordinated Fe^{3+}. Fit to the data has been carried out by Voigt profile analysis (after Mack et al. 2005).

investigating the role of iron in mullite formation from kaolin using Mössbauer spectroscopy and Rietveld refinement (Soro et al. 2003) and by extended X-ray absorption fine structure (EXAFS) measurements (Ocana et al. 2000).

Cameron (1977b), referring to reflectance spectra of Faye and Harris (1969) from titaniferous varieties of andalusite, stressed the idea of preferred octahedral incorporation of Fe^{3+} and Ti^{4+} in mullite. The pleochroism from lilac (moderate iron/titanium ratios) to pale yellow (high iron/titanium ratios) was ascribed to Fe^{2+} → Fe^{3+}, Ti^{3+} → Ti^{4+}, and Ti^{3+} + Fe^{3+} → Ti^{4+} + Fe^{2+} charge transfer. For charge transfer to take place, it is considered necessary that most of the titanium and iron substitute for octahedral aluminum in mullite.

Electron paramagnetic resonance studies on iron-doped mullite have been carried out by Schneider and Rager (1986). Different Fe^{3+} centers could be distinguished in the mullite electron paramagnetic resonance (EPR) spectra: Signals with g_{eff}-values near 6.8 and 5.1 belonging to a common electron paramagnetic resonance center designated as center I, and a signal near g_{eff} = 4.2, which was designated as center II (Fig. 1.3.11). The temperature dependence of the spectra was found to be similar for all mullites: The intensity of the signals near g_{eff} = 4.2 (center II), and 5.1 and 6.8 (center I) increases with decreasing temperature. The distortion factor $\lambda \approx 0.12$ of center I indicates a rather axially symmetric crystal field, while center II has a λ-value of about 1, which is attributed to a completely orthorhombic crystal field. Mullite electron paramagnetic resonance centers II and I were attributed by Schneider and Rager (1986) to isolated Fe^{3+} ions at tetrahedral and octahedral sites, respectively. Bond distances and angles of the octahedron of 3/2-mullite as determined by Saalfeld and Guse (1981) yield approximate axial symmetry, thus confirming the assignment of center I to the octahedral position, which in turn corresponds with a low λ-value. Contrary to the appearance of the iron-rich sinter-mullite patterns, the electron paramagnetic resonance spectra of iron-poor fused-mullite exhibit signals corresponding to center II ($g_{eff} \approx 4.2$), but not those of center I. This is interpreted in terms of nearly exclusive iron incorporation at tetrahedral sites though in very low concentrations. Though Mössbauer spectroscopy and electron paramagnetic resonance analysis prove that iron essentially occurs as Fe^{3+} in mullite, there is evidence for the presence of traces of Fe^{2+} in titanium-rich mullites grown from a melt in a reducing atmosphere (Schneider and Rager, 1984, Rager et al. 1993).

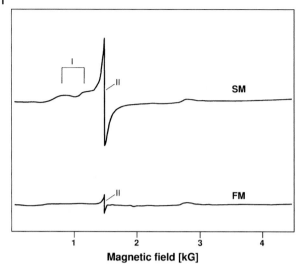

Fig. 1.3.11 Electron paramagnetic resonance (EPR) spectra of iron-substituted mullite. Electron paramagnetic resonance centers I and II are assigned to octahedrally and tetrahedrally coordinated Fe^{3+} in mullite, respectively. SM: Sinter-mullite (0.22 wt.% Fe_2O_3); FM: Fused-mullite (0.10 wt.% Fe_2O_3). Note that the sinter-mullite contains Fe^{3+} at octahedral and tetrahedral sites, while the fused-mullite contains Fe^{3+} preferentially at tetrahedral sites (from Schneider et al. 1994).

The temperature-dependent iron distribution between mullite and a coexisting silicate melt was investigated between 1300 and 1670 °C by Schneider (1987a). The maximum iron content of the mullites strongly decreases from about 10.5 wt.% (6 mol%) Fe_2O_3 at 1300 °C to about 2.5 wt.% (1.5 mol%) Fe_2O_3 at 1670 °C (Fig. 1.3.12). Mullite dissolution in the temperature field under consideration is too low to fit a simple eutectic melting in the system $3Al_2O_3 \cdot 2SiO_2$-"$3Fe_2O_3 \cdot 2SiO_2$" with limited mutual solubility of the compounds. Therefore, a temperature-controlled iron distribution between crystal and melt has to be taken into account (Schneider, 1987a). Further studies on the iron distribution between mullites and coexisting iron-rich silicate glasses were performed between 1350 and 1670 °C by Schneider (1989). They found a reciprocal dependence between iron incorporation into mullite and the annealing temperature, though the amount of iron incorporation was also correlated with the bulk iron content of the samples.

With respect to the iron distribution mechanism two temperature regions could be distinguished:

- A temperature field extending between about 1500 and 1670 °C with small standard deviations of the iron oxide contents of mullite and of the coexisting glass phase. Iron exchange processes between mullite and coexisting glass phase in this temperature region were discussed using Nernst iron distribution patterns (Fig. 1.3.13). Results were interpreted as follows:

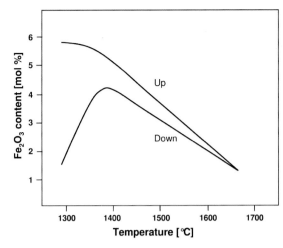

Fig. 1.3.12 Temperature-dependent evolution of iron incorporation in mullite coexisting with a silicate melt. The experiment used a starting mullite with 10.5 wt.% Fe_2O_3. Up: Heating up experiments; down: Back annealing experiments. The strong deviation of "up" and "down" curves at temperatures below about 1400 °C is explained by slow diffusion velocities, which do not allow the iron equilibrium distribution to be established (from Schneider et al. 1994).

- Iron content ratios between mullite (c^s), and coexisting silicate melts (c^l), produce linear curves indicating equilibrium iron distribution;
- In relation to the iron content of the glass phase the amount of iron incorporated into mullite increases with decreasing temperature;
- The iron oxide distribution c^s/c^l ratios increase with increasing iron contents of the bulk samples;
- With increasing annealing temperatures the iron distribution curves approach each other and cross at about 1670 °C. However, it is not clear whether iron incorporation then becomes independent of the bulk iron content of the samples, or whether iron incorporation into mullite is then reciprocally correlated with the bulk iron content of the samples;
- Between 1500 and 1670 °C saturation with iron is reached in any of the mullites.

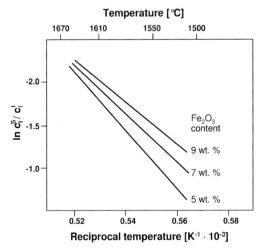

Fig. 1.3.13 Arrhenius plot (Nernst diagram) of the iron distribution between mullite (c^s) and coexisting glass (c^l) for samples with Fe_2O_3 bulk compositions of 5, 7 and 9 wt.%. A ln c^s/c^l versus $1/T$ diagram is given (from Schneider et al. 1994).

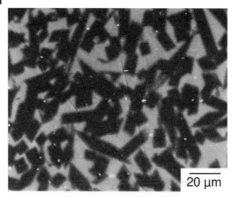

20 µm

50 µm

Fig. 1.3.14 Electron microprobe photographs of mullite powder compacts coexisting with an iron-rich glass, corresponding to a bulk Fe_2O_3 content of 7.5 wt.%. Above: Annealed at 1670 °C (7 days); below: Annealed at 1350 °C (22 days). Note the patchy iron distribution in mullite of the "low temperature" sample, indicated by darker ("low" iron contents in the middle of mullite crystals) and lighter areas ("high" iron contents at the mullite crystal rims, from Schneider et al. 1994).

- A temperature field extending below about 1500 °C with widespread iron oxide composition ranges in mullite and between mullite and the coexisting glass phase. This indicates non-equilibrium of iron distribution in mullite and between mullite and the coexisting silicate glass. Electron microprobe analyses show that iron is enriched at the outer rims of mullite, whereas the inner crystals' areas contain less iron (Fig. 1.3.14). Obviously the velocity of diffusion species was too slow below 1500 °C to establish equilibrium iron distribution throughout the crystals (see also Fig. 1.3.12 and Schneider, 1987a).

Things become even more complex if not only iron but multiple foreign cation incorporation in mullite is taken into account. Schneider (1987b) described the temperature-dependent solubility of iron, titanium and magnesium in mullites coexisting with silica-rich partial melts in a used schamotte refractory brick. The mullites of the starting material had relatively high mean titania (TiO_2) but lower iron oxide (Fe_2O_3) and very low magnesia (MgO) contents. Cation removal from mullite to the coexisting glass phase was observed at high temperature though the onset of exsolution is lowest for magnesium and highest for titanium. Cation exsolution curves follow exponential laws with an exsolution rate being considerably higher for Fe_2O_3 than for TiO_2 and MgO (Fig. 1.3.15). The study documents a higher mobility of the iron ion than those of titanium and magnesium.

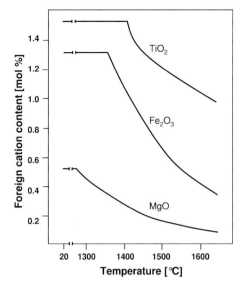

Fig. 1.3.15 Iron, titanium and magnesium incorporation in mullite coexisting with a silicate melt in a used schamotte brick. The Fe_2O_3, TiO_2 and MgO contents are shown versus increasing annealing temperatures. Note that the onset temperature of foreign cation exsolution from mullite follows the sequence $T_{Mg} < T_{Fe} < T_{Ti}$ (from Schneider et al. 1994).

1.3.1.6 Cobalt Incorporation

Preliminary data on cobalt incorporation in mullite were published by Schneider (1990). Cobalt-doped mullite produces electron paramagnetic resonance spectra with signals near $g_{eff} = 4.9$ and 2.2. These electron paramagnetic resonance signals are attributed to the occurrence of Co^{2+} in octahedral coordination. Owing to the low CoO content of about 1 wt.%, neither significant incorporation-induced changes of the lattice constants nor microchemical interdependences between Al_2O_3 and SiO_2 on the one hand and CoO on the other hand can be detected. No information on the incorporation of Co^{3+} into the mullite structure is available as yet.

1.3.1.7 General Remarks on Transition Metal Incorporation

Dependent on synthesis temperatures and atmospheres, mullite incorporates Ti^{3+}, Ti^{4+}, V^{3+}, V^{4+}, Cr^{3+}, Mn^{3+}, Fe^{2+}, Fe^{3+} and Co^{2+}, though in strongly differing amounts. The upper solubility limit is controlled by radii and oxidation states of transition metal ions, with the highest degrees of incorporation being observed for V^{3+}, Cr^{3+} and Fe^{3+} followed by Ti^{4+}. Only very low amounts of Fe^{2+} and Co^{2+} ions can enter the mullite structure (Schneider, 1990).

The transition metal ions are preferably incorporated into the $M(1)O_6$ octahedra in mullite. Two differently distorted octahedral sites have been revealed by spectroscopic methods in the case of Ti^{3+} and Fe^{3+}. Fe^{3+} also enters the oxygen tetrahedra in small amounts, while Cr^{3+} is possibly distributed over the $M(1)O_6$ oxygen octahedra and octahedral interstitial lattice sites, or, alternatively forms Cr^{3+}-Cr^{3+} clusters in the octahedral chains in the **c** direction (see Section 1.3.1.3). Mn^{3+}-substituted mullites are characterized by major changes of lattice spacings and associated increased lattice strains. Both observations are explained by Jahn-Teller distortion of Mn^{3+} in an octahedral environment (Schneider, 1990).

The relatively large cation sizes and high octahedral but lower tetrahedral crystal-field splitting parameters of most transition metal cations may explain their preference for octahedral coordination. Fe^{3+} and Mn^{2+} ions in the middle of the 3d transition metal series are exceptional: these cations have a stable d^5 electron configuration, with a spherically symmetrical charge distribution similar to those of noble gases. Consequently Mn^{2+} and Fe^{3+} exhibit no site preference and their incorporation behavior is mainly controlled by the sizes of the cations. Actually a small amount of Fe^{3+} does enter the oxygen tetrahedra in mullite at high temperature. On the other hand, Mn^{2+} is obviously too large to be tetrahedrally incorporated in mullite (Schneider, 1990).

Entry of the relatively large transition metal cations into the oxygen octahedra in place of aluminum should produce greater expansion of b than of a, because the most elastic bond of the octahedron (M(1)-O(D)) lies about 30° to either side of **b** (Fig. 1.1.14). Greater **b** than a expansion has actually been observed for V^{3+}, Mn^{3+} and Fe^{3+} substituted mullites (Table 1.3.2). The extreme **b** lengthening (and **a** shortening) of Mn^{3+}-substituted mullites is believed to be due to repulsion of singly occupied d_{z2} electron orbitals of Mn^{3+} by the electrons of the O(D) ligands along the octahedral M(1)-O(D) bonds (Jahn-Teller distortion, Fig. 1.3.9). The intense **a** expansion of Ti^{4+}-doped mullites may be explained by substitution of tetrahedral Si^{4+} by Al^{3+}, which occurs along with the octahedral incorporation of Ti^{4+} for charge compensation. The anomalously high **a** expansion of chromium- substituted mullites can possibly be explained by partial entry of Cr^{3+} into interstitial structural channels parallel to the crystallographic **c** axis. Possibly the unique tendency of Cr^{3+} to occupy interstitial lattice positions in mullite can be explained by its ability to produce metastable octahedral oxygen environments in glass and silicate matrices (Schneider, 1990).

While the structural position of 3d transition metals in mullite seems to be controlled essentially by electron configuration and size of cations, the amount of cation incorporation is also dependent on the oxidation state of the ions. Obviously maximum incorporation can be observed if the transition metal occurs as an M^{3+} ion. This can be understood, taking into account that octahedral transition metal incorporation is associated with removal of Al^{3+} from the structure (Fig. 1.3.4). The entry of cations with deviating oxidation states (e. g.: Ti^{4+} and V^{4+}) is less favorable, even if their radii are closer to that of Al^{3+}, since it requires simultaneous tetrahedral substitution of Si^{4+} by Al^{3+} in order to compensate for the excess positive charge.

1.3.2
Other Foreign Cation Incorporation

Besides transition metals the mullite structure is able to incorporate a variety of other foreign cations, although in variable concentrations. Synthesis experiments on the incorporation of Ga^{3+} in mullite were described by Schneider (1986a). A maximum of about 12 wt.% Ga_2O_3 was determined. According to microchemical relationships (Fig. 1.3.4), and because Ga^{3+}-doped mullites display strongest incorporation-induced expansion parallel to the crystallographic **b** axis (Fig. 1.1.14), it has been suggested that Ga^{3+} substitutes Al^{3+} favorably at octahedral lattice positions.

In an early work Scholze (1956) described the solubility of boron in aluminum silicon mullite. Grießer (2005) re-studied the potential boron incorporation into mullite at 950 and 1300 °C. According to this work mullite has a strong tendency to incorporate B^{3+}. B_2O_3 contents up to about 20 mol% have been reported (Fig. 1.3.16). Incorporation of B^{3+} in mullite probably occurs interstially or at tetrahedral sites. Substitution of Al^{3+} by B^{3+} is charge neutral, while substitution of Si^{4+} produces excess negative charge and therefore is less probable. Incorporation of boron in mullite should cause a reduction of lattice spacings, which was, in fact, is observed. In contrast to boron incorporation into mullite no significant silicon incorporation into boron aluminate $Al_{6-x}B_xO_9$ has been described (see Section 1.1.3.14). The excess positive charge produced by Si^{4+} for Al^{3+} replacement may be one reason for this. Thus, there exists a clear solubility gap between mullite and boron aluminate of the composition $Al_{6-x}B_xO_9$. This is understood in terms that although both end members belong to the mullite family, they are distinguished by structural details that do not allow complete and continuous solid solution (see also Section 1.1.3).

Microchemical studies have shown that alkali or alkaline earth cations can enter the mullite structure, though in small quantities (should not be mixed up with the alkali aluminates with mullite-type structure, see Section 1.1.3.7). Owing to its large cation size, Na^+ ($^{[8]}Na^+$: 1.16 Å) can only be incorporated into the thermally expanded mullite structure at very high temperature (up to about 0.4 wt.% Na_2O, Fig. 1.3.17). The Na_2O content of mullite rapidly decreases with decreasing temperature (Schneider, 1984). Mullite incorporates up to about 0.5 wt.% MgO (Schneider, 1985). Magnesium incorporation decreases with temperature and is

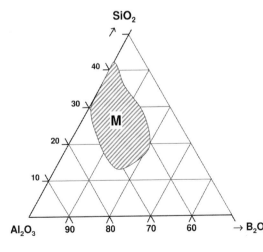

Fig. 1.3.16 Boron incorporation in mullite (M). Note the extended mixed crystal range with B_2O_3 contents of up to 20 mol% being incorporated into mullite at 1300 °C (after Grießer, 2005).

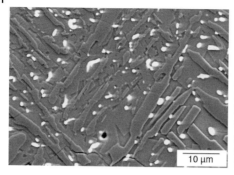

Fig. 1.3.17 Scanning electron micrograph of a fused-mullite refractory brick consisting of mullite (matrix), α-alumina (acicular phases), zirconia (light grains) and silicate glass (at grain junctions). The elongated crystals of α-alumina are the result of rapid crystallization at the outer rim of the refractory brick.

10 µm

reciprocally correlated with the aluminum content of the phase. These observations have been interpreted in terms of an interstitial incorporation of Na^+ and of a substitution of octahedral Al^{3+} ions by Mg^{2+} ($^{[6]}Mg^{2+}$: 0.72 Å, Schneider, 1984, 1985). In a recent study on the thermal decomposition of muscovite, Rodriguez-Navarro et al. (2003) reported more than 1.5 wt.% of MgO into the newly formed mullite. Similarly high incorporation of MgO were described by Rubie and Brearley (1987) and Worden et al. (1987) in mullites derived from the thermal breakdown of magnesium-rich muscovites. No explanation of these high magnesium oxide contents in mullite with correlated charge compensation mechanisms has yet been provided.

Zirconium enters the mullite structure in low amounts (≤0.8 wt.% ZrO_2, Schneider, 1986b). Schneider suggested that zirconium incorporation increases with temperature rather than with the bulk ZrO_2 content, and that "high" zirconium contents can be preserved in mullite only if the material is quenched rapidly to room temperature. It has also been stated that zirconium incorporation increases with the Al_2O_3 content of mullite. Only a very small amount of zirconium (< 0.1 wt.% ZrO_2) is incorporated into 3/2-type mullite, whereas slightly higher zirconium contents (< 0.5 wt.% ZrO_2) can enter 2/1-type mullite. The conclusion of both observations is that the large Zr^{4+} cations ($^{[8]}Zr^{4+}$: 0.84 Å) is incorporated at thermally expanded structural O(C) oxygen vacancies in a similar way to Na^+ (Schneider, 1984). Charge excesses or deficiencies produced by Na^+, Zr^{4+} and Mg^{2+} entry into mullite may be compensated by $Al^{3+} \rightarrow Si^{4+}$, or by $Si^{4+} \rightarrow Al^{3+}$ substitution, respectively.

Caballero and Ocana (2002) investigated the incorporation of tin into mullite synthesized from aerosols. They stated that Sn^{4+} substitutes Al^{3+} at octahedral sites. The authors provided no direct data on the maximum SnO_2 content in mullite, but believed that it is in the same range than that of other foreign tetravalent cations.(i.e. about 3 mol%). Our own synthesis experiments yielded significantly lower SnO_2 incorporation into mullite.

Tomsia et al. (1998), investigating functionally graded molybdenum/mullite composites, mentioned that more than 2 wt.% of molybdenum enters the mullite structure. However, they gave no information on the charge and structural sites of molybdenum in mullite.

Europium-doped mullites have achieved increasing research interest in recent years, owing to their potential application in fluorescent lamps, since Eu^{2+}-doped mullites emit in the blue region. Photoluminescence emission and excitation may also serve as a tool to provide information on the local structural environment of europium in mullite, because of associated electron transitions. Effects are strongly influenced by the crystal fields of the ligands surrounding the europium centers (see Kutty and Nayak, 2000). Piriou et al. (1977) analyzed Eu^{3+}-incorporation into mullite by emission spectroscopy. They described the "unusual" character of the spectra, which they attributed to the strong and anisotropic field arising from the presence of the Eu^{3+} bonds. Kutty and Nayak (2000) studied europium-doped 3/2-mullite (0.41 wt.% Eu_2O_3) and 2/1-mullite (0.67 wt.% Eu_2O_3). They demonstrated that Eu^{3+} in mullite is easily reduced to Eu^{2+} at high temperature and low oxygen partial pressure, and that this process is reversible. Their conclusion was that Eu^{2+} and Eu^{3+} are both incorporated into the relatively large oxygen vacancies in mullite. No model explaining the mechanisms necessary for charge compensation of the interstitial europium ions was provided by the authors. Possibly it is effected by simultaneous Al^{3+} for Si^{4+} substitution. Charge excesses or deficits, caused by $Eu^{2+} \rightarrow Eu^{3+}$ oxidation or $Eu^{3+} \rightarrow Eu^{2+}$ reduction can also be explained by elimination or formation of O(C) oxygen vacancies.

1.4
Mullite-type Gels and Glasses
M. Schmücker and H. Schneider

Instead of natural raw materials like andalusite, sillimanite and refractory-grade bauxite, synthetic precursors have been increasingly used to produce mullite ceramics of high chemical purity, high sinterability and low mullitization temperatures (below 1250 °C).

Since mullite formation mechanisms are strongly affected by the constitution and structure of precursors, knowledge of their temperature-dependent development allows us to design the microstructure and properties of the final mullite ceramics. As an example, mullite ceramics designated for structural applications at moderate temperatures should have a microstructure with small crystal size and a minimum amount of pores, while small amounts of glassy phase can be accepted. In contrast, mullite ceramics for high temperature structural applications have to be glass-free, and a greater crystal size is favored. Moreover, mullite matrices of fiber-reinforced ceramics have to be processed at relatively low temperature (<1350 °C) to avoid fiber degradation (see Section 7.2.1.2), while optical window materials are fabricated at high temperatures (>1600 °C) to achieve dense, optically transparent ceramics. Thus, the precursors must be designed with respect to mullite formation rate, sinterability, composition or additives in order to obtain a final product which best fits the specific applications.

It is generally accepted that two types of mullite precursors exist. One type displays direct mullitization from the amorphous state at temperatures as low as about 950 °C, while the other type shows mullitization above 1200 °C by reaction of transient spinel-type alumina with silica. The former precursor type has been designated as single phase (Hoffman et al. 1984), polymeric (Yoldas, 1990) or type I (Schneider et al. 1993c) while the latter one is called diphasic, colloidal, type II, or NM (no mixing, Okada and Otsuka, 1986, Schneider et al. 1993b). The type of mullite precursor depends on the starting materials and on synthesis conditions. Colloidal suspensions of aluminum and silicon compounds lead to diphasic precursors, whereas true solutions of salts or organometallic compounds give rise to single phase precursors. However, homogeneous solutions of the starting compounds can also produce diphasic mullite precursors. For instance, ethanol diluted admixtures of tetraethyloxysilane plus aluminum-*sec*-butylate can produce single phase or diphasic gels, depending on the process route (see below and Schneider et al.1993c, Voll, 1995). This special case of diphasic mullite precursor has been been designated as rapid hydrolysis gel or type III precursor (Okada and Otsuka, 1986, Schneider et al. 1993c). They are amorphous up to about 950 °C, then form spinel-type alumina plus silica, and mullitization is observed above about 1200 °C.

In the following an overview on the different mullite precursors and their thermal evolution is given. The main emphasis of this section is on the structure (atomic arrangement, short- range-order) of the non-crystalline aluminum silicate phases, structural development before mullitization, correlation between precursor homogeneity and crystallization process, and the mechanisms of mullite formation.

1.4.1
Type I (Single Phase) Mullite Precursors and Glasses

Non-crystalline aluminosilicate precursors transforming directly to mullite at temperatures below 1000 °C have been designated as single phase (e. g. Hoffman et al. 1984). In more recent studies, however, the single phase character of these precursors was a point of controversy: Huling and Messing (1992) mentioned that spinodal phase separation in the amorphous state preceedes mullite crystallization, while Okada et al. (1996) and Schmücker et al. (2001) showed that diphasic precursors may also directly transform to mullite at low temperatures under special circumstances. None the less, cation mixing at the atomic level, i.e. occurrence of Al-O-Si sequences, can in general be assumed for all these precursors. Thus, in the following, to avoid misunderstandings, this type of mullite precursor will be designated as type I, according to Schneider's nomenclature (Schneider et al. 1993c).

1.4.1.1   Preparation of Type I Mullite Precursors and Glasses
Various kinds of type I mullite precursors can be distinguished, depending on their synthesis routes: Chemical methods, sol-gel route, co-precipitation or spray hydrolysis. Aluminosilicate glasses with mullite composition can also be regarded

as type I precursors, since their structural short-range order and crystallization behavior is virtually the same as that of the gels. The preparation of mullite glasses is difficult, since the glass-forming ability of aluminosilicate melts is low, and hence extremely high quenching rates are required to suppress crystallization. Recently it has been shown that ultra homogeneous type I mullite precursors can also be produced by vapor deposition techniques, such as high frequency sputtering (Schmücker et al. 2001, Taake, 1999). In addition, non-crystalline aluminosilicate powders with atomic short-range orders similar to that of mullite gels and glasses can be prepared by mechanical amorphization of mullite carried out via long-term ball milling (Schmücker et al. 1998). Thus, starting materials of ultra homogeneous non-crystalline type I mullite precursors can be liquids (solutions, melts), vapor phases or crystalline solids, as indicated in Fig. 1.4.1.

Chemically derived type I mullite precursors   A number of routes to process type I mullite precursors have been described in the literature. In general, solutions containing the aluminum and silicon species have to be prepared, which, in a second step, are transformed into solids by precipitation or condensation. An overview of methods and techniques leading to type I mullite precursors is given by Schneider et al. (1994a) and Voll (1995).

Ossaka (1961) was among the first to describe a precursor forming mullite below 1000 °C. This precursor was prepared by dissolving sodium silicate and potassium aluminate in sulfuric acid. Addition of hexamethylene tetramine led to aluminosilicate precipitates which were filtered and washed to remove residual alkali and amine components. The main disadvantages of this process are a poor yield of aluminosilicate precipitate and the occurrence of residual alkali ions. A sol-gel method starting with an organometallic silicon source was introduced by Hoffman et al. (1984) for the preparation of type I mullite precursors. They used a solution of tetraethyloxysilane ($Si(OC_2H_5)_4$, TEOS) and aluminum nitrate nonahydrate ($Al(NO_3)_3 \cdot 9H_2O$, ANN) in ethanol, which had been gelled at 60 °C for several days by a hydrolysis-polymerization reaction. The mullite precursor derived by this method has been designated "single-phase xerogel". A similar process was used by Okada and Otsuka (1986). They emphasized that the aluminosilicate gel transforms directly to mullite in the case of slow hydrolysis ("SH xerogel"). Rapid hydrolysis, on the other hand, produces a gel that forms transition alumina as the first crystalline phase (see below, type III mullite precursors). The same results are

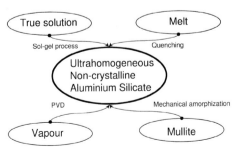

Fig. 1.4.1 Flow chart showing starting materials and processes leading to ultra-homogeneous type I non-crystalline mullite precursors and glasses.

achieved if the gels are aged for more than one month (Okada and Otsuka, 1990), or if gelation is carried out at temperatures below 60 °C (Okada et al. 1996).

Mullite precursor preparation by hydrolysis of organometallic aluminum and silicon compounds is also described in the literature. Alkoxides such as aluminum-isopropylate, aluminum-*sec*-butylate, tetraethyloxysilane (TEOS) or tetramethyloxysilane (TMOS) have been used as starting materials. The preparation of these type I precursors from alkoxides of aluminum and silicon is difficult and very sensitive to the reaction conditions. This is due to the different rates of hydrolysis and polycondensation of the starting compounds, which may cause demixing effects. The problems arising from different reactivities of aluminum and silicon species during the sol-gel process can be minimized by:

− Very slow hydrolysis, e. g. by ambient humidity (Yoldas 1990, Okada and Otsuka, 1986, Colomban, 1989)
− Prehydrolysis of silicon (Voll, 1995)
− Reduction of the hydrolysis rate of aluminum alkoxides using β-diketone (e. g. acetylacetone), as a chelating agent (Heinrich and Raether, 1992)

Spray drying or spray pyrolysis provides an alternative to the hydrolysis-based sol-gel process. Spraying small droplets of a solution containing aluminum and silicon species into a hot reaction chamber causes simultaneous evaporation of the solvents, thermal decomposition and polymerization of the compounds. Thereby the rapid reaction process successfully suppresses demixing effects. The small droplets produced by atomizers or ultrasonicators, yield a precursor powder with particles of spherical shape in the (sub-)micrometer range. Kanzaki et al. (1985) were the first to describe ultra homogeneous type I mullite precursors prepared by spray pyrolysis. Tetraethyloxysilane (TEOS) and aluminum nitrate dissolved in water-methanol solution were atomized and subsequently sprayed into a furnace heated at temperatures between 350 and 650 °C.

Melt-derived type I mullite glasses   Gani and McPherson (1977a) showed that aluminosilicate glasses with mullite composition are suitable precursors for mullite ceramics. However, since the glass-forming ability of aluminosilicate melts with alumina contents >20 % is low, extremely high cooling rates are required to suppress the crystallization of melts during the cooling process. Essentially, two methods of mullite glass preparation have been applied. In the first route small melt droplets produced by flame spheroidisation (Takamori and Roy, 1973), plasma spraying (Gani and McPherson, 1977b), or melt atomization (Morikawa et al. 1982) are quenched in water or oil. In an alternative method, the melt is splat cooled between two rollers made of steel (MacDowell and Beall, 1969), titanium (Risbud et al. 1987) or aluminum (Schmücker et al. 1997). Mullite crystallization in the latter case can be suppressed only if particle sizes or flake thicknesses are below about 20 µm (Gani and McPherson, 1977b, Schmücker et al. 1995). Critical particle sizes or glass-flake thicknesses are controlled by the thermal conductivity of the supercooled melt rather than by the cooling medium. Fig. 1.4.2 shows a rapidly solidified aluminosilicate flake produced by roller quenching of the melt. Thin

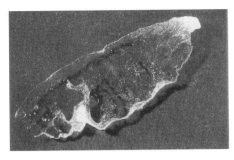

Fig. 1.4.2 Rapidly solidified aluminosilicate glass prepared by roller quenching. Thin and transparent areas are vitreous, while thicker flake areas are partially crystalline (mullite) due to insufficient cooling rates.

areas are transparent and non-crystalline, while thicker areas appear white due to light scattering by micron-sized mullite crystals formed during cooling.

1.4.1.2 Temperature-induced Structural Evolution of Type I Mullite Precursors and Glasses

Wet (solution) stage of type I mullite precursors Little information exists about the structural evolution during hydrolysis and gelation ("wet stage") of type I mullite precursors: Fukuoka et al. (1993) investigated the aluminum coordination of sols and wet gels derived from tetraethyloxysilane (TEOS) plus various aluminum sources by means of ^{27}Al NMR spectroscopy. According to these studies the aluminum ions are octahedrally coordinated in aluminosilicate sols and wet gels derived from aluminum nitrate nonahydrate (ANN) and from boehmite sol, while octahedrally and tetrahedrally bound Al^{3+} is found when aluminum formoacetate is used. Four-, five-, and sixfold coordinated Al^{3+}, on the other hand, occur in gels derived from aluminum di-(butoxide)-ethylacetoacetic ester chelate. Pouxviel and Boilot (1989) investigated the early gelation stages of aluminum silicon esters. Small angle X-ray scattering (SAXS) reveals aggregation of 5 to 8 Å sized elementary clusters. In the case of precursors derived from tetraethyloxysilane (TEOS) plus aluminum *sec*-butylate using acetylacetone as chelating agent, primary particles of 21 to 34 Å form and the condensation is a reaction-limiting cluster-cluster aggregation (Pouxviel et al. 1987). Particles of similar size (about 20 Å) are observed in aluminum nitrate/tetraethyloxysilane (TEOS)-derived sols by means of dynamic light scattering (Jaymes and Douy, 1996; for more details see Section 6.2.1, which deals with mullite fibers).

Dried gel stage of type I mullite precursors The development of dried mullite gels prior to crystallization was investigated by Schmücker and Schneider (1999). These investigations focus on the development of primary particles, volatilization of water and organic groups, condensation of the network, and on the evolution of aluminum oxygen polyhedra. By means of high resolution scanning electron microscopy (SEM), submicron-sized particles are identified in dried or calcined aluminum silicate gels (Fig. 1.4.3). The micrographs show a uniform microstructure of the gel consisting of 20 to 50 nm spherulites. This microstructure remains

Fig. 1.4.3 Scanning electron micrograph of single phase type I gel showing spherical particles about 20 to 50 nm in size.

unchanged up to 800 °C, indicating no change in particle size and agglomeration behavior. The findings agree well with density data based on refractive index determinations, which reveal a density gain of only 7 % after calcination at 900 °C (Okuno et al. 1997). Similar results have been reported by Li and Thomson (1990) on the basis of surface area measurements.

Dehydration and condensation of type I mullite precursors The temperature-dependent removal of water, hydroxyl groups and organic residuals from type I mullite precursors has been studied by means of Fourier transform infrared (FTIR) spectroscopy and mass spectrometry (Voll et al. 1998, Mackenzie et al. 1996). Fig. 1.4.4 shows the analytically determined water loss of tetraethyloxysilane (TEOS)/aluminum *sec*-butylate-derived single phase mullite gels as a function of calcination temperatures together with the integral absorbances of water and hydroxyl groups. Up to annealing temperatures of about 600 °C mullite gels lose virtually all molecular water which is weakly bound at the gel surface and in open pores, whereas the stronger, structurally bound hydroxyl groups are less affected. At temperatures above 700 °C the thermal energy is high enough for dehydroxylation and for subsequent recombination of hydroxyl groups to water ($2OH^- \rightarrow H_2O + 0.5 O_2$). Some of the newly formed water is trapped in nanopores giving rise to a relative increase of the water content. Temperatures above 800 °C are required for the diffusion of water molecules through the gel network, with the consequence that at 900 °C the mullite precursors are almost water-free.

Investigations of the temperature-dependent decomposition and combustion of organic residuals performed by means of in-situ mass spectrometry reveal that organic species can be retained in mullite gels up to about 900 °C. Released residual organic species formed by prolonged heating below about 250 °C are predominantly straight chains or cyclic hydrocarbons, whereas heating at about 350 °C leads to their conversion into aromatic species. Prolonged heating above 350 °C gradually destroys the aromatic species by oxidation (Mackenzie et al. 1996).

^{29}Si NMR spectroscopic investigations have been used to describe the temperature-induced condensation of the network of mullite gels (Schneider et al. 1992, Mackenzie et al. 1996). Fig. 1.4.5 shows a series of ^{29}Si NMR spectra of calcined

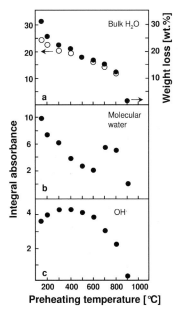

Fig. 1.4.4 Temperature-induced dehydroxylation of single phase (type I) mullite gels: (a) Analytical H_2O contents. Open circles: Data from moisture evolution analysis (MEA), filled circles: Data derived from thermobalance (TG). (b) Integral absorbance values of the infrared H_2O combination band centered at 5160 cm^{-1}. (c) Integral absorbance values of the infrared (Si,Al)-OH combination band at 4540 cm^{-1} (after Voll et al. 1998).

aluminosilicate gels prepared by slow hydrolysis of tetraethyloxysilane (TEOS) and aluminum *sec*-butylate. The NMR spectra contain a major resonance centered at about -88 ppm and two broad shoulders at about -55 and -110 ppm, respectively. The spectral region around -88 ppm is typical for tetrahedrally coordinated silicon in aluminosilicates such as mullite or mullite glass (Risbud et al. 1987, Schneider et al. 1992). The broad shoulder at -110 to -115 ppm is usually very weak in type I mullite gels and glasses, and indicates the presence of a silica-rich phase, while the -55 ppm signal is attributed to residual organo-silicon species and to their transient combustion products (Mackenzie et al. 1996). The temperature-dependent increase of the -110 ppm signal may reflect gradual demixing of the gel into silica-rich and alumina-rich domains (e.g. Huling and Messing, 1992, see below). The major resonance, on the other hand, shows a slight but significant up-field shift from about -84 ppm at 250 °C to -91 ppm at 650 °C. According to Mägi et al. (1984) and Engelhardt and Michel (1987) an up-field peak shift in the ^{29}Si-NMR spectrum of aluminosilicates is either due to a decreasing number of next-nearest aluminum atoms around silicon or due to an increasing number of bridged (Al,Si)-O-(Si,Al) oxygen atoms. Since the major resonance position of the gel approaches the position in corresponding aluminosilicate glasses (e.g. Schmücker et al. 1997) it can be assumed that the temperature-induced upfield shift reflects a gradual condensation process of the gel network accompanied by the evaporation of volatile compounds.

The low increase in the gel density during calcination (see below) indicates that the overall porosity is not dramatically reduced by the precursor network condensation, and a rigid oxide skeleton is formed in an early stage of the precursor evolution. Calculations of radial distribution functions (RDF) have provided fur-

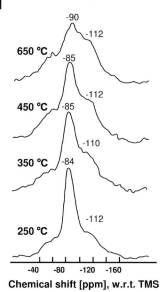

Fig. 1.4.5 ^{29}Si NMR spectra of single phase (type I) mullite gels heat-treated at different temperatures prior to crystallization (after MacKenzie et al. 1996).

ther evidence for a temperature-induced condensation of the network of aluminosilicate gels. As-prepared mullite type I precursors have RDF patterns with a prominent peak at 1.8 Å, and broad peaks of low intensity near 2.9, 3.2 and 4.2 Å (Fig. 1.4.6). On the basis of the ionic radii of Al^{3+}, Si^{4+} and O^{2-}, the peak near 1.8 Å was asssigned to T-O (T = Al^{3+}, Si^{4+}) atomic pairs, whereas those at 2.9, 3.2 and 4.2 Å are associated with O-O, T-T (i.e. T(1)-O(1)-T(2)), and T-O(2) (i.e. T(1)-O(1)-T(2)-O(2)) pairs, respectively. The maxima near 3.2 and 4.2 Å gradually become more intense with heat treatment. This again was interpreted as an increasing condensation of the precursor network (Okuno et al. 1997).

^{27}Al NMR spectroscopy has also been used as a tool to provide information on the mode and distribution of AlO polyhedra in aluminosilicate gels. Ultra homogeneous type I mullite precursors display ^{27}Al NMR spectra with three peaks centered near 0, 30 and 60 ppm (Fig. 1.4.7). In the as-dried gel the 0 ppm signal is stronger than the 60 ppm signal, while the 30 ppm peak is very weak. With heat

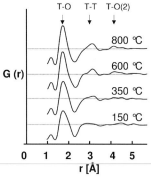

Fig. 1.4.6 Pair distribution functions of single phase (type I) gels heat-treated at different temperatures prior to crystallization (after Okuno et al. 1997).

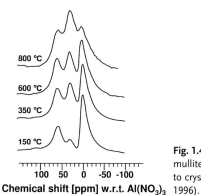

800 °C

600 °C

350 °C

150 °C

100 50 0 -50 -100
Chemical shift [ppm] w.r.t. Al(NO₃)₃

Fig. 1.4.7 ^{27}Al NMR spectra of single phase (type I) mullite gels heat-treated at different temperatures prior to crystallization (after Schmücker and Schneider, 1996).

treatment, a strong increase of the 30 ppm signal intensity is observed, approaching a spectrum very similar to that of aluminosilicate glasses. ^{27}Al NMR data on glasses of the system Al_2O_3–SiO_2 was first published by Risbud et al. (1987). They assigned the ^{27}Al NMR signals near 0 ppm and 60 ppm to octahedral ($Al^{[6]}$) and tetrahedral ($Al^{[4]}$) aluminum, respectively. The 30 ppm resonance was attributed to fivefold coordinated aluminum ($Al^{[5]}$) because of its intermediate position between those of $Al^{[6]}$ and $Al^{[4]}$, and because it exhibits a chemical shift similar to the isotropic shift of andalusite[4]. A large number of NMR spectroscopic studies carried out on aluminosilicate gels and glasses refer to the paper of Risbud et al. and mention the presence of $Al^{[4]}$, $Al^{[5]}$ and $Al^{[6]}$ in these materials. On the contrary, Schmücker and Schneider (1996, 2002) believe that the 30 ppm ^{27}Al NMR signal has to be attributed to tetrahedral triclusters. However, the final assignment of the 30-ppm resonance in the ^{27}Al NMR spectrum is still a point of discussion (see 1.4.4.2).

Gerardin et al. (1994) and Schneider et al. (1994b) pointed out that there is a correlation between the 30 ppm NMR signal and mullite formation. If aluminum sites corresponding to the 30 ppm signal are predominant, mullite formation is preferred to transition alumina crystallization or, in other words, the intensity of the 30 ppm peak in sol-gel-derived mullite precursors is assumed to correlate with their degree of structural homogeneity. Taylor and Holland (1993), on the other hand, reported that aluminum is tetrahedrally coordinated in very homogeneous gels, while diphasic gels are characterized by high amounts of octahedrally coordinated aluminum. Fivefold coordinated aluminum was believed to occur in interfacial regions between homogeneous and less homogeneous domains.

Metastable immiscibility in the system SiO_2–Al_2O_3 Demixing occurring in noncrystalline materials of the system Al_2O_3–SiO_2 has been well established for many

---

4) Andalusite contains chains of AlO_6 octahedra running parallel to the **c** axis. These octahedral chains are connected by double chains consisting of SiO_4 tetrahedra and AlO_5 polyhedra (Burnham and Buerger, 1961).

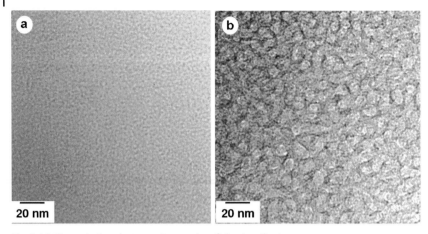

Fig. 1.4.8 Transmission electron micrographs of aluminosilicate glasses. (a) After rapid quenching; (b) After rapid quenchnig and calcination below crystallization temperature (890 °C). Note the occurrence of unmixing zones (Courtesy M. Schmücker).

years. MacDowell and Beall (1969) and Galakhov (1976) were among the first reporting the existence of a metastable immiscibility region deduced from microstructural investigations of rapidly quenched aluminum silicate glasses. They obtained a miscibility gap extending from about 15 to 70 mol% Al_2O_3, and estimated the upper consolute temperature as near 1650 °C. Their results, however, are not without problems, since phase separation was heavily disturbed by simultaneous mullite crystallization. Jantzen et al. (1981) studied ultrarapid quenching from melts and the kinetics of demixing. They obtained a consolute temperature as low as about 725 °C at a composition of about 46 mol% Al_2O_3. Fig. 1.4.8 shows the microstructure of a rapidly solidified aluminosilicate glass with mullite composition. The as-quenched glass appears featureless (Fig. 1.4.8a) while annealing at 890 °C produces significant segregation effects (Fig. 1.4.8b).

Ban et al. (1996) emphasize that mullite crystals formed in glasses and ultrahomogeneous gels at temperatures below 1000 °C are supersaturated in Al_2O_3, irrespective of the bulk chemical composition of the starting material. Ban et al. assume that the unusual crystallization behavior of mullite is caused by phase separation in the amorphous state, and that the composition of mullite formed at 950 °C corresponds to the Al_2O_3-rich demixing zones existing at this temperature. Rapid mullite crystallization at temperatures above 950 °C, however, disturbs the observation of metastable phase separation. It is therefore extremely difficult to determine the high temperature region of the immiscibility gap by experimental methods.

There have been several approaches for calculating the immiscibility region from thermodynamic data. Risbud and Pask (1977) calculated the immiscibility gap for the pseudo-binary system silica-mullite using a regular solution model. Ban et al. (1996) also assumed a regular solution model but considered the binary system Al_2O_3–SiO_2. Takei et al. (2000), on the other hand, used thermodynamic

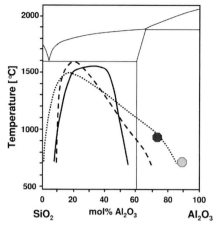

Fig. 1.4.9 Metastable immiscibility in the system Al_2O_3–SiO_2. Curves refer to results of Risbud and Pask (1977; full line), Ban et al. (1996; dotted line), and Takei et al. (2000; dashed line). Data points correspond to mullite compositions formed at 980 °C (dark) and 700 °C (light), respectively (after Fischer et al. 1994).

parameters derived from molecular dynamics simulations. The calculations confirmed the existence of metastable immiscibiltity in the Al_2O_3–SiO_2 system. Miscibility gaps display critical compositions near 20 mol% Al_2O_3 and consolute temperatures ranging between 1500 and 1700 °C. The shape of gaps is asymmetric and tail towards the Al_2O_3-rich composition side of the system. Hillert and Jonsson (1992) simulated the miscibility gap with the CALPHAD method. Their upper consolute point at about 1600 °C fitted well with the other models. However, its critical composition, near 50 mol% Al_2O_3, is obviously too Al_2O_3-rich.

The calculated immiscibility regions are plotted in Fig. 1.4.9 together with typical compositions of mullite formed at 980 °C and 700 °C (Fischer et al. 1994), respectively. The results calculated by Ban et al. (1996) fit well with the crystallization temperature and the composition data determined for mullite. However, the extension of the immiscibility gap towards SiO_2 may correspond to calculations of Risbud or Takei rather than to the curve of Ban, since no phase separation could be observed in silica glasses with 5 mol% Al_2O_3 prior to mullitization (Schmücker, unpublished results). Djuric and Mihajlov (1996) working on theoretical criteria and using data along the liquidus curves of mullite and α-alumina suggested immiscibility regions at both ends of the phase diagram. This observation has not been confirmed by other workers.

1.4.1.3 Mechanisms of Mullite formation From Type I Precursors and Glasses

Takei et al. (1999) studied mullite crystallization in commercial aluminosilicate glass fibers that were relatively SiO_2-rich (49 mol% Al_2O_3) and Al_2O_3-rich (69 mol% Al_2O_3), respectively. Mullite formation from the amorphous state is believed to take place in three steps.

- In the first stage, below 1000 °C, mullite nucleation is the dominant process, characterized by activation energies ranging between about 850 kJ mol^{-1} (SiO_2-rich) and 1000 kJ mol^{-1} (Al_2O_3-rich). The higher activation energy of the Al_2O_3-rich system is surprising, since the viscosity of aluminosilicate glass decreases with the Al_2O_3 content and the correlated mobility of atomic species increases as well. Takei et al. explained this discrepancy by a phase separation occurring in the liquid state of the Al_2O_3-rich glass fibers.

- In the second stage, between about 1000 and 1200 °C, mullite nucleation and growth are controlling mechanisms. Activation energies range between about 1200 kJ mol^{-1} (SiO$_2$-rich) and 1100 kJ mol^{-1} (Al$_2O_3$-rich). The activation energies of this stage have been explained by the slow diffusion of species.

- In the third stage, above about 1200 °C, the main effect is mullite grain growth via coalescence. Processes in this mullitization stage proceed more rapidly because of the higher temperatures, and as a result the activation energies become lower (SiO$_2$-rich: about 700 kJ mol^{-1}, Al$_2O_3$-rich: about 650 kJ mol^{-1}).

Wei and Rongti (1999) studied the crystallization kinetics of other aluminosilicate glass fibers with 67 mol% Al_2O_3 (79 wt.%). They observed first mullite crystallization at about 1250 °C, with maximum mullite precipitation at 1280 °C. and a rather low activation energy of crystallization (about 650 kJ mol^{-1}). Takei et al. (2001) and Okada et al. (2003) investigated the mullite formation process from type I single-phase gels of stoichiometric composition (60 mol% Al_2O_3) and from aluminosilicate glasses with Al_2O_3 contents varying between 15 and 50 mol%. The activation energies of mullite nucleation and growth have been calculated to be about 1200 kJ mol^{-1}, or from 900 to 1300 kJ mol^{-1}, respectively. Diffusion-controlled mechanisms of mullite formation are assumed for these gels and glasses. Tkalcec et al. (1998) starting from a gel of stoichiometric mullite composition (60 mol% Al_2O_3) obtained similar activation energies of mullitization (>1050 kJ mol^{-1}). They discuss a two-step mullite formation model, attributed to phase separation in the precursor, and suggest that mullitization is controlled by phase separation rather than by nucleation and growth of mullite. From literature data it can be concluded that the nucleation density of mullite is extremely high (see also Fig. 1.4.24, below) and that crystal growth occurs isotropically. The kinetic results of Takei et al., Okada et al. and Tkalcec et al. are in a clear contrast to the data published by Li and Thomson (1990) with activation energies between about 300 and 350 kJ mol^{-1} and without any induction branch of the transformation curves, while all other studies observe distinct nucleation periods. A possible explanation is that Li and Thomson measured only the activation energies of nucleation, whereas the other studies measured those of nucleation and growth (see Okada et al. 2003). Similar reasons may account for the low activation energy published by Wei and Pongti (1999).

A time-temperature correlation of the incubation time (τ) was used to estimate the activation energy of nucleation (about 980 kJ mol^{-1}) which is somewhat smaller than the activation energy determined for subsequent crystal growth (about 1100 kJ mol^{-1}). Johnson et al. (2001) stated that aluminosilicate glass samples are fully nucleated by the time they reach 850 °C, and hence crystallization occurs with a constant number of nuclei. James et al. (1997), on the other hand, reviewing the nucleation rates of various silicate glasses, found that the maximum nucleation temperature (T_M) is strongly correlated with the liquidus temperature (T_L), thus leading to T_M/T_L ratios scattered in a remarkably narrow range of 0.54 to

0.58. Using these data, the nucleation temperature of mullite glasses has been determined to range between 895 and 980 °C, suggesting that the formation of mullite at temperatures above 900 °C is not only due to the growth of pre-existing nuclei but involves both nucleation and growth. Inconsistencies in Johnson's findings are caused by the moderate quenching rates, which may have caused mullite nucleation during cooling also.

1.4.2
Type II (Diphasic) Mullite Precursors

Although in type I (single phase) mullite precursors mullitization takes place below 1000 °C this material is not without problems: As diffusion in mullite is very sluggish, sintering-induced densification of ceramics becomes difficult. Diphasic mullite precursors of type II can provide the solution to this problem.

These mullite precursors consist of transitional alumina and non-crystalline silica (type II, Schneider et al. 1993c) and thus allow densification prior to mullitization by viscous flow sintering. The particle size of constituents of diphasic mullite precursors range from a few to several hundreds of nanometers. The precursors are usually prepared by sol-gel or precipitation techniques, but well homogenized and heavily ground powders of alumina and silica phases may also come into this category.

1.4.2.1 Synthesis of Type II Mullite Precursors

Most preliminary work on diphasic mullite precursors via sol-gel techniques originates from the pioneering study of Hoffman et al. (1984). They synthesized diphasic mullite precursors with aqueous silica and boehmite sols as starting materials. Gelation was carried out by gradual solvent evaporation. In alternative approaches admixtures of boehmite sol plus alcoholic tetraethyloxysilane (TEOS) solutions or silica sol plus aqueous aluminum nitrate solutions, have been employed. Okada and Otsuka (1986) used demixed alcoholic solutions of tetraethyloxysilane (TEOS) plus aluminum chloride and gelled them after slow hydrolysis. More recently, Voll (1995) prepared diphasic mullite precursors starting with alcoholic solutions of tetraethyloxysilane (TEOS) and aluminum *sec*-butylate. Tetraethyloxysilane (TEOS) solutions are prehydrolyzed by addition of water under strongly basic conditions (pH = 13) to induce silica self-condensation. After a short aging time the silica sol, containing a considerable amount of excess water, is put into the aluminum-bearing solution, leading to rapid hydrolysis of aluminum *sec*-butylate and subsequent formation of pseudo-boehmite colloids. Transmission electron microscopy of the dried gel reveals the existence of relatively large spherical silicaparticles and nanometer-sized pseudo-boehmite aggregates intimately embedded in a non-crystalline silica matrix (Fig. 1.4.10).

A novel preparation method for type II mullite precursors has been developed by Sacks et al. (1991) by means of transient viscous sintering (TVS) techniques (see Section 4.2.6). According to their concept the powder particles of the mullite pre-

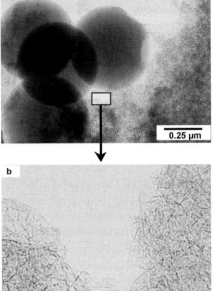

Fig. 1.4.10 Transmission electron micrographs of a diphasic mullite precursor consisting of relatively large spherical silica particles and nanometer-sized pseudo-boehmite aggregates, intimately embedded in a noncrystalline silica matrix. (a) Overview, (b) Detail showing pseudo-boehmite aggregates.

cursor are "microcomposites" consisting of α-alumina cores enveloped by amorphous silica layers. These microcomposites are prepared by dispersing the submicron alumina particles in an alcoholic solution of tetraethyloxysilane. In the next step silica is precipitated at the surface of the alumina particles by the addition of ammoniated water. Compacts of the composite particles are sintered to almost full density at about 1300 °C by viscous flow of the amorphous silica layer. On the other hand, the relatively long diffusion paths necessary for full mullitization (about 200 nm) still require temperatures of about 1600 °C. Based on the idea of microcomposite powders, Bartsch et al. (1999) improved the transient viscous sintering process by preparing "nanocomposite" mullite precursors consisting of γ-alumina particles coated by nanometer-thick silica layers (Fig. 1.4.11). These nanocomposites exhibit a similar densification behavior by transient viscous flow as the α-alumina/silica microcomposites but mullite forms at temperatures as low as 1300 °C. This enhancement of sintering has been explained by diffusional distances required for mullitization which are about one order of magnitude smaller in the latter case.

1.4.2.2 Temperature-induced Structural Evolution of Type II Mullite Precursors

The transformation of diphasic type II mullite precursors involves two steps:
– The development of spinel-type transition alumina ("γ-alumina") phases below about 1200 °C
– Mullite formation by reaction of γ-alumina and silica above about 1200 °C.

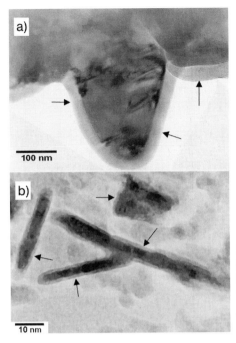

Fig. 1.4.11 Transmission electron micrograph of a diphasic mullite precursor prepared by coating (a) α-alumina powder and (b) γ-alumina powder with amorphous silica according to the transient viscous sintering technique (see also Section 4.2.6. Courtesy M. Bartsch).

The spinel-type transition alumina phase Diphasic mullite precursors prepared by sol-gel methods typically consist of pseudo-boehmite (γ-AlOOH) plus amorphous silica (e.g. Hoffman et al. 1984). By dehydration above about 500 °C pseudo-boehmite converts to γ-alumina, the latter transforming into the structurally related δ- and θ-alumina at elevated temperatures (Wefers and Misra, 1987). The composition of the γ-alumina phases has been a point of controversy for many years. The spinel-type transition phase occurring during the kaolinite-mullite transformation was found to incorporate considerable amounts of silicon. This derives from the analyses of leached samples that reveal aluminum-silicon spinels with compositions approaching that of mullite (e.g. Chakraborty and Ghosh 1978, Srikrishna et al. 1990). Brown et al. (1985), on the other hand, using ^{29}Si NMR spectroscopy established virtually silicon-free γ-alumina as the transient phase in the kaolinite-mullite reaction sequence. Contradictory results have also been reported for the spinel-type transition phase in gel-derived mullite precursors. Low and McPherson (1989), from infrared spectroscopic investigations, assumed a composition corresponding to that of 2/1-mullite (i.e. 33 mol% SiO_2), while Wei and Halloran (1988a) and Komarneni and Roy (1986) on the basis of analytical transmission electron microscopy (TEM) and ^{29}Si NMR spectroscopy concluded that the spinel phase is essentially pure aluminum oxide. Okada and Otsuka (1986) compared the IR spectrum of pure γ-alumina with that of the gel-derived spinel phase and reported a high silicon incorporation, with an Al_2O_3/SiO_2 ratio in the spinel phase of about 6 to 1.

Schneider et al. (1994c) re-examined the composition of the transition spinel

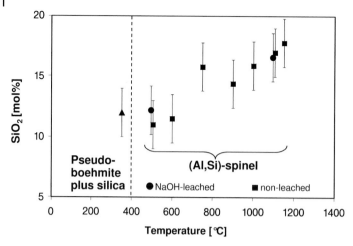

Fig. 1.4.12 SiO_2 content of pseudo-boehmite plus silica phase assemblages and of spinel-type transition alumina ("γ-alumina") determined by the energy dispersive X-ray (EDX) technique. As leaching of the spinel phase has no influence on the composition, it is concluded that silicon has been incorporated in the spinel phase (after Schneider et al. 1994c).

phase that develops in mullite precursors derived from tetraethyloxysilane (TEOS) plus aluminium *sec*-butylate (see above) by combining spectroscopic methods with analytical transmission electron microscopic results on non-leached and leached samples that had been calcined between 350 and 1150 °C. The dried precursor powder consists of relatively large spherical silica particles and much finer grained agglomerates of pseudo-boehmite embedded in a silica matrix (Fig. 1.4.10). Above about 350 °C the pseudo-boehmite and silica admixture converts completely to aluminum-silicon spinel with about 12 mol% SiO_2. Up to 750 °C the composition of the spinel phase remains constant, and above this temperature the SiO_2 content gradually increases up to about 18 mol% at 1150 °C (Fig. 1.4.12), obviously triggered by partial dissolution of the larger SiO_2 spherules.

Mechanisms of mullite formation from type II precursors   The kinetics and mechanisms of mullite crystallization from diphasic type II precursors in the system Al_2O_3–SiO_2 have been studied by several research groups. Wei and Halloran (1988b) were the first to describe the kinetics of mullite formation from pseudo-boehmite and tetraethyloxysilane (TEOS) with an activation energy of the overall process of about 1070 kJ mol⁻¹. These authors believe that mullite forms by a direct solid-state reaction between transition alumina and the silica-rich non-crystalline phase by an interface- or, alternatively, a diffusion-controlled process. Mullite formation is preceded by an incubation period, corresponding to nucleation. Mullite nucleation has an activation energy of about 990 kJ mol⁻¹, which is very close to that of the overall process (see above). Scanning electron microscopic analyses reveal a nucleation density of the type II precursors of about 2×10^{11} cm⁻³ which is signicantly smaller than that of type I single-phase mullite precursors (about 10^{17} cm^{-3}, see Fig. 1.4.24 below). The growth of mullite grains follows a $t^{-0.63}$ law. Li and Thomson (1991) used the same starting materials and achieved similar activation energies of mullitization (about 1040 to 1080 kJ mol^{-1}).

Huling and Messing (1991) started from aluminum nitrate nonahydrate and tetraethyloxysilane (TEOS) for their precursor synthesis. They give an activation energy of mullite formation of about 1030 kJ mol^{-1}. Ivankovic et al. (2003) investigating the influence of the particle size of the aluminum sources (boehmite, γ-alumina, aluminum nitrate nonahydrate) on the reaction kinetics, suggested that the mullitization is a two-step process: In a first step Al$_2$O$_3$-rich 2/1-mullite (66 mol% Al$_2$O$_3$) forms, which, in a second step, is transformed to stoichiometric mullite (60 mol% Al$_2$O$_3$) with the activation energy of the overall reaction ranging between about 900 and 1150 kJ mol^{-1} (see also Tkalcec et al. 2003). Finally Boccaccini et al. (1999), starting from boehmite and fumed silica nanopowders obtained activation energies of about 880 kJ mol^{-1}. The latter authors suggest that mullitization is characterized first by an induction period during which mullite nucleation takes place, followed by the main transformation regime, in which mullite nucleation and growth occur simultaneously.

The activation energies of mullitization in type I (single phase) and type II (diphasic) gels and glasses of mullite composition are similar in value. This may indicate that similar mechanisms of mullitization are active. In diphasic precursors dissolution of γ-alumina in the coexisting silica phase and reprecipitation of mullite are rate-controlling steps, being more probable than direct solid-state reactions between transitional alumina and silica as quoted by Wei and Halloran (1988b). Acceleration of mullite formation and associated reduction of the activation energy below 900 kJ mol^{-1} is possible if precursors of higher reactivity are used, as in the case of Boccaccini et al. (1999). Their diphasic mullite precursor was prepared by admixing nanometer-sized fumed silica (aerosol) with boehmite sol. Microcomposites consisting of sub-micron α-alumina particles enveloped by nanometer-thick silica glass coatings behave quite similarly to diphasic gels (see Sacks et al. 1996). These systems have typical activation energies of mullitization of 1040 kJ mol^{-1}. Sacks et al. suggest that mullite formation in this case proceeds via solution of aluminum in amorphous silica with subsequent nucleation and growth of mullite.

A suitable way to further drastically reduce the activation energies of diphasic type II mullite precursors is doping of the starting compounds with foreign atoms. In this way, the viscosity of the amorphous silica is reduced drastically. This causes higher diffusion rates, and as a consequence, induces accelerated nucleation and crystal growth of mullite. Hong et al. (1996) and Hong and Messing (1997, 1999) studied the mullite formation kinetics of sol-gel-derived diphasic gels, doped with phosphorus oxide (P$_2$O$_5$), boria (B$_2$O$_3$) and titania (TiO$_2$, see Table 1.4.1). Low activation energies for enhanced mullitization have also been observed by Hildmann et al. (1996) in commercial fibers consisting of γ-alumina plus silica-rich amorphous phase containing a small amount of boria (see Table 1.4.1). The fact that the nucleation barrier of mullite is clearly lowered by boria addition goes along

Table 1.4.1 Activation energy of mullite formation from precursors and glasses.

Composition [wt. %]	Starting compounds	Investigation technique	Effective activation energy [kJ mol⁻¹]	Reference
Single-phase (type I) precursors and glasses				
Al_2O_3: 72% SiO_2: 28%	Al nitrate nona-hydrate + TEOS	DXRD	300	Li and Thomson (1990)
Al_2O_3: 49% SiO_2: 51%	Commercial glass fiber	XRD	≈ 1200	Takei et al. (1999)
Al_2O_3: 69% SiO_2: 31%	Commercial glass fiber	XRD	≈ 1100	Takei et al. (1999)
Al_2O_3: 79% SiO_2: 21%	Arc-furnace produced from α-Al_2O_3 + quartz	DTA	≈ 650	Wei and Rongti (1999)
Al_2O_3: 23–72% SiO_2: 77–28%	Al nitrate nona-hydrate + TEOS	DTA	≈ 1200	Takei et al. (2001)
Al_2O_3: 72–87% SiO_2: 28–13%	Al nitrate nona-hydrate + TEOS	DTA	≈ 900–1300	Okada et al. (2003)
Al_2O_3: 72% SiO_2: 28%	Al nitrate nona-hydrate + TEOS	DSC	≈ 1050	Tkalcec et al. (1998)
Diphasic precursors (type II) and single phase/diphasic (type III) precursors				
Al_2O_3: 46–84% SiO_2: 54–16%	Boehmite + TEOS	DXRD	≈ 1040–1080	Li and Thomson (1991)
Al_2O_3: 63–77% SiO_2: 37–23%	Pseudoboehmite + TEOS	XRD	≈ 1070	Wei and Halloran (1988b)
Al_2O_3: 72% SiO_2: 28%	Al nitrate nona-hydrate, γ-Al_2O_3, boehmite, TEOS	XRD	≈ 890–1060	Tkalcec et al. (2003)
Al_2O_3: 73% SiO_2: 27%	Al nitrate nona-hydrate + TEOS	DTA	≈ 1030	Huling and Messing (1991)
Al_2O_3: 72% SiO_2: 28%	Boehmite sol + silica nanopowder	DTA	≈ 880	Boccaccini et al. (1999)
Al_2O_3: 72% SiO_2: 28%	Al nitrate nona-hydrate, γ-Al_2O_3, boehmite, TEOS	XRD	≈ 950–1090	Ivankovic et al. (2003)
Hybrid precursors (types I and III)				
Al_2O_3: 70–73% SiO_2: 30–27%	Al nitrate nona-hydrate, silica sol, boehmite, TEOS	DTA	≈ 930–1090	Huling and Messing (1991)

with the observation of a significantly increased nucleation density of boria-doped systems (about 10^{15} cm⁻³) in comparison to that in undoped systems (about 2×10^{11} cm⁻³). In the case of boria addition Hong and Messing argue that the boron aluminate $9Al_2O_3 \cdot 2B_2O_3$ is intermediately formed. Since this phase has a crystal structure belonging to the mullite family (Ihara et al. 1980, Garsche et al. 1991) it

Table 1.4.1 (continued).

Composition [wt. %]	Starting compounds	Investigation technique	Effective activation energy [kJ mol⁻¹]	Reference
Foreign cation oxide-doped diphasic precursors				
Al$_2$O$_3$: 70% SiO$_2$: 28% B$_2$O$_3$: 2%	Commercial fiber (Nextel 440, 3M)	XRD	≈ 900	Hildmann et al. (1996)
Al$_2$O$_3$: 3/2 plus Na$_2$O	Al nitrate plus Na metasilicate	XRD	≈ 730	Campos et al. (2002)
Al$_2$O$_3$: 27% SiO$_2$: 53% ZnO: 10% B$_2$O$_3$: 4% Li$_2$O: 0.7% ZrO$_2$: 2.5–3.0% TiO$_2$: 1.7–2.5%	Al(OH)$_3$, quartz sand, H$_3$BO$_3$, ZnO, TiO$_2$, ZrO$_2$, Li$_2$O	DTA	≈ 465	Tkalcec et al. (2001)
Al$_2$O$_3$: 73% SiO$_2$: 24% TiO$_2$: 3%	Boehmite, silica sol, TiO$_2$	DTA	≈ 1020	Hong and Messing (1997)

(D)XRD: (Dynamic) X-ray diffraction, DTA: Differential thermal analysis, DSC: Differential scanning calorimetry.

may provide sites for topotactical or epitactical nucleation of mullite. Unfortunately Hong and Messing gave no experimental evidence supporting this interesting approach. The crystallization of mullite from multicomponent silicate glasses of the system Li$_2$O–ZnO–Al$_2$O$_3$–B$_2$O$_3$–ZrO$_2$–SiO$_2$ with different ZrO$_2$/TiO$_2$-ratios has been described by Tkalcec et al. (2001). Mullitization at temperatures as low as 800 °C with activation energies below 500 kJ mol⁻¹ indicates the high reactivity of the low-viscosity precursor caused by the presence of foreign cations (see Table 1.4.1). Campos et al. (2002) described a similar although less significant trend due to the occurrence of sodia (Na$_2$O, see Table 1.4.1).

Sundaresan and Aksay (1991) re-examined the kinetics of mullite formation from type II (diphasic) precursors published by Wei and Halloran. They emphasize that the reported time-dependent growth rate is not consistent with interface- or diffusion-controlled transformations but is in excellent accord with a dissolution and precipitation mechanism. In this scenario the alumina particles dissolve in the silica phase and mullite nuclei form when the vitreous aluminosilicate phase exceeds a critical concentration. The solution plus precipitation process can be illustrated in a free enthalpy versus composition diagram (Fig. 1.4.13). It turns out that the equilibrium composition of the amorphous phase is richer in Al$_2$O$_3$ when coexisting with alumina (SA) than when coexisting with mullite (SM). Mullite nucleation occurs if the critical nucleation concentration (CNC) is reached at a composition beyond SM. According to Sundaresan and Aksay, mullite nucleation within the non-crystalline silica implies that dissolution of alumina is the rate-

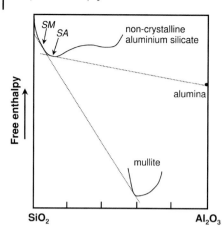

Fig. 1.4.13 Schematic free enthalpy diagram of the system $Al_2O_3–SiO_2$ (after Sundaresan and Aksay, 1991). SM and SA are the compositions of the non-crystalline aluminosilicate phase coexisting with mullite and alumina, respectively.

controlling step in mullite growth. Diffusion of aluminum ions through the vitreous phase as the rate-controlling step is ruled out: If that were the case a compositional gradient would be expected within the silica-rich phase, giving rise to mullite formation at the alumina/silica interface rather than in the bulk of the non-crystalline silica-rich phase.

Mullite as an interfacial product is observed in sapphire/silica reaction couples heated at temperatures above 1678 °C (Aksay and Pask, 1975). As a conclusion from experimental observations, Sundaresan and Aksay (1991) predict a change of the rate-limiting effects at some point between 1350 and 1650 °C, such that dissolution is rate-controlling below this temperature and diffusion is rate-controlling above. Microscopic evidence for the solution/precipitation mechanism and for a change of mullite formation mode from dissolution- to diffusion-controlled was given for the first time by Schmücker et al. (1994) investigating the mullitization of α-alumina/quartz powder admixtures. The starting powders used in this study were coarse (>100 nm) compared with the alumina and silica particles occurring in gel-derived mullite precursors, and hence mullitization temperatures are shifted to about 1500 °C. Since quartz grains form (metastable) viscous melt layers at their peripheries upon (rapid) heating at temperatures above about 1300 °C, the mullitization reaction takes place between α-alumina and amorphous silica rather than between α-alumina and quartz (Fig. 1.4.14a). Upon heating the powder admixtures to 1550 °C, mullite crystallites form randomly within the non-crystalline silica-rich phase (Fig. 1.4.14b).

The compositional evolution of the vitreous phase as determined by energy dispersive X-ray (EDX) analyses is shown in Fig. 1.4.15. About 4 mol% Al_2O_3 is incorporated into the viscous silica melt at about 1450 °C. Firing temperatures of 1500 °C produce a bimodal compositional distribution with maxima at about 4 and 2.5 mol% Al_2O_3. The obvious reduction of the Al_2O_3 content of the non-crystalline phase with temperature is attributed to mullite crystallization from this phase. The composition of the silica-rich melt coexisting with mullite (SM) is about 97.5 mol% SiO_2 and 2.5 mol% Al_2O_3, while the critical concentration of mullite nucleation

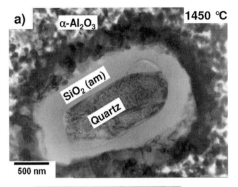

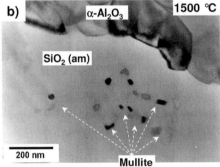

Fig. 1.4.14 Transmission electron micrographs of heat-treated α-alumina/quartz powder compacts. (a) 1450 °C – Formation of a metastable viscous melt layer at the periphery of quartz grains; (b) 1500 °C – Formation of randomly oriented mullite crystallites within the liquid siliceous phase (arrows, Courtesy M. Schmücker).

(CNC) is about 4 mol% Al_2O_3 at 1550 °C (Fig. 1.4.16). The microstructure of quartz/alumina powder admixtures heat-treated at 1600 °C indicates a change in the mullite formation mechanism: Although silica has converted completely to cristobalite, the extension of previous outer melt zones can be recognized by ori-

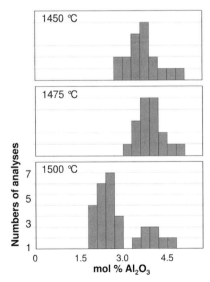

Fig. 1.4.15 Composition of aluminosilicate melts formed at the periphery of quartz grains during reaction sintering of mullite. The primarily existing silica melt incorporates up to 4 mol% Al_2O_3 prior to mullite nucleation. After mullite nucleation (1500 °C) the Al_2O_3 content of the liquid phase is about 2.5 mol% (after Schmücker et al. 1994).

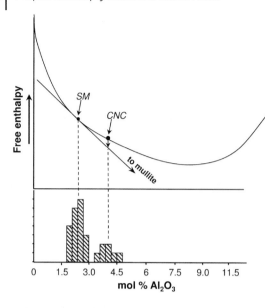

Fig. 1.4.16 Plot of the free energy versus Al_2O_3 content of aluminosilicate melts formed at the periphery of quartz grains during reaction sintering of mullite. The bimodal compositional distribution of the aluminosilicate phase occurring during reaction sintering of quartz plus α-alumina (see Figs. 1.4.14 and 1.4.15) suggests that the silicate melt coexisting with mullite (i.e. SM, see Fig. 1.4.13) contains about 2.5 mol% Al_2O_3, while the critical nucleation concentration (CNC) is about 4 mol% Al_2O_3.

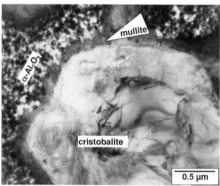

Fig. 1.4.17 Transmission electron micrograph of an α-alumina/quartz powder compact sintered at 1600 °C The picture indicates a different reaction mechanism with respect to firing conditions at 1500 °C (Fig. 1.4.14) No mullite crystals occur in the peripheral amorphous silica zone enveloping the cristobalite grains, but a mullite layer forms at the α-alumina/silica contact instead (Courtesy M. Schmücker).

entational contrasts of the cristobalite. It is noteworthy that no mullite crystals are incorporated in the outer cristobalite zone, but a mullite layer forms at the α-alumina/silica contact (Fig. 1.4.17), which is consistent with the model of Sundaresan and Aksay suggesting diffusion-controlled mullite growth at high temperatures.

1.4.3
Type III (Single Phase/Diphasic) Mullite Precursors

Type I single phase mullite precursors are non-crystalline and convert to Al_2O_3-rich mullite at about 950 °C (see Section 1.4.1), while diphasic type II precursors typically consist of poorly crystalline aluminum hydroxides and oxides and non-crystalline silica, which react to mullite above 1200 °C (see Section 1.4.2). A further

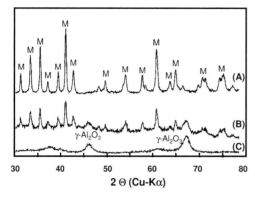

Fig. 1.4.18 X-ray diffraction traces of mullite precursors prepared from tetraethyloxysilane (TEOS) and aluminum nitrate nonahydrate solutions gelled at (A) 60 °C; (B) 40 °C and (C) 20 °C, respectively, and after firing at 1000 °C. At low gelation temperatures the precursor converts into transition alumina ("γ-alumina") plus amorphous silica rather than into mullite.

non-crystalline mullite precursor type has been described in the literature and was designated as rapid hydrolysis gel (Okada and Otsuka, 1986) or type III gel (Schneider et al. 1993c). The crystallization temperature of type III aluminosilicate gels corresponds to type I mullite precursors. However, transitional alumina ("γ-alumina") forms at a first stage, while mullite crystallizes by reaction of the aluminum and silicon compounds only at about 1200 °C. The latter reaction is similar to that of diphasic type II precursors (see Section 1.4.2). There exist gradual transitions between type I and type III precursors, resulting in mullite, mullite plus γ-alumina, or γ-alumina crystallization (Voll, 1995, Taake, 1999; see also Fig. 1.4.18).

1.4.3.1 Synthesis of Type III Mullite Precursors

Type III mullite gels can be prepared by various routes. Okada and Otsuka (1986) and Hyatt and Bansal (1990) used tetraethyloxysilane (TEOS) and aluminum nitrate nonahydrate solutions as starting compounds. Slow hydrolysis at 60 °C was reported to produce type I single phase mullite precursors (see above), but type III gels can be achieved by slow hydrolysis if the sol is aged at room temperature instead of 60 °C (Okada et al. 1996, Taake, 1999). The mullite gels gradually change from type III to type I on increasing the aging temperature from room temperature to 60 °C (Fig. 1.4.18). Voll (1995) prepared type III mullite precursors starting with alcoholic solutions of tetraethyloxysilane (TEOS) and aluminum *sec*-butylate by a similar method to that described for the preparation of diphasic gels. The tetraethyloxysilane (TEOS) solution is pre-hydrolyzed but mild basic conditions (pH 7-10) are used in order to prevent silica self-condensation.

1.4.3.2 Temperature-induced Structural Evolution of Type III Mullite Precursors

The structural development of tetraethyloxysilane (TEOS) plus aluminum nitrate nonahydrate-derived type III gels synthesized by aging at room temperature and calcined between 900 and 1200 °C has been investigated by Schmücker and Hoffbauer (unpublished results) with ^{29}Si NMR spectroscopy (Fig. 1.4.19). The 900 °C precursors show a single resonance centered at about 90 ppm which is typical for

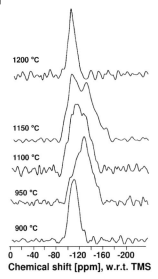

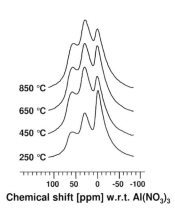

Fig. 1.4.19 ^{29}Si NMR spectra of type III mullite gels heat-treated between 900 and 1200 °C. Gels calcined at 900 °C (amorphous) and 1200 °C (after mullite formation) show a single resonance centered at about 90 ppm, typical of tetrahedrally coordinated silicon in aluminosilicates. Between 950 and 1100 °C a signal centered at about –110 ppm is observed with a shoulder in the –90 ppm region. The corresponding silicon sites are attributed to vitreous silica (–110 ppm) and to a minor extent to an incorporation of silicon into the spinel-type transient phase (–90 ppm).

Fig. 1.4.20 ^{27}Al NMR spectra of type III mullite gels heat-treated at different temperatures prior to crystallization (250 to 850 °C). The structural development corresponds closely to type I gels (Fig. 1.4.7). NMR peaks near 0 and 60 ppm have been attributed to aluminum with sixfold (Al$^{[6]}$), and fourfold (Al$^{[4]}$) coordination, while that near 30 ppm was assigned either to fivefold aluminum (Al$^{[5]}$) coordination, or to tetrahedral triclusters.

tetrahedrally coordinated silicon in mullite and in type I single phase aluminosilicate gels and glasses (see Section 4.1.1). Obviously the silicon environments of type I and type III gels are very similar, and significant phase separation in the amorphous state prior to crystallization can be ruled out. Between 950 and 1100 °C a ^{29}Si NMR signal centered at –110 ppm appears with a shoulder in the –90 ppm region. The corresponding silicon sites are attributed to a spinel-type transition phase with minor silicon incorporation (–90 ppm), and to coexisting vitreous silica (–110 ppm). Above 1100 °C the spinel phase and silica gradually react to mullite, indicated by a gradual increase of the –90 ppm resonance. Aluminum nuclear magnetic resonance (^{27}Al NMR) spectroscopy indicates a slightly higher amount of sixfold coordinated aluminum (Al$^{[6]}$) in type III than in type I gels. It was argued that Al$^{[6]}$ may facilitate formation of γ-alumina phases (Schneider et al. 1993c). Aluminum nuclear magnetic resonance spectra of type III mullite gels in the noncrystalline stage (150–900 °C) are shown in Fig. 1.4.20. The spectral development is virtually the same as that observed in single-phase type I mullite gels (see Fig. 1.4.7), supporting the suggestion of a high structural similarity between non-crystalline type I and type III mullite gels.

1.4.3.3 Mechanisms of Mullite Formation From Type III Mullite Precursors

Mullite formation as a reaction between γ-alumina and non-crystalline silica proceeds in type III precursors in a similar way as in type II materials above about 1200 °C (see Section 1.4.2.2).

1.4.4
General Remarks on the Structure and Crystallization Behavior of Mullite Precursors and Glasses

1.4.4.1 Mullite Precursors: Similarities and Differences

In order to understand the differences between type I single phase and type III mullite precursors and to check possible demixing effects in non-crystalline aluminosilicate gels and glasses, Schmücker et al. (2001) studied non-crystalline aluminosilicate materials with a well-defined degree of chemical inhomogeneity ranging from nanometer to almost atomic scale. For that purpose aluminosilicate films consisting of thin (about 30 nm) to ultrathin (about 2 nm) alumina and silica sublayers were physical vapor deposited by two-source evaporation using a jumping electron beam physical vapor deposition technique (EB-PVD). Four series of non-crystalline physical vapor deposited (PVD) films with bulk compositions ranging between 50 and 63 mol% Al_2O_3 are produced by varying the electron beam jumping frequencies (Tab. 1.4.2). Transmission electron microscopic cross sections of the vapor-deposited films show periodic contrasts perpendicular to the deposition direction. The contrast periodicities correspond reasonably with the calculated thicknesses of the respective alumina/silica double layers. Line scans confirm that the physical vapor deposited films actually consist of Al_2O_3- and SiO_2-rich sublayers (Fig. 1.4.21). Silicon nuclear magnetic resonance spectra (^{29}Si NMR) ob-

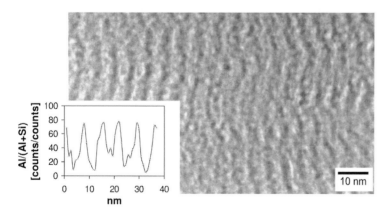

Fig. 1.4.21 Cross section transmission electron micrograph of physical vapor deposited aluminosilicate double layers (series 2, see text) in high magnification with the energy dispersive X-ray (EDX) line scan perpendicular to the observed contrast modulations. The EDX profile yields evidence for periodical chemical variations (Courtesy M. Schmücker).

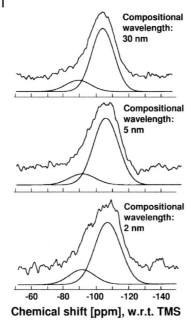

Fig. 1.4.22 ^{29}Si NMR spectra of physical vapor deposited aluminosilicate double layers with nominal thicknesses of 30, 5 and 2 nm, respectively. The three spectra are very similar, showing a main resonance at –110 ppm and a shoulder centered at –90 ppm, indicating the occurrence of "free" silica and atomically "mixed" alumina and silica (after Schmücker et al. 2001).

tained from series with about 30, 5, and 2 nm silica/alumina double-layer thicknesses are shown in Fig. 1.4.22: All spectra are very similar and exhibit a resonance in the –110 ppm region. The peak profiles, however, are slightly asymmetric indicating a resonance of minor intensity in the –90 ppm region.

Deconvolution of the asymmetric resonances into two signals centering at about –110 and –90 ppm, respectively, results in a peak area ratio of about 9/1. This means that virtually pure silica layers (–110 ppm resonance) occur in all PVD aluminosilicate films. The minor fraction of silicon sites surrounded by aluminum (–90 ppm resonance) is interpreted in terms of interfacial silicon, taking into account that the multilayer thickness is in the nanometer range. The occurrence of virtually aluminum-free silica sub-layers in all samples indicates the absence of any significant atomic mixing during deposition. The 30 nm thick alumina/silica double layers calcined at 1000 °C consist of transitional alumina. Transitional alumina plus small amounts of mullite appear in couples of 9 and 5 nm thickness. Only mullite forms in the 2 nm thick alumina/silica system. Above 1200 °C, couples with thicknesses of 30, 9, and 5 nm yield mullite by reaction of transitional alumina with silica. While the crystallization behavior of the 2 nm thick couple corresponds to that of type I single phase mullite precursors, the 30, 9, and 5 nm double layer systems behave like diphasic type II mullite precursors. This is a remarkable result, since all starting alumina/silica double layers are diphasic though on a nanometer scale. The different crystallization behavior has been explained by interdiffusion-induced chemical homogenization between adjacent alumina and silica layers prior to crystallization. Obviously at 1000 °C the complete homogenization necessary for mullitization occurs in reaction couples being about

2 nm thick. Relationships are completely different if the double layers become thicker. In that case zones of atomic mixing are separated by much broader areas of alumina and silica. It is assumed that the extension of the homogenized zones (about 1 to 2 nm) is below the critical size of a mullite nucleus, starting from the suggestion that stable growing mullite nuclei should be at least several unit cell dimensions in size.

To learn more about the crystallization of mullite precursors, the microstructures of type I and type III gels prepared either from tetraethyloxysilane (TEOS) plus aluminum *sec*-butylate (Voll, 1995) or aluminum nitrate nonahydrate solutions (Taake, 1999), respectively, were re-examined by transmission electron microscopy (Schmücker, unpublished results). Fig. 1.4.23 reveals the significant morphological differences between type I and type III mullite gels: Irrespective of the

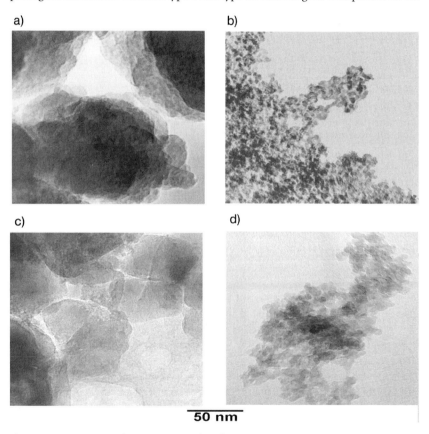

Fig. 1.4.23 Transmission electron micrographs of (a), (c) Single phase type I gels, and (b), (d) Diphasic type III gels. Starting materials are tetraethyloxysilane (TEOS) plus aluminum *sec*-butylate solutions (a,b) and tetraethyloxysilane) plus aluminum nitrate nonahydrate solutions (c,d). In contrast to type I gels, type III gels consist of primary particles smaller than 10 nm (Courtesy M. Schmücker).

Table 1.4.2 Experimental conditions of electron beam physical vapor deposition (EB-PVD) runs.

	Series 1	Series 2	Series 3	Series 4
Film thickness [μm]	70	75	30	29
Average deposition rate [nm s^{-1}]	190	125	75	50
Jumping beam frequency [Hz]	7	14	14	25
Nominal thickness of Al$_2$O$_3$–SiO$_2$ double layer [nm]	27	9	5.5	2

From Schmücker et al. (2001).

starting compounds, primary particles of 5 to 10 nm can be resolved in type III gels (Fig. 1.4.23b,d). In contrast, only faint contours of nanometer-sized primary particles become visible in tetraethyloxysilane (TEOS) plus aluminum *sec*-butylate-derived type I gels (Fig. 1.4.23a) indicating intense particle aggregation. Tetraethyloxysilane (TEOS) plus aluminum nitrate nonahydrate-derived type I gels (Fig. 1.4.23c), on the other hand, consist of agglomerated particles of about 30 to 100 nm (see also Fig. 1.4.3). Small-angle X-ray scattering reveals particles sizes of 7 nm and 39 nm for type III and type I mullite gels, respectively (Okada, unpublished results), in good accordance with microscopic data.

Microstructural analyses of type I and type III mullite gels suggest that their crystallization behavior is influenced by their particle sizes. Obviously, gels with intense aggregation of primary particles or with primary particles several tens of nanometers in size tend to transform directly into mullite, while particulate gels with primary particles below 10 nm form transitional alumina as the first crystalline phase. This finding is explained in terms of critical crystal nucleus sizes, which are assumed to be larger than 10 nm for mullite but smaller than 10 nm for γ-alumina. Actually, there is experimental evidence, from mullite crystal size data reported for early mullitization stages, that stable mullite crystallites exceed 10 nm. Interestingly, crystallite sizes determined by various methods all range from about 10 to 40 nm (Table 1.4.3). Transitional alumina crystals, on the other hand, are well known to be as small as 3 nm (e.g. Wefers and Misra, 1987). Thus homogeneous aluminosilicate gels with particles greater than a stable mullite nucleus (about 10 nm) can directly convert to mullite, while gels consisting of smaller particles should form transitional alumina.

1.4.4.2 The Coordination of Aluminum in Mullite Precursors and Glasses

The structure of gels and glasses in the system Al$_2$O$_3$–SiO$_2$, and the coordination of aluminum has been a point of interest for many years. In the 1960s, Lacy (1963) assumed on the basis of charge balance considerations that tetrahedral triclusters,

Table 1.4.3 Mullite crystallite sizes in early stages of crystallization.

Mullite crystal size [nm]	Method	Reference
20	XRD	Takamori and Roy (1973)
26–42	XRD	Tkalcec et al. (1998)
≈ 15	SEM	Takei et al. (1999)
25–40	XRD	Thom (2000)
10–20	TEM	Bartsch et al. (1999)
12.6	TEM	Johnson et al. (2001)
≈ 20	TEM	This chapter, Figure 1.4.14
20–50	TEM	This chapter, Figure 1.4.24

XRD = X-ray diffractometry
SEM = Scanning electron microscopy
TEM = Transmission electron microscopy

i.e. three tetrahedra linked together by one common oxygen atom, are formed, if silicon is partially replaced by aluminum in melts and glasses. Years later, however, with the introduction of magic angle spinning (MAS) nuclear magnetic resonance (NMR) spectroscopy another structural model became popular. In the pioneering work of Risbud et al. (1987) three ^{27}Al NMR signals are observed, which center at 0, 60 and 30 ppm. These resonances have been attributed to AlO polyhedra in octahedral (Al[6]), tetrahedral (Al[4]) and fivefold (Al[5])coordination, respectively (see above). In another approach Meinhold et al. (1993) attributed the 30 ppm peak to tetrahedrally bound aluminum (Al[4]) with elongated Al-O bonds rather than to Al[5]. This suggestion was obtained on the basis of ^{27}Al NMR line shape analyses together with radial distribution functions (RDFs) and extended X-ray absorption fine structure (EXAFS) data. The assignment of the aluminum 30 ppm NMR signal is still a point of controversy: Bodart et al. (1999) supported the idea of the presence of Al[5] on the basis of recent multiple quantum ^{27}Al NMR studies, while Peeters and Kentgeris (1997), for example, provided evidence for the model of distorted tetrahedra. According to McManus et al. (2001), the experimental data do not allow the questionable resonance to be assigned unambiguously.

Schmücker and Schneider (1996) reactivated the tricluster model of Lacy. Starting from the observation that mullite forms within the bulk of aluminosilicate gels or glasses in extremely high nucleation densities (Fig. 1.4.24), structural short-range-order similarities are believed to occur in mullite and the non-crystalline counterparts. Since no Al[5] occurs in mullite but tetrahedral triclusters instead, it was argued that (Si,Al)O$_4$-triclusters rather than AlO$_5$ polyhedra exist in aluminosilicate gels and glasses. There is also evidence for short-range-order similarities between non-crystalline aluminosilicates and mullite from similar ^{27}Al NMR spectra (Fig. 1.4.25). It has been suggested that the two different tetrahedrally bound aluminum (Al[4]) sites in mullite cause a splitting of the ^{27}Al NMR signal into two peaks at about 60 ppm and 43 ppm, the latter being attributed to tricluster-forming aluminum (Al*) sites (Merwin et al. 1991). The intense upfield shift of the Al*

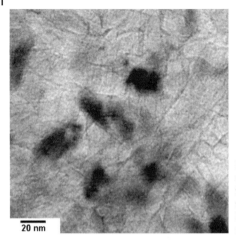

Fig. 1.4.24 Mullite crystals formed after calcination of aluminosilicate glass. Note the numerous crystallites with sizes of 20 to 30 nm (dark areas) which account for a high nucleation density (Courtesy M. Schmücker).

20 nm

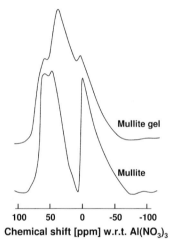

Mullite gel

Mullite

100   50   0   -50   -100
Chemical shift [ppm] w.r.t. Al(NO₃)₃

Fig. 1.4.25 ^{27}Al NMR spectra of a single phase (type I) mullite gel. The spectrum of mullite is given for comparison (after Schmücker and Schneider, 1996).

resonance in mullite is close to the 30 to 35 ppm signal occurring in non-crystalline aluminosilicates, and therefore the latter is attributed to triclustered AlO_4 tetrahedra by Schmücker and Schneider (1996).

Table 1.4.4 presents the mean coordination numbers of different aluminosilicate glasses on the basis of ^{27}Al NMR spectroscopy and on pair distribution function studies. From the results of NMR spectroscopy, two structural models have been developed: Model 1 with 20% of the total aluminum in fourfold ($Al^{[4]}$), 55% in five-fold ($Al^{[5]}$) and 25% in six-fold coordination ($Al^{[6]}$) and model 2 with 75% of the total aluminum in fourfold ($Al^{[4]}$) and 25% in six-fold coordination ($Al^{[6]}$). Coordination numbers calculated on the basis of model 1 (with $Al^{[5]}$) are higher than values derived from pair distribution functions. On the other hand, it turns out that the coordination numbers calculated from model 2 (without $Al^{[5]}$) agree reasonably with the values resulting from the pair distribution functions (PDFs). In a

Table 1.4.4 Mean cation coordination numbers derived from models 1 and 2 (see the text) compared with values from pair distribution function (PDF) data.

Glass composition [mol% Al$_2$O$_3$]	Model 1 (AlO$_4$, AlO$_5$, AlO$_6$)	Model 2 (AlO$_4$, AlO$_6$)	Calculated from PDF
60	4.80	4.38	4.3
50	4.70	4.33	4.2
35	4.54	4.26	4.1

After Schmücker et al. (1999).

further step Schmücker and Schneider (1996) fitted the first pair distribution function (PDF) maximum of the glasses to either four gaussian functions according to model 1 (Si$^{[4]}$-O, Al$^{[4]}$-O, Al$^{[5]}$-O, Al$^{[6]}$-O) or with three gaussian functions according to model 2 (Si$^{[4]}$-O, Al$^{[4]}$-O, Al$^{[6]}$-O), using interatomic distances calculated from ionic radii and the site occupancies derived from nuclear magnetic resonance (NMR) spectra. The calculations clearly show that pair distribution function data can be fitted well for all three glass compositions if AlO$_4$ and AlO$_6$ polyhedra in addition to SiO$_4$ are proposed. On the other hand, if AlO$_5$ is included in the fitting procedure, the shapes of the PDFs and the calculated distance distributions match only poorly (Fig. 1.4.26).

A further fitting strategy has been used by Schmücker and Schneider (1996) for the first maxima of type I gels calcined at different temperatures (see Fig. 1.4.6). The maxima are fitted to two normal functions without parameter constraints

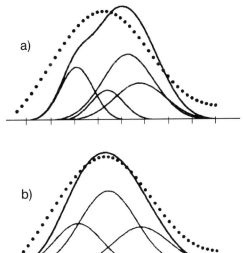

a)

b)

Fig. 1.4.26 Simulation of the first pair distribution function (PDF) maximum of mullite glass by (a) Four, and (b) Three gaussian functions corresponding to (Si$^{[4]}$-O, Al$^{[4]}$-O, Al$^{[5]}$-O, Al$^{[6]}$-O) distances, and (Si$^{[4]}$-O, Al$^{[4]}$-O, Al$^{[6]}$-O) distances, respectively Dotted lines: Measured data. Full lines: Calculated curves (after Schmücker et al. 1999).

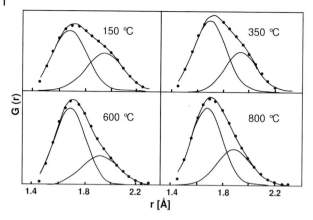

Fig. 1.4.27 Simulation of the first pair distribution function (PDF) maximum of mullite gels heat-treated at different temperatures (after Schmücker and Schneider, 1996).

(Fig. 1.4.27). The following conclusions can be drawn from the positions and intensities of the calculated normal functions: Position of function 1 (maximum at 1.69 Å) corresponds exactly to the mean tetrahedral (Al,Si)-O distance in mullite (Saalfeld and Guse, 1981). Its relative intensity increases with the calcination temperature, which implies a gradual increase of the relative amount of tetrahedrally coordinated aluminum. The intensity of the second normal function decreases with annealing temperature, and the peak maximum position shifts from 1.95 to 1.89 Å. The 1.89 Å distance is typical for $Al^{[6]}$-O bond lengths in crystalline aluminosilicates (Saalfeld and Guse, 1981), while the elongated 1.95-Å distance in the 150 °C and 350 °C samples may rather be attributed to Al-OH bonds. The ratio of $(Si,Al)^{[4]}$ to $Al^{[6]}$ changes slightly from about 60/40 at 150 °C to about 70/30 at 600 to 800 °C. Thus, the ratio of tetrahedrally to octahedrally coordinated cations in the calcined gels is very similar to that of mullite (67/33). The increase of fourfold-coordinated cations with the calcination temperature corresponds to the tendency noted in the NMR spectral development (Fig. 1.4.7), provided the 30 ppm signal is attributed to $Al^{[4]}$.

The assignment of the 30 ppm resonance to distorted AlO_4 tetrahedra and Lacy's charge balance considerations are self-consistent. Schmücker et al. (1997) show that a reciprocal dependence exists between the 30 ppm signal intensity and the Na_2O content in glasses and gels of the system Na_2O-Al_2O_3-SiO_2, suggesting that tricluster formation and incorporation of alkali ions are competitive mechanisms to achieve charge neutrality (Fig. 1.4.28; see also Taake, 1999). It has also been demonstrated that sodium addition to aluminosilicate glasses or melts has a strong influence on mullite nucleation (Schmücker and Schneider, 2002). Fig. 1.4.29 shows reaction couples consisting of Al_2O_3-rich mullite single crystals $(Al_2O_3/SiO_2 \approx 2/1)$ and aluminosilicate glasses with and without sodium addition after firing at 1650 °C. While stoichiometric 3/2-mullite ($3Al_2O_3 \cdot 2SiO_2$) nucleates

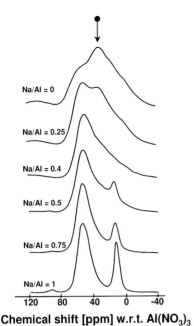

Chemical shift [ppm] w.r.t. Al(NO₃)₃

Fig. 1.4.28 ²⁷Al MAS NMR spectra of alumino-silicate glasses containing 10 mol% Al₂O₃ and varying amounts of Na₂O, given as Na/Al ratio. Note the decreasing intensity of the 30 ppm resonance (arrow) with increasing sodium content (after Schmücker et al. 1997).

within the pure Al₂O₃–SiO₂ melt (see also Schmücker et al. 2002), 3/2-mullite forms epitactically at the surface of the parent 2/1-mullite crystal if the aluminum silicate melt contains sodia (Na₂O). To explain these findings, it was suggested that sodium addition to Al₂O₃–SiO₂ glasses suppresses the population of tetrahedral triclusters and as a consequence the difference in the structural short-range-order of mullite and the aluminosilicate melt is high. Therefore, the nucleation barrier of mullite in Na₂O–Al₂O₃–SiO₂-melts is higher than in undoped Al₂O₃–SiO₂ melts. Thus, mullite crystallization in the bulk of the silicate melt becomes unfavorable if sodium is present, and epitactic growth of mullite needles from the surface of the single-crystal substrate occurs instead.

1.4.4.3 The Origins of Mullite Crystallization

High density nucleation within the bulk of non-crystalline aluminosilicates indicates that the energy barrier for mullitization is low. This has been attributed to short-range-order similarities between mullite and its non-crystalline counterparts. According to Schmücker and Schneider (1996) mullite and the non-crystalline aluminosilicates both consist of the same type of cation-oxygen polyhedra (SiO₄-tetrahedra, AlO₄-tetrahedra, AlO₆-octahedra, triclustered (Si,Al)O₄-tetrahedra), which display similar polyhedral distribution frequencies. In terms of the classical nucleation theory it is argued that a high degree of structural similarity, on the one hand, reduces the surface energy (σ) of the nucleating phase, but, on the other hand, lowers the driving force of transformation (ΔG). However, since

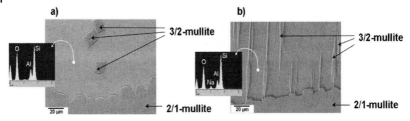

Fig. 1.4.29 Cross-section scanning electron micrograph of (a) Mullite single crystal / SiO_2-Al_2O_3 glass and (b) Mullite single crystal / Na_2O-Al_2O_3-SiO_2 glass reaction couples. Both systems have been heat-treated at 1650 °C (100 h). Note that newly formed mullite crystallites appear in the bulk of the pure aluminosilicate glass (a), but grow epitactically from the substrate surface into the sodium silicate melt (b). Inserts: EDX spectra of the vitreous phases (Courtesy M. Schmücker).

the surface energy dominates the free enthalpy of transformation (σ^3 versus ΔG^2)[5], in total the activation energy of nucleation should be lowered.

Analogous correlations between the structural arrangments of glasses and crystals and nucleation behavior have been reported by several authors. Müller et al. (1993), reviewing literature data, stated that silicate glasses displaying "homogeneous" nucleation have short-range-order similarities with their crystalline phase, in contrast to glasses typically transforming by heterogeneous nucleation events. The same was reported by Mastelaro et al. (2000) and Schneider et al. (2000) investigating $Na_2Ca_2Si_3O_9$-, $CaSiO_3$-, $CaMgSi_2O_6$-, and $PbSiO_3$-glasses by means of extended X-ray absorption fine structure (EXAFS) and silicon nuclear magnetic resonance (^{29}Si NMR) spectroscopy.

Mullitization of type I single phase aluminosilicate glasses and gels at temperatures below 1000 °C is obviously the result of rapid nucleation and very short diffusion distances. It is noteworthy that the close structural relation between crystalline and non-crystalline materials does not only affect the mullite nucleation but also enhances subsequent crystal growth, which then may be considered as polyhedral rearrangement rather than a diffusion-controlled reaction over a distance of several nanometers. In contrast, diphasic mullite precursors require dissolution of alumina and subsequent diffusion of the aluminum species in the silica phase prior to mullite nucleation. Depending on reaction temperature, dissolution (below about 1600 °C) or diffusion (above about 1600 °C) are rate-controlling steps, leading to mullite formation either in the bulk of the siliceous melt or at the alumina/silica interface. Only minor Al_2O_3 supersaturation of the amorphous silica phase is required to induce mullite nucleation (Fig. 1.4.16) which again shows that the nucleation barrier of mullite is low. Mullite crystal growth in diphasic precursors requires relatively long-distance diffusion of aluminum species

5) Derived from classical nucleation theory
where the activation energy of nucleation
(E_A) is given by $E_A \propto \sigma^3 / T \Delta G^2$.

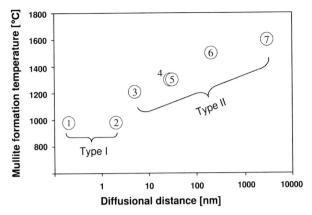

Fig. 1.4.30 Diagram showing the degree of alumina to silica segregation plotted versus the mullitization temperature. Data point 1 corresponds to aluminosilicate glasses or gels with compositional homogeneity on the atomic level. Data points 2 to 4 are from vapor deposited alumina/silica multilayers with layer thickness of 2, 5, and 30 nm, respectively. Data points 5 to 7 refer to literature data obtained from diphasic gels (Wei and Halloran, 1988 a,b), silica-coated alumina particles (Wang and Sacks, 1996), and alumina and silica powder admixtures (Albers, 1994).

through the silica-rich melt or, alternatively at high temperatures, aluminum and silicon interdiffusion through the interfacial mullite layer.

The interdependence between diffusion distance and mullitization temperature is depicted in Fig. 1.4.30. Data point 1 corresponds to type I aluminosilicate glasses or gels with compositional homogeneity on an atomic level. The assumed "diffusional distance" is 0.2 nm. Data points 2 to 4 come from vapor-deposited alumina/silica multilayers (see Section 1.4.4.1) with layer thickness of 2, 5, and 30 nm, respectively. Data points 5 to 7, finally, refer to literature data obtained for diphasic gels (data point 5), silica-coated alumina particles (data point 6) and alumina/silica powder admixtures (data point 7). Obviously, the transformation temperature of type I mullite-forming precursors is not affected by nanometer-sized segregation. In this case homogenization takes place prior to crystallization (see Section 1.4.4.1).

In contrast, data for true type II diphasic materials display mullite formation at significantly higher temperatures, since diffusional distances become larger. The following factors may have a crucial influence on the mullitization process of type II diphasic precursors:

- *Dissolution of alumina*: Small (highly curved) alumina particles display higher solubility in the non-crystalline silica phase than coarser particles owing to the Gibbs-Thompson effect. Moreover, the solubility of metastable transient alumina phases (γ-alumina) is higher than that of the stable α-alumina.
- *Critical nucleation concentration (CNC)*: The smaller the silica zones, the more

rapid the critical Al_2O_3 concentration is achieved necessary for mullite nucleation.

- *Growth of mullite crystals:* The diffusional distance of aluminum species in a silica-rich melt or the aluminum-silicon interdiffusion through the interfacial mullite layer, respectively, are directly controlled by the degree of alumina/silica segregation.

References

Abrahams, I., Bush, A. J., Hawkes, G.E and Nunes, T. (1999). Structure and oxide ion conductivity mechanism in $Bi_2Al_4O_9$ by combined X-ray and high-resolution neutron powder diffraction and ^{27}Al solid state NMR. J. Solid State Chem. **147**, 631–636.

Aksay, I. A. and Pask, J. A. (1975). Stable and metastable equilibria in the system SiO_2–Al_2O_3. J. Am. Ceram. Soc. **58**, 507–512.

Albers, W. (1994). Reaktionsintern von α-Al_2O_3 und verschiedenen SiO_2-Modifikationen zu Mullit. Ph. D. Thesis, Univ. Hannover.

Alonso, J.A., Casais, M. T., Martínez-Lope, M. J. and Rasines, I. (1997a). High oxygen pressure preparation, structural refinement, and thermal behavior of RMn_2O_5 (R = La, Pr, Nd, Sm, Eu). J. Solid State Chem. **129**, 105–112.

Alonso, J.A., Casais, M. T., Martínez-Lope, M. J., Martínez, J. L. and Fernández-Díaz, M.T. (1997b). A structural study from neutron diffraction data and magnetic properties of RMn_2O_5 (R = La, rare earth). J. Phys. Cond. Matter **9**, 8515–8526.

Anderson, J.S. and Kahn, A. S. (1975) quoted in: Levin, E. M. and McMurdie, H. F.: Phase diagrams for ceramists. The American Ceramic Society 1975 Supplement, p. 16, Fig. 1470.

Angel, R. J. and Prewitt, C. T. (1986). Crystal structure of mullite: A re-examination of the average structure. Amer. Min. **71**, 1476–1482.

Angel, R. J. and Prewitt, C. T. (1987).The incommensurate structure of mullite by Patterson synthesis. Acta Cryst. **B 42**, 116–126.

Angel, R. J., McMullan, R. K. and Prewitt, C.T. (1991). Substructure and superstructure of mullite by neutron diffraction. Amer.Min. **76**, 332–342.

Angerer, P. (2001). Alkalialuminate und Alkaligallate mit Mullitstruktur. Ph. D. Thesis Univ. Hannover.

Baerlocher, C., Hepp, A. and Meier, W. M. (1976). DLS-76, a program for the simulation of crystal structures by geometric refinements. ETH Zürich, Switzerland.

Balzar, D. and Ledbetter, H. (1993): Crystal structure and compressibility of 3:2 mullite. Amer. Min. **78**, 1192–1196.

Ban, T. and Okada, K. (1992). Structure refinement of mullite by the Rietveld method and a new method for estimation of chemical composition. J. Am. Ceram. Soc. **75**, 227–230.

Ban, T., Hayashi, S., Yasumori, A. and Okada, K. (1996). Calculation of metastable immiscibility region in the Al_2O_3–SiO_2 system. J. Mater. Res. **11**, 1421–1427.

Bärnighausen, H. (1980). Group-subgroup relations between space groups: A useful tool in crystal chemistry. Communications in mathematical chemistry MATCH, **9**, 209–233.

Bartsch, M., Saruhan, B., Schmücker, M. and Schneider, H. (1999). Novel low-temperature processing route of dense mullite ceramics by reaction sintering of amorphous SiO_2 coated γ–Al_2O_3 particle nanocomposites. J. Am. Ceram. Soc. **82**, 1388–1392.

Basso, R., Palenzona, A., Zefiro, L. (1989): Crystal structure refinement of a Sr-bearing term related to copper vanadates and arsenates of adelite and descloizite groups. N. Jb. Miner. Mh. 1989, 300–308.

Bauchspiess, R., Schneider, H. and Kulikov, A. (1996). EXAFS studies of Cr-doped mullite. J. Europ. Ceram. Soc. **16**, 203–209.

Baudin, C., Osendi, M. I. and Moya, J. S. (1983) Solid solution of TiO_2 in mullite. J. Mater. Sci. Lett. **2**, 185–187.

Baur, W. H. (1994). Rutile-type derivatives. Z. Krist. **209**, 143–150.

Baur, W. H. and Fischer, R. X. (2000). Zeolite-Type Crystal Structures and their Chemistry. Zeolite Structure Codes ABW to CZP. Subvolume B in Landolt-Börnstein, New Series, Group IV, Microporous and other Framework Materials with Zeolite-Type Structures, Springer, Berlin, Vol. 14 2000.

Baur, W. H. and Fischer, R. X. (2002). Zeolite-Type Crystal Structures and their Chemistry. Structure Codes DAC to LOV. Subvolume C in Landolt-Börnstein, New Series, Group IV, Microporous and other Framework Materials with Zeolite-Type Structures, Springer, Berlin, Vol. 14 2002.

Becker, H.-D. and Schneider, H. (2005). High temperature crystal field spectroscopy of chromium-doped mullite. Unpublished results.

Birkenstock, J., Fischer, R. X. and Messner, T. (2004). BRASS, the Bremen Rietveld Analysis and Structure Suite. University of Bremen, Germany.

Birx, J. and Hoppe, R. (1991). $Na_2Li_3CoO_4$, das erste quaternaere Oxocobalt(III) der Alkalimetalle. Z. Anorg. Allg. Chemie **597**, 19–26.

Boccaccini, A. R., Khalil T. K. and Bücker M. (1999). Activation energy for the mullitization of a diphasic gel obtained from fumed silica and boehmite sol. Mater. Sci. Lett. **38**, 116–120.

Bock, O. and Müller, U. (2002). Symmetrie-verwandtschaften bei Varianten des Perowskit-Typs. Acta Cryst., B**58**, 594–606.

Bodart, P. R., Parmentier, J., Harris R. K. and Thompson, D. P. (1999). Aluminium environments in mullite and an amorphous sol-gel precursor examined by ^{27}Al triple quantum MAS NMR. J. Phys. Chem. Solids **60**, 223–228.

Böhm, H. (1977). Eine erweiterte Theorie der Satellitenreflexe und die Bestimmung der modulierten Struktur des Natriumnitrits. Habilitation Thesis, Univ. Münster.

Bohn, D. (1979). Gleichgewichte zwischen Korund, Mullit und Tialit im System Al_2O_3–SiO_2–Fe_2O_3–TiO_2. Diploma Thesis, Univ. Aachen.

Bowen, N. L., and Greig, J. W. (1924). The system Al_2O_3 – SiO_2. J. Am. Ceram. Soc. **7**, 238–254.

Boysen, H.; Frey, F. and Jagodzinski, H. (1984). Diffuse scattering and disorder in crystals. Rigaku **1**, 3–14.

Brown, I. W. M., MacKenzie, K. J. D., Bowen, M. E. and Meinhold, R. H. (1985). Outstanding problems in the kaolinite-mullite reaction sequence investigated by ^{29}Si and ^{27}Al solid-state nuclear magnetic resonance: II. High-temperature transformations of metakaolinite. J. Am. Ceram. Soc. **68**, 298–301.

Brownell, J. (1958). Subsolidus relations between mullite and iron oxide. J. Am. Ceram. Soc. **41**, 226–230.

Brunauer, G., Boysen, H., Frey, F., Hansen, T., Kriven, W. (2001). High temperature crystal structure of a 3:2 mullite from neutron diffraction data. Z. Krist. **216**, 284–290.

Burnham, C. W. (1963a). Crystal structure of mullite. Annual Report Geophys. Lab. Carnegie Inst. Wash., **63**, 223–227.

Burnham, C. W. (1963b). Refinement of the crystal structure of sillimanite. Z. Krist. **118**, 127–148.

Burnham, C. W. (1964a). Composition limits of mullite and the sillimanite – mullite solid solution problem. Carnegie Inst. Wash. Yearb. **63**, 223–227.

Burnham, C. W. and Buerger, M. J. (1961). Refinement of the crystal structure of andalusite. Z. Krist. **115**, 269–290.

Burns, P. C. and Hawthorne, F. C. (1995). Rietveld refinement of the crystal structure of olivenite: A twinned monoclinic structure. Canad. Min. **33**, 885–888.

Butler, B. D. and Welberry, T. R. (1994). Analysis of diffuse scattering from the mineral mullite. J.Appl.Cryst. **27**, 742–754.

Caballero, A. and Ocana, M. (2002). Synthesis and structural characterization by X-ray absorption spectroscopy of tin-doped mullite solid solutions. J. Am. Ceram. Soc. **85**, 1910–1914.

Cameron, W. E. (1976a). A mineral phase intermediate in composition between sillimanite and mullite. Amer. Min. **61**, 1025–1026.

Cameron, W. E. (1976 b) Coexisting sillimanite and mullite. Geol. Mag. **113**, 497–514.

Cameron, W. E. (1977a). Mullite: A substituted alumina. Amer. Min. **62**, 747–755.

Cameron, W. E. (1977b). Nonstoichiometry in sillimanite: Mullite compositions with sillimanite-type superstructures. Phys. Chem. Minerals **1**, 265–272.

Cameron, W. E. (1977c). Composition and cell dimensions of mullite. Amer. Ceram. Soc. Bull. **56**, 1003–1011.

Campos, A. L., Silva, N. T. Melo, F.C.L., Oliveira, M.A.S. and Thim G. P. (2002). Crystallization kinetics of orthohombic mullite from diphasic gels. J. Non-Cryst. Solids **304**, 19–24.

Cesbron, F. P., Ginderov, D., Girand, R., Pelisson, P. and Pilard, F. (1987). La nickelanstiute Ca (Ni, Zn) (AsO₄) (OH): nouvelle espèce minérale du district cobalto-nickelifère de Bon-Azzer, Marvc. Canad. Min. **25**, 401–407.

Chakraborty A. K. and Ghosh D. K. (1978). Reexamination of kaolinite-to-mullite reaction series, J. Am. Ceram. Soc. **61**, 170–173.

Chater, R. and Gavarri, J.-R. (1985). Evolution structurale entre 2 et 300 K de l'oxyde $MnSb_2O_4$: Propriétés élastiques et magnétiques anisotropes. J. Solid State Chem. **59**, 123–131.

Chater, R., Gavarri, J. R. and Hewat, A. (1985). Structures isomorphes MeX_2O_4 – Evolution structurale entre 2 K et 300 K de l'antimonite $FeSb_2O_4$: Elasticité et ordre magnétique anisotropes. J. Solid State Chem. **60**, 78–86.

Clark, L.A., Pluth, J.J., Steele, I., Smith, J.V. and Sutton, S. R. (1997). Crystal structure of austinite, $CaZn(AsO_4)OH$. Min. Mag. **61**, 677–683.

Colomban, Ph. (1989). Structure of oxide gels and glasses by infrared and Raman scattering, J. Mater. Sci. **24**, 3011–3020.

Cooper, M.A. and Hawthorne, F.C. (1995). The crystal structure of mottramite, and the nature of $Cu \leftrightarrows Zn$ solid solution in the mottramite-descloizite series. Canadian Mineral. **33**, 1119–1124.

Cordsen, A. (1978). A crystal-structure refinement of libethenite. Canad. Min. **16**, 153–157.

Cowley, J. M. and Moodie, A. F. (1957). The scattering of electrons by atoms and crystals. I. A new theoretical approach. Acta Cryst. **10**, 609–619.

Daniel, V. and Lipson, H. (1944). The dissociation of an alloy of copper, iron and nickel. Further X-ray work. Proc. Royal Soc. London **A128**, 378–387.

Darken, L. S. and Gurry, R. W. (1945). The system iron-oxygen. I. The wustite field and related equilibria. J. Am. Ceram. Soc. **28**, 1398–1412.

Darken, L. S. and Gurry, R. W. (1946). The system iron-oxygen. II. Equilibrium and thermodynamics of liquid oxide and other phases. J. Am. Ceram. Soc. **29**, 798–816.

Deville, S. C. and Caron, D. (1865). Cited in: Pask, J.A. (1990). Critical review of phase equilibria in the $Al_2O_3 - SiO_2$ system. Ceramic Trans. **6**, 1–13.

De Wolff, P. M. (1974). The pseudo-symmetry of modulated crystal structures. Acta Cryst. **A30**, 777–785.

Ďurovič, S. (1962a). A statistical model of the crystal structure of mullite. Kristallografiya **7**, 339–349.

Ďurovič, S. (1962b). Isomorphism between sillimanite and mullite. J. Am. Ceram. Soc. **45**, 157–162.

Ďurovič, S. (1969). Refinement of the crystal structure of mullite. Chemicke Zvesti **23**, 113–128.

Ďurovič, S. and Fejdi, P. (1976). Synthesis and crystal structures of germanium mullite and crystallochemical parameters of D-mullites. Silikaty **2**, 192.

Duvigneaud, P. H. (1974). Existence of mullite without silica. J. Am. Ceram. Soc. **57**, 224.

Engelhardt, G. and Michel, D. (1987). High Resolution Solid State NMR of Silicates and Zeolites. John Wiley and Sons, New York.

Epicier, T. (1991). Benefits of high-resolution electron microscopy for the structural characterization of mullites. J. Am. Ceram. Soc. **74**, 2359–2366.

Epicier, T., O'Keefe, M. A. and Thomas, G. (1990). Atomic imaging of 3:2 mullite. Acta Cryst. **A46**, 948–962.

Faye, G. H. and Harris, D. C. (1969). On the origin of colour and pleochroism in andalusite from Brazil. Canad. Min. **10**, 47–56.

Fischer, R. X. and Baur, W. H. (2004). The standardization of zeolite-type crystal structures. Proceedings of the 14th International Zeolite Conference, In press.

Fischer, R. X., Kahlenberg, V., Voll, D., Mac Kenzie, K. J. D., Smith, M. E. and Schneider, H. (2005). Cation coordination in synthetic boron aluminate, $Al_{6-x}B_xO_9$ ($x = 2$), studied by ¹¹B and ²⁷Al MAS NMR spectroscopy and Rietveld analyses. Abstracts Jahrestagung Deutsche Gesellschaft für Kristallographie (DGK), Köln, 2005.

Fischer, R. X. and Messner, T. (2004). STRUPLO, a new version of the crystal structure drawing program. University of Bremen, Germany.

Fischer, R. and Pertlik, F. (1975). Verfeinerung der Kristallstruktur des Schafarzikites, FeSb$_2$O$_4$. Tschermaks Mineral. Petrogr. Mitt. **22**, 236–241.

Fischer, R. X. and Schneider, H. (1992). Crystal chemistry of iron containing germanium andalusites, Fe$_x$Al$_{8-x}$Ge$_4$O$_{20}$. Z. Krist. **201**, 19–36.

Fischer, R. X. and Schneider, H. (2000). Crystal structure of Cr-mullite. Amer. Min. **85**, 1175–1179.

Fischer, R. X., Schneider, H. and Schmücker, M. (1994). Crystal structure of Al-rich mullite. Amer. Min. **79**, 983–990.

Fischer, R. X., Schneider, H. and Voll, D. (1996). Formation of aluminum rich 9 : 1 mullite and its transformation to low alumina mullite upon heating. J. Europ. Ceram. Soc. **16**, 109–113.

Fischer, R. X., Schmücker, M., Angerer, P. and Schneider, H. (2001). Crystal structures of Na and K aluminate mullites. Amer. Min. **86**, 1513–1518.

Foster Jr., P. A. (1959). The nature of alumina in quenched cryolite-alumina melts. J. Electrochem. Soc. **106**, 971–975.

Freimann, S. (2001). Bestimmung der Sauerstoff-Leerstellenverteilung in 2:1-Mullit mit Hilfe der videographischen Methode. Ph. D. Thesis, Univ. Hannover.

Freimann, S. and Rahman, S. (2001). Refinement of the real structures of 2:1 and 3:2 mullite. J. Europ. Ceram. Soc. **21**, 2453–2461.

Freimann, S., Thoke, S. and Rahman, S. H. (1996). Bestimmung der Nahordnungsvektoren in Mullit mit Hilfe der videographischen Rekonstruktion. 4. Jahrestagung DGK Marburg, Abstracts 121.

Fukuoka, M., Onoda, Y., Inoue, S., Wada, K., Nukui, A. and Makashima, A. J. (1993). The role of precursors in the structure of SiO$_2$–Al$_2$O$_3$ sols and gels by the sol-gel-process. Sol-Gel Sci. Tech. **1**, 47–53.

Gani, M.S.J. and McPherson, R. (1977a). Crystallization of mullite from Al$_2$O$_3$–SiO$_2$ glasses. J. Austr. Ceram. Soc. **13**, 21–24.

Gani, M.S.J. and McPherson, R. (1977b). Glass formation and phase transformation in plasma prepared Al$_2$O$_3$–SiO$_2$ powders. J. Mater. Sci. **12**, 999–1009.

Garsche, M., Tillmanns, E., Almen, H., Schneider, H. and Kupčik. V. (1991). Incorporation of chromium into aluminium borate 9Al$_2$O$_3$ · 2B$_2$O$_3$ (A$_9$B$_2$). J. Europ. Min. Soc. **3**, 793–808.

Gavarri, J. R. (1981). Sur les composes isomorphes MeX$_2$O$_4$: evolution structurale par diffraction de neutrons de NiSb$_2$O$_4$ et ZnSb$_2$O$_4$ entre 5 et 300 K: anisotropie, rigidite. Compt. Rend. Hebd. Seances Acad. Sci. **292**, 895–898.

Gavarri, J. R., Vigouroux, J. P., Calvarin, G. and Hewat, A. W. (1981). Structure de SnPb$_2$O$_4$ à quatre températures: relation entre dilatation et agitation thermiques. J. Solid State Chem. **36**, 81–90.

Gavarri, J.-R., and Weigel, D. (1975). Oxydes de Plomb. I. Structure cristaline du minium Pb$_3$O$_4$, à température ambiante (293 K). J. Solid State Chem. **13**, 252–257.

Gavarri, J. R., Weigel, D. and Hewat, A. W. (1978). Oxydes de plomb. IV. Evolution structurale de l'oxyde Pb$_3$O$_4$ entre 240 et 5° K et mécanisme de la transition. J. Solid State Chem. **23**, 327–339.

Gelato, L. M. and Parthé, E., J. (1987). STRUCTURE TIDY – a computer program to standardize crystal structure data. J. Appl. Cryst. **20**, 139–143.

Gelsdorf, G., Müller-Hesse, H. and Schwiete, H. E. (1958). Einlagerungsversuche an synthetischem Mullit und Substitutionsversuche mit Galliumoxyd und Germaniumoxyd. Teil II. Arch. Eisenhüttenwesen **29**, 513–519.

Gelsdorf, G., Müller-Hesse, H. and Schwiete, H. E. (1961). Untersuchungen zur Frage der Mischkristallbildung in Mulliten und mullitähnlichen Verbindungen. Sprechsaal **94**, 502–508.

Gerardin, C., Sundaresan, S., Benziger, J. and Navrotsky, A. (1994). Structural investigation and energetics of mullite formation from sol-gel-precursors. Chem. Mater. **6**, 160–168.

Giaquinta, D. M., Papaefthymiou, G. C., Davis, W. M. and zur Loye, H.-C. (1992). Synthesis, structure, and magnetic properties of the layered bismuth transition metal oxide solid solution $Bi_2Fe_{4-x}Ga_xO_9$. J. Solid State Chem. **99**, 120–133.

Giaquinta, D. M., Papaefthymiou, G. C. and zur Loye, H.-C. (1995). Structural and magnetic studies of $Bi_2Fe_{4-x}Al_xO_9$. J. Solid State Chem. **114**, 199–205.

Green, C. R. and White, J. (1974). Solid solubility of TiO_2 in mullite in the system Al_2O_3–TiO_2–SiO_2. Trans. Brit. Ceram. Soc. **73**, 73–75.

Grigor'ev, A. P. (1976). Crystallization of the intermediate phases in the series sillimanite-mullite. Neorg. Mater. **12**, 519–521.

Grießer, K. J. (2004). Untersuchungen im System Al_2O_3–SiO_2–B_2O_3 "Bormullite". Master Thesis, Univ. Vienna.

Guse, W. (1974). Compositional analysis of Czochralski grown mullite single crystals. J. Crystal Growth **26**, 151–152.

Guse, W. and Mateika, D. (1974). Growth of mullite single crystals ($2Al_2O_3$ .SiO_2) by the Czochralski method. J. Crystal Growth **22**, 237–240.

Hahn (2002) International Tables for Crystallography, Volume A. Space Group Symmetry. Hahn, T. (ed.). Kluwer Academic Publishers.

Hammonds, K. D., Bosenick, A., Dove, M. T. and Heine, V. (1998). Rigid unit modes in crystal structures with octahedrally coordinated atoms. Amer. Min. **83**, 476–479.

Hariya, Y., Dollase, W. A. and Kennedy, G. C. (1969). An experimental investigation of the relationship of mullite to sillimanite. Amer. Min. **54**, 1419–1441.

Hawthorne, F. C. and Faggiani, R. (1979). Refinement of the structure of descloizite. Acta Cryst. **B35**, 717–720.

Heinrich, T. and Raether, F. (1992). Structural characterization and phase development of sol-gel derived mullite and its precursors. J. Non-Cryst. Solids **147**, 152–156.

Hildmann, B., Schneider, H. and Schmücker, M. (1996). High temperature behaviour of polycrystalline alumo-silicate fibers with mullite bulk composition: II. Kinetics of γ-Al_2O_3 -mullite transformation. J. Europ. Ceram. Soc. **16**, 287–292.

Hill, R. J. (1976). The crystal structure and infrared properties of adamite. Amer. Min. **61**, 979–986.

Hiroi, Y., Grew, E. S., Motoyoshi, Y., Peacor, D. R., Rouse, R. C., Matsubara, S., Yokoyama, K., Miyawaki, R., McGee, J. J., Su, S. C., Hokada, T., Furukawa, N. and Shibasaki, H. (2002). Ominellite, $(Fe,Mg)Al_3B$-SiO_9 (Fe^{2+} analogue of grandidierite), a new mineral from porphyritic granite in Japan. Amer. Min. **87**, 160–170.

Hoffman, D. W., Roy, R. and Komarneni, S. (1984). Diphasic xerogels, a new class of materials: phases in the system Al_2O_3–SiO_2. J. Am. Ceram. Soc. **67**, 468–471.

Hong, S.-H. and Messing, G. L. (1997). Mullite transformation kinetics in P_2O_5–B_2O_3– and TiO_2-doped aluminosilicate gels. J. Am. Ceram. Soc. **80**, 1551–1559.

Hong, S.-H. and Messing G. L. (1999). Anisotropic grain growth in boria-doped diphasic mullite gels. J. Europ. Ceram. Soc. **19**, 521–526.

Hong, S.-H., Cermignani, W. and Messing, G. L. (1996). Anisotropic grain growth in seeded and B_2O_3-doped diphasic mullite gels. J. Europ. Ceram. Soc. **16**, 133–141.

Hübner, J. S. and Sato, M. (1970). The oxygen fugacity-temperature relationships of manganese oxide and nickel oxide buffers. Amer. Min. **55**, 934–952.

Huling, J. C. and Messing, G. L. (1991). Epitactic nucleation of spinel in aluminium silicate gels and effect on mullite crystallization. J. Am. Ceram. Soc. **74**, 2374–2381.

Huling, J. C. and Messing, G. L. (1992). Chemistry-crystallization relations in molecular mullite gels. J. Non-Cryst. Solids **147/148**, 213–221.

Hyatt, M. J. and Bansal, N. P. (1990). Phase transformations in xerogels of mullite composition. J. Mater. Sci. **25**, 2815–2821.

Ihara , M., Imai, K, Fukunaga, J. and Yoshida N. (1980). Crystal structure of boraluminate $9Al_2O_3$ $2B_2O_3$. Yogyo Kyokaisi **88**, 77–84.

Ikeda, K., Schneider, H., Akasaka, M. and Rager, J. (1992). Crystal-field spectroscopic study of Cr-doped mullite. Amer. Min. **77**, 251–257.

Ivankovic, H., Tkalcec, E., Nass R. and Schmidt H. (2003). Correlation of the precursor type with densification behavior and microstructure of sintered mullite ceramics. J. Europ. Ceram. Soc. **23**, 283–292.

Jagodzinski, H. (1949). Eindimensionale Fehlordnung in Kristallen und ihr Einfluß auf die Röntgeninterferenzen. I. Berechnung des Fehlordnungsgrades aus Röntgenintensitäten. Acta Cryst. **2**, 201–207.

Jagodzinski, H. (1964a). Allgemeine Gesichtspunkte für die Deutung diffuser Interferenzen von fehlgeordneten Kristallen. Advances in Structure Research by Diffraction Methods, Vol. I, R. Brill. (ed.) Vieweg Braunschweig.

Jagodzinski, H. (1964b). Diffuse disorder scattering by crystals. Advanced Methods of Crystallography, G. N. Ramachandran (ed.). London, New York Academic Press.

James, P. F., Iqbal, Y., Jais, U. S., Jordery, S. and Lee, W. E. (1997). Crystallization of silicate and phosphate glasses. J. Non-Cryst. Solids **219**, 17–29.

Jantzen, C. M., Schwalm, D., Schelten, J. and Herrman, H. (1981). The SiO_2–Al_2O_3 system, Part 1. Later stage spinoidal decomposition and metastable immiscibility. Phys. Chem. Glasses **22**, 122–137.

Jaymes I. and Douy, A. (1996). New aqueous mullite precursor synthesis. Structural study by ^{27}Al and ^{29}Si NMR spectroscopy. J. Europ. Ceram. Soc. **16**, 155–160.

Johnson, B. R., Kriven, W. M. and Schneider, J. (2001). Crystal structure development during devitrification of quenched mullite. J. Europ. Ceram. Soc. **21**, 2541–2562.

Kahlenberg, V., Fischer, R. X. and Baur, W. H. (2001). Symmetry and structural relationships among ABW-type materials. Z. Krist. **216**, 489–494.

Kanzaki, S., Tabata, H., Kumazawa, T. and Ohta, S. (1985). Sintering and mechanical properties of mullite. J. Am. Ceram. Soc. **68**, C-6–C-7.

Keller, P., Lissner, F. and Schleid, T. (2003). The crystal structure of arsendescloizite, $PbZn(OH)[AsO_4]$, from Tsumeb (Namibia). N. Jb. Min. Mh. 2003, 374–384.

Kharisun, Taylor, M. R., Bevan, D.J.M. and Pring, A. (1998). The crystal chemistry of duftite, $PbCuAsO_4(OH)$ and the β-duftite problem. Min. Mag. **62**, 121–130.

Kolitsch, U. (2001). Refinement of pyrobelonite, $PbMn^{II}VO_4(OH)$, a member of the descloizite group. Acta Cryst. E**57**, i119–i121.

Komarneni, S. and Roy, R. (1986). Application of compositionally diphasic xerogels for enhanced densification, the system Al_2O_3–SiO_2. J. Am. Ceram. Soc. **69**, C155–C156.

Korekawa, M. (1967). Theorie der Satellitenreflexe. Habilitation Thesis Univ. München.

Korekawa, M., Nissen, H.-U., and Philipp, D. (1970). X-ray and electronmicroscopic studies of a sodium-rich low plagioclase. Z. Krist. **131**, 418–436.

Kunath, G., Losso, P. Steuernagel, S., Jäger, C. and Schneider H. (1992). ^{27}Al satellite transition spectroscopy (SATRAS) of polycrystalline aluminium borate, $9Al_2O_3$ $2B_2O_3$ (A_9B_2). Solid State NMR **1**, 261–266.

Kunze, G. (1959). Fehlordnungen des Antigorites. Z. Krist. **111**, 190–221.

Kutty, T.R.N. and Nayak, M. (2000). Photoluminescence of Eu^{2+}-doped mullite (xAl_2O_3 $ySiO_2$; x/y = 3/2 and 2/1) prepared by the hydrothermal method. J. Mater. Chem. Phys. **65**, 158–165.

Kvick, Å., Pluth, J.J., Richardson, Jr, J. W. and Smith, J. V. (1988). The ferric ion distribution and hydrogen bonding in epidote: a neutron diffraction study at 15K. Acta Cryst. B**44**, 351–355.

Lacy, E. D. (1963). Aluminium in glasses and melts. Phys. Chem. Glasses **4**, 234–238.

Li, D. X. and Thomson, W. J. (1990). Mullite formation kinetics of a single phase gel. J. Am. Ceram. Soc. **73**, 964–969.

Li, D. X. and Thomson, W. J. (1991). Mullite formation from non-stoichiometric diphasic precursors. J. Am. Ceram. Soc. **74**, 2382–2387.

Low, I. M. and McPherson, R. (1989). The origins of mullite formation. J. Mater. Sci. **24**, 926–936.

MacDowell, J. F. and Beall, G. H. (1969). Immiscibility and crystallization in Al_2O_3-SiO_2 glasses. J. Am. Ceram. Soc. **52**, 17–25.

Mack, D. E., Becker, K. D. and Schneider, H. (2005). High temperature Mössbauer study of Fe-subtituted mullite. Amer. Min., in press.

Mackenzie, K. J. D., Meinhold, R. H., Patterson, J. E., Schneider, H. , Schmücker, M. and Voll, D. (1996). Structural evolution in gel-derived mullite precursors. J. Europ. Ceram. Soc. **16**, 1299–308.

Mägi, M., Lipmaa, E., Samoson, A., Engelhardt, G. and Grimmer, A. R. (1984). Solid state high resolution silicon-29 chemical shifts in silicates. J. Phys. Chem. **88**, 1518–1522.

Mastelaro, V. R., Zanotto, E. D., Lequeux, N.

and Cortès, R. (2000). Relationship between short-range order and ease of nucleation in $Na_2Ca_2Si_3O_9$, $CaSiO_3$ and $PbSiO_3$ glasses. J. Non-Cryst.Solids **262**, 191–199.

Mazza, D., Vallino, M. and Busca, G. (1992). Mullite-type structures in the systems Al_2O_3–Me_2O (Me = Na,K) and Al_2O_3–B_2O_3. J. Am. Ceram. Soc. **75**, 1929–1934.

McConnell, J. D. C. and Heine, V. (1985). Incommensurate structure and stability of mullite. Phys. Rev. **B 31**, 6140–6142.

McManus, J., Ashbrook, S. E., MacKenzie, K.J.D. and Wimperis, S. (2001). ^{27}Al multiple quantum MAS and ^{27}Al $\{^1H\}$ CPMAS NMR study of amorphous aluminosilicates. J. Non-Cryst. Solids **282**, 278–290.

McNear, E., Vincent, M. G. and Parthé, E. (1976). The crystal structure of vuagnatite, $CaAl(OH)SiO_4$. Amer. Min. **61**, 831–838.

Meinhold, R. H., Slade, R. C. T. and Davies, T. W. (1993). High field ^{27}Al MAS NMR studies of the formation of metakaolinite by flash calcination of kaolinite. J. Appl. Magn. Reson. **4**, 141–155.

Mellini, M. and Merlino, S. (1979). Versiliaite and apuanite: derivative structures related to schafarzikite. Amer. Min. **64**, 1235–1242.

Merwin, L. H., Sebald, A., Rager, H. and Schneider, H. (1991). ^{29}Si and ^{27}MAS NMR spectroscopy of mullite. Phys. Chem. Minerals **18**, 47–52.

Moore, P. B. and Smyth, J. R. (1968). Crystal chemistry of the basic manganese arsenates: III. The crystal structure of eveite, $Mn_2(OH)(AsO_4)$. Amer. Min. **53**, 1841–1845.

Moore, J. M., Waters, D. J. and Niven, M. L. (1990). Werdingite, a new borosilicate mineral from the granulite facies of the western Namaqua-land metamorphic complex, South Africa. Amer. Min. **75**, 415–420.

Morikawa, H., Miwa, S., Miyake, M., Marumo, F. and Sata, T. (1982) Structural analysis of SiO_2–Al_2O_3 glasses. J. Am. Ceram. Soc. **65**, 78–81.

Muan, A. (1957). Phase equilibria at liquidus temperatures in the system iron oxide-Al_2O_3–SiO_2 in air atmosphere. J. Am. Ceram. Soc. **40**, 121–133.

Müller, E., Heide, K. and Zanotto, E. D. (1993). Molecular structure and nucleation in silicate glasses. J. Non-Cryst. Solids **155**, 56–66.

Müller-Buschbaum, H. and Chales de Beaulieu, D. (1978). Zur Besetzung von Oktaeder- und Tetraederpositionen in $Bi_2Ga_2Fe_2O_9$. Z. Naturforsch. **B33**, 669–670.

Murthy, M. K. and Hummel, F. A. (1960) X-ray study of the solid solution of TiO_2, Fe_2O_3, and Cr_2O_3 in mullite ($3Al_2O_3$ $2SiO_2$). J. Am. Ceram. Soc. **43**, 267–273.

Nakajima, Y., Morimoto, M and Watanabe, E. (1975). Direct observation of oxygen vacancy in mullite, $1.86\ Al_2O_3 \cdot SiO_2$ by high resolution electron microscopy. Proc. Jpn. Acad. Sci. **51**, 173–178.

Nguyen, N., Legrain, M., Ducouret, A. and Raveau, B. (1999). Distribution of Mn^{3+} and Mn^{4+} species between octahedral and square pyramidal sites in $Bi_2Mn_4O_{10}$-type structure. J. Mater. Chem. **9**, 731–734.

Niizeki, N. and Wachi, M. (1968). The crystal structures of $Bi_2Mn_4O_{10}$, $Bi_2Al_4O_9$ and $Bi_2Fe_4O_9$. Z. Krist. **127**, 173–187.

Niven, M. L., Waters, D. J. and Moore, J. M. (1991). The crystal structure of werdingite, $(Mg,Fe)_2Al_{12}(Al,Fe)_2Si_4(B,Al)_4O_{37}$, and its relationship to sillimanite, mullite and grandidierite. Amer. Min. **76**, 246–256.

Nyfeler, D., Hoffmann, C., Armbruster, T., Kunz, M. and Libowitzky, E. (1997). Orthorhombic Jahn-Teller distortion and Si-OH in mozartite, $CaMn^{3+}O[SiO_3OH]$; A single-crystal X-ray, FTIR, and structure modeling study. Amer. Min. **82**, 841–848.

Ocana, M., Cabellero, A., Ganzalez-Carreno, T. and Serna, C. J. (2000). Preparation by pyrolysis of aerosols and structural characterization of Fe-doped mullite powders. Mater. Res. Bull. **35**, 775–788.

Okada K. and Otsuka, N. (1986). Characterization of the spinel phase from SiO_2-Al_2O_3 xerogels and the formation process of mullite. J. Am. Ceram. Soc. **69**, 652–656.

Okada K. and Otsuka, N. (1990). Preparation of transparent mullite films by dip-coating method. Ceramic Trans. **6**, 425–430.

Okada, K., Aoki, C., Ban, T., Hayashi, S. and Yasumori, A. (1996). Effect of aging temperature on the structure of mullite precursor prepared from tetraethoxysilane and aluminum nitrate in ethanol solution. J. Europ. Ceram. Soc. **16**, 149–153.

Okada, K., Kaneda, J., Kameshima, Y., Yasumori, M. and Takei, T. (2003). Crystallization kinetics of mullite from polymeric Al_2O_3 – SiO_2 xerogels. Mater. Lett. **4304**, 1–5.

Okuno, M., Shimada, Y., Schmücker, M., Schneider, H., Hoffbauer, W. and Jansen, M. (1997). LAXS and Al-NMR studies on the temperature-induced changes of non-crystalline single phase mullite precursors. J. Non-Cryst. Solids **210**, 41–47.

Oschatz and Wächter (1847). Cited in: Litzow, K. (1984). Keramische Technik. Vom Irdengut zum Porzellan. Callwey, München.

Ossaka, J. (1961). Tetragonal mullite-like phase from co-precipitated gels. Nature **191**, 1000–1001.

Padlewski, S., Heine, V. and Price, G. D. (1992a). Atomic ordering around the oxygen vacancies in sillimanite. A model for the mullite structure. Phys. Chem. Minerals **18**, 373–378.

Padlewski, S., Heine, V. and Price, G. D. (1992b). The energetics of interaction between oxygen vacancies in sillimanite: A model for the mullite structure. Phys. Chem. Minerals **19**, 196–202.

Park, H. and Barbier, J. (2001). $PbGaBO_4$, an orthoborate with a new structure-type. Acta Cryst. E**57**, 82–84.

Park, H., Barbier, J. and Hammond. R. P. (2002). Crystal structure and polymorphism of $PbAlBO_4$. Solid State Sci. **5**, 565–571.

Park, H., Barbier, J., Hammond, R. P. (2003a). Crystal structure and polymorphism of $PbAlBO_4$. Solid State Sci. **5**, 565–571.

Park, H., Lam, R., Greedan, J. E. and Barbier, J. (2003b). Synthesis, crystal structure and magnetic properties of $PbMBO_4$ (M = Cr, Mn, Fe). A new structure type exhibiting one-dimensional magnetism. Chem. Mater. **15**, 1703–1712.

Parmentier, J. Vilminot, S. and Dormann, J.-L. (1999). Fe- and Cr-substituted mullites: Mössbauer spectroscopy and Rietveld structure refinement. Solid State Sci. **1**, 257–265.

Parthé, E. and Gelato, L. M., (1984). The standardization of inorganic crystal-structure data. Acta Cryst. A**40**, 169–183.

Paulmann, C. (1996). Study of oxygen vacancy ordering in mullite at high temperatures. Phase Transitions **59**, 77–90.

Paulmann, C., Rahman, S. H. and Strothenk, S. (1994). Interpretation of mullite HREM images along [010] and [100]. Phys. Chem. Minerals **21**, 546–554.

Peacor, D. R., Rouse, R. C. and Grew, E. S. (1999). Crystal structure of boralsilite and its relation to a family of boroaluminosilicates, sillimanite, and andalusite. Amer. Min. **84**, 1152–1161.

Peeters, M.P.J. and Kentgens, A.P.M. (1997). A ^{27}Al MAS, MQMAS and off-resonance nutation NMR study of aluminium containing silica-based sol-gel materials, Solid State NMR, **9**, 203–217.

Perez Y Jorba, M. (1968). Les systemès GeO_2–Al_2O_3 et GeO_2–Fe_2O_3 avec les systèmes correspondants à base de silice. Silicates Ind. **33**, 11–17.

Perez Y Jorba, M. (1969). Contribution a l'étude de phases formées par l'oxyde de germanium avec quelques oxydes d'éléments trivalents. Rev. Int. Hts. Temp. et Réfract. **6**, 283–298.

Permer, L., Laligant, Y. and Ferey, G. (1993). Crystal structure of $(Pb_{2.8}Fe_{1.2})Cu_4O_{1.6}$ $(VO_4)_4(OH)_2$; structural relationships with mineral gamagarite. Europ. J. Solid State Inorg. Chem. **30**, 383–392.

Perrotta, A. J. and Young Jr., J. E. (1974). Silica-free phases with mullite-type structures. J. Am. Ceram. Soc. **57**, 405–407.

Pertlik, F. (1975). Verfeinerung der Kristallstruktur von synthetischem Trippkeit, $CuAs_2O_4$. Tschermaks Mineral. Petrogr. Mitt. **22**, 211–217.

Pertlik, F. (1989). The crystal structure of čechite, $Pb(Fe^{2+}, Mn^{2+})(VO_4)(OH)$ with Fe > Mn. A mineral of the descloizite group. N. Jb. Mineral. Mh. 1989, 34–40.

Pilati, T., Demartin, F. and Gramaccioli, C. M. (1997). Transferability of empirical force fields in silicates: Lattice-dynamical evaluation of atomic displacement parameters and thermodynamic properties for the Al_2OSiO_4 polymorphs. Acta Cryst. B**53**, 82–94.

Piriou, B., Rager, H. and Schneider, H. (1996). Time resolved fluorescence spectroscopy of Cr^{3+} in mullite. J. Europ. Ceram. Soc. **16**, 195–201.

Pouxviel, J. C. and Boilot, J. P. (1989). Gels from a double alkoxide $(BuO)_2$-Al-O-Si-$(OEt)_3$, J. Mater. Sci. **24**, 321–327.

Pouxviel, J. C., Boilot, J. P., Lecomte, A. and Dauger, A. (1987). Growth process and structure of aluminosilicate gels. J. Phys. **48**, 921–925.

Puebla, E.G., Rios, E.G., Monge, A. and Rasines, I. (1982). Crystal growth and structure of diantimony(III) zinc oxide. Acta Cryst. B**38**, 2020–2022.

Qurashi, M.M. and Barnes, W.H. (1963). The structures of the minerals of the descloizite and adelite group. IV: Descloizite and conichalcite (part 2). The structure of conichalcite. Canad. Min. **7**, 561–577

Rager, H., Schneider, H. and Graetsch, H. (1990). Chromium incorporation in mullite. Amer. Min. **75**, 392–397.

Rager, H., Schneider, H. and Bakhshandeh, A. (1993). Ti^{3+} centres in mullite. J. Europ. Min. Soc. **5**, 511–514.

Rahman, S.H. (1991). Die videographische Methode: Ein neues Verfahren zur Simulation und Rekonstruktion fehlgeordneter Kristallstrukturen. Habilitation Thesis Univ. Hannover.

Rahman, S.H. (1992). Real structure image reconstruction from the random phase random amplitude model. Xth European Congress on Electron Microscopy Granada Spain. Vol. I. 453–454.

Rahman, S.H. (1993a). The videographic method: A new procedure for the simulation and reconstruction of real structures. Acta Cryst. A**49**, 56–68.

Rahman, S.H. (1993b). Interpretation of mullite HREM images using the potential-exchange method. Z. Krist. **203**, 67–72.

Rahman, S.H. (1994a). The real crystal structure of mullite. In: H.Schneider, K.Okada and J.A.Pask. Mullite and Mullite Ceramics. Wiley and Sons, Chichester, 4–31.

Rahman, S.H. (1994b). Videographic reconstructions and simulation of the real Cu$_3$Au structure at various temperatures. Z. Krist. **209**, 315–321.

Rahman, S.H. and Rodewald, M. (1992). HRTEM-Kontrastsimulation von Defektstrukturen nach dem Multi-slice Verfahren. Optik Suppl. 4, Vol. 88, 17.

Rahman, S.H. and Rodewald, M. (1995). Simulation of short range order in FCC-alloys. Acta Cryst. A**51**, 153–158.

Rahman, S.H. and Weichert, H.-T. (1990). Interpretation of HREM images of mullite. Acta Cryst. B**46**, 139–149.

Rahman, S.H., Strothenk, S., Paulmann, C. and Feustel, U. (1996). Interpretation of mullite real structure via inter-vacancy correlation vectors. J. Europ. Ceram. Soc. **16**, 177–186.

Rahman, S., Feustel, U. and Freimann, S. (2001). Structure description of the thermic phase transformation sillimanite-mullite. J. Europ. Ceram. Soc. 21, 2471–2478.

Razumovskii, S.N., Tunik, T.A., Fisher, O.N. and Schmitt-Fogelevich, S.P. (1977). X-ray analysis of solid solutions of ferric oxide in mullite. Ogneupory **18**, 46–48.

Rehak, P., Kunath-Fandrei, G., Losso, P., Hildmann, B., Schneider H. and Jäger, C. (1998). Study of the Al coordination in mullites with varying Al:Si ratio by ^{27}Al NMR spectroscopy and X-ray diffraction. Amer. Min. **83**, 1266–1276.

Risbud, S.H. and Pask, J.A. (1977). Calculated thermodynamic data and metastable immiscibility in the system SiO$_2$–Al$_2$O$_3$. J. Am. Ceram. Soc. **60**, 419–424.

Risbud, S.H., Kirkpatrick, R.J., Taglialavore, A.P. and Montez, B. (1987) Solid state NMR evidence of 4-, 5-, and 6-fold aluminium sites in roller-quenched SiO$_2$-Al$_2$O$_3$ glasses. J. Am. Ceram. Soc. **70**, C10–C12.

Rodriguez-Navarro, C., Cultrone, G., Sanchez-Navas, A. and Sebastian, E. (2003). TEM study of mullite growth after muscovite breakdown. Amer. Min. **88**, 713–724.

Ronchetti, S., Piana, M., Delmastro, A., Salis, M. and Mazza, D. (2001). Synthesis and characterization of Fe and P substituted 3:2 mullite. J. Europ. Ceram. Soc. 21, 2509–2514.

Rossman, G.R., Grew., E.S. and Dollase, W.A. (1982). The colors of sillimanite. Amer. Min. **67**, 749–761.

Rossouw, C.J. and Miller, P.R. (1999). Location of interstitial Cr in mullite by incoherent channeling patterns from characteristic X-ray emission. Amer. Min. **84**, 965–969.

Rubie, D.C. and Brearley, A.J. (1987). Metastable melting during the breakdown muscovite + quartz at 1 kbar. Bull. Mineral. **110**, 533–549.

Rymon-Lipinski, T., Hennicke, H.W. and Lingenberg, W. (1985). Zersetzung von 9Al$_2$O$_3$ · 2B$_2$O$_3$ bei hohen Temperaturen. Keram. Z. **37**, 450–453.

Saalfeld, H. (1962). A modification of Al$_2$O$_3$ with sillimanite structure. Trans. VIIIth Int. Ceramic Congress Copenhagen, pp 71–74.

Saalfeld, H. (1979). The domain structure of 2:1-mullite (2 Al_2O_3 1SiO_2). N. Jb. Miner. Abh. **134**, 305–316.

Saalfeld, H. and Gerlach, H. (1991). Solid solution and optical properties of (Al, Ge)-mullites. Z. Krist. **195**, 65–73.

Saalfeld, H. and Guse, W. (1981). Structure refinement of 3:2-mullite (3 Al_2O_3 · 2SiO_2). N. Jb. Min. Mh. 1981, 145–150.

Saalfeld, H. and Klaska, K. H. (1985). A Pb/Nd-stabilized mullite of the composition $Al_{5.03}Ge_{0.97}Pb_{0.15}Nd_{0.06}O_{9.71}$. Z. Krist. **172**, 129–133.

Sacks, M. D., Bozkurt, N. and Scheiffele, G. W. (1991). Fabriction of mullite and mullite matrix composites by transient viscous sintering of composite powders. J. Am. Ceram. Soc. **74**, 2828–2837.

Sacks, M. D., Wang, K., Scheiffele, G. W. and Bozkurt, N. (1996). Activation energy for mullitization of α-alumina/silica microcomposite particles. J. Am. Ceram. Soc. **79**, 571–573.

Sadanaga, R., Tokonami, M. and Takéuchi, Y. (1962). The structure of mullite, 2Al_2O_3 · SiO_2, and relationship with the structures of sillimanite and andalusite. Acta Cryst. **15**, 65–68.

Schmücker, M. and Schneider, H. (1996). A new approach on the coordination of Al in non-crystalline gels and glasses of the system $SiO_2–Al_2O_3$, Ber. Bunsenges. Phys. Chem. **100**, 1550–1555.

Schmücker, M. and Schneider, H. (1999). Structural development of single phase mullite gels. J. Sol-Gel Sci. Tech. **15**, 191–199.

Schmücker, M. and Schneider, H. (2002). New evidence for tetrahedral triclusters in alumino silicate glasses, J. Non-Cryst. Solids **311**, 211–215.

Schmücker, M., Albers, W. and Schneider, H. (1994). Mullite formation by reaction sintering of quartz and α-alumina – a TEM study. J. Europ. Ceram. Soc.**14**, 511–515.

Schmücker M., Schneider, H., Poorteman, M., Cambier, F. and Meinhold, R. (1995). Formation of Al_2O_3-rich glasses in the system $SiO_2–Al_2O_3$. J. Europ. Ceram Soc.**15**, 1201–1205.

Schmücker, M., Mackenzie, K.J.D., Schneider, H., and Meinhold, R. (1997). NMR-studies on rapidly solidified $SiO_2–Al_2O_3$ -and

$SiO_2–Al_2O_3–Na_2O$ glasses. J. Non-Cryst. Solids **217**, 99–105.

Schmücker, M., Schneider, H., Mackenzie, K.J.D. (1998). Mechanical amorphization of mullite and recrystallization. J. Non-Cryst. Solids **226**, 99–103.

Schmücker, M., Schneider, H., Mackenzie, K.J.D. and Okuno, M. (1999). Comparative Al NMR and LAXS studies on rapidly quenched aluminosilicate glasses J. Europ. Ceram. Soc, **19**, 99–103.

Schmücker, M., Hoffbauer, W. and Schneider H. (2001). Constitution and crystallization behaviour of ultrathin physical vapor deposited (PVD) Al_2O_3/SiO_2 laminates. J. Europ. Ceram. Soc. **21**, 2503–2507.

Schmücker, M., Hildmann, B. and Schneider, H. (2002). The mechanisms of 2/1- to 3/2-mullite transformation. Amer. Min. **87**, 1190–1193.

Schmücker., M., MacKenzie, K.J.D., Smith, M. E., Carroll, D. E. and Schneider, H. (2005). AlO_4/SiO_4 distribution in tetrahedral double chains of mullite. J. Am. Ceram. Soc., in press.

Schneider, H. (1981). Infrared spectroscopic investigations on andalusite- and mullite-type structures in the systems $Al_2GeO_5–Fe_2GeO_5$ and $Al_2GeO_5–Ga_2GeO_5$. N. Jb. Min. Abh. **142**, 111–123.

Schneider, H. (1984). Solid solubility of Na_2O in mullite. J. Am. Ceram. Soc. **67**, C-130–C-131.

Schneider, H. (1985). Magnesium incorporation in mullite. N. Jb. Min. Mh., 491–496.

Schneider, H. (1986a).Formation, properties and high-temperature behaviour of mullite. Habilitation Thesis Univ. Münster.

Schneider, H. (1986b). Zirconium incorporation in mullite. N. Jb. Min. Mh. **4**, 172–180.

Schneider, H. (1987a). Temperature-dependent iron solubility in mullite. J. Am. Ceram. Soc. **70**, C43–C45.

Schneider, H. (1987b). Solubility of TiO_2, Fe_2O_3 and MgO in mullite. Ceramics Intern. **13**, 77–82.

Schneider, H. (1989). Iron exchange processes between mullite and coexisting silicate melts at high temperature. Glastechn. Ber. **62**, 193–198.

Schneider, H. (1990). Transition metal distribution in mullite. Ceramic Trans. **6**, 135–158.

Schneider, H. and Eberhard, E. (1990). Thermal expansion of mullite. J. Am. Ceram. Soc. **73**, 2073–2076.

Schneider, H. and Rager, H. (1984) Occurrence of Ti^{3+} and Fe^{2+} in mullite. J. Am. Ceram. Soc. **67**, C248–C250.

Schneider, H. and Rager, H. (1986) Iron incorporation in mullite. Ceramics Intern. **12**, 117–125.

Schneider, H. and Rymon-Lipinski, T. (1988). Occurrence of pseudotetragonal mullite. J. Am. Ceram. Soc. **71**, C162–C164.

Schneider, H. and Vasudevan, R. (1989) Structural deformation of manganese substituted mullites: X-ray line broadening and lattice parameter studies. N. Jb. Min. Mh., 165–178.

Schneider, H. and Werner, H. D. (1981). Synthesis and crystal chemistry of andalusite-, kyanite-, and mullite-type structures in the system Al_2GeO_5–Fe_2GeO_5. N. Jb. Min. Abh. **140**, 153–164.

Schneider, H. and Werner, H. D. (1982). Synthesis and crystal chemistry of andalusite- and mullite-type structures in the system Al_2GeO_5–Ga_2GeO_5. N. Jb. Min. Abh. **143**, 223–230.

Schneider, H. and Wohlleben, K. (1981). Microchemical composition and cell dimensions of mullites from refractory-grade South American bauxites. Ceramics Intern. **7**, 130–136.

Schneider, H., Merwin, L., Sebald, A. (1992). Mullite formation from non-crystalline solids. J. Mater. Sci. **27**, 805–812.

Schneider, H., Fischer, R. X. and Voll, D. (1993a). Mullite with lattice constants $a > b$. J. Am. Ceram. Soc. **76**, 1879–1881.

Schneider, H., Schmücker, M., Ikeda, K., and Kaysser, W. A. (1993b). Optically translucent mullite ceramics. J. Am. Ceram. Soc. **76**, 2912–2916.

Schneider, H., Saruhan, B., Voll, D., Merwin, L., and Sebald, A. (1993c). Mullite precursor phases. J. Europ. Ceram. Soc. **11**, 87–94.

Schneider, H., Okada, K, and Pask, J. A. (1994a) Mullite and Mullite Ceramics. John Wiley and Sons, Chichester.

Schneider, H., Voll, D., Saruhan, B., Sanz, J., Schrader, G., Rüscher, C. and Mosset, A. (1994b) Synthesis and structural characterization of non-crystalline mullite precursors, J. Non-Cryst. Solids **17**, 262–271.

Schneider, H., Voll, D., Saruhan, B., Schmücker, M., Schaller, T. and Sebald, A. (1994c). Constitution of the γ-alumina phase in chemically produced mullite precursors. J. Europ. Ceram. Soc. **13**, 441–448.

Schneider, J., Mastelaro, V. R., Panepucci, H. and Zanotto, E. D. (2000). ^{29}Si MAS-NMR studies of Q^n structural units in metasilicate glasses and their nucleating ability. J. Non-Cryst. Solids **273**, 8–18.

Scholze, H. (1956). Über Aluminiumborate. Z. Allg. Anorg. Chemie **284**, 272–277.

Schryvers, D., Srikrishna, K., O'Keefe, M. A. and Thomas, G. (1988). An electron microscopy study of the atomic structure of a mullite in a reaction-sintered composite. J. Mater. Res. **3**, 1355–1366.

Schwerdtfeger, G. and Muan, A. (1975) quoted in: Levin, E. M. and McMurdie, H. F.: Phase Diagrams for Ceramists. The American Ceramic Society, Supplement, p. 8, Fig. 4156.

Shannon, R. D. (1976). Revised effective ionic radii and systematic studies of interatomic distances in halides and chalcogenides. Acta Cryst. A**32**, 751–767.

Soro, N., Aldon, L., Olivier-Fourcade, J., Jumas, J. C., Laval, J. P. and Blanchart. P. (2003). Role of iron in mullite formation from kaolins by Mössbauer spectroscopy and Rietveld refinement. J. Am. Ceram. Soc. **86**, 129–134.

Stephenson, D. A. and Moore, P. B. (1968). The crystal structure of grandidierite, (Mg, Fe)Al_3SiBO_9. Acta Cryst. B**24**, 1518–1522.

Srikrishna, K., Thomas, G., Martinez, R., Corral, M. P., De Aza, S. and Moya, J. S. (1990). Kaolinite-mullite reaction series: A TEM study. J. Mater. Sci. **25**, 607–612.

Sundaresan, S. and Aksay, I. A. (1991). Mullitization of diphasic aluminosilicate gels. J. Am. Ceram. Soc. **74**, 2388–92.

Taake, C. (1999). Synthese und strukturelle Charakterisierung amorpher Festkörper im System SiO_2/Al_2O_3. Ph. D. Thesis, Univ. Hannover.

Takamori, T. and Roy, R. (1973). Rapid crystallization of SiO_2–Al_2O_3 glasses, J. Am. Ceram. Soc. **56**, 639–644.

Takei, T., Kameshima, Y., Yasumori, A. and Okada, K. (1999). Crystallization kinetics of mullite in alumina-silica fibers. J. Am. Ceram. Soc. **82**, 2876–2880.

Takei, T., Kameshima, Y., Yasumori, A. and Okada, K. (2000). Calculation of metastable immiscibility region in the Al_2O_3–SiO_2 system using molecular dynamics simulation. J. Mater. Res. **15**,186–193.

Takei, T., Kameshima, Y., Yasumori, A. and Okada, K. (2001). Crystallization kinetics of mulite from Al_2O_3–SiO_2 glasses under non-isothermal conditions. J. Europ. Ceram. Soc. **21**, 2487–2493.

Taylor, A. and Holland, D. (1993).The chemical synthesis and reaction sequence of mullite. J. Non-Cryst. Solids **152**, 1–7

Thom, M. (2000). Zusammensetzungsänderungen von Mullit-Mischkristallen. Diploma Thesis, Univ. Köln.

Thomas, H.H. (1922). On certain xenolithic tertiary minor inclusions in the island of Mull (Argyllshire). J. Geol. Soc. London **78**, 229–260.

Tkalcec, E., Nass, R., Schmauch, J., Schmidt, H., Kurajica, S., Bezjak, A. and Ivankovic, H. (1998). Crystallization kinetics of mullite from single phase gel determined by isothermal differential scanning calorimetry. J. Non-Cryst. Solids **223**, 57–72.

Tkalcec, E., Kurajica, S. and Ivankovic, H. (2001). Isothermal and non-isothermal crystallization kinetics of zinc-aluminosilicate glasses. Thermochim. Acta **378**, 135–144.

Tkalcec, E., Ivankovic, H., Nass, R. and Schmidt, H. (2003). Crystallization kinetics of mullite formation in diphasic gels containing different alumina components. J. Europ. Ceram. Soc. **23**, 1465–1475.

Toman, K. (1978). Ordering in olivenite – adamite solid solutions. Acta Cryst. B**34**, 715–721.

Tomsia, A.P., Saiz, E., Ishibashi, h., Diaz, M., Requena, J. and Moya, J.DS. (1998). Powder processing of mullite/Mo functionally graded materials. J. Europ. Ceram. Soc. **18**, 1365–1371.

Trömel, G., Obst, K.-H., Konopicky, K., Bauer, H. and Patzak, I. (1957). Untersuchungen im System SiO_2–Al_2O_3. Ber. Dtsch. Keram. Ges. **34**, 397–402.

Vernadsky, W.I. (1890).Cited in: Pask, J.A. (1990). Critical review of phase equilibria in the Al_2O_3 – SiO_2 system. Ceramic Trans. **6**, 1–13.

Voll, D. (1995). Mullitprecursoren. Synthese, temperaturabhängige Entwicklung der strukturellen Ordnung und Kristallisationsverhalten. Ph. D. Thesis, Univ. Hannover

Voll, D., Beran, A. and Schneider, H. (1998). Temperature-dependent dehydration of sol-gel-derived mullite precursors: A FTIR spectroscopy study. J. Europ. Ceram. Soc. **18**, 1101–1106.

Voll, D., Lengauer, C., Beran, A. and Schneider, H. (2001). Infrared band assignment and structural refinement of Al–Si, Al–Ge, and Ga–Ge mullites. J. Europ. Min. Soc. **13**, 591–604.

Voll, D., Beran, A. and Schneider, H. (2005). Local structure of Bi_2 (Al_4Fe_{4-x}) O_9 mixed crystals. Unpublished results.

Wada, H., Sakane, K., Kitamura, T., Sunai, M. and Sasaki, N. (1993). Thermal expansion of aluminium borate. Mater. Sci. Lett. **12**, 1735–1737.

Wada, H., Sakane, K., Kitamura, T. Kayahara, Y., Kawahara, A. and Sasaki, N. (1994). Thermal conductivity of aluminium borate. J. Ceram. Soc. Jp. **102**, 695–701.

Wang, K. and Sacks, M. (1996). Mullite formation by endothermic reaction of α-Al_2O_3/silica microcomposite particles. J. Am. Ceram. Soc. **79**, 12–16.

Wefers, K. and Misra, C. (1987). Oxides and hydroxides of aluminium. Alcoa Technical Paper No. 19, Rev., Alcoa Lab, Pittsburgh PA.

Wei, W.-Ch., and Halloran, J.W. (1988a). Phase transformation of diphasic aluminosilicate gels. J. Am. Ceram. Soc. **71**, 166–172.

Wei, W.-Ch. and Halloran, J.W. (1988b). Transformation kinetics of diphasic aluminosilicate gels, J. Am. Ceram. Soc. **71**, 581–587.

Wei, P. and Rongti, L. (1999). Crystallization kinetics of the aluminum silicate glass fiber. J. Mater. Sci. Eng. A**271**, 298–305.

Weiss, Z., Bailey, S.W. and Rieder, M. (1981). Refinement of the crystal structure of kanonaite, $(Mn^{3+}, Al)^{[6]}(Al,Mn^{3+})^{[5]}$ $O[SiO_4]$. Amer. Min. **66**, 561–567.

Wells, A.F. (1984) Structural Inorganic Chemistry. Clarendon Press, 5th Edition, Oxford.

Werding, G. and Schreyer, W. (1992). Synthesis and stability of werdingite, a new phase in the system MgO – Al_2O_3 – B_2O_3 – SiO_2 (MABS), and another new phase in the

ABS-system. J. Europ. Min. Soc. **4**, 193–207.

Wilhelmi, K. A. (1975) quoted in: Levin, E. M. and McMurdie, H. F. Phase Diagrams for Ceramists. The American Ceramic Society 1975 Supplement, p. 6, Fig. 4152.

Witzke, T., Steins, M., Doering, T. and Kolitsch, U. (2000). Gottlobite, $CaMg(VO_4, AsO_4)(OH)$, a new mineral from Friedrichroda, Thuringia, Germany. N. Jb. Min. Mh., 444–454.

Worden, R. H., Champness, P. E. and Droop, G.T.R. (1987). Transmission electron microscopy of pyrometamorphic breakdown of phengite and chalorite. Min. Mag. **51**, 107–121.

Yamashita, T., Fujino, T., Masaki, N. and Tagawa, H. (1981). The crystal structures of alpha- and beta-$CdUO_4$. J. Solid State Chem. **37**, 133–139.

Yang, H., Hazen, R. M., Finger, L. W., Prewitt, C. T. and Downs, R. T. (1997). Compressibility and crystal structure of sillimanite, Al_2SiO_5, at high pressure. Phys. Chem. Minerals **25**, 39–47.

Ylä-Jääski, J. and Nissen, H.-U. (1983). Investigation of superstructure in mullite by high resolution electron microscopy and electron diffraction. Phys. Chem. Minerals **10**, 47–54.

Yoldas, B. E. (1990). Mullite formation from aluminium and silicon alkoxides. Ceramic Trans. **6**, 255–263.

Yue, R., Wang, Y., Chen, C. and Xu, C. (1998). Interface reaction of Ti and mullite ceramic substrate. Appl. Surf. Sci. **126**, 255–264.

Zha, S., Cheng, J., Liu, Y., Liu, X. and Meng, G. (2003). Electrical properties of pure and Sr-doped $Bi_2Al_4O_9$ ceramics. Solid State Ionics **156**, 197–200.

2
Basic Properties of Mullite

2.1
Mechanical Properties of Mullite
H. Schneider

2.1.1
Strength, Toughness and Creep

Only a small amount of information is available on the response of mullite single crystals under mechanical load. Dokko et al. (1977) provided stress-strain curves of mullite single crystals under compression about 3° off the crystallographic **c** axis. Single crystals subjected to a compressive stress of about 480 MPa at 1400 °C for 200 h display no detectable plastic strain, which is consistent with the high brittleness of mullite. The high compressive strength of mullite single crystals and the low diffusion-controlled creep of polycrystalline mullite ceramics (see Section 4.3.7) indicate that the mobility of dislocations in mullite is low. Dokko et al. ascribe the high creep resistance of mullite to the "complexity" of the crystal structure (probably they mean the real structure of mullite, see Section 1.2). Menard and Doukhan (1978) studied plastic deformation in natural sillimanite, which can contribute to our understanding since sillimanite is closely related to mullite, although it has no "complex" real structure as mullite does (see Sections *1.1.3.7* and *1.1.3.10*). Sillimanite shows low-energy dislocations and gliding ($hk0$) planes parallel to the crystallographic **c** axis. This allows a mobility of dislocations that does not involve crossing the AlO_6 chains, and supports the suggestion that in all members of the mullite family (see Section 1.1.3) these structural units are the most stable. A detailed analysis of the plastic deformation of mullite single crystals in comparison to sillimanite is not available yet, but is highly necessary for an understanding of whether its behavior is determined by the average crystal structure alone, or if the real structure has a controlling influence as well. A suitable model of the deformation mechanism of mullite would allow to predict the response of mullite ceramics under mechanical load.

2.1.2
Elastic Moduli and Compressibility

The elastic constants of mullite single crystals and of directionally solidified poly-crystalline mullite fibers determined by Brillouin and resonant ultrasound spectroscopy (RUS) are given in Table 2.1.1 (Schreuer et al. 2005, Hildmann et al. 2001, Palko et al. 2002, Vaughan and Weidner, 1978). All data are in good agreement except those of Hildmann et al., which display slight deviations.

Hildmann et al. (2001) believe that the high longitudinal elastic coefficients c_{33} of mullite (about 350 GPa) and sillimanite (about 390 GPa) are caused by the "stiff" load-bearing tetrahedral chains parallel to the crystallographic **c** axis, while the "soft" octahedral chains stabilize the tetrahedral chains against tilting (see, for example, Fig. 1.1.13, Section 1.1). The c_{33} value of mullite is lower than that of sillimanite, because the mean tetrahedral bond strength is reduced by partial re-placement of Si^{4+} by the larger Al^{3+} ions. The longitudinal stiffness coefficients c_{11} and c_{22} describing relationships perpendicular to the **c** axis in the (001) plane are significantly lower than c_{33}. This has been explained by the occurrence of the sequence of "soft" octahedra with "stiff" tetrahedra in the (001) plane, the former having a stiffness-reducing influence. The observation that the stiffness parallel to the **b** axis is lower (c_{22}: about 235 GPa) than that parallel to **a** (c_{11}: about 290 GPa) has been explained by the long and elastic octahedral Al-O(D) bond (i. e. Al-O1),[1] which is more effective on **b** than on **a** (see, for example, Fig. 1.1.13, Section 1.1 and Section 2.2.2). Shear stiffness coefficients of mullite increase from c_{55} (about 75 GPa) to c_{66} (about 80 GPa) to c_{44} (about 110 GPa), indicating increasing re-sistance against shear deformation within the lattice planes (010), (001) and (100). Hildmann et al. (2001) on the basis of a model developed for sillimanite by Vaughan and Weidner (1978) proposed an explanation for the shear behavior of mullite. They pointed out that the lattice plane with the highest shear resistance (100) is built up of an oxygen-oxygen network, diagonally braced along <011> (the so-called "Jägerzaun" configuration). This oxygen network can be sheared by com-pression and elongation only along the short oxygen-oxygen interaction lines, which, however, is unlikely (Fig. 2.1.1). Since such oxygen-oxygen networks do not occur in the (010) and (001) planes, shear deformations are easier, and as a con-sequence c_{55} and c_{66} coefficients have lower numerical values than c_{44}. Hildmann et al. further quoted that the shear coefficients of mullite are all slightly lower than those of sillimanite. This can be understood by the reduction of the mean tetra-hedral bond strength of mullite caused by the partial substitution of silicon by aluminum and the occurrence of oxygen vacancies (see above).

---

1) Different designations of the oxygen posi-tions of mullite have been described in litera-ture: 01, 02, 03, 04 which corresponds to O(A,B), O(D), O(C) and O(C*), (Table 1.1.7). The first setting follows strict geometrical rules and allows an easy description of mem-bers of the mullite family being in group-subgroup relationship to the aristotype with the highest symmetry (Tables 1.1.2 to 1.1.10, 1.1.17 and Fig. 1.1.4). Since the second set-ting has frequently been cited in the crystal chemical literature of mullite it has been used in the following.

Table 2.1.1 Comparison of elastic properties of mullite single crystals at room temperature.

Property		$2Al_2O_3 \cdot SiO_2$[a] (RUS)	$2Al_2O_3 \cdot SiO_2$[b] (RUS)	$2.5Al_2O_3 \cdot SiO_2$[c] (Brillouin spectroscopy)	$2.5Al_2O_3 \cdot SiO_2$[d] (Brillouin spectroscopy)
ϱ	[g cm^{-3}]	3.125	3.096	3.11	3.10
c_{11}	[GPa]	279.5	291.3	281.9	280.0
c_{22}		234.9	232.9	244.2	245.0
c_{33}		360.6	352.1	363.6	362.0
c_{44}		109.49	110.3	111.7	111.0
c_{55}		74.94	77.39	78.2	78.1
c_{66}		79.89	79.90	79.2	79.0
c_{12}		103.1	112.9	105.1	105.0
c_{13}		96.1	96.22	100.3	99.2
c_{23}		135.6	121.9	142.3	135.0
s_{1111}	[TPa^{-1}]	4.357	4.314	4.325	4.366
s_{2222}		6.015	5.995	5.832	5.670
s_{3333}		3.615	3.540	3.647	3.567
s_{2323}		2.283	2.267	2.238	2.252
s_{1313}		3.336	3.230	3.197	3.201
s_{1212}		3.129	3.129	3.157	3.164
s_{1122}		−1.587	−1.801	−1.511	−1.525
s_{1133}		−0.565	−0.556	−0.602	−0.628
s_{2233}		−1.839	−1.584	−1.866	−1.697
g_{11}	[GPa]	26.11	11.60	30.6	24.0
g_{22}		21.16	18.83	22.1	21.1
g_{33}		23.21	33.00	25.9	26.0
c^{iso}_{11}	[GPa]	284.8	286.4	290.9	289.5
c^{iso}_{44}		111.2	110.6	114.9	112.7
K	[GPa]	166.5	167.5	171.0	169.4

RUS = Resonant ultrasound spectroscopy
ϱ = Density
c_{ij} = Elastic stiffness coefficients
s_{ijkl} = Elastic compliance coefficients
g_{ij} = Cauchy relations
c^{iso}_{ij} = Isotropic elastic stiffness coefficients are averages of Voigt and Reuss models
K = Bulk modulus

a) Schreuer et al. 2005; b) Hildmann et al. 2001; c) Palko et al. 2002; d) Kriven et al. 1999.

Kriven et al. (1999), Palko et al. (2002) and Schreuer et al. (2005) provided data on the temperature dependence of the elastic properties of single-crystal mullite from room temperature to 1200 and 1400 °C, respectively. Palko et al. described a linear development of the elastic longitudinal and shear moduli over the whole

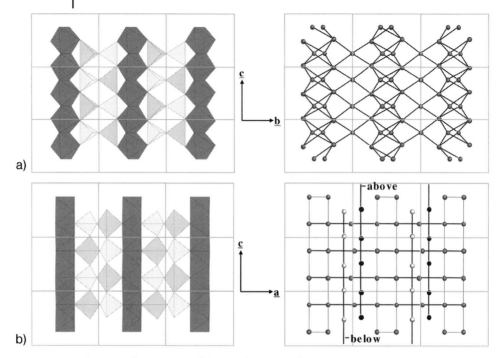

Fig. 2.1.1 Schematic view of the crystal structure of sillimanite drawn to explain the elastic behavior of mullite. (a) Projection down the **a** axis ([100]). The picture shows layers of octahedral and tetrahedral chains in the (100) plane (left) with the diagonally braced oxygen network (right) being characterized by strong oxygen-oxygen interactions, responsible for the relatively high resistance against shear deformation in (100); (b) Projection down the **b** axis ([010]). The picture shows layers of octahedral and tetrahedral chains in the (010) plane (left) with the unbraced oxygen-oxygen network being weaker than the braced one in (100), which explains the lower resistance against shear deformation (after Hildmann et al. 2001 from Vaughan and Weidner, 1978).

temperature range, while Schreuer et al. with their more sensitive resonant ultrasound spectroscopy (RUS) technique at about 1000 °C identified discontinuous decreases of the shear resistances especially in the (100) and (010) planes (c_{44} and c_{55}), and less pronounced ones in the (001) plane (c_{66}), accompanied by a rapidly increasing ultrasound attenuation (Figs. 2.1.2a,b). The discontinuities correspond with the anomalies of other properties, such as thermal expansion and heat capacity (see Section 2.2). A possible interpretation of the slight, anomalous softening of shear resistances especially in (100) and (010) (corresponding to a decrease of the c_{44} and c_{55} shear stiffnesses) is discussed in Section 2.5.2.

Another remarkable feature of the elastic behavior of mullite derived from resonant ultrasound spectroscopy is a frequency- and temperature-dependent max-

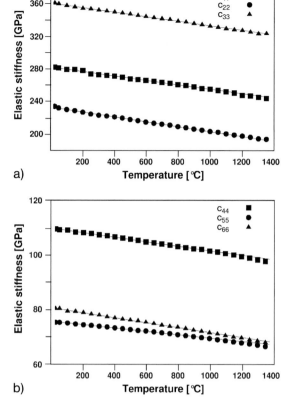

a)

b)

Fig. 2.1.2 Temperature dependence of the elastic moduli of single crystal mullite. (a) Linear moduli c_{11}, c_{22} and c_{33}. (b) Shear moduli c_{44}, c_{55} and c_{66}. Note the slight anomalies of the elastic shear moduli c_{44} and c_{55} above 1000°C (see also Section 2.5.2 according Schreuer et al. 2005).

imum in the signal width of the mechanical resonances that is observed at about 70 °C in the investigated frequency range from 300 kHz to 1.7 MHz (Fig. 2.1.3). This observation is considered to be a characteristic signature of relaxation attenuation due to interaction between elastic waves and point defects of the crystal (e. g. Nowick and Berry, 1972, Weller, 1996). The altenuation above 1000 °C has been

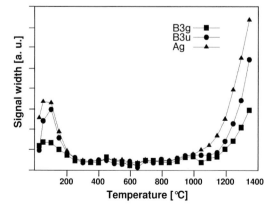

Fig. 2.1.3 Temperature-dependent development of the signal width of resonant ultrasound spectra of mullite single crystal. Note that signal broadenings appear near room temperature and above about 1000 °C, which is associated with a weakening of the structure in these temperature ranges (a. u. = arbitrary units, according to Schreuer et al. 2005).

ascribed to a wealeing of the structure due to the possible phase transformation (see Section 2.5.2.2).

The bulk elastic properties of mullite single crystals published by Hildmann et al. (2001) and Palko et al. (2002) agree well with those of Pentry et al. (1972) and Ledbetter et al. (1998), but there are substantial differences with other literature data (e.g. Ismail et al. 1987, Kelly et al. 1997) obtained from mullite ceramics. Obviously the latter data, reported from polycrystalline mullite ceramics, are affected by the occurrence of grain boundaries and possible existing grain boundary phases, which influence the inherent behavior of mullite.

No direct experimental data on the compressibility of mullite are available yet, while results have been published on the structurally related phase sillimanite (Yang et al. 1997, see also Section 1.1.3). Vaughan and Weidner (1978) using Brillouin scattering, and Yang et al. (1997) with high-pressure single crystal X-ray diffraction techniques both obtained isothermal compressibilities of about 170 GPa. The highest compressibilities occur parallel to the crystallographic **b** axis, and the lowest parallel to the **c** axis (axial compression ratios are: $\sigma_a : \sigma_b : \sigma_c = 1.22 : 1.63 : 1.00$). This again supports the suggestion that the rigid tetrahedra control the mechanical and thermal behavior of mullite along the **c** direction, and that the long and elastic Al-O(D) bond of the oxygen octahedron is responsible for the high **b**-axis compressibility (Fig. 2.1.4). The occurrence of the oxygen vacancies, and the predominance of aluminum oxygen tetrahedra (AlO$_4$) at the expense of silicon oxygen tetrahedra (SiO$_4$) in mullite, may decrease the **c** axis stiffness and increase the compressibility with respect to that of sillimanite. Actually the elastic c_{33} constant of mullite is about 10% lower than that of sillimanite (see Hildmann et al. 2001).

2.1.3
Microhardness of Mullite

Kollenberg and Schneider (1989) and Kriven et al. (2004) measured the microhardness of mullite single crystals of 2/1-composition parallel to the **b** (i.e [010]) and **c**

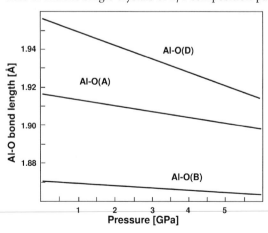

Fig. 2.1.4 Pressure-dependent compressibility of the aluminum oxygen octahedron of sillimanite which is closely related to mullite structurally (see seections 1.1.3.7 and 1.1.3.10). Shown are aluminum-oxygen bond lengths dependent on applied pressure. Note that the relatively long and elastic Al-O(D) (i.e. Al–O2) bond displays the highest compressibility (see also Table 1.1.17). (after Yang et al. 1997).

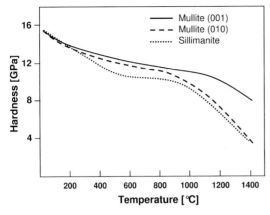

Fig. 2.1.5 Microhardness of (010) and (001) faces of mullite single crystals measured up to 1400 °C. Note that the temperature-dependent microhardness develops in three stages. Microhardness data of sillimanite are given for comparison (after Kriven et al. 2004 and Pitchford et al. 2001).

axes (i. e. [001]) up to 1000 and 1400 °C, respectively. According to Kriven et al. the microhardness versus temperature plots display sigmoidal shapes suggesting three sequential stages (Fig. 2.1.5):

- A first stage between room temperature and about 300 °C with a pronounced decrease in microhardness from the initial value of about 16 to 13 GPa.
- A second stage between about 300 and 1000 °C with little microhardness reduction from about 13 to 10 GPa.
- A third stage above 1000 °C with a strong microhardness decrease to a mean value of about 6 GPa at 1400 °C.

Sillimanite, which is closely related to mullite structurally (see Section 1.1.3), displays a sigmoidal microhardness versus temperature behavior similar to mullite (Fig. 2.1.5, Pitchford et al. 2001).

The complexity of the indentation makes its structural interpretation difficult. Below 300 °C the significant microhardness increase accompanies a reduction of the thermal expansion of mullite (first stage, see Section 2.2.2). The latter was explained by the fact that the shapes of the potential curves of solids from about room temperature to the absolute zero temperature (−273 °C) gradually become symmetric, which results in zero or near-zero expansion. Low to zero expansion, on the other hand means that bond lengths are shortest and bond strength highest. This has the consequence of increasing the hardness. The small decrease of microhardness above 300 and up to about 1000 °C (second stage) is "normal", caused by a temperature-induced decrease of the elastic moduli, which in turn is caused by a gradual decrease of the bond-length-correlated Al,Si-O bond strengths. The reasons for the discontinuous change of the microhardness above about 1000 °C (third stage) are not yet fully understood. It may be due to thermally activated lattice gliding and enhanced atomic mobility in an indentation-produced amorphous phase (see below). Moreover, it may be influenced by a phase transformation, corresponding to the anomalous developments of the elasticity, thermal expansion and heat capacity of mullite.

Kollenberg and Schneider (1989) and Kriven et al. (2004) identified no hardness

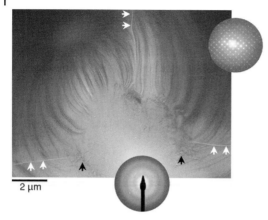

Fig. 2.1.6 Microindentation-induced amorphization of mullite. Transmission electron micrograph of the indented region of an (001) face of mullite. The indentation core has become amorphous, while the outer zone is crystalline with typical bend contours. Black arrows indicate dislocation networks at the interface between amorphous and crystalline areas. White arrows indicate radial cracks (Courtesy M. Schmücker).

anisotropy below 1000 °C, in spite of the anisotropic character of the mullite structure and of most mechanical and thermal properties (Fig. 2.1.5; see, for example, Sections 1.1.3.7, 2.1 and 2.2). A possible explanation of the near-isotropic deformation of mullite below 1000 °C is the role of an indentation-produced amorphous phase (see below), which should be equally effective in all lattice directions. At temperatures above 1000 °C, however, the microhardness becomes anisotropic. At 1400 °C the microhardness on (010) (about 8 GPa) is about twice as high as on (001) (about 4 GPa). At a first glance this is astonishing, since the stiff character of tetrahedral double chains parallel to the crystallographic **c** axis (see Section 2.1.2) implies the highest resistance against indentation in this lattice direction. Kriven et al. believe that lattice gliding has to be considered. Indentation on (001) activates gliding preferentially parallel to (*hk*0) faces, whereas indentation on (010) induces gliding along (*h*0*l*). The observation that gliding parallel to (*hk*0) is more easily activated than that parallel to (*h*0*l*), has been associated with the stable aluminum oxygen octahedral chains in mullite, running parallel to the **c** axis. Indentation on (010) requires lattice gliding parallel to (*h*0*l*) which involves breakage of octahedral chains. This, obviously, is an unfavorable process, associated with a high resistance to deformation. In the case of (001) indentation, lattice gliding occurs preferentially parallel to (*hk*0), where no octahedral chains have to be broken. The resistance to lattice gliding and the associated microhardness are therefore lower.

Schmücker et al. (2003), on the basis of transmission electron microscopy studies demonstrated that micro-indented mullite single crystals display a complex sequence of deformation effects: Directly under the indenter, in the region of the highest compressive stress, mullite becomes amorphous. Further out towards the undeformed areas there are regions of high plastic deformation in the form of dislocation networks, radial microcracks and bend contours (Fig. 2.1.6). Indentation-induced amorphization in mullite to some extend is comparable to the amorphization effects produced by dynamic shock or intense ball milling.

2.1.4
Mechanical Response to Dynamic Stress

The structural response of mullite to extremely high stress rates as realized at high dynamic shock loading has been investigated by Braue et al. (2005a). They describe the formation of shock-wave-produced intragranular aluminosilicate glass lamellae in mullite, similar to the "diaplectic glasses" reported from shock-loaded framework silicates such as quartz, feldspar and cordierite. The glass lamellae occur in {$hk0$} planes. The {$hk0$} deformation planes are suitable, since the stable octahedral chains running parallel to the crystallographic **c** axis are thus preserved (see above).

2.2
Thermal Properties of Mullite
H. Schneider

2.2.1
Thermochemical Data

2.2.1.1   Enthalpy, Gibbs Energy and Entropy
Although mullite is a phase of outstanding importance in high-temperature technology, no systematic overview of the thermochemical properties has yet been published. Experimentally measured enthalpies ΔH_f of mullite formation from the oxides on the basis of calorimetry measurements are given in Table 2.2.1 (after Shornikov et al. 2003). Values range between 21.0 kJ mol^{-1} and 38.5 kJ mol^{-1} at room temperature (298 K). The calculated enthalpy of mullite formation from the oxides at room temperature from data of Barin et al. (1973) and Barin et al. (1993) is 23.32 kJ mol^{-1} and agrees well with the experimental results. The calculated enthalpy at high temperature decreases significantly to a value of about 17 kJ mol^{-1} at 1900 K (1627 °C), while the measured data, depending on the experimental technique, vary in a very wide range (see Table 2.2.1). The formation enthalpies ΔH_f of 3/2-mullite from kyanite, andalusite and sillimanite calculated by Holm and Kleppa (1966) from oxide melt solution calorimetry are 52.75 kJ mol^{-1}, 45.97 kJ mol^{-1} and 36.68 kJ mol^{-1}, respectively.

The calculated Gibbs energy ΔG_f value of mullite formation from α-alumina plus silica according to Barin et al. (1993) is 16.65 kJ mol^{-1} at room temperature (298 K, Fig. 2.2.1). It decreases with temperature to reach a negative value above about 750 K (477 °C). This indicates that mullite formation at temperatures above about 750 K is the energetically favorable process. Experimentally determined values of ΔG_f at 1823 K (1550 °C) according to Rein and Chipman (1965) and Kay and Taylor (1960) are −23.45 and −24.28 kJ mol^{-1}, respectively, while the calculated value at 1823 K based on data of Barin et al. (1993) is −30.0 kJ mol^{-1}. Calculated values of ΔG_f for mullite formation from γ-alumina and silica are negative throughout the temperature range of 298 K (25 °C) to 2000 K (1727 °C), indicating

Table 2.2.1 Experimentally measured enthalpies and entropies of mullite formation from the oxides at room temperature and at high temperature.

Temperature		Enthalpy	Entropy	Method of determination
[K]	[°C]	ΔH_f [kJ mol^{-1}]	ΔS_f [J mol^{-1} K^{-1}]	
298	25	29.5	79.0	Solution calorimetry
298	25	38.5	–	Solution calorimetry
298	25	30.7	–	Bomb calorimetry
298	25	21.0	–	Solution calorimetry
968	695	22.8	–	Solution calorimetry
968	695	27.3	–	Solution calorimetry
973	700	40.7	–	Solution calorimetry
1617	1344	−4.1	7.2	Knudsen mass spectroscopy
1773	1500	111.0	75.5	Heteregeneous equilibria in slags
1800	1527	−4.35	10.45	Heterogeneous equilibria in slags
1933	1660	79.5	69	Knudsen mass spectroscopy
1950	1677	64.5	58	Knudsen mass spectroscopy

Data are taken from Shornikov et al. (2003). Note that Shornikov and coworkers gave their thermodynamic data on the basis of 0.6 Al$_2$O$_3$ · 0.4 SiO$_2$ units, while the present values refer to 3Al$_2$O$_3$ · 2SiO$_2$ units. This means that Shornikov's data have to be multiplied by five. For references see Shornikov et al. (2003).

the stability of mullite with respect to γ-alumina plus silica over the whole temperature field. Shornikov et al. (2003) describe separate linear relationships between temperature and ΔG_f below and above about 1600 K (1327 °C).

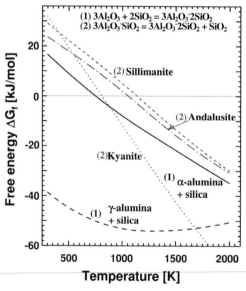

Fig. 2.2.1 Free energies of mullite formation (ΔG_f) from oxides and from aluminosilicates. Curves have been calculated from data of Barin (1987).

The occurrence of linear temperature-Gibbs energy relationships has been taken as evidence for the considerable energetic stability of mullite at high temperature caused by constant entropies. According to Shornikov et al. (2003) the Gibbs free energy becomes positive below 950 K (677 °C), which means that the phase can not form under equilibrium conditions. In spite of this thermodynamic condition, decomposition of mullite below 950 K is unlikely, because of the high activation energy of the process. Calculated values of ΔG_f of mullite formation from the aluminosilicates kyanite, andalusite and sillimanite on the basis of the thermo-chemical data of Barin et al. (1993) are also shown in Fig. 2.2.1. The graph demon-strates that from a theoretical point of view andalusite and sillimanite are stable with respect to mullite plus silica below about 1050 K (777 °C) and 1130 K (857 °C), while kyanite should decompose to mullite plus silica above about 810 K (537 °C). Although the experimentally measured transformation temperatures of the alumi-nosilicates all are much higher, the sequence of decomposition temperatures also follows the same sequence, i.e. kyanite < andalusite < sillimanite (Schneider and Majdic, 1979, 1980, 1981).

Shornikov and Archakov (2002) estimated the enthalpy of fusion ΔH_{fus} of mul-lite on the basis of enthalpies and entropies of melt formation in the system Al_2O_3-SiO_2, determined experimentally by Knudsen mass spectrometry. Their value of 372.5 kJ mol^{-1} is in satisfactory agreement with the previous estimates of Berezh-noi (1970, i.e. 428.5 kJ mol^{-1}) and Bubkova (1992, i.e. 320 kJ mol^{-1}, both publica-tions cited in Shornikov and Archakov, 2002), but differs considerably from the value given by Barin and Knacke (1973, i.e. 188.5 kJ mol^{-1}). It also differs from the estimate of Nordine and Weber (pers. comm., i.e. 248 kJ mol^{-1}), derived from undercooling measurements on liquids with mullite composition. The endother-mic character (positive ΔH_{fus}) and the high temperature of melting again account for the very high thermal stability of mullite.

On the basis of free energy of formation values of Rain and Chipman (1965) and the high temperature heat content data of Pankratz et al. (1963), Holm and Kleppa (1966) calculated the standard entropy of 3/2-mullite to be $S°_{298K} = 269.76$ J mol^{-1} K^{-1}). This result agrees well with the values given in the JANAF Thermochemical Tables ($S°_{298K} = 275.07$ J mol^{-1} K^{-1}, JANAF 1965–1968), and by Barin et al. (1993, $S°_{298K} = 274.89$ J mol^{-1} K^{-1}, 1600–2023 K), Robie and Heming-way (1995, $S°_{298K} = 275.00$ J mol^{-1} K^{-1}, 1400–1800 K), Chase (1998, $S°_{298K} = 274.25$ J mol^{-1} K^{-1}, 1700–2180 K) and by Kirschen et al. (1999, $S°_{298K} = 274.90$ J mol^{-1} K^{-1}, 1600–2180 K). It is, however, in serious disagreement with the value of Pankratz et al. (1963, $S°_{298K} = 254.56$ J mol^{-1} K^{-1}), calculated on the basis of low temperature heat capacity measurements. Holm and Kleppa attributed this discrepancy to the entropy of aluminum-silicon disorder in mullite, since the value of Pankratz and coworkers did not consider the configurational entropy associated with distributing the aluminum and silicon atoms over the tetrahe-drally coordinated lattice positions. If all the tetrahedrally coordinated aluminum atoms mix randomly with all silicon atoms, this entropy contribution amounts to $-(16/3)R[(5/8)\ln(5/8) + (3/8)\ln(3/8)] = 29.31$ J mol^{-1} K^{-1}, (298 K). Holm and Kleppa stated that this value is clearly too large. Better agreement is obtained if

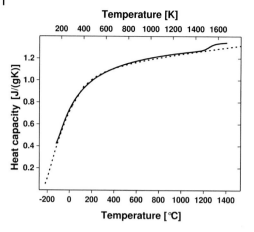

Fig. 2.2.2 Experimental heat capacity curves of mullite plotted versus temperature. Solid line: Hildmann and Schneider (2004); dashed line: Pankratz et al. (1963).

partial ordering of aluminum and silicon atoms according to $^{[4]}Al^{[6]}Al[^{[4]}Al_{1/4}$ $^{[4]}Si_{3/4}]O_{39/8}$ is considered. In a random mixing approximation, this model gives a configurational entropy of $(-8/3)R((1/4)\ln(1/4) + (3/4)\ln(3/4)\ln3/4) =$ 12.56 J mol^{-1} K^{-1}. Holm and Kleppa concluded that some of the tetrahedrally co-ordinated aluminum atoms in the mullite structure do not randomly exchange with silicon. Entropies of mullite formation from the oxides (ΔS_f) determined on an experimental basis, were provided by Shornikov et al. (2003, see Table 2.2.1). It is difficult to evaluate the temperature dependence of entropies, owing to the considerable spread of values. However, Shornikov et al. believe that the entropies at high temperature are constant at about 60 J mol^{-1} K^{-1}.

2.2.1.2 Heat Capacity

Experimentally determined heat capacities (c_p, at constant pressure) of mullite were first published by Pankratz et al. (1963), while Barin and Knacke (1973) calculated heat capacity data. In a recent study Hildmann and Schneider (2004) determined the heat capacity of mullite in a temperature range between −125 and 1400 °C. Their heat capacity master curve fits well with previously published data (Fig. 2.2.2). Heat capacities range between 0.415 J g^{-1} K^{-1} at −125 °C, 0.78 J g^{-1} K^{-1} at room temperature and 1.25 J g^{-1} K^{-1} at 1000 °C. Hildmann and Schneider identified a step-like, yet unknown, heat capacity increase above about 1100 °C. The anomaly has been ascribed to a temperature-induced phase transformation of mullite (see Section 2.5).

2.2.2
Thermal Expansion

One of the most important properties of mullite is its low thermal expansion and the associated excellent resistance to sudden temperature changes (thermal shock). Consequently intense work has been performed to quantify the thermal expansion behavior of mullite.

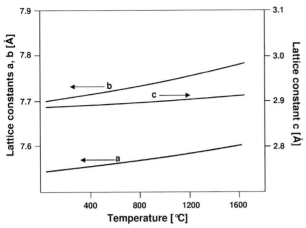

Fig. 2.2.3 Temperature-dependent evolution of the lattice constants *a*, *b* and *c* of mullite measured by in situ high temperature X-ray diffractometry of a "low" alumina (72 wt. % Al_2O_3, 28 wt. % SiO_2) mullite ceramic (after Schneider and Schneider, 2005).

Table 2.2.2 Thermal expansion coefficients of mullite.

Mullite sample	Temperature range (°C)	Expansion coefficients [$\times 10^{-6}$ °C^{-1}]				Reference
		a axis $\alpha(a)$	b axis $\alpha(b)$	c axis $\alpha(c)$	Mean $\bar{\alpha}^{a}$	
Undoped	300–1000	4.1	6.0	5.7	5.3	Brunauer et al. (2001)
		3.9	7.0	5.8	5.6	Schneider and Eberhard (1990)
	1000–1600	6.8	9.3	6.3	7.5	Brunauer et al. (2001)
	800–1300	5.4	7.1	6.5	6.3	Schneider and Schneider (2005)
	1300–1600	5.3	10.4	5.9	7.2	Schneider and Schneider (2005)
Cr-doped	300–1000	3.6	5.9	5.2	4.9	Brunauer et al. (2001)
		3.1	6.2	5.6	5.0	Schneider and Eberhard (1990)
	1000–1600	5.8	11.0	6.1	7.6	Brunauer et al. (2001)

a) Mean thermal expansion coefficients are given by the equation:
$$\bar{\alpha} = (\alpha\,(a) + \alpha\,(b) + \alpha\,(c))/3$$

Structure-resolved thermal expansion coefficients of mullite, undoped and doped with 10 wt. % Fe_2O_3 and Cr_2O_3, were measured by Schneider and Eberhard (1990) up to 900 °C, and Brunauer et al. (2001) and Schneider and Schneider (2005) up to 1600 °C using in situ high temperature powder X-ray and neutron diffractometry. Schreuer et al. (2005) provided data on the basis of single-crystal dilatometer measurements between about −175 and 1425 °C. The expansion parallel to the crystallographic **a** axis is low, while it is higher parallel to **b** and **c** (Table 2.2.2). At temperatures below about 300 °C the thermal expansion is very low and non-linear, and coefficients $\alpha_{[010]}$ and $\alpha_{[001]}$ are similar, while above 300 °C up to about 1000 °C it increases more or less linearly, with the expansion coefficient $\alpha_{[010]}$ being slightly higher than $\alpha_{[001]}$ and especially $\alpha_{[100]}$ (see Fig. 2.2.3 and Table 2.2.2). Between about 1000 and 1300 °C expansion coefficients change discontinuously, which has been attributed to a phase transformation of mullite (see above and Section 2.5.2).

With increasing Al_2O_3 contents (*x*-values) of mullite $\alpha_{[100]}$ and $\alpha_{[001]}$ coefficients increase, while $\alpha_{[010]}$ decreases. Schneider and Eberhard believe that the composition-controlled change of expansion coefficients is associated with Al* and O(C) oxygen vacancy distributions. This approach yields further evidence by the suggestion that oxygen vacancies form favorably along **b** with the Al_2O_3 content of mullite and increase the expansion in this lattice direction (see Section 1.3). Iron (Fe^{3+}) and chromium (Cr^{3+}) incorporation into mullite reduces the overall expansion, essentially by decreasing the α_a coefficient. The substitution of Al^{3+} by Fe^{3+} and Cr^{3+} may cause an expansion of the oxygen octahedra (MO_6) with an almost full pre-stressing of the relatively short and inelastic M−O(A) bonds. This has the effect that these bonds, which lie close to the **a** and **c** axes, can not expand further and behave rigidly upon heat treatment, whereas the longer and more elastic M-O(D) bonds lying close to **b** are still able to expand.

Schreuer et al. (2005) on the basis of mullite single crystal dilatometer measurements found that below about 300 °C the expansion of mullite gradually becomes lower and approaches zero near −100 °C. The low expansion of mullite at low temperature is consistent with the fact that atoms near to the absolute zero temperature (0 K, −273 °C) vibrate symmetrically around their balance points (symmetrical potential curves), and thus do not contribute to the thermal expansion. With an increase in temperature the potential curve of mullite becomes more and more asymmetric, causing an increase in the thermal expansion.

Since no conclusive high temperature single crystal structure refinement of mullite is available so far, the principles of expansion up to 1000 °C are deduced from the behavior of sillimanite, which is closely related to mullite (see Section 1.1.3). Winter and Ghose (1979) demonstrated that the thermal expansion of sillimanite is mainly controlled by the elongation of the long and elastic Al–O(D) (i. e. Al1–O2, see Section 1.1.3) bonds of the octahedra. Since the Al–O(D) bonds form an angle of about 30° with the **b** axis but 60° with the **a** axis, this has the effect of a stronger **b**- than **a**-axis lengthening. The suggestion of Winter and Ghose that **b**-axis expansion is reinforced by clockwise and counter-clockwise rotations of the polyhedra into the structural channels of sillimanite seems to be less probable. This is derived from the theory of rigid unit modes (RUM, Hammonds et al. 1998), which says that the stiff tetrahedral networks in silicates show little tendency for polyhedra rotations, if they are stabilized by octahedra as in the case of sillimanite. We believe that the structural expansion of mullite follows the same rules as for sillimanite.

Brunauer et al. (2001) performed high-temperature X-ray and neutron diffracto-metry of mullite from room temperature up to 1600 °C, while Schneider and Schneider (2005) carried out in situ high-temperature X-ray diffraction studies on mullite in a mirror-heated furnace up to 1600 °C. Both groups found the strongest expansion parallel to **b** followed by **c** and **a** (Table 2.2.2) They also observed an expansion anomaly above 1000 °C, with a related increase of the lattice expansion mainly parallel to the crystallographic **b** axis, while effects parallel to **a** and **c** are less significant. The discontinuous development of lattice spacings has been asso-ciated with a high-temperature phase transformation of mullite, which is dis-cussed separately in Section 2.5.2.

2.2.3
Thermal Conductivity

The thermal diffusivity D_{th} ("temperature conductivity") of mullite single crystals determined by Hildmann and Schneider (2005) with the laser flash technique is shown in Fig. 2.2.4. The thermal diffusivity is anisotropic being much higher par-allel to the crystallographic **c** axis ([001]) than parallel to **a** [(100)] or **b** [(010)], i.e. $D_{th[001]} \gg D_{th[100]} > D_{th[010]}$. The thermal diffusivity strongly decreases from about 0.032 cm^2 s^{-1} ($D_{th[001]}$) and 0.022 cm^2 s^{-1} ($D_{th[100]}$, $D_{th[010]}$) at room temperature to reach nearly constant values above about 700 °C ($D_{th[001]}$: about 0.013 cm^2 s^{-1}, $D_{th[100]}$, $D_{th[010]}$: about 0.008 cm^2 s^{-1}). There is no significant change of the aniso-tropy with temperature.

The thermal conductivity λ is related to the thermal diffusivity D_{th} by the equation:

$$\lambda = D_{th} \, \rho \, c_p \tag{1}$$

where ρ is the density and c_p the heat capacity at constant pressure.

The thermal conductivity displays a similar variation with temperature as does the thermal diffusivity ($\lambda_{[001]} \gg \lambda_{[100]}$, $\lambda_{[010]}$, Fig. 2.2.5). According to Hildmann and Schneider (2005) mullite is characterized by a low mean thermal conductivity of about 6 W m^{-1} K^{-1} at room temperature. With an increase in temperature it fur-ther decreases to reach a nearly constant value of about 3.5 W m^{-1} K^{-1} above about 800 °C. These data correspond well with those on the basis of mullite ceramics published by Dreyer (1974) and Kingery et al. (1976), although the shape of the temperature-dependent thermal conductivity curve may be different. The thermal conductivity may be further decreased by doping mullite with foreign cations (e.g. Cr^{3+} and Fe^{3+}), which produce additional scattering centers in the structure.

Thermal diffusivity and conductivity values reflects the anisotropy of the mullite structure: Along the stiff and strongly bonded tetrahedral chains running parallel to the crystallographic **c** axis ([001]) the thermal conductivity should be higher than parallel to the **a** ([100]) and **b** axes ([010]) in the (001) plane, where there exists a sequence of "strongly" bonded tetrahedra and the more "weakly" bonded octahe-dra (see also Section 2.1.2). This actually is the case (Fig. 2.2.5).

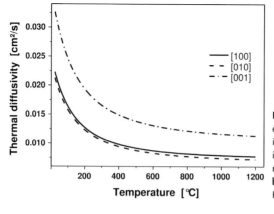

Fig. 2.2.4 Temperature-dependent evolution of the thermal diffusivity D_{th} ("temperature conductivity") of single crystal mullite measured along the **a** ([100]), **b** ([010]) and **c** ([001]) axes (after Hildmann and Schneider, 2005).

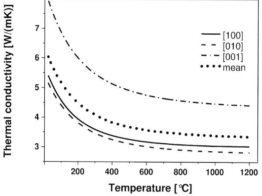

Fig. 2.2.5 Temperature-dependent evolution of the thermal conductivity λ of mullite single crystal determined for directions parallel to the **a** ([100]), **b** ([010]) and **c** ([001]) axes. The mean value of the thermal conductivity has been included for comparison (after Hildmann and Schneider, 2005).

2.2.4
Atomic Diffusion

2.2.4.1 Oxygen Diffusion

The anisotropic character of the mullite structure and the number of oxygen vacancies, increasing in going from 3/2-mullite (1 oxygen vacancy per 4 unit cells) to 2/1-mullite (1 oxygen vacancy per 2.5 unit cells) implies a structural control of oxygen diffusion. However, oxygen ^{18}O tracer diffusion and oxygen ^{18}O to ^{16}O isotope exchange measurements carried out on mullite single crystals along the crystallographic axes **b** ([010]) and **c** ([001]) in the range 1100 to 1600 °C yield only a small variation of diffusion velocities (Ikuma et al. 1999, Fielitz et al. 2001a,b; Fig. 2.2.6). This militates against a simple oxygen hopping from one vacancy to the other, which should favor **c** axis ([001]) diffusion. Further evidence against a vacancy-hopping-controlled oxygen diffusion can be derived from a comparison of diffusion data for mullite and zirconia. In calcium-stabilized zirconia, where oxygen diffusion definitely takes place by migration of structural vacancies, the diffusion coefficients are eight orders of magnitude higher than in mullite at 1200 °C.

Also, migration enthalpies in zirconia are much lower (below 1.3 eV) than in mullite (about 4.5 eV, Fielitz et al. 2001a). Since oxygen in mullite obviously does not diffuse over initially existing vacancies, new mobile vacancies have to be formed by thermal activation. It is not clear whether these are the pre-existing immobile vacancies, which are set in motion at high temperature, or if new vacancies are formed. The diffusion data of Fielitz and coworkers are in some contrast to the earlier results of Fischer and Janke (1969) and of Rommerskirchen et al. (1994), who, on the basis of electromotive force measurements (EMF), predicted an ionic conductivity that at 1400 °C was supposed to be higher than that of zirconia.

Oxygen diffusion in polycrystalline mullite is more complex than in single crystals, since it displays simultaneous volume and grain boundary migration. Fielitz et al. (2003a) performed gas/solid ^{18}O to ^{16}O oxygen exchange experiments and determined grain boundary diffusion coefficients that are five orders of magnitude higher than the mullite lattice diffusivities (Fig. 2.2.7). The influence of composition and microstructure on the oxygen diffusivity has been evaluated for low (72 wt. % Al_2O_3) and high alumina (78 wt. % Al_2O_3) polycrystalline mullite ceramics. A faster grain boundary diffusion with a lower activation enthalpy (about 360 kJ mol^{-1}) in the low than in the high alumina material (about 550 kJ mol^{-1}) has been reported. Fielitz et al. (2003a) explain the different diffusivities of these ceramics by their different microstructures. In low alumina ceramics some silica-rich glass coexists with mullite at grain triple junctions, whereas mullite/mullite grain boundaries are more or less glass-free (Figs. 2.2.8a,b). The high alumina materials are virtually glass-free. Impurity atoms like alkalis and iron are concentrated in the glassy phase in the low alumina material, whereas they are more homogeneously distributed over the grain boundaries in high alumina mullite. Fielitz et al. (2003a) believe that in the latter case the impurities act as oxygen traps and thus hinder oxygen migration in the ceramics. As impurities are concentrated in grain triple junctions in the low alumina material mullite/mullite grain bounda-

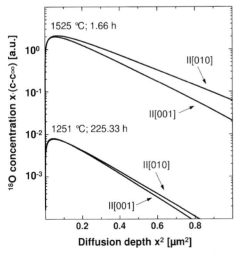

Fig. 2.2.6 Temperature-dependent oxygen ^{18}O diffusion profiles of single crystal mullite parallel to the **b** ([010]) and **c** ([001]) axes. The curves represent best fits of the experimental data by using Fick's law modified by Crank (1975) (a. u. = arbitrary units, after Fielitz et al. 2001a).

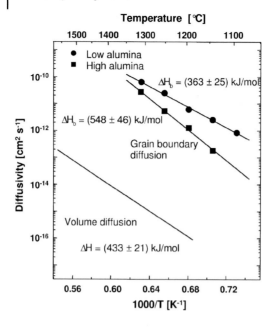

Fig. 2.2.7 Temperature-dependent grain boundary oxygen ^{18}O diffusion coefficients in "low" (72 wt. % Al_2O_3, 28 wt. % SiO_2) and "high" alumina (78 wt. % Al_2O_3, 22 wt. % SiO_2) mullite ceramics. Data are shown in comparison to single crystal mullite volume diffusion. Activation energies of diffusion are also given (after Fielitz et al. 2003a).

ries are more or less impurity-free and enhanced oxygen diffusion occurs in these ceramics.

The results of Fielitz et al. agree with transmission electron microscopy observations of Kleebe et al. (1996), although their interpretation of data is different. Kleebe and coworkers describe thin grain boundary films around the crystals in mullite ceramics of low alumina bulk composition. Although these grain boundary films tend to be nano-sized (<1 nm), their actual thickness depends on the impurity content of the materials, the wetting behavior of the grain boundary phase on specific faces of mullite and on the bulk SiO_2 content. According to Kleebe et al. (1996) diffusion is enhanced in these grain boundary films. As a result of the oxygen diffusion experiments it can be stated that high alumina polycrystalline mullite coatings are more suitable for oxidation protection of non-oxide substrates (e. g. silicon nitride or silicon carbide) than low alumina mullite coatings.

2.2.4.2 Silicon Diffusion

Data on silicon diffusion are rare in literature, because diffusivities are very slow and appropriate silicon tracer isotope measurements are expensive. In spite of these difficulties silicon tracer diffusion in single-crystal 2/1-mullite has been measured by Fielitz et al. (2003b) parallel to the crystallographic **b** ([010]) and **c** axes ([001]). A thin aluminosilicate layer, enriched in ^{30}Si silicon, was deposited for this purpose on mullite (010) and (001) faces by argon ion beam sputtering. Fielitz et al. (2003b) found no significant structure control of silicon diffusion in mullite (Fig. 2.2.9). Isotropic Si^{4+} diffusion is consistent with its relatively high activation

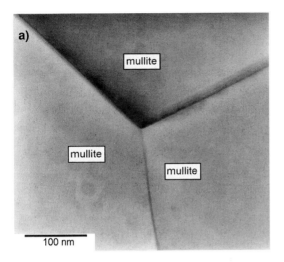

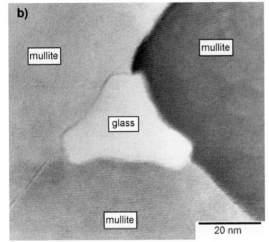

Fig. 2.2.8 Transmission electron micrographs of (a) "High" alumina (78 wt. % Al$_2$O$_3$, 22 wt. % SiO$_2$) and (b) "Low" alumina (72 wt. % Al$_2$O$_3$, 28 wt. % SiO$_2$) mullite ceramics. Note the occurrence of glass pockets in triple point junctions in the "low" but not in the "high" alumina material (Courtesy P. Fielitz).

energy ($\approx$ 600 kJ mol^{-1} between 1300 and 1525 °C). The diffusivity of silicon is about two orders of magnitude slower than that of oxygen, which has been explained by the strong covalent Si-O bonding in mullite.

2.2.4.3 Aluminium Diffusion

Preliminary investigations of the research group of P. Fielitz yielded relatively high diffusities of aluminium in the order of that of oxygen.

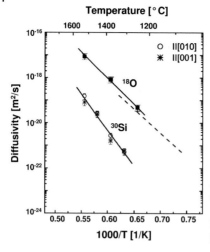

Fig. 2.2.9 Temperature-dependent oxygen ^{18}O and silicon ^{30}Si diffusion coefficients of single-crystal mullite parallel to the **b** ([010]) and **c** ([001]) axes. Dashed line: oxygen ^{18}O diffusion in quasi single crystal mullite (approx. 3/2-composition) (after Fielitz et al. 2001a and 2003b).

2.2.5
Grain Growth

Electron micrographs were used by Konopicky et al. (1965) for an early and qualitative study of the crystal growth of mullite produced by decomposition of argillaceous refractory raw materials. According to Konopicky et al. mullite development is controlled by the size and structural imperfections of the starting materials at relatively low temperature (below 1300 °C), while it depends on the content of coexisting impurities (i. e. fluxes) at high temperature (above 1300 °C). A reciprocal dependence between acicular shape and size of these newly formed mullites has been reported.

The kinetics of densification and grain growth of mullite ceramics has been studied by Ghate et al. (1975), who believe that silicon (Si^{4+}) diffusion in particular has a rate-controlling influence on mullite grain growth. Huang et al. (2000) investigated the grain growth of dense mullite ceramics above 1550 °C. These ceramics display slow and isotropic mullite grain growth below the eutectic temperature (1590 °C) in the system Al$_2$O$_3$–SiO$_2$. Above this temperature limit strong and anisotropic grain growth occurs, which is explained by the enhanced liquid-phase-assisted diffusion.

Qualitative analyses of the temperature-induced mullite grain growth in commercial aluminosilicate fibers (Nextel 440, Nextel 720, Sumitomo Altex 2k, see section 6) have been provided by Schmücker et al. (1996, 2001) and Schneider et al. (1998). A quantitative study of mullite grain growth kinetics has recently been carried out between 1300 and 1700 °C by Schmücker et al. (2004) on laboratory-produced mullite fibers with stoichiometric composition (about 72 wt. % Al$_2$O$_3$, about 28 wt. % SiO$_2$) in comparison to commercial Nextel 550 and 720 aluminosilicate fibers. At 1300 °C the nano-sized mullites still display the as-received mosaic-type microstructure, which is typical for mullites formed from diphasic (type II) mullite precursors (see Schneider et al. 1998, and Section 1.4.2). Schmücker et al.

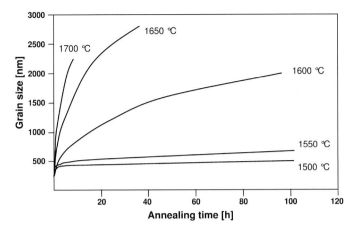

Fig. 2.2.10 Temperature- and time-dependent grain growth curves of mullite in aluminosilicate fibers (after Schmücker et al. 2005a).

(2005a) proved that mullite grain growth is very slow up to 1600 °C, although recrystallization and elimination of intragranular pores have been observed. Above about 1600 °C, however, strong grain growth occurs (Fig. 2.2.10). The driving force for the grain growth of mullite is the reduction of grain surface areas by reduction of the grain surface "curvatures" which decreases during grain growth. The diffusion-controlled process follows the parabolic law:

$$D^{1/n} - D_o^{1/n} = k\,t \tag{2}$$

where D is the grain size after heat treatment, D_o the starting grain size, t the duration of heat treatment, k the reaction constant, and n the grain growth exponent.

Grain growth exponents n of mullite have been determined from $\ln D$ versus $\ln t$ plots (Fig. 2.2.11). Ideally n is 2. In practice, n is often higher, typically with values up to 4, as in the case of α-alumina. Things are different for mullite. Up to about 1600 °C slow grain growth of mullite occurs with grain growth exponents up to 13. Above this temperature grain growth exponents decrease drastically displaying "normal" values between 2.5 and 3. The high grain growth exponents of mullite in the "low" temperature regime correspond to a low grain boundary mobility. This can be explained by foreign atom segregation (solute drag), agglomerations of pores (pore drag) and second phases (second-phase drag) at the phase boundaries of mullite (see also Malow and Koch, 1997). The microstructural studies of Schmücker et al. (2005a), however, yield no evidence for any of these grain growth retarding mechanisms. Therefore, as yet unknown structure-controlled processes have to be taken into account for the different grain growth exponents in the "low" and "high" temperature ranges.

Temperature-induced grain growth is an inherent effect, which on the one hand

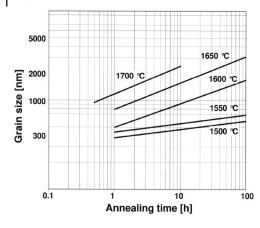

Fig. 2.2.11 Logarithmic time- and temperature-dependent evolution of the grain size of mullite in aluminosilicate fibers (after Schmücker et al. 2005a).

decreases the strength and may cause formation of critical flaws, but on the other hand improves the creep resistance of ceramics. A possible way to reduce grain growth in mullite ceramics is based on the segregation of secondary phases at the grain boundaries of mullite. The efficiency of grain growth inhibitors improves as their particle size decreases and volume fraction increases, and with their homogeneous distribution over the mullite's grain boundaries. Decorating phases suitable for reducing mullite grain growth should not react with mullite under either processing or application conditions (temperature, atmosphere). Ceria (CeO_2) has a high potential to serve as a grain growth inhibitor in mullite ceramics (see Mechnich et al. 1999).

2.2.6
Wetting Behavior

Braue et al. (2005b) performed reactive wetting experiments in mullite-glass systems. Single crystal substrates cut parallel to (100), (010) and (001) and a polycrystalline material are brought in contact with two glasses – a highly reactive yttrium aluminosilicate glass (42 wt. % Y_2O_3, 25 wt. % Al_2O_3, 33 wt. % SiO_2, designated as YAS glass) and a less reactive boron sodia aluminosilicate glass (19 wt.% B_2O_3, 4.5 wt. % Na_2O, 2 wt. % Al_2O_3, 74 wt. % SiO_2, designated as BS glass). Apparent contact angles between substrate and glass (Fig. 2.2.12) have been measured from 1100 to 1600 °C. Independently of the glass composition, the apparent contact angles (Θ_{app}) decrease on the single crystal faces in the following sequence: $\Theta_{(010)} > \Theta_{(100)} > \Theta_{(001)}$. The angles of polycrystalline samples roughly correspond to the average value of the single-crystal specimens (Figs. 2.2.13 and 2.2.14). The anisotropic character of reaction wetting is stronger in the case of the highly reactive yttrium aluminosilicate glass/mullite than in the less reactive boron sodia aluminosilicate glass/mullite pairs.

Reactive wetting of mullite by silicate melts is complex and cannot be adequately explained by the classical wetting theory. This is because the composition of the liquid in the interfacial region changes continuously with the duration of the pro-

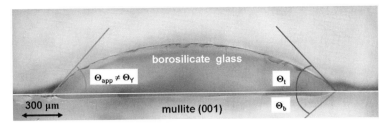

Fig. 2.2.12 Optical micrograph from a cross-section through a boron sodium aluminosilicate glass (BS) deposited on a mullite (001) substrate, revealing the flat morphology of the liquid/solid interface established after reactive wetting. The white line indicates the mullite substrate surface prior to the sessile drop experiment (Θ = wetting (contact) angle. Courtesy W. Braue).

cess owing to gradual dissolution of the substrate, local chemical reactions and precipitation of devitrification products. Therefore, the local surface energies can differ considerably from the surface energies of the starting mullite-glass systems. It has been suggested that the mullite-liquid interfacial energy (γ_{SL}) is reduced by possibly occurring reactions, leading to a drop of the contact angle.

Although the wetting anisotropy depends on the composition and reactivity of the glass system used (see Figs. 2.2.13 and 2.2.14 and above) the anisotropy of surface energies on (100), (010) and (001) faces seems to control the wetting behavior. While the solid-liquid interfacial energy γ_{SL} is most likely reduced during any interfacial reaction, its orientation dependence remains unclear. Braue et al. (2005b) suggest that the solid-vapor interfacial energy γ_{SV} represents the dominant orientation-dependent term in the wetting behavior of mullite. For a wetting system with an apparent wetting angle $\Theta < 90°$, increasing the solid-vapor energy γ_{SV} but keeping other factors constant results in a decrease of apparent contact angle. The following orientational dependence of the solid-vapor interfacial energies γ_{SV} can be anticipated from Figs. 2.2.13 and 2.2.14: $\gamma_{SV(010)} < \gamma_{SV(100)} < \gamma_{SV(001)}$.

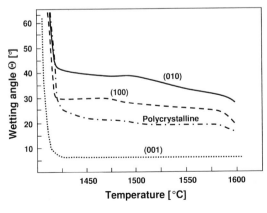

Fig. 2.2.13 Apparent wetting (contact) angles Θ of single crystal mullite substrates cut parallel to (100), (010) and (001) and of a polycrystalline mullite ceramic. The experiments were performed with a highly reactive yttrium aluminosilicate glass (YAS) at temperatures up to 1600 °C (after Braue et al. 2005b).

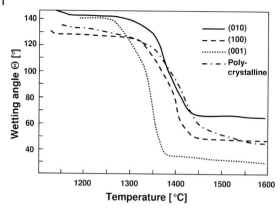

Fig. 2.2.14 Apparent wetting (contact) angles Θ of single crystal mullite substrates cut parallel (100), (010) and (001) and of a polycrystalline mullite ceramic. The experiments were performed with a low reactive boron sodium aluminosilicate glass (BS) at temperatures up to 1600°C (after Braue et al. 2005b).

The anisotropy of surface energies can be rationalized in terms of the anisotropy of the crystal structure of mullite. Rigid tetrahedral chains extending parallel to the crystallographic **c** axis, which are stabilized against tilting by continuous chains of edge-sharing AlO_6 octahedra, give rise to a high stiffness of the structure in this direction (c_{33}) and to a relative high surface energy on (001). Perpendicular to the **c** axis the mullite structure exhibits no such chains. Consequently, stiffness coefficients parallel to the **a** and **b** axes (c_{11} and c_{22}) are lower (see Section 2.1.2), indicating that the anisotropy of the surface energies γ_{SV} does indeed follow the order $\gamma_{SV(010)} < \gamma_{SV(100)} < \gamma_{SV(001)}$.

2.3
Miscellaneous Properties
H. Schneider

2.3.1
Optical and Infrared Properties

The basic optical data of mullite have been summarized e.g. by Tröger (1982). The refractive indices are $\alpha = 1.630–1.670$, $\beta = 1.636–1.675$ and $\gamma = 1.640–1.691$, corresponding to a birefringence of 0.010 to 0.029 and an optical angle ($2V_\gamma$) between 45 and 61°. The wide range of optical properties corresponds to the variation of the Al_2O_3 content and the incorporation of foreign cations, especially Fe^{3+}, Cr^{3+} and Ti^{4+} (see Section 1.3.1). Pure and Ti^{4+}-doped mullites are colorless, while Fe^{3+} and Cr^{3+}-substituted mullites are reddish-brown and green in color, respectively. Fe^{3+}-plus Ti^{4+}-doped mullites show faint pleochroism owing to charge-transfer reactions between the transition metal cations.

The preparation of optically translucent mullite ceramics was first reported by Prochazka and Klug (1983) and further studies to improve the transmittance have been performed in following years (for references see Aksay et al. 1991). The optical transmittance of mullite up to a wavelength of 5 μm is very good (at a

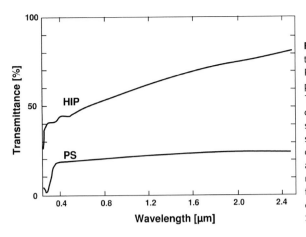

Fig. 2.3.1 Optical transmittance of mullite ceramics. PS: mullite ceramic pressureless sintered at 1700 °C, HIP: mullite ceramic pressureless sintered at 1700 °C and subsequently hot isostatically pressed at 1600 °C and 1.2 kbar. Starting material: commercial fused-mullite (Al_2O_3 content: 76 wt. %) (after Schneider et al. 1993a).

wavelength of 3 µm, for example, it is as high as 95 %). Above a wavelength of 5 µm, mullite displays an intrinsic cut-off of the transmittance. Schneider et al. (1993a) investigated the optical transmittance of pressureless-sintered and of a hot isostatically pressed mullite ceramics in the near infrared (NIR), visible light (VIS), and in the ultraviolet (UV) spectral ranges. In pressureless sintered specimens the transmittance is low in the visible and near infrared ranges (about 20 %). In optically translucent pore-free mullite ceramics a transmittance of 40 % has been measured in the visible range and up to 80 % in the near infrared range (Fig. 2.3.1). The ultraviolet transmittance of both pressureless sintered and mullite ceramics is low.

2.3.2
Electrical Properties

Mullite is an electrical insulator suitable as a substrate material for electronic devices (see Section 4.6.2). In spite of this fact, no comprehensive study exists on the electrical conductivity of mullite. Böhm and Schneider (unpublished results) compared the electrical conductivities σ of undoped and chromium-doped (1 wt. % Cr_2O_3) mullite single crystals, with the electrical field vector being parallel to the crystallographic **a** ($E//[100]$) and **c** ($E//[001]$) axes, respectively. The electrical resistivity of undoped mullite measured along the crystallographical **c** axis (electrical field parallel [001]) is about 2.9×10^7 Ω cm at 500 °C. At this temperature chromium-doped mullite shows a lower value with about 1.9×10^7 Ω cm parallel **c**. However, a strong amisotropy exists with a value of about 5.5×10^8 Ω cm along to the **a** axis (electrical field parallel [100], 500 °C). The anisotropy of resistivity and conductivity (Fig. 2.3.2) is also documented by the different activation energies of conduction (parallel to the **c** axis: 0.85 eV; parallel to the **a** axis: 1.5 eV; both values are for chromium-doped mullite). The electrical conductivity of chromium-doped mullite single crystals is slightly higher than that of undoped mullite.

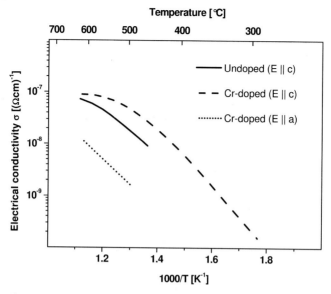

Fig. 2.3.2 Temperature-dependent electrical conductivity (σ) of mullite, undoped and chromium-doped (1 wt. % Cr_2O_3), in direction of the crystallographical **a** ([100]) and **c** axes ([001]). Note that the electrical conductivity of mullite parallel to **c** is one order of magnitude higher that parallel to **a**, while the activation energies A of the conductivity behave in the opposite way (according to Böhm and Schneider, unpublished results).

Data on the electrical conductivity of polycrystalline mullite has also been measured by Böhm and Schneider (unpublished results). They state that the electrical conductivity of polycrystalline mullite is more than one order of magnitude higher than that of single crystals parallel to the **c** axis, probably because the conductivity is enhanced by grain boundary effects. According to Chaudhuri et al. (1999) the electrical resistivity of mullite ceramics drops from about $10^{13}\ \Omega$ cm at room temperature, and $10^8\ \Omega$ cm at 500 °C, to $10^4\ \Omega$ cm at 1400 °C, i.e. a change of nine orders of magnitude. Doping mullite with transition metals reduces the resistivity of mullite to about $10^{11}\ \Omega$ cm at room temperature, and to about 0.2×10^4 at 1400 °C. Chaudhuri et al. (1999) mentioned that the ability to lower the resistivity of mullite is highest for Ti^{4+} but is lowest for Fe^{3+}. Since the transition metals essentially substitute Al^{3+} at octahedral lattice sites, the strong decrease in resistivity and associated increase of the electrical conductivity in titanium-doped mullite can be explained by hopping of electrons according to the equation:

$$Ti^{4+} + e^- \rightarrow Ti^{3+} \tag{3}$$

where e^- represents an electron

There is an old discussion on whether mullite at high temperatures displays ionic conductivity caused by oxygen vacancy hopping. Fischer and Janke (1969) and Rommerskirchen et al. (1994) mentioned that mullite ceramics feature ionic conductivities superior to that of calcia-stabilized zirconia solid electrolytes at temperatures above 1400 °C. On the other hand oxygen diffusion in mullite is rather slow (see Section 2.2.4), so there is little evidence for a high conductivity of mullite.

2.4
Structure-controlled Formation and Decomposition of Mullite
H. Schneider and M. Schmücker

2.4.1
Temperature-induced Formation

This section includes information on the structure-controlled formation of mullite from aluminosilicates involving epitactic or topotactic processes. Formation of mullite from the oxides or from other phases with no structure control are not described here (but see Section 4.1).

2.4.1.1 Formation from Kaolinite and Related Minerals
The series of reactions that make kaolinite and related minerals transform to 3/2-mullite have been intensively studied for several decades, because of their crystallo-chemical aspects and their importance for ceramic technology. An enormous number of publications exist dealing with the thermal decomposition of kaolinite and the formation of mullite. This review deals only to those papers that directly refer to the structure-controlled transformation process.

When kaolinite is heated, it transforms to mullite by several steps according to the following reaction scheme:

$$2(Al_2(OH)_4[Si_2O_5]) \rightarrow\ >550\ °C \rightarrow 2(Al_2Si_2O_7) + 4H_2O \tag{4}$$
$$\text{kaolinite} \qquad\qquad\qquad\qquad \text{metakaolin}$$

$$0.375(Si_8[Al_{10.67}\square_{5.33}]O_{32}) + SiO_2 \tag{5a}$$
$$\text{"}\gamma\text{-alumina"} \qquad\qquad \text{amorphous silica}$$

$$\rightarrow\ \geq 980\ °C$$

$$0.188(Al_8[Al_{13.33}\square_{2.66}]O_{32} + 4SiO_2 \tag{5b}$$
$$\text{"}\gamma\text{-alumina"} \qquad\qquad \text{amorphous silica}$$

$$\rightarrow\ \geq 1000\ °C \rightarrow 0.66(3Al_2O_3 \cdot 2SiO_2) + 2.68SiO_2 \tag{6}$$
$$\text{3/2-mullite} \qquad\qquad \text{amorphous silica}$$

$\square$ = vacancy

where equations (5a) and (5b) lead to alternative "γ-alumina" (spinel-type) phases containing or not containing silicon. Details are given below; the composition of the phases is from Bulens et al. (1978).

Brindley and Nakahira (1959a,b,c), Brindley and McKinstry (1961) and Gehlen (1962) were the first to suggest that a series of topotactic conversions occur in which each successive product is closely related structurally to the "low-temperature" phase. The differential thermal analysis (DTA) curves of heat-treated kaolinite (Chakraborty and Gosh, 1978a), with a strong endothermic peak at about 550 °C, a strong exothermic peak at about 950 °C and a weak exothermic peak at about 1250 °C, show that the phase dehydroxylates at about 550 °C to form metakaolin (Insley and Ewell, 1935, Roberts, 1945, Grimshaw et al. 1945). The removal of the hydroxyl groups is accompanied by a reorganization of the octahedral sheets of kaolinite to a tetrahedral configuration in metakaolin (see Iwai et al. 1971). The kaolinite-metakaolin transformation proceeds very slowly, and metakaolin has an extreme defect structure. About 20 vol.% of the metakaolin structure consists of lattice vacancies produced by the temperature-induced water release (Freund, 1967). However, the two-dimensional sheets of SiO_4 tetrahedra seem to stabilize the metakaolin network.

Metakaolin decomposes at about 950 °C to a γ-alumina (spinel-type) structure, non-crystalline free silica and some minor amount of mullite. Reaction details remained a subject of controversy. Different mechanisms were envisaged: Brindley and Nakahira (1959b,c), Chakraborty and Ghosh (1977, 1978a,b) and Srikrishna et al. (1990) proposed that metakaolin decomposes upon heating to an aluminum silicon spinel and additional silica. Compositions of the spinel phase proposed by Brindley and Nakahira, Chakraborty and Ghosh, and by Srikrishna et al. are: $2Al_2O_3 \cdot 3SiO_2$,i.e. $^{[4]}[Si_8]^{[6]}[Al_{10.67}\square_{5.33}]O_{32}$ ($\square$ = vacancy) and $3Al_2O_3 \cdot 2SiO_2$ i.e. $^{[4]}[Si_{4.92}Al_{3.08}]^{[6]}[Al_{11.69}\square_{4.31}]O_{32}$. Another possibility, discussed by Leonard (1977), Bulens et al. (1978), Percival et al. (1974) and Brown et al. (1985) on the basis of diffraction techniques, microscopy, spectroscopy and analytical methods suggest the formation of pure transition alumina (γ-alumina) with the composition: $^{[4]}[Al_8]^{[6]}[Al_{13.33}\square_{2.66}]O_{32}$ and additional silica (see also Section 1.4.2). On the other hand, Okada et al. (1986a) and Sonuparlak et al. (1987) described a γ-alumina phase with a small amount of silicon incorporation. The crystallite size of the spinel phase is very small, falling between 75 and 125 Å. Electron diffraction studies show that the (111) plane of the spinel phase is parallel to (001) of metakaolin (Comer, 1961). The formation of γ-alumina spinel, and the degree of silicon incorporation into this phase has also been the subject of many investigations studying the crystallization paths of mullite precursors (see also Section 1.4.2). Similar to the relationships during thermal decomposition of kaolinite, there are contradictory understandings of the degree of silicon incorporation into γ-alumina. Low and McPherson (1989) postulated a high degree of silicon incorporation into γ-spinel, corresponding to 2/1-mullite composition ($^{[4]}[Al_4Si_4]^{[6]}[Al_{12}\square_4]O_{32}$), while Wei and Halloran (1988), Hyatt and Bansal (1990) and Lee and Wu (1992) suggested that the γ-spinel was virtually pure alumina. Okada and Otsuka (1986) and Schneider et al. (1994) determined silicon incorporations up to 18 mol% SiO_2

($^{[4]}[Si_2Al_6]^{[6]}[Al_{12.67}\square_{3.33}]O_{32}$), and discussed temperature-dependent silicon incorporation mechanisms.

Mullite forms from large well-crystallized kaolinites at temperatures above 1000 °C, frequently with acicular morphology (Johnson and Pask, 1982). The mullites can have preferred orientation with the **c** axis parallel to <110> of the spinel phase (Brindley and Nakahira, 1959c, Comer, 1960, 1961). From disordered kaolinite, mullite crystallizes above about 1200 °C. In that case mullite initially has a rounded-up shape with random orientation. At higher firing temperatures (about 1300 °C), via a needle-like form, the mullite crystallites sometimes develop a hexagonal orientation on the former kaolinite sheets (Campos et al. 1976). Srikrishna et al. (1990) performed a transmission electron microscopy study on the transformation of kaolinite single crystals of high chemical purity and a high degree of structural order. They found that the newly formed mullite crystallites grew in an oriented way with respect to the kaolinite plates, with the **c** axis of mullite being perpendicular to the plates. Lee et al. (1999) presented a new approach to the mullite formation mechanism from kaolinite: According to them there is close structural control of the kaolinite to metakaolinite and metakaolinite to γ-alumina transformations. In their experiments mullite starts to crystallize above 940 °C, but no structural relationship with metakoalinite or the spinel phase has been found. Lee et al. believed that the breakdown of metakaolinite triggered mullitization. In this early stage mullitization is slow, owing to the coexistence of the spinel phase with mullite, the former acting as a silicon trap. These results of Lee and coworkers are in contrast to the earlier observations of Comoforo et al. (1948), Brindley and Nakahira (1959c), and Comer (1960, 1961), who described an oriented transformation. Their suggestion that nucleation and growth of mullite from chemically pure and well-crystallized kaolinite proceeds at a lower temperature than that from disordered kaolinite may give evidence for a preservation of the aluminum oxygen octahedral chains of the spinel phase during mullite formation in the case of ordered kaolinite. The following orientational relationships have been considered: $[010]_K//[010]_{MK}//[110]_\gamma//[001]_M$ (K = kaolinite, MK = metakaolinite, γ = γ-alumina, M = mullite).

Another interpretation of the orientational relationships between kaolinite and mullite is a structural breakdown of the intermediate γ-alumina and epitactical growth of mullite on kaolinite residuals, with the octahedral chains in kaolinite serving as nuclei for mullite (see the relationships during the thermal decomposition of muscovite, below). The hypothesis that mullite growth occurs preferentially along the ordered aluminum oxygen octahedral chains of the spinel phase agrees with the suggestion that these chains are most stable elements in the aluminosilicate minerals. Mullite formed at about 1000 °C has been designated as transitional mullite or primary mullite and its composition was proposed to be $SiO_2 \cdot Al_2O_3$ (Brindley and Nakahira, 1959c). This is certainly not correct, since it has been found that these mullites are richer in Al_2O_3 than stoichiometric mullite (see also Section 2.5.1).

The thermal decomposition of pyrophyllite ($4SiO_2 \cdot Al_2O_3 \cdot H_2O$) was studied by Heller (1962). Pyrophyllite decomposes to an anhydrous phase and converts to mullite plus silica at around 1000 °C. No intermediate phase such as γ-spinel was detected prior to mullitization in this case. McConville (1999) performed a comparative study on the temperature-dependent microstructural developments of natural kaolinite, illite and smectite. The breakdown of kaolinite occurred below 600 °C, while illite and smectite were stable up to 800 and 900 °C, respectively. The materials all contained high-impurity contents (especially sodium, calcium, titanium and iron) and therefore formed a large amount of glassy phase, from which mullite crystallized randomly.

MacKenzie et al. (1987) studied the thermal history of the sheet silicate muscovite. Muscovite first dehydroxylated into a phase of low crystallinity, where aluminum is predominantly fivefold coordinated. On heat treatment above about 1100 °C a feldspathoid phase and mullite form. The presence of iron enhanced the transformation process. MacKenzie and co-workers provided no information on the structural relationship between muscovite and mullite. Structural aspects of the high-temperature breakdown of muscovite to mullite have been investigated by Rodriguez-Navarro et al. (2003). Above 900 °C dehydroxylation of muscovite was followed by local melting. Within these melt pockets mullite crystals develop. At a later stage of annealing skeletal mullite crystals are frequent, formed by coalescence of smaller individuals (Figs. 2.4.1 and 2.4.2). The newly formed mullites are acicular with their **c** axes running parallel to <010> (or parallel to the symmetrically equivalent <310> directions) in (001) of muscovite. Rodriguez-Navarro et al. suggested that the edge-sharing AlO_6 octahedral chains in mullite run parallel to those

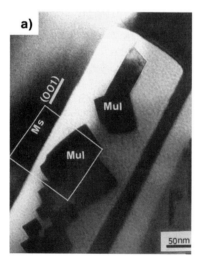
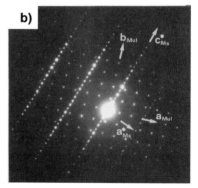

Fig. 2.4.1 Transmission electron micrograph of muscovite partially transformed to mullite. (a) Mullite (Mul) crystallization in melt pockets between (001) muscovite (Ms) basal planes; (b) Diffraction pattern of the framed area in (a). Note the existence of orientational relationships with [001]$_{Mul}$ running parallel to [010]$_{Ms}$ (Courtesy C. Rodriguez-Navarro).

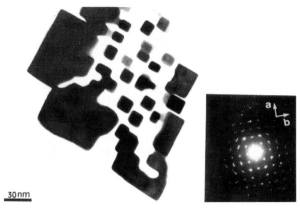

Fig. 2.4.2 Skeletal mullite crystal grown in a melt pocket during thermal decomposition of muscovite. The pseudo-crystal is cut perpendicular to the crystallographic **c** axis and has been formed by coalescence of smaller mullite individual crystals. The diffraction pattern (inset) shows slight rotations (misorientations) of individual crystallites (Courtesy C. Rodriguez-Navarro).

in muscovite. Their conclusion was that the octahedral units served as nuclei or templates for an eptactical growth of mullite on muscovite (see also the orientational relationships during the thermal decomposition of kaolinite to mullite, above). Although Rodriguez-Navarro et al. could not identify direct contacts between muscovite and mullite, this approach is reasonable, since the rarely occurring primary muscovite-mullite contacts could have been lost by secondary dissolution. A topotactic transformation, on the other hand, is less probable, because of the different crystal structures of muscovite and mullite. Interestingly, mullite formed from muscovite has the stoichiometric composition in spite of its relatively low crystallization temperature (below about 1100 °C), whereas mullites crystal-

Table 2.4.1 Data for the high temperature transformation of andalusite and sillimanite powders (≤ 40 μm) to 3/2-mullite plus silica

	Kyanite	Andalusite	Sillimanite[a]
Beginning of transformation [°C]	1150	1250	1300
Completion of transformation [°C]	1300	1500	1700
Width of transformation interval [°C]	150	250	400
Reference	Schneider and Majdic (1980)	Schneider and Majdic (1979)	Schneider and Majdic (1981)

The data for the kyanite high transformation are given for comparison.
a) The transformation temperature of sillimanite is rather low, probably due to its relatively high K_2O content (1.36 wt. %).

lized from kaolinite, and also from gels and glasses in the same temperature range, are characterized by much higher Al_2O_3 contents (Okada and Otsuka, 1988, see also Section 2.5.1). The reason possibly is the fitting of nuclei on the specific muscovite faces, which in the case of 3/2-mullite obviously is better than for 2/1-mullite.

2.4.1.2 Formation from Andalusite and Sillimanite

The polymorphic aluminosilicate minerals andalusite and sillimanite (all of the composition $Al_2O_3 \cdot SiO_2$ or Al_2SiO_5), transform to 3/2-mullite plus silica on heating at high temperature under oxidizing conditions according to the equation:

$$3Al_2SiO_5 \rightarrow 3Al_2O_3 \cdot 2SiO_2 + SiO_2 \qquad (7)$$

andalusite, 3/2-mullite   amorphous silica,
sillimanite   cristobalite

Going from andalusite to sillimanite, an increase of the transformation temperature and of the width of the transformation interval is observed (Table 2.4.1).

Structural transformation mechanisms Sillimanite and andalusite are structurally closely related to mullite, both belonging to the mullite family (see Section 1.1.3), and epitactic or topotactic transformations to mullite can be considered. Kyanite, does not belong to the mullite structural family and thus is not included in this section.

Andalusite. The decomposition of andalusite (A) to 3/2-mullite (M) is structure-controlled with the (201) lattice planes of mullite lying parallel to (011) and (101) of andalusite. The **c** axes of andalusite and mullite are parallel, and the **a** and **b** axes of the two phases are interchanged (due to the different unit cell settings, see Section 1.1.3 and Pannhorst and Schneider 1978). Hülsmans et al. (2000a,b) performed a detailed scanning and transmission electron microscopy study on the decomposition reaction of andalusite to 3/2-mullite plus silica. According to this study the reaction starts at the (001) plane of andalusite and proceeds rapidly along the $[001]_A$ direction. No glassy or crystalline transition phase occurs at the andalusite/ mullite phase boundary (Fig. 2.4.3). The overall reaction front is parallel to $(001)_{A,M}$, although in a nanoscale view it exhibits a zig-zag shape parallel to $(011)_A$ and $(201)_M$. These lattice planes of andalusite and mullite fit well together, especially at transformation temperature. The results of Hülsmans et al. agree perfectly with the earlier single crystal X-ray diffraction studies of Pannhorst and Schneider.

The definite orientational relationships, the excellent fit of the andalusite and mullite lattices and the lack of heavy lattice distortions across the andalusite-mullite phase boundary (Pannhorst and Schneider, 1978, Hülsmans et al. 2000a) suggest a topotactical transformation. The schematic view of the $(011)_A/(201)_M$ phase boundary (Fig. 2.4.4) shows that the aluminum oxygen (AlO_6) octahedral chains continue without any distortion upon the transition from andalusite to mullite.

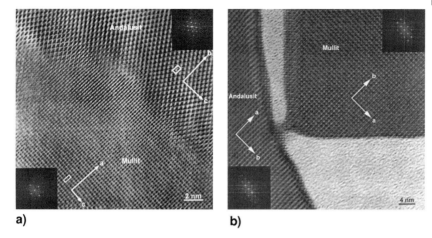

Fig. 2.4.3 High resolution transmission electron micrographs showing the andalusite -mullite interface region in a partially transformed andalusite. (a) Transformation in a view down the crystallographic **a** axis of andalusite ([100]$_A$) and the **b** axis of mullite ([010]$_M$), respectively. Note that the contact between andalusite and mullite is not planar, and that both phases are in a direct, well-fitting contact (from Hülsmans et al. 2000a); (b) Transformation in a view down the crystallographic **c** axis of andalusite ([001]$_A$) and mullite ([001]$_M$), respectively. Insets in both figures are diffractograms of the imaged regions (Courtesy A. Hülsmans).

Thus the fivefold coordinated aluminum (Al) in andalusite corresponds to in-plane aluminum (Al) and silicon (Si) positions in the tetrahedral double chains in mullite, whereas the tetrahedrally coordinated silicon (Si) positions in andalusite correspond to tricluster aluminum (Al*) sites in mullite. Hülsmans et al. (2000a) starting from earlier suggestions of Sadanaga et al. (1962) and Pannhorst and Schneider (1978) concluded that the octahedral chains along the crystallographic **c** axis in andalusite are preserved during the transformation into mullite. Preservation of the AlO_6 octahedral chains is reasonable because these structural units are similar in andalusite and mullite, differing only little in the degree of distortion.

The double chains consisting of SiO_4 tetrahedra and AlO_5 bipyramids, which connect the octahedral chains in andalusite, are decomposed and reconstructed during the transformation process. The reconstruction requires diffusion processes, which control the transformation velocity. The silica liquid, which is formed along with the andalusite to mullite transformation, is exsolved into the small channels occurring between the newly formed elongated mullite crystallites. Most of the silica liquid is rapidly transported from the reaction front to the former (001) andalusite surface by means of capillary forces, forming a continuous surface layer. According to Hülsmans et al. (2000a) the exsolution of silica has no strong retarding effect on the transformation process.

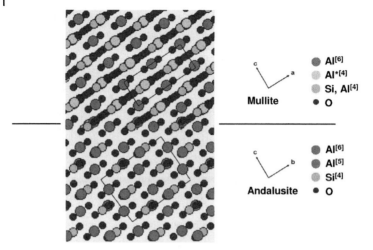

Fig. 2.4.4 Schematic structural presentation of the andalusite –
mullite interface region in a view down the **a** axis of andalusite
([100]_A) and the **b** axis of mullite ([010]_M), respectively. Note
that the aluminum oxygen octahedral chains running parallel
to the **c** axes ([001]) in both phases continue without disrup-
tion from andalusite to mullite (after Hülsmans et al. 2000a).

Transformation mechanisms perpendicular to the octahedral aluminum oxygen
chains (i. e. perpendicular to the **c** axis of andalusite and mullite, respectively) are
complex. Mullite nucleation starts epitactically on the andalusite's surface. Mullitization then proceeds by dissolution of andalusite in the coexisting silica liquid,
transport of aluminum to the mullite nuclei and further growth of mullite. The
dissolution-precipitation process perpendicular to **c** is much slower than the topotactical transformation parallel to **c**. The approach of Hülsmans et al. (2000a,b)
further supports the earlier suggestion that the octahedral chains in mullite-type
phases are so firmly built that they are preserved during transformation.

With respect to the transformation of andalusite (A) to mullite (M) it is interesting to look at the andalusite (A) to sillimanite (S) conversion, the latter phase
being closely related to mullite structurally (see Section 1.1.3). The same orientational relationships as for the andalusite to mullite transformation have been described for the andalusite to sillimanite reaction pair (i. e.: [100]_A//[010]_S, [010]_A//
[100]_S, [001]_A//[001]_S; note that the [100] and [010] are interchanged in andalusite
and sillimanite/mullite, see Vernon, 1987). Cesare et al. (2002) provided new information on the andalusite to sillimanite transformation, investigating the inter-
growth of both phases in xenoliths from volcanic rocks. They basically confirmed
the earlier findings but stated that the orientation of the newly formed sillimanite
deviates slightly from that predicted by Vernon by rotating the sillimanite lattice by
about 2.5° around the **a** axis of andalusite. Two different orientational relationships
are described: (011)_A//(101)_S and (032)_A//(302)_S (Fig. 2.4.5). Cesare et al. give no
interpretation of the andalusite to sillimanite transformation mechanism. How-

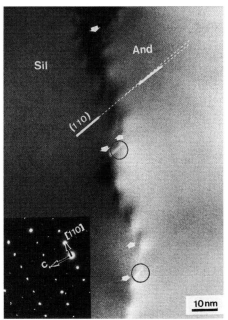

Fig. 2.4.5 High resolution transmission electron micrograph of the andalusite (And)-sillimanite (Sil) interface region in a partially transformed andalusite. Note that the interface is not planar, with the arrows marking areas with $(032)_{And}/(302)_{Sil}$ and $(011)_{And}/(101)_{Sil}$ orientations. Mismatch at the interface is documented by dislocations (circles). The inset is a diffractogram of the imaged region (Courtesy B. Cesare).

ever, a preservation of the octahedral chains as for the andalusite to mullite and sillimanite to mullite reaction pairs can be taken into account. Besides the topotactical andalusite to sillimanite transformation, andalusite dissolution with random sillimanite precipitation from silicate melts has been described as well. Owing to the high activation energy of the transformation and the small entropy difference between reaction products the absence or presence of impurities is crucial. In impurity-free systems a topotactical, basically "solid-state", transformation may be the most favorable transformation path energetically. In the presence of impurities, such as alkalis, coexisting melts form, and andalusite dissolution with subsequent random mullite re-precipitation is more likely.

Sillimanite. The transformation of sillimanite to 3/2-mullite plus silica has been described as a multiple-step reaction. Heat treatment of sillimanite just below the transformation interval to 3/2-mullite yields slightly expanded lattice constants and a symmetry change from *Pbnm* (natural) to *Pbam* (heat-treated). The structural changes have been explained by a continuous change of composition from sillimanite to 3/2-mullite (Hariya et al. 1969), or alternatively by increasing aluminium/silicon disorder in sillimanite at constant composition (Beger et al. 1970, Chatterjee and Schreyer, 1972, Navrotsky et al. 1973, Guse et al. 1979). Winter and Ghose (1979) made an attempt to detect spatial aluminum/silicon disorder in sillimanite, and linear curves of the isotropic equivalent temperature factor B of each atom versus absolute temperature were fitted. At absolute zero (0 K) these curves should extrapolate to a temperature factor of $B = 0$, since thermal motion ceases. For sillimanite, such extrapolations resulted in B values deviating from zero. This was interpreted as due to aluminum/silicon disorder in sillimanite. Lattice parameters and heat of solution measurements of sillimanite carried out in an oxide melt under pressure by Navrotsky et al. (1973) indicate an aluminum/silicon disordering in several steps prior to the transformation to mullite.

Guse et al. (1979) found that the thermal transformation of sillimanite to 3/2-mullite is a reaction with a high degree of orientation, with the crystallographic axes of the newly formed 3/2-mullite being parallel to those of sillimanite. Starting from the fact that the structures of sillimanite and mullite, especially their aluminum oxygen octahedral chains, are very similar in their internal symmetry and relative orientations in the respective unit cells, Sadanaga et al. (1962) suggested that the octahedra are maintained during phase transformation as in the case of andalusite. The initial reaction stages of sillimanite prior to the decomposition to mullite were studied by Rahman et al. (2001) with X-ray diffraction and electron microscopy techniques. Rahman et al. first observed aluminum/silicon disordering and slight tilting and/or rotation of tetrahedral and octahedral units. Simultaneously, partial occupation of the mullite-specific Al* positions occurs. The results agree with the data derived from infrared spectroscopy (Rüscher, 2001). In the next step an alumina-rich "mullite-type" matrix containing isolated sillimanite areas is formed by interdiffusion of aluminum and silicon (Fig. 2.4.6). Finally complete conversion of sillimanite to 3/2-mullite plus silica is observed. The excess silica melt is precipitated in (110)-faceted rectangular channels running parallel to **c** of mullite, which probably are "residuals" of the above-described isolated sillimanite areas.

Cameron (1976a,b) studied the formation conditions of coexisting sillimanites and mullites in α-alumina/sillimanite/mullite xenoliths from the Bushveld intrusion in South Africa. Cameron found evidence that earlier formed mullite in the core of the xenolith may have exsolved sillimanite lamellae after settling near the floor of the magma chamber during a very slow cooling process. The contact plane of the two phases is (100). The transformation process is so sluggish that a concentration gradient is set up in the sillimanite host, resulting in a second generation of sillimanite. In spite of the precisely orientated sillimanite lamellae, Cameron suggested that the exsolution takes place by nucleation and growth but not topotac-

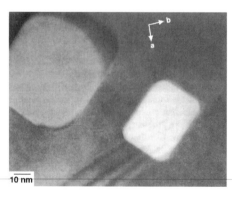

Fig. 2.4.6 High resolution transmission electron micrograph of single crystal sillimanite partially transformed to mullite in a view down the crystallographic **c** axis ([001]). The rectangles are residuals of sillimanite within the newly topotactically formed mullite (Courtesy S. Rahman).

tically with preservation of structural elements. On the basis of these observations it is concluded that in the presence of a fluid phase ("wet" reaction conditions), dissolution of mullite, ion diffusion and epitactical nucleation and growth of silli-manite take place. Under "dry" reaction conditions dissolution is probably un-favorable energetically and a transformation mechanism with conservation of structural elements proceeds instead. Similar transformation mechanisms may also be active for andalusite, assuming "dry" or "wet" reaction conditions, re-spectively.

Heat treatment of sillimanite single crystals up to 1650 °C, carried out by Gype-sova and Durovič (1977), yielded no mullite but the formation of α-alumina to-gether with silica melt. Gypesova and Durovič interpreted these phases as metast-able reaction products. Similar processes of andalusite (Schneider, unpublished results) and kyanite (Saalfeld, 1977) lead to the conclusion that the furnace atmos-phere, especially the presence of alkali, are responsible for this type of decomposi-tion reaction.

Nature of the silica phase formed during the thermal decomposition of kyanite, andalusite and sillimanite The heat-induced transformation of andalusite and sil-limanite to 3/2-mullite produces X-ray amorphous (liquid) silica as an additional reaction product, while that of kyanite yields cristobalite (Schneider and Majdic, 1979, 1980, 1981). This is a remarkable result, since the transformation intervals of the three aluminosilicates lie well within the cristobalite stability field. It seems reasonable to assume that cristobalite nucleation takes place more easily during the thermal decomposition of kyanite than during that of andalusite or sillimanite. Single crystal investigations by Saalfeld (pers. comm.) actually showed that cristo-balite can display a definite structural relationship to kyanite with the cubic [111] direction of cristobalite lying parallel to **c** of the newly formed mullite. Thus the nucleation energy may be lowered to such a degree that cristobalite crystallization is favored relative to the formation of a silica melt.

Kinetics of the thermal transformation of kyanite, andalusite and sillimanite to mullite Kinetic studies on the high-temperature transformation of the aluminosi-licates kyanite, andalusite and sillimanite (all having the composition $Al_2O_3 \cdot SiO_2$) to 3/2-mullite ($3Al_2O_3 \cdot 2SiO_2$) plus silica (SiO_2, glass or cristobalite) were performed by Wilson (1969) and Schneider and Majdic (1979, 1980, 1981; Fig. 2.4.7). All transformation curves initially have steep slopes with high reaction rates, but reach saturation after extended annealing times. After 180 min heat treatment, complete transformation to mullite plus silica occurs at about 1320 °C for kyanite, 1440 °C for andalusite and 1660 °C for sillimanite (data from Schneider and Majdic, 1979, 1980, 1981). Simultaneously, the transformation in-tervals become wider and the activation energies increase (kyanite: about 420 kJ mol^{-1}, andalusite: about 585 kJ mol^{-1}, sillimanite: about 710 kJ mol^{-1}, data from Wilson, 1969, see also Section 2.2.1.1).

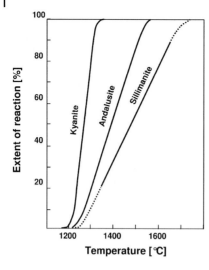

Fig. 2.4.7 Kinetics of the thermal transformation of sillimanite and andalusite to mullite plus silica. Reaction rates of kyanite are given for comparison. The reaction time is 60 min (after Schneider and Majdic, 1981).

2.4.1.3 Formation from X-sialon

X-sialon (stoichiometric composition: $Si_{12}Al_{18}O_{39}N_8$) transforms to mullite at high temperature and oxidizing environments, according to the general equation:

$$Si_{12}Al_{18}O_{39}N_8 + 9\ O_2 \rightarrow 3(3Al_2O_3 \cdot 2SiO_2) + 6\ SiO_2 + 8\ NO_2 \qquad (8)$$
X-sialon (X)   mullite (M)

It has been suggested that X-sialon (x) belongs to the mullite structure family[2] and has the same backbone of edge-sharing AlO_6 octahedral chains (parallel to $[010]_X$) as mullite (M, parallel to $[001]_M$, see Section 1.1.3) topotactical transformations between the two phases can be considered. The idea of this approach again is that the octahedral chains are so firmly built that they are preserved during transformation, whereas the other structural units decompose and rearrange. Transmission electron microscopy studies by Schmücker and Schneider (1999) on X-sialon partially transformed into mullite actually provide evidence for this approach. The (100) plane of X-sialon, which contains $[010]_X$ lies parallel to the (110) of mullite, which contains $[001]_M$. The orientational relationship $(100)_X // (110)_M$ makes sense, since three (110) lattice spacings of mullite ($d(110) = 5.386$ Å, $\times$ 3 = 16.158 Å) fit with two (100) lattice spacings of X-sialon ($d(100) = 7.86$ Å $\times$ 2 = 15.720 Å), (Fig. 2.4.8).

2.4.1.4 General Remarks

The described mullite formations all display specific orientational relationships between the mother phases and the product mullite. In andalusite, sillimanite, X-

---

2) Since no exact structural data are avaible from X-sialon it has not been included into the discussion of the mullite-type family of crystall structures (see Section 1.1).

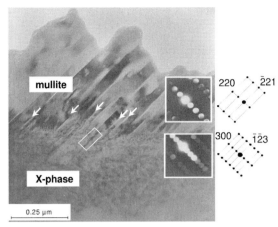

Fig. 2.4.8 Transmission electron micrograph of an X-sialon ceramic, heat treated in air (1250 °C, 1 h). The micrograph shows an X-sialon grain partially transformed to mullite. Convergent beam electron diffraction patterns (insets) show parallel lattice planes of X-sialon and newly formed mullite, with all mullite crystallites being similarly oriented. The arrows indicate channels with glass (Courtesy M. Schmücker).

sialon and mullite the common backbones are the chains of edge-sharing AlO_6 octahedra. These chains are so firmly built that they are preserved or serve as nucleation sites during the mullite formation process. Preservation of the edge-sharing octahedral chains is also suggested for other transformations between members of the mullite family (see Section 1.1.3) not investigated yet. For example the preservation of the aluminum oxygen (AlO_6) octahedral chains during the high-temperature transformation of the boron aluminate $Al_{6-x}B_xO_9$ to its high temperature phase $Al_{18}B_4O_{33}$ (A_9B_2) phase is most propable. The important role of the octahedral units during mullitization has also been demonstrated during the thermal decomposition of kaolinite and muscovite to mullite. In these cases, however, the octahedral chains serve as nuclei during epitactic mullite formation.

2.4.2
Pressure-induced Decomposition

2.4.2.1 Decomposition to Sillimanite
Mullite single crystals treated hydrothermally (e.g. at 800 °C and 2 kbar) decompose to sillimanite plus α-alumina probably following a dissolution-reprecipitation process. Transformations are initiated on linear surface defects of mullite, with the newly formed sillimanite often having two mirror-imaged orientations with respect to the mother phase (Eils et al. 2005, Fig. 2.4.9). With the progress of the transformation the lens-shaped sillimanite individual crystals can coalesce to sheets covering the whole mullite surface. These observations suggest that in the presence of a fluid phase ("wet" reaction conditions), dissolution of mullite, ion diffusion and epitactical nucleation and growth of sillimanite takes place. Under "dry" reaction conditions dissolution is probably energetically unfavorable and a topotactical solid-state transformation mechanism with conservation of structural elements is more likely to proceed.

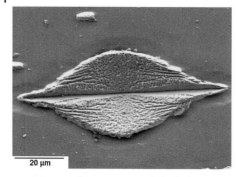

Fig. 2.4.9 Scanning electron micrograph of a mullite single crystal surface parallel to (001), hydrothermally treated at 800 °C and 2 kbar. The lens-shaped newly formed sillimanite agglomeration is probably produced by a dissolution-precipitation process, and the sillimante crystals display two mirror image orientations with respect to the mother phase mullite.

20 µm

2.4.2.2   Decomposition to γ-Alumina plus Silica

Under "static" conditions mullite transforms, depending on pressure and temperature, to the aluminosilicates sillimanite, andalusite or kyanite plus α-alumina, which in turn at very high pressure decompose to α-alumina and stishovite (a sixfold-coordinated high-pressure silica phase). Under dynamic shock waves with extreme thermodynamic conditions equilibrium reactions are not realized, and the aluminosilicate phase fields are overjumped. Details of the mechanisms of the shock-induced disproportionation of mullite can be derived from the shock behavior of the structurally related phase andalusite (Schneider and Hornemann, 1977, see also Section 1.1.3). Infrared and X-ray diffractometry studies on shock-loaded andalusite indicate a progressive shock-induced transformation into poorly crystallized γ-alumina plus non-crystalline silica. The newly formed γ-alumina shows orientational relationships with the andalusite mother phase, which may be due to a preferred nucleation and growth of γ-alumina on specific faces of andalusite (Schneider and Hornemann, 1977). Sillimanite from the impact crater of Nördlingen (Southern Germany) and experimentally shock-loaded basically display similar transformation mechanisms (Stöffler, 1970, Schneider and Hornemann, 1981, see also Braue et al. 2005a).

2.5

Mullite-mullite Phase Transformations

H. Schneider and M. Schmücker

2.5.1

Compositional Transformations

Since mullite, $Al_2[Al_{2+2x}Si_{2-2x}]O_{10-x}$, forms in a wide chemical composition range ($0.2 < x < 0.85$, see Section 1.1.3), it is interesting to look at the mechanisms of the various mullite-mullite transformations with change of the x-values (Al_2O_3 contents).

Voll et al. (2005) investigated the development of composition and structural state of mullites obtained from non-crystalline single-phase (type I) precursors and glasses (see Section 1.4.1). Voll et al. distinguished two mullite crystallization routes (Fig. 2.5.1a,b).

- A first group of mullites with Al_2O_3 contents up to about 81 wt. % (72 mol%) obtained from non-crystalline precursors and glasses by annealing at 900 °C. The composition of these mullites corresponds approximately to 5/2-mullite ($5Al_2O_3 \cdot 2SiO_2$). Increasing the temperature to 1100 °C makes the Al_2O_3 content of mullite slightly decrease, while above 1100 °C a strong Al_2O_3 decrease has been observed. Stoichiometric mullite composition (72 wt. % i.e. 60 mol% Al_2O_3) is achieved above 1300 °C (Fig. 2.5.1a).

- A second group of mullites, obtained by annealing non-crystalline precursors and glasses between 920 and 940 °C. These mullites also have Al_2O_3 contents up to about 81 wt. % (72 mol%). From crystallization temperature to about 1000 °C they display a strong Al_2O_3 decrease to values near 76 wt. % (65 mol%). Above this temperature and to about 1100 °C little compositional change or even a re-increase of the Al_2O_3 content of mullite has been observed. Exceeding 1100 °C the Al_2O_3 content of mullite decreases again and reaches stoichiometric composition (72 wt. % i.e. 60 mol%) above 1300 °C (see also Okada and Otsuka, 1988, Nass et al. 1995, Gerardin et al. 1994). A small amount of α-alumina typically appears in the X-ray diffraction diagram of these samples between 920 to 950 °C. The amount of the coexisting α-alumina slightly increases up to 1050 °C, but completely disappears at higher temperatures (Fig. 2.5.1b).

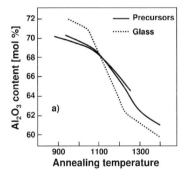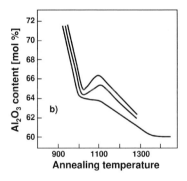

Fig. 2.5.1 Evolution of the Al_2O_3 content of precursor- and glass-derived type I (single phase) mullite as a function of temperature. Mullites formed above about 900 °C are very Al_2O_3-rich (about 81 wt. % i.e 72 mol%). With increasing temperature the Al_2O_3 content drops to achieve the stoichiometric composition (72 wt. % i.e. 60 mol% Al_2O_3). (a) Continuous evolution of the Al_2O_3 content of mullite. Note that no transitional alumina coexists with the early stages of mullite. (b) Discontinuous evolution of the Al_2O_3 content of mullite. Note that small amounts of transitional alumina coexist with the early stages of mullite.

We believe that the non-occurrence or occurrence of local Al_2O_3 enrichments in the precursor decides whether the first or the second mullite-mullite transformation mode is activated. In the first case, with no local Al_2O_3 enrichment, the Al_2O_3 content of mullite decreases slowly and continuously with temperature. In the second case, nano-sized Al_2O_3-rich domains are believed to occur in the precursors, giving rise to nucleation of alumina below 1000 °C. The alumina precipitates act as sinks for aluminum, thus enhancing the exsolution of aluminum from mullite. Above 1050 °C, however, the highly reactive nano-sized alumina particles become unstable and dissolve in the coexisting silica glass with subsequent crystallization of a second generation of mullite having slightly higher Al_2O_3 contents than that formed at temperatures below 1050 °C. Above 1100 °C both types of mullite gradually approach the composition of stoichiometric 3/2-mullite.

The composition of mullite in 2/1-mullite/silicate glass reaction couples annealed in the stability field of stoichiometric 3/2-mullite changes continuously from 2/1- (78 wt. % i.e. 66 mol% Al_2O_3) to 3/2-mullite composition (72 wt. % i.e. 60 mol% Al_2O_3), as indicated by chemical analyses and the appearance of diffraction patterns with specific satellite reflections (Fig. 2.5.2, Schmücker et al. 2002). A crucial point for the mullite-mullite transformation process is whether there exists a coexisting silicate liquid, which at high temperature has the function of a diffusion medium and a "cation reservoir". The coexisting silicate liquid at high temperature can act either as a source or as a sink for aluminum (see also Schneider and Pleger, 1993). Schmücker and Schneider (2002) have shown that in reaction couples consisting of 2/1-mullite and aluminosilicate glass the 3/2-mullite can also secondarily form in the coexisting glass during cooling of the system. Schmücker and Schneider found that the mechanism of this mullite formation depends on the degree of mullite solution in the glass phase at high temperature and on whether sodia (Na_2O) is present in the glass or not. In Na_2O-free glasses the secondary 3/2-mullite forms randomly in the coexisting glass, whereas in reaction couples of sodia/silica glass/mullite it forms epitactically on the mullite substrate surface (Fig. 2.5.3).

The different nucleation mechanisms of the newly formed, secondary 3/2-mullite have been explained in terms of tetrahedral triclusters existing in the Al_2O_3–SiO_2 but not in Na_2O–Al_2O_3–SiO_2 gels and glasses. Oxygen-deficient tetrahedral triclusters in aluminosilicate glasses were first postulated by Lacy (1963) as a charge balance mechanism occurring in competition to network-modifying cations such as Na^+. According to this approach, mullite and aluminosilicate glasses consist of the same type of structural units, i.e. $(Al,Si)O_4$ tetrahedra, $(Al,Si)O_4$ tetrahedral triclusters and AlO_6 octahedra, although in partially ordered or random arrangements, respectively. Owing to the structural similarities between mullite and the Na_2O-free aluminosilicate gels and glasses there is only a small energy barrier for mullite nucleation and thus mullite can easily form in the bulk of the glass. In the case of Na_2O-containing systems, a lower degree of structural similarities exist between mullite and glass, due to a lack of triclusters, and hence the nucleation barrier is higher. In that case an epitactical nucleation of the newly formed 3/2-mullite on the substrate is more favorable than in the glass bulk (for more details see Schmücker and Schneider, 2002).

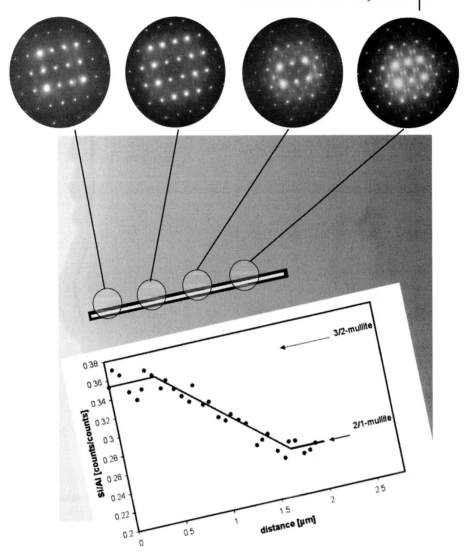

Fig. 2.5.2 Aluminum/silicon line scan over the 2/1- to 3/2-mullite transformation area, using the scanning transmission electron microscopic technique. The line scan indicates that there is a gradual change between the compositions of 2/1- and 3/2-mullite. The diffraction patterns of the transformation area (top of the figure) also show a gradual and continuous development of superstructure reflections from 2/1-mullite (sharp and intense) to 3/2-mullite (weak and diffuse) (Courtesy M. Schmücker).

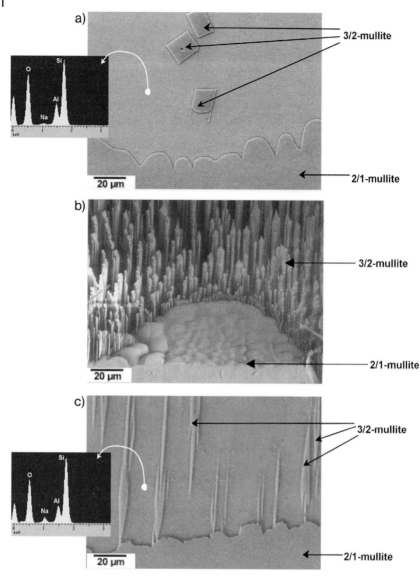

Fig. 2.5.3 Cross section scanning electron micrographs of glass/mullite reaction couples heat-treated at 1650 °C for 100 h. (a) Silica (SiO$_2$) glass/mullite system. Note the formation of secondary mullite in the bulk of the glass. (b) and (c) Sodia (Na$_2$O)/silica (SiO$_2$) glass/mullite system. Note the epitactical growth of newly formed 3/2-mullite needles on the mullite substrate. Both reaction couples display cellular mullite substrate surfaces due to partial dissolution (Courtesy M. Schmücker).

Schmücker et al. give no evidence for the existence of intermediate "stable" and "unstable" mullites in the 2/1- to 3/2-composition series. They even rule out a nanometer-scale phase separation into alumina-rich and silica-rich mullites as suggested by Dabbs et al. (1999) for gel-derived mullites. Other studies, however, mention that specific mullite compositions occur more frequently than others. Of special importance are the 3/2- and 2/1- (= 4/2) compositions. However, relatively stable mullites with higher Al_2O_3/SiO_2 ratios may also occur: The investigations of Voll et al. (2005) have shown that mullites with 5/2-composition (81 wt. % i.e. 72 mol% Al_2O_3) preferentially form from type I precursors and glasses at about 950 °C (see Fig. 2.5.1), while Bauer et al. (1950) and Bauer and Gordon (1951) grew mullite single crystals with 3/1- (= 6/2) composition (83 wt. % i.e. 75 mol% Al_2O_3) from melts. Finally Fischer et al. (1996) described an extremely aluminum-rich mullite with 9/1-composition (= 18/2, with 93 wt. % i.e. 90 mol% Al_2O_3; see below). The average structure of mullite provides no evidence for the occurrence of such specific stable compositions. Thus, if they exist they should be associated with the real structure of mullite, i.e. with favorable oxygen vacancy distributions (see Sections 1.1.3.7 and 1.2).

It is interesting to have a look at the temperature-dependent development of the extremely Al_2O_3-rich mullites described by Schneider et al. (1993b) and Fischer et al. (1996), (Fig. 2.5.4). Initially a small amount of mullite appears together with transitional alumina and non-crystalline phases at temperatures as low as 600 °C. These mullites have Al_2O_3 contents of about 93 wt. % (88 mol%) and coexist with transitional alumina. Upon heating, the Al_2O_3 content of mullite further increases up to 96 wt. % (92 mol%) at about 1000 °C, which is the highest Al_2O_3 content of mullite detected so far. Above this temperature limit the Al_2O_3 content of mullite again drops to that of stoichiometric mullite (72 wt. % i.e. 60 mol% Al_2O_3) at 1400 °C. Together with the decrease of the Al_2O_3 content of mullite above 1000 °C, transformations of the coexisting transitional alumina phases are observed. This observation suggests that the aluminum exsolution from mullite is not only temperature-controlled, but depends on the character of the coexisting transitional alumina, which may or may not act as a sink for the aluminum.

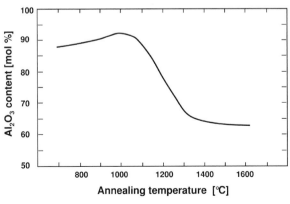

Fig. 2.5.4 Evolution of the Al_2O_3 content of extremely aluminum-rich mullite, synthesized by a specific sol-gel process. Note that the first mullite crystallization takes place below 800 °C. The Al_2O_3 content of this mullite at 1000 °C is about 93 wt. % (88 mol%), which is the highest content so far observed (after Fischer et al. 1996).

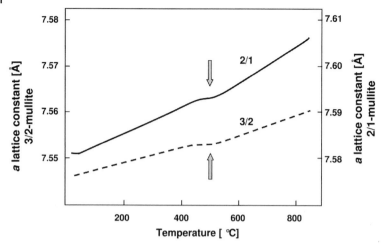

Fig. 2.5.5 Thermal expansion along the crystallographic **a** axis of chemically produced 3/2-mullite ($\approx$ 72 wt. % Al_2O_3) and of fused 2/1-mullite ($\approx$ 76 wt. % Al_2O_3) measured by X-ray diffraction. Note the occurrence of a slight expansion discontinuity between 400 and 500 °C, indicated by arrows (after Schneider et al. 1993c).

2.5.2
Structural Transformations

2.5.2.1 Transformation at about 450 °C

High accuracy high temperature X-ray Guinier powder and high resolution single crystal X-ray (Bond) techniques indicate that lattice constant expansion curves of mullite are discontinuous between about 400 and 500 °C (Schneider et al. 1993c). The temperature-induced expansion discontinuities are strongest parallel to **a** (Fig. 2.5.5), while there is no significant change parallel to **b** and **c**. The slopes of the *a* lattice constant curves are greater above the expansion discontinuity than below, and the effects are reversible without any hysteresis. Schneider et al. ascribed the slight increase of the **a** expansion above the expansion discontinuity to deformations, rotations and tiltings of the aluminum oxygen (AlO_6) octahedra, and to a geometrical deformation of the aluminum oxygen (Al^*O_4) tetrahedra.

2.5.2.2 Transformation above 1000 °C

Investigating the high temperature heat capacity c_p of mullite, Hildmann and Schneider (2004) found a weak step-like increase of the heat capacity c_p above about 1100 °C ($\Delta c_p \approx 0.03$–0.10 J g^{-1} K^{-1}). This heat-capacity anomaly is reproducible under various heating conditions and reversible both on heating and cooling (Fig. 2.5.6). The onset of processes varies in a wide temperature range, and seems to depend on the heating velocity and whether the material is a single crystal or

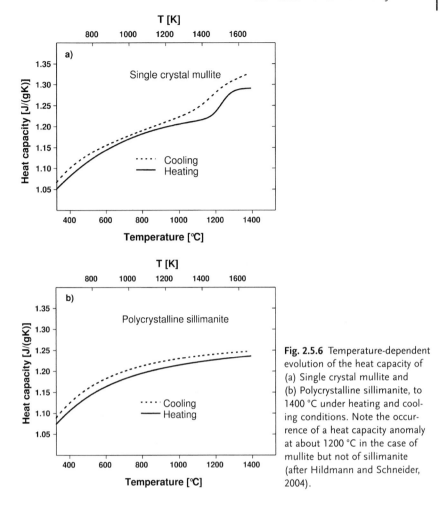

Fig. 2.5.6 Temperature-dependent evolution of the heat capacity of (a) Single crystal mullite and (b) Polycrystalline sillimanite, to 1400 °C under heating and cooling conditions. Note the occurrence of a heat capacity anomaly at about 1200 °C in the case of mullite but not of sillimanite (after Hildmann and Schneider, 2004).

polycrystalline (single crystal: 1020 to 1100 °C, polycrystalline ceramic: >1250 °C) but not on the composition of the mullite sample. Interestingly, sillimanite, which is closely related to mullite (see Section 1.1.3), displays no such heat-capacity anomaly. The shape of the heat-capacity anomaly is similar to the effects of silicate glasses at the glass transition point.

Brunauer et al. (2001) carried out in situ high-temperature X-ray and neutron diffraction studies on undoped and chromium-doped mullite (10.3 wt. % Cr_2O_3) up to 1600 °C. In spite of a strong spread of values they identified a discontinuous variation in the lattice constants above about 1000 °C. Schneider and Schneider (2005) performing X-ray in situ high-temperature diffraction studies on mullite ceramics in a mirror-heated furnace confirmed the existence of an expansion anomaly, although at higher temperatures (above 1200 °C). They describe a main increase of the lattice expansion parallel to the crystallographic **b** axis, while effects

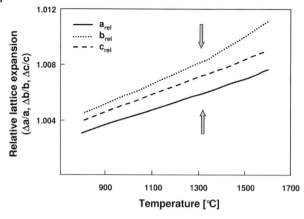

Fig. 2.5.7 Relative temperature-induced changes of the lattice parameters *a*, *b* and *c* referred to room temperature values. Note the occurrence of expansion anomalies at about 1300 °C indicated by arrows (after Schneider and Schneider, 2005).

parallel to **a** and **c** are less significant (Fig. 2.5.7). An expansion anomaly at about 1200 °C has also been identified by Schreuer et al. (2005) using single-crystal dilatometry.

Finally Schreuer et al. (2005) describe anomalous developments of the elastic shear constants c_{44} and c_{55}, and to a lesser degree of c_{66}, at about 1100 °C, while the linear stiffness coefficient constants c_{11}, c_{22} and c_{33} are not affected (Fig. 2.5.8, see also Figs. 2.1.2 a and b). A more detailed view of processes can be derived from Rietveld structure refinements carried out by Schneider and Schneider (2005). In the transformation region, extending from 1000 to about 1350 °C, the expansion of structural units of mullite develops discontinuously: While the AlO_6 octahedra contract in volume, that of the T^*O_4 tetrahedra increase. Above about 1350 °C, the volume of the AlO_6 octahedron re-increases and the T^*O_4 tetrahedra slightly con-

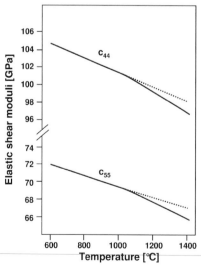

Fig. 2.5.8 Temperature-dependent evolution of the shear moduli c_{44} and c_{55} of single crystal mullite in the 600 to 1400 °C area. Note the occurrence of anomalies at about 1000 °C (after Schreuer et al. 2005).

tract. A further important result of the in situ high temperature structure refinement of mullite is that above 1000 °C, the dimensions and volumes of TO_4 and T^*O_4 tetrahedra and displacements of oxygen atoms converge. This indicates that TO_4 and T^*O_4 tetrahedra gradually become similar in dimensions and degree of deformation.

At present it is difficult to understand the high-temperature transformation mechanism of mullite. Hildmann and Schneider (2004) interpreted the heat capacity anomaly as being caused by an onset of tetrahedral aluminum and oxygen atoms hopping between adjacent T and T* and O(C) and O(C)* structural sites. Hildmann and Schneider suggest that the tetrahedral aluminum T and T* and O(C) and O(C)* oxygen atoms at temperatures below 1000 °C occupy structural positions that are locally partially ordered ("static state"). Above about 1000 °C dynamic site-exchange processes of T and T* and O(C) and O(C)* atoms, respectively, are initiated ("dynamic state"). During cooling, the mullite local structural arrangements are frozen in. This leads back to partially ordered structural ("static") arrangements again, which, however, may not be completely identical with the initial state, and depending on the samples and their cooling histories a wide range of "transformation" temperatures may be possible. The discontinuous decrease of the elastic shear constants c_{44} and c_{55} and to a lower degree of c_{66} but not of the linear constants c_{11}, c_{22} and c_{33} may account for rotational gliding in the (100) and (010) lattice planes during the high temperature transformation of mullite.

Molecular dynamics (MD) simulations have been carried out comparing mullite with sillimanite, in order to check the possible occurrence of a high-temperature phase transformation in mullite (Lacks et al. 2005). At high temperatures the simulations of mullite actually yield an anomaly in the heat capacity as found experimentally, while sillimanite shows no measurable effect. According to Lacks et al. the heat capacity jump coincides with an onset of aluminum atom diffusion. Paulmann (1996), carrying out in situ single crystal diffractometry up to 1400 °C, found slight weakening and broadening of satellite reflections, which can be attributed to the increase of thermal vibrations. There is, however, no indication of a change of the oxygen vacancy arrangement, which means that the composition-dependent ordering scheme of oxygen vacancies of mullite persists during the eventual phase transformation up to the melting point of mullite.

In summary it must be stated that, in spite of the experimental data and of the results of MD simulations the high temperature transformation in mullite is not yet fully understood.

2.6
Spectroscopy of Mullite and Compounds with Mullite-related Structures
K. J. D. MacKenzie

A variety of spectroscopic techniques has been used to provide structural information about mullite and related solid compounds, including solid-state nuclear magnetic resonance (NMR), electron paramagnetic resonance (EPR), infrared (IR) and

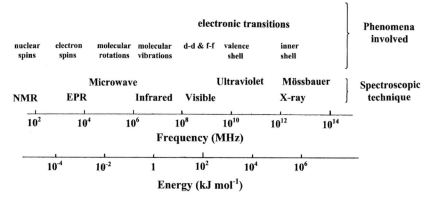

Fig. 2.6.1 Frequency and energy ranges of the various spectroscopic techniques used in studies of mullite and mullite-related compounds.

Mössbauer spectroscopies. All these methods are based on the principle that under certain conditions materials absorb or emit energy. The energy supplied to the sample is usually electromagnetic radiation, which spans a very large range of frequencies (i. e. energy), from radio frequencies (MHz), at which nuclear and electron spins are modulated, beyond the visible and ultraviolet region to X-rays, which interact with the inner-shell electrons (Figure 2.6.1).

The stimulation of a particular event within the structure is detected either as the adsorption of part of the incident radiation at the resonance frequency, or by the emission of radiation of different frequency resulting from the return of the structure from an excited state to its original state. The energy of IR radiation stimulates stretching or bending molecular vibrations within a solid, whereas the much smaller energies in the microwave region are sufficient to interact with the spins of nuclei (NMR) or electrons (EPR) in an applied magnetic field. Mössbauer spectroscopy is similar to NMR spectroscopy insofar as it is concerned with the transitions taking place inside atomic nuclei, but in this case the incident radiation is a highly monochromatic beam of gamma-rays produced by decay of radioactive elements such as ^{57}Fe or ^{119}Sn whose energy is varied by making use of the Doppler effect.

2.6.1
Solid-state Nuclear Magnetic Resonance (NMR) Spectroscopy

2.6.1.1 Brief Principles of Solid-state NMR Spectroscopy
For a comprehensive description of the theory and practice of solid state NMR techniques and its application to inorganic materials, the reader should consult a modern text such as MacKenzie and Smith (2002). The following is a very brief outline of the important principles.

Many nuclei possess a quantized property called spin. When such nuclei are placed in a strong magnetic field (typically 2-18 T) the energy levels between the various spin states are split (the Zeeman effect) and their magnetic moment precesses about the magnetic field axis at a frequency characteristic of that particular nucleus (the Larmor frequency). The Larmor frequencies are typically in the radio frequency (rf) range. If the spin system is then irradiated with a pulse of plane-polarized rf radiation at the Larmor frequency, it inclines as a result of the interaction between its magnetic moment and the magnetic field associated with the rf pulse. It follows from this that NMR spectroscopy is specific to a particular nucleus, the spectrometer being tuned to its characteristic Larmor frequency. At the end of the pulse, the spin system dephases in the xy plane to return to its equilibrium position along the z axis with a characteristic relaxation time T_1. The resulting change in magnetization induces a small decaying voltage (called a Free Induction Decay or FID) in a detector coil surrounding the sample. The FID, which contains encoded frequency information, is recorded in the time domain and then Fourier-transformed into the frequency domain to yield a typical NMR spectrum. The frequency distribution of the detected signal displays small but significant differences from the applied frequency, resulting from shielding of the nucleus from the applied magnetic field by the surrounding species. These frequency shifts, reported as differences in parts per million (ppm) from the peak position of a standard substance, provide structural information regarding the atomic environment of the nucleus under investigation. A large number of nuclei have nuclear characteristics suitable for NMR spectroscopy; those of greatest importance for mullite investigations are ^{27}Al, ^{29}Si and ^{17}O, although others such as ^{23}Na and ^{11}B have proved useful in studies of mullite-related structures.

A difficulty in applying NMR spectroscopy to solids arises from broadening of the spectra due to various nuclear interactions. Many of these broadening effects can be removed or significantly reduced by inclining the sample at an angle of 54.7° to the applied magnetic field axis (the "magic angle") and spinning the sample rapidly at this angle. This process, known as magic angle spinning (MAS) is capable in many cases of narrowing the broad solid-state NMR spectral lines sufficiently to resolve different sites in a structure. More advanced techniques for improving spectral resolution which have been applied to mullite studies include double angle spinning (DOR), dynamic angle spinning (DAS) and multiple quantum (MQ) techniques. For details of these methods the reader is referred to MacKenzie and Smith (2002).

2.6.1.2 NMR Spectroscopic Structural Studies of Aluminosilicate Mullite

All the constituent elements of aluminosilicate mullite (Al, Si, O) have nuclides suitable for NMR spectroscopy. The utility of MAS NMR spectroscopy for observing ^{29}Si and ^{27}Al has made MAS NMR spectroscopy a favored tool for studying aluminosilicate structures, including mullite. Furthermore, since NMR does not depend on long-range order or crystallinity of the samples, it is superior to X-ray diffraction for studying amorphous gel precursors of mullite and mullite glasses.

^{27}Al is a quadrupolar nucleus of spin 5/2 and 100% abundance; its MAS NMR

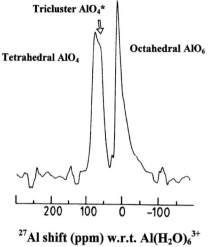

Tricluster AlO₄*

Tetrahedral AlO₄

Octahedral AlO₆

$$^{27}\text{Al shift (ppm) w.r.t. Al(H}_2\text{O)}_6^{3+}$$

Fig. 2.6.2 7-T ^{27}Al MAS NMR spectrum of crystalline 3:2 mullite (from Merwin et al. (1991) by permission of the copyright owner).

spectrum can readily be observed, but for best resolution of overlapping peaks arising from multiple sites, fast MAS speeds (>12 kHz) and magnetic fields as strong as possible (>11.7 T) are preferable (MacKenzie and Smith 2002). Additionally, special techniques for removing higher-order quadrupolar broadening effects, DOR, DAS, MQ and satellite transition spectroscopy (SATRAS) have all been used in ^{27}Al NMR studies of mullite and related structures.

The typical ^{27}Al MAS NMR spectrum of well-crystallized mullite (Fig. 2.6.2) shows two resonances; one, in the tetrahedral Al-O region, is split into two closely overlapping peaks, at about 64 and 46 ppm (Turner et al. 1987, Merwin et al. 1991). Note that all these ^{27}Al chemical shifts are quoted with respect to Al(H$_2$O)$_6^{3+}$. The resonance at 64 ppm has been assigned to the regular tetrahedral Al sites, while the peak at 46 ppm is identified with the T* sites associated with the tricluster units. An intense narrow resonance at about 0 ppm arises from the Al in the octahedral columns.

Additional detail in the overlapping tetrahedral sites has been sought by the SATRAS method (Kunath-Fandrei et al. 1994, Rehak et al. 1998) in which the MAS spinning sidebands on each side of the central transition (Fig. 2.6.3a) reveal details which suggest the presence of three tetrahedral peaks. These are the "regular" sites with an isotropic chemical shift of 69 ppm, and two types of distorted tetrahedral tricluster sites T* and T' with shifts of 55 and 48 ppm respectively. These results were used to fit the central transition line shape by a simulated curve (Fig. 2.6.3b) allowing quantitative information to be determined about the relative Al site occupancies of mullites with varying Al:Si ratios. These studies indicate that despite small differences in the partitioning of the Al over the various sites, which can be correlated with structural details, the overall line shape of the ^{27}Al spectrum is essentially identical for 2:1 and 3:2 mullites (Rehak et al. 1998).

Bodart et al. (1999) have used ^{27}Al triple-quantum MAS NMR spectroscopy in an

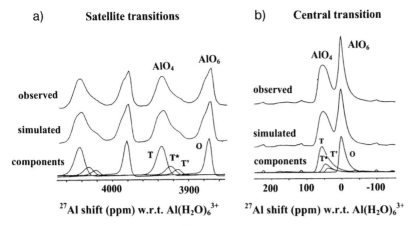

a)   **Satellite transitions**   b)   **Central transition**

Fig. 2.6.3 (a) Satellite transitions of the ^{27}Al MAS NMR spectrum of crystalline mullite, simulated with one octahedral and three tetrahedral resonances. (b) Central transition of the same mullite spectrum, simulated with the spectral components derived from the satellite transition (adapted from Rehak et al. (1998) by permission of the copyright owner).

attempt to confirm the presence of three tetrahedral resonances in 2:1 mullite. The resulting two-dimensional 3QMAS spectrum (Fig. 2.6.4) clearly resolves two tetrahedral and one octahedral site but was unable to distinguish the T' site. Boudart et al. suggest that the overlap in the two tetrahedral tricluster sites is too great for resolution in their MQ experiments. The 3QMAS mullite spectrum indicates a broad distribution of quadrupolar interactions related to the various arrangements of atoms and vacancies within the mullite structure.

Although its natural abundance is much less (4.7 %), the NMR-active spin $^1/_2$ nucleus ^{29}Si is readily observable at moderate MAS speeds and magnetic fields. The ^{29}Si NMR spectrum of crystalline mullite (Fig. 2.6.5a) contains a principal resonance at −86.8 ppm, attributed to Si in an environment with a sillimanite-like Al/Si ordering scheme, and a second resonance at −94.2 ppm, tentatively attrib-

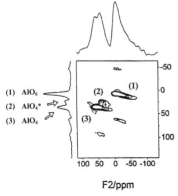

Fig. 2.6.4 ^{27}Al triple-quantum MAS NMR spectrum of 2:1 mullite showing resolution of three Al sites but no evidence of the tetrahedral T' site suggested by SATRAS (from Bodart et al. (1999) by permission of the copyright owner).

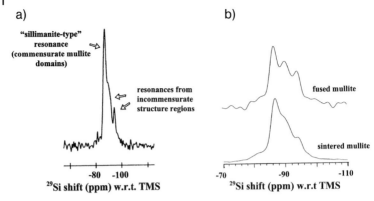

Fig. 2.6.5 (a) 7-T ^{29}Si MAS NMR spectrum of crystalline 3:2 mullite [from Merwin et al. (1991) by permission of the copyright owner]. (b) 11.7-T ^{29}Si MAS NMR spectra of crystalline 3:2 mullites showing differences related to their provenance and thermal history (from MacKenzie et al. (unpublished data).

uted to Si in an environment with Al/Si ordering more typical of mullite (i.e. with Al in the T* positions) (Merwin et al. 1991). However, this site identification would require the resonance at –94.2 ppm to be considerably larger than observed. An alternative suggestion proposed by Merwin et al. (1991) attributes the resonance at –86.8 ppm to local domains of commensurate mullite structure, where the number of resonances in the spectrum equals the number of physically non-equivalent nuclei per unit cell. The resonance at –94.2 (and probably also the shoulder at –90 ppm) is attributed to incommensurate mullite regions arising from the modulation of two sets of ordering with different symmetries. In one ordering scheme the O atoms and vacancies are ordered on the bridging oxygen site, while a second scheme orders some of the tetrahedral Al and Si atoms in a similar manner to the C1 sillimanite structure (McConnell and Heine, 1985). The ^{29}Si MAS NMR spectra of 3/2- and 2/1-mullites are reported to be identical (Merwin et al. 1991).

A different interpretation of the ^{29}Si spectrum of mullite was advanced by Jaymes et al (1995), who found that the spectra of samples derived from aluminosilicate gels could all be fitted by four peaks; a broad peak appearing as a shoulder at –80 ppm was ascribed to tricluster silicon sites near oxygen vacancies, a sharp peak at –86 ppm was assigned to sillimanite-type sites, while smaller peaks at –90 and –94 ppm were ascribed to the replacement of Al in the second coordination sphere by one and two Si atoms respectively. A more recent ^{29}Si MAS NMR study of a series of well-characterized highly crystalline synthetic mullites (Schmücker, MacKenzie et al., unpublished results) resulted in spectra which can be satisfactorily be fitted according to Jaymes et al. (1995) and suggested that the relative populations of the various Si sites may be influenced by the origin and thermal history of the samples (Fig. 2.6.5b). The relationship between these spectral variations and crystalline parameters such as the degree of Al-Si ordering, calculated according to

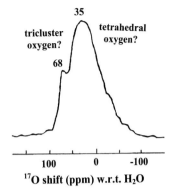

Fig. 2.6.6 7-T ^{17}O MAS NMR spectrum of mullite from an aluminosilicate gel fired at 1300 °C (from Jaymes et al. (1994) by permission of the copyright owner).

various mixing models, suggested that the Si/Al distribution within the tetrahedral chains of mullite is not completely random and there is some tendency towards ordering in favor of the sillimanite-type structure (Schmücker et al. 2005b).

^{17}O, the NMR-active quadrupolar spin-5/2 nuclide of oxygen, is of such low natural abundance (0.037%) that it can only be detected in samples specially enriched in this isotope. For this reason, ^{17}O NMR studies of mullite and related phases are limited in number, and the structural implications for mullite are not yet well understood. The ^{17}O MAS NMR spectrum of mullite crystallized at 1300 °C from an aluminosilicate gel enriched in ^{17}O by hydrolyzing with $H_2^{17}O$ has been reported by Jaymes et al (1994) to contain a major broad resonance at 35 ppm and a smaller shoulder peak at 68 ppm (Fig. 2.6.6), but these authors did not attempt to relate these spectral features to the mullite structure. However, on the basis of subsequent ^{17}O studies of Ga-Ge mullite (Meinhold and MacKenzie, 2000), the resonance at 35 ppm may be associated with all the oxygen atoms coordinated to four Si or Al atoms; this includes all the cations except those of the T* sites, which are tri-coordinated. The much lower area of the 68-ppm peak suggests that this may arise from the tri-coordinated oxygens.

2.6.1.3 NMR Spectroscopic Studies of Amorphous Materials of Mullite Composition

NMR spectroscopy has provided valuable information about the constitution of amorphous mullite-related materials, including precursor compounds, mullite glasses and mechanochemically treated mullite. A common feature of these materials is the appearance in the ^{27}Al NMR spectrum of a resonance at about 30 ppm, in addition to the normal octahedral and tetrahedral resonances (Fig. 2.6.7). The 30-ppm resonance is commonly observed in heated aluminosilicate minerals such as kaolinite (Gilson et al. 1987, Sanz et al. 1988, Lambert et al. 1989, Rocha and Klinowski 1990, Guo et al. 1997, Yao et al. 2001), monophasic gel precursors of mullite (Schneider et al. 1992, 1994a, MacKenzie et al. 1996a, Ikeda et al. 1996, Miller and Lakshmi 1998, Zhao et al, 2002, Lugmair et al. 2002), glasses of mullite composition (Risbud et al. 1987, Sato et al. 1991, Schmücker et al. 1995) and

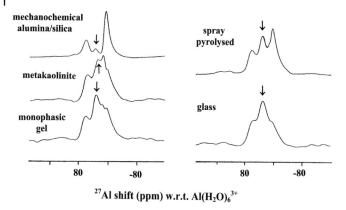

mechanochemical alumina/silica

spray pyrolysed

metakaolinite

glass

monophasic gel

80   -80   80   -80

^{27}Al shift (ppm) w.r.t. $Al(H_2O)_6^{3+}$

Fig. 2.6.7 9.4-T 27Al MAS NMR spectra of amorphous aluminosilicates of mullite composition derived from mechanochemical processing of alumina/silica mixture, thermal dehydroxylation of the clay mineral kaolinite, aluminosilicate gel, spray pyrolyzed alumina/silica mixture and a rapidly quenched aluminosilicate glass. All spectra show the characteristic resonance at about 30 ppm (arrowed) (from McManus et al. (2001) by permission of the copyright owner).

mullite in which the crystallinity has been destroyed by high-energy grinding (Schmücker et al. 1998) (for examples of these spectra see MacKenzie and Smith 2002). Although the 30-ppm resonance is frequently attributed to five-fold-coordinated Al, an alternative proposal advanced by Schmücker and Schneider (1996) suggests that this resonance arises from distorted tetrahedral Al associated with the T* tricluster sites, and is thus the precursor of the resonance at about 46 ppm in crystalline mullite. Schmücker and Schneider (1996) suggest that this peak is shifted upfield in the amorphous aluminosilicate because in these environments it lacks the steric constraints of the crystalline structure. Evidence for this viewpoint has been provided by the simulation of large-angle X-ray scattering (LAXS) profiles using Al site distributions derived from ^{27}Al NMR spectra (Schmücker at al. 1999) and from the observation of the behavior of the 30-ppm Al NMR signal in aluminosilicate glasses where the presence of Na^+ provides a charge compensation mechanism which competes with tricluster formation (Schmücker et al. 1997).

Recently, more advanced NMR techniques such as multiple-quantum (MQMAS NMR) and off-resonance nutation studies have been invoked in an attempt to shed light on the questions surrounding the origin of the 30-ppm ^{27}Al NMR resonance in these amorphous materials. A ^{27}Al MQMAS NMR study of kaolinite heated to various temperatures (Rocha, 1999) came to a similar conclusion to a previous ^{27}Al SATRAS study (Massiot et al. 1995) that the three overlapping resonances in metakaolinite are due to Al(IV), Al(V) and Al(VI). The MQMAS resolution of these peaks yielded a value of 5.0 MHz for the nuclear quadrupole coupling constant of the Al(V) resonance (isotropic chemical shift 37 ppm). The corresponding coupling constant for the Al(IV) resonance (isotropic chemical shift 63 ppm) was

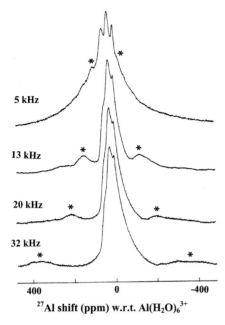

5 kHz

13 kHz

20 kHz

32 kHz

$$\underset{400}{\rule{0pt}{0pt}} \qquad \underset{0}{\rule{0pt}{0pt}} \qquad \underset{-400}{\rule{0pt}{0pt}}$$

^{27}Al shift (ppm) w.r.t. Al(H$_2$O)$_6^{3+}$

Fig. 2.6.8 Effect of the MAS speed on the shape of the 9.4-T ^{27}Al NMR central transition of the spectrum of metakaolinite. The spinning side bands are indicated by asterisks (from Rocha (1999) by permission of the copyright owner).

5.4 MHz (Rocha, 1999). This study also revealed substantial changes in the shape of the ^{27}Al metakaolinite spectrum at very fast MAS speeds up to 32 kHz (Fig. 2.6.8), attributed to the proportionately greater narrowing at higher speeds of the sites with larger quadrupole couplings.

Triple-quantum ^{27}Al MAS NMR spectra of an amorphous gel precursor of 2:1 mullite composition (Bodart et al. 1999) has yielded values of the composite quadrupolar coupling constant λ for all three sites, including that attributed to Al(V). The composite quadrupolar coupling constant, also known as the quadrupolar product P_Q, is a function containing both the nuclear quadrupolar coupling constant χ and the asymmetry parameter η:

$$\lambda = \chi\sqrt{(1 + \eta^2/3)} \tag{9}$$

Inspection of the λ values of the three sites in the amorphous precursor and comparison with the values obtained by the same authors for the octahedral and two tetrahedral sites in crystalline 2/1-mullite reveals a large degree of overlap; although Bodart et al. (1999) interpret these results as militating against the existence of triclusters in the amorphous precursor, the evidence is not clear-cut. A range of amorphous mullite precursors including aluminosilicate glass, metakaolinite, aluminosilicate gel and flame-sprayed aluminosilicate was investigated by triple-quantum ^{27}Al MAS NMR spectroscopy and cross-polarization between the ^{27}Al and ^{1}H nuclei (McManus et al. 2001), yielding very similar values of P_Q (i.e. λ) for the three sites in all these compounds, but not permitting the questionable resonance at about 30 ppm to be unambiguously assigned. A different conclusion was reached by Peeters and Kentgens (1997) who interpreted the results of their ^{27}Al MQMAS NMR and off-resonance nutation NMR study of amorphous aluminosilicate gels dehydrated at 200 °C as evidence that the peak at 30 ppm arises from tetrahedral species in a very distorted environment experiencing large quadrupole-induced shifts, rather than the Al(V) species. It should therefore be observed that the methods available to date have been unable unambiguously to resolve this question.

2.6.1.4 NMR Spectroscopic Studies of Mullite Formation from Minerals

Notwithstanding the debate surrounding site assignments in amorphous mullite precursors, solid-state MAS NMR spectroscopy has been used extensively in studies of mullite formation from a range of minerals, providing valuable insights into the reaction sequences and formation and constitution of intermediate phases. The most studied of these reactions is the thermal decomposition of the 1:1 layer lattice clay mineral kaolinite, $Al_2Si_2O_5(OH)_4$. The earliest ^{29}Si and ^{27}Al MAS NMR studies of the reaction sequence (Meinhold et al. 1985, MacKenzie et al. 1985a, Brown et al, 1985), although hampered by low MAS speeds and magnetic fields which prevented them to observe all the aluminum intensity, especially that of the 30-ppm resonance, exploited the NMR spectroscopy results to provide information about the constitution of the X-ray amorphous dehydroxylated intermediate (metakaolinite) and its transformation to mullite above 980 °C via a cubic spinel phase. The NMR spectroscopy results suggested the latter is more closely related to γ-alumina than a previously-suggested Al-Si spinel, although a later NMR study of this phase after leaching with alkali (MacKenzie et al. 1996b) indicates the incorporation of up to 3.9 wt.% of SiO_2. The earlier mechanistic conclusions were essentially confirmed by later NMR studies using faster MAS speeds and higher magnetic fields (Sanz et al. 1988, Lambert et al. 1989, Rocha and Klinowski, 1990), which, however, provided additional detail about changes in the aluminum environment during the thermal reactions. A [27]Al MQMAS study of the transformation of metakaolinite to a mixture of poorly crystalline mullite and γ-alumina (Rocha, 1999) indicated a distribution of both quadrupole interactions and chemical shifts, but contributed little further information about the reaction mechanism.

^{29}Si and ^{27}Al MAS NMR spectroscopy has been used to study the formation of mullite from kaolinite under a variety of different reaction conditions, including flash calcination (Slade and Davies, 1991, Meinhold et al. 1993), the influence of lithium nitrate mineralizer on the thermal decomposition of kaolinite (Rocha et al. 1991), the effect of water vapor on the thermal reactions of kaolinite (Temuujin et al. 1999), the effect of various reducing and oxidizing reaction atmospheres on mullite formation from kaolinite (MacKenzie et al. 1996c) and the effect of mechanochemical treatment (high-energy grinding) on the thermal reactions of kaolinite and its mixtures with aluminum hydroxides (Ashbrook et al. 2000, Temuujin et al. 2000).

The 2:1 layer lattice aluminosilicate minerals such as pyrophyllite, $Al_2Si_4O_{10}$ $(OH)_2$, also form mullite on heating. The intermediate dehydroxylated phase (the counterpart of metakaolinite) is of particular interest, since it contains Al(V), independently verifiable by X-ray powder diffraction and containing the expected quadrupolar ^{27}Al MAS NMR line shape rather than the broad 30-ppm peak as in metakaolinite (Fitzgerald et al. 1989). ^{29}Si and ^{27}Al MAS NMR spectroscopy has been used to study the thermal decomposition of pyrophyllite (Frost and Barron, 1984, MacKenzie et al. 1985b, Sánchez-Soto et al. 1993) and the effect of grinding on the thermal decomposition reactions of pyrophyllite (Sánchez-Soto et al. 1997). ^{27}Al and ^{29}Si MAS NMR spectroscopy has also been used to study the thermal decomposition reactions of the 2:1 layer lattice silicates montmorillonite, $(Al,Mg,Fe)_4$ $(Si,Al)_8O_{20}(OH)_4(1/2Ca,Na)_{0.7} \cdot nH_2O$, (Brown et al. 1987) and muscovite mica, $(K_2Al_4(Si_6Al_2)O_{20}(OH)_4$, (MacKenzie et al. 1987), both of which form mullite as one of their high-temperature products. The formation of mullite by thermal decomposition of topaz $(Al_2SiO_4(F,OH)_2)$ has also been studied by solid-state MAS NMR spectroscopy (Day et al. 1995), as has the formation of mullite by thermal decomposition of the amorphous aluminosilicate mineral allophane (MacKenzie et al. 1991) and a related tubular form imogolite (MacKenzie et al. 1989).

2.6.1.5 NMR Spectroscopic Studies of Other Compounds with Mullite Structure

Gallium and germanium mullites   Analogies between the chemistry of aluminum and gallium and between silicon and germanium have led to the successful preparation of several gallium and germanium compounds with the mullite structure and their investigation by multinuclear MAS NMR spectroscopy (Meinhold and MacKenzie, 2000). The ^{27}Al NMR spectrum of the mullite-structured phase $Al_6Ge_2O_{13}$ (Fig. 2.6.9a) shows no unexpected features, containing a narrow octahedral resonance at 5 ppm and a broadened tetrahedral resonance at 61 ppm (but no resolvable analogy of the T* tetrahedral resonance that occurs in aluminosilicate mullite). A similar ^{27}Al MAS NMR spectrum was reported by Jäger et al. (1992) for the central transition of $Al_6Ge_2O_{13}$ mullite, but observations of the satellite transitions revealed a weak shoulder upfield of the main tetrahedral resonance in the position expected for the T* resonance. Attempts by Meinhold and MacKenzie (2000) to observe the ^{73}Ge NMR spectrum were unsuccessful; although this is an NMR-active nucleus, it is extremely intractable and difficult to observe (MacKenzie and Smith, 2002).

The mullite-structured compound $Ga_6Ge_2O_{13}$ has also been successfully prepared and investigated by ^{71}Ga MAS NMR spectroscopy (Fig. 2.6.9b) (Meinhold and MacKenzie, 2000). The broad, featureless nature of the spectrum is thought to be due to the presence of disorder, with its associated range of electric field gradients (EFGs). ^{69}Ga MAS NMR spectra were also acquired from these samples, but all showed a broad, uninformative peak at about 120 ppm. These spectra all had large half-widths resulting from the high quadrupole moment of this nuclide and the presence of large EFGs (Meinhold and MacKenzie, 2000).

The ^{17}O MAS NMR spectra have been reported of an isotopically-enriched gel of initial composition $Ga_6Ge_2O_{13}$ heated to various temperatures (Meinhold and

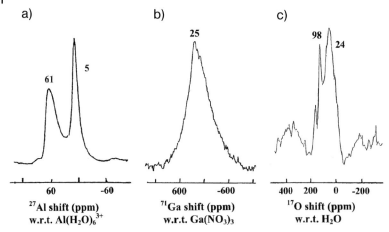

Fig. 2.6.9 (a) 11.7-T ^{27}Al MAS NMR spectrum of Al$_6$Ge$_2$O$_{13}$; (b) 11.7-T ^{71}Ga MAS NMR spectrum of Ga$_6$Ge$_2$O$_{13}$; (c) 11.7-T ^{17}O MAS NMR spectrum of Ga$_6$Ge$_2$O$_{13}$ crystallized from an isotopically-enriched gel (from Meinhold and Mackenzie (2000) by permission of the copyright owner).

MacKenzie, 2000).The spectra (Fig. 2.6.9c) are similar in appearance to the ^{17}O spectra of gel-derived Al$_6$Si$_2$O$_{13}$ reported by Jaymes et al. (1994), and contain a broad resonance at 24–29 ppm, with a narrower feature at 98 ppm. On the basis of the ^{17}O shift differences expected to result from Ga-for-Al and Ge-for-Si substitutions, the broad peak at 24 ppm was tentatively ascribed to the oxygens coordinated to all the octahedral and tetrahedral cations other than the T* cations. The smaller 98-ppm peak may therefore arise from the oxygens associated with the tricluster site, but its observed downfield shift is explicable only if the geometry of this site is slightly different from that of aluminosilicate mullite (Meinhold and MacKenzie, 2000). Alternatively, this minor resonance may arise from an impurity phase such as Ga$_4$GeO$_8$.

Mullite-structured sodium aluminate   The compound NaAl$_9$O$_{14}$ has a mullite-like structure (Fischer et al. 2001) containing aluminum in one octahedral and two tetrahedral sites. ^{27}Al MAS and DOR NMR spectroscopy at several magnetic fields ranging from 8.45 to 16.9 T enabled the spectra to be narrowed sufficiently at the highest field to resolve the two tetrahedral sites (Fig. 2.6.10a) and provide an estimate of their relative populations, confirming the Al site distribution expected from the Rietveld X-ray structure refinement (MacKenzie et al. 2001). Since ^{23}Na is a tractable spin-3/2 quadrupolar NMR nucleus with 100% natural abundance, the crystallographic environment of this element was also investigated by ^{23}Na MAS and DOR NMR spectroscopy (Fig. 2.6.10b), yielding evidence of the presence of two non-equivalent Na sites of approximately equal occupation (MacKenzie et al. 2001). This result is at variance with the average *Pbam* structure deduced from the

a)　　　　　　　　　b)

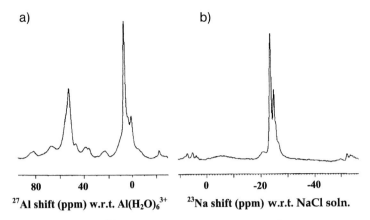

^{27}Al shift (ppm) w.r.t. Al(H$_2$O)$_6^{3+}$　　^{23}Na shift (ppm) w.r.t. NaCl soln.

Fig. 2.6.10 (a) 11.7-T ^{27}Al double rotation (DOR) spectrum of crystalline NaAl$_9$O$_{14}$; (b) 11.7-T ^{23}Na DOR spectrum of crystalline NaAl$_9$O$_{14}$ resolving the two non-equivalent Na sites (from MacKenzie et al. (2001) by permission of the copyright owner).

Rietveld structure refinement, and suggests that NMR spectroscopy is able to distinguish small variation in the Na environments which are not resolved in the averaged X-ray structure.

Mullite-structured aluminum borate　The synthetic aluminum borate Al$_4$B$_2$O$_9$ has a structure closely related to mullite and to the mineral boralsilite. A Rietveld X-ray analysis of this structure has been augmented by ^{27}Al and ^{11}B NMR studies at a variety of magnetic fields from 8.45 T to 16.5 T. ^{11}B is a favorable NMR quadrupolar nucleus with a spin of 3/2 and a high natural abundance (80.42%), allowing facile discrimination of trigonal BO$_3$ and tetrahedral BO$_4$ units (MacKenzie and Smith, 2002). The ^{11}B MAS NMR spectrum of Al$_4$B$_2$O$_9$ (Fig. 2.6.11a) reveals the presence of both tetrahedral and trigonal B–O units in the approximate ratio 25:75, confirming the X-ray structural analysis. The ^{27}Al MAS NMR spectra (Fig. 2.6.11b) show complex line shapes arising from overlapping resonances. Experiments at different fields indicate the presence of aluminum in at least three octahedral sites and three tetrahedral sites, the ratio of octahedral to tetrahedral Al site occupancy being approximately 50:50. Identification of the overlapping Al resonances was confirmed by MQMAS NMR spectroscopy.

X-phase sialon　Sialons are silicon aluminum oxynitride compounds, which can be prepared in a variety of compositions and structural types. The various sialon phases have excellent physical properties, making them suitable as engineering ceramics, high-speed cutting tools and refractories for metal smelting activities. One of the sialons, X-phase, has a composition variously described as Si$_3$Al$_6$O$_{12}$N$_2$ and Si$_{12}$Al$_{18}$O$_{39}$N$_8$. Its structure is similar to that of mullite, and it is sometimes thought of as a solid solution of Si$_3$N$_4$ in mullite. The ^{29}Si NMR spectrum of X-

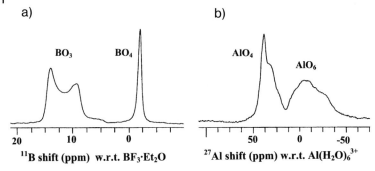

Fig. 2.6.11 (a) 14.1-T ^{11}B MAS NMR spectrum of crystalline Al$_4$B$_2$O$_9$ showing resolution of the characteristic quadrupolar line shape of the more distorted trigonal BO$_3$ unit and the single tetrahedral BO$_4$ resonance; (b) 14.1-T ^{27}Al MAS NMR spectrum of crystalline Al$_4$B$_2$O$_9$ ((MacKenzie and Smith (to be published)).

sialon acquired with a long recycle delay time of 3000 s (Fig. 2.6.12a) consists of a broad envelope with maximum intensity at −66 to −68 ppm containing at least three shoulders at −57, −76 and −90 ppm (Sheppard et al. 1997). The resonance at −57 coincides with the major peak at −56.5 ppm reported by Smith (1994), suggesting the presence of SiO$_2$N$_2$ units in the structure, in which the nitrogen preferentially occupies those tricordinate sites in the mullite structure with Si as nearest neighbors. The other resonances in the ^{29}Si spectrum arise from a range of Si-O-N environments, especially SiO$_3$N, containing at least two Si environments. The ma-

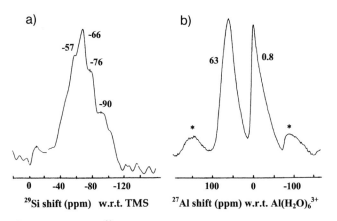

Fig. 2.6.12 (a) 11.7-T ^{29}Si MAS NMR spectrum of X-sialon, Si$_{12}$Al$_{18}$O$_{39}$N$_8$ showing shoulders corresponding to a range of Si-O-N environments. (b) 11.7-T ^{27}Al MAS NMR spectrum of X-sialon showing the characteristic tetrahedral and octahedral Al resonances. Asterisks denote the spinning side bands (from Sheppard et al. (1997) by permission of the copyright owner).

jor Si-O resonance of mullite at −88 to −90 ppm is also present in the ^{29}Si spectrum. The ^{27}Al NMR spectrum (Fig. 2.6.12b) contains distinct signals from both tetrahedral and octahedral Al sites (62–63 ppm and 0.5 to 0.8 ppm, respectively) but no characteristic Al-N or Al-N-O resonances which occur at more positive chemical shifts (112 ppm for Al-N, MacKenzie and Smith 2002).

X-sialon can conveniently be prepared by silico-thermal reaction of a clay such as pure kaolinite with elemental silicon and additional γ-alumina to achieve the desired composition. The course of the reaction, which is carried out in an atmosphere of purified nitrogen at >1400 °C, has been studied by ^{27}Al and ^{29}Si MAS NMR spectroscopy (Sheppard et al. 1997), and found to proceed by the progressive conversion of Si_3N_4 to SiO_2N_2 and SiO_3N units and a change in the tetrahedral/octahedral Al ratio from 0.5 (characteristic of γ-alumina) through about 1.1 (as in mullite) to the typical X-sialon value of 1.3. The effect of a number of metal oxide additives on the sintering and densification of X-sialon has also been studied by ^{27}Al and ^{29}Si MAS NMR spectroscopy (Sheppard and MacKenzie, 1999), and the particular interaction of MgO, Y_2O_3 and Fe_2O_3 sintering additives with X-sialon has also been studied using ^{25}Mg and ^{89}Y NMR spectroscopy and ^{57}Fe Mossbauer spectroscopy respectively (MacKenzie et al. 2000).

Oxidation of X-sialon is of particular concern if the ceramic is to be used at high temperatures in air. The oxidation products are mullite and silica, and a ^{27}Al and ^{29}Si MAS NMR study has revealed details of the reaction, including the formation of a surface layer of amorphous SiO_2, undetected by X-ray diffraction, at an early stage of the oxidation (MacKenzie et al. 1998). It is worth noting that the oxidation products of the other sialons, including β-sialon and O-sialon are also predominantly mullite (MacKenzie et al. 1998). Oxidation of X-sialon to mullite under humid conditions has been studied by ^{27}Al, ^{29}Si and ^{17}O MAS NMR spectroscopy using an atmosphere of ^{17}O-labelled water (Kiyono et al. 2001). The ^{17}O NMR spectra indicate that the oxygen atoms from the water vapor are implicated in the formation of both the mullite and silica oxidation products.

Other minerals structurally related to mullite   The three polymorphs of the 1/1-aluminosilicate Al_2SiO_5 are sillimanite, kyanite and andalusite. Of these, sillimanite is structurally the most similar to mullite, containing Al in both tetrahedral and octahedral sites, as reflected in the ^{27}Al NMR spectrum acquired at 18.8 T (Fig. 2.6.13a), in which both sites show quadrupolar line shapes (MacKenzie et al. unpublished results). The ^{29}Si NMR spectrum of sillimanite (Fig. 2.6.13b) shows a single narrow resonance at −86.4 ppm (Sherriff and Grundy, 1988) to −85.9 ppm (MacKenzie et al. unpublished results), indicative of an ordered Al-Si distribution. By contrast, kyanite contains four equally populated octahedral aluminum sites which can only be distinguished by ^{27}Al MAS NMR spectroscopy with difficulty (Alemany et al. 1991, 1999) and ^{29}Si resonances at −82 and −83 ppm (Hartman and Sherriff, 1991).

The third polymorph, andalusite, contains Al in both fivefold and sixfold coordination with oxygen. The large nuclear quadrupole coupling constant of the octahedral site in andalusite makes it necessary to use high magnetic fields and very

a)

b)

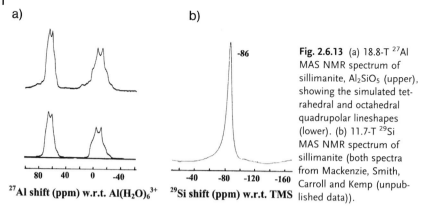

Fig. 2.6.13 (a) 18.8-T ^{27}Al MAS NMR spectrum of sillimanite, Al_2SiO_5 (upper), showing the simulated tetrahedral and octahedral quadrupolar lineshapes (lower). (b) 11.7-T ^{29}Si MAS NMR spectrum of sillimanite (both spectra from Mackenzie, Smith, Carroll and Kemp (unpublished data)).

^{27}**Al shift (ppm) w.r.t. Al(H$_2$O)$_6^{3+}$** ^{29}**Si shift (ppm) w.r.t. TMS**

fast MAS speeds to narrow the ^{27}Al spectrum sufficiently to observe the quadrupolar line shape of the octahedral resonance (Dec et al. 1991). A ^{27}Al MAS NMR study of andalusite has been reported at the very high magnetic field strength of 18.8 T and fast MAS speed of 34 kHz, allowing the relative amounts of Al(V) and Al(VI) to be determined (Alemany et al. 1991, 1999). The ^{29}Si spectrum of andalusite is unremarkable, containing a resonance at –80 ppm (Magi et al. 1984).

Grandidierite, $(Mg,Fe)Al_3SiBO_9$, is an unusual boron aluminosilicate mineral containing aluminum in two octahedral sites and one five-coordinated site. Although it has a similar c axis translation to sillimanite, it is probably more closely related to andalusite. All the constituent elements of this mineral have NMR-active nuclides, and their NMR spectra have been determined. The ^{27}Al spectrum (Fig. 2.6.14a) displays a complex line shape which is best interpreted by making MAS measurements at several different field strengths (Smith and Steuernagel, 1992). The ^{11}B spectrum (Fig. 2.6.14b) confirms that the boron is present in trigonal BO_3 units, while the ^{29}Si spectrum (Fig. 2.6.14c) shows a single resonance with a chemical shift of –79 to –82 ppm, corresponding to Q^0 aluminosilicate units (Smith and Steuernagel, 1992, MacKenzie and Meinhold, 1997). The magnesium in this structure is unusual in being in fivefold coordination, and shows a broad ^{25}Mg NMR resonance with a quadrupolar line shape (Fig. 2.6.14d) (MacKenzie and Meinhold, 1997).

2.6.2
Electron Paramagnetic Resonance (EPR) Spectroscopy

This technique is closely related to NMR spectroscopy, but detects changes in the electron spin configuration rather than the nuclear spin. EPR depends on the presence of unpaired electrons such as those that occur in many transition metal ions, and has therefore been used mainly for studies of transition metal-doped mullites. EPR spectrometers operate at microwave frequencies in the GHz range and the spectra are normally obtained at constant frequency and varying magnetic fields. The measured parameter is the gyromagnetic ratio (g), the value of which

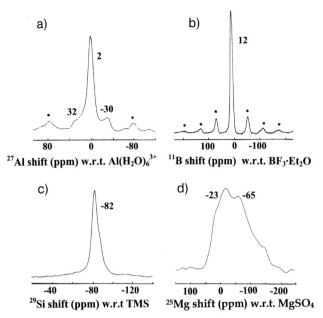

Fig. 2.6.14 11.7-T Multinuclear MAS NMR spectra of grandidierite, $(Mg,Fe)Al_3SiBO_9$. (a) ^{27}Al MAS NMR spectrum. Asterisks denote spinning side bands. (b) ^{11}B MAS NMR spectrum. Asterisks denote spinning side bands. (c) ^{29}Si MAS NMR spectrum. (d) ^{25}Mg spectrum obtained using a Hahn echo pulse sequence, showing a pronounced quadrupolar lineshape (from MacKenzie and Meinhold (1997) by permission of the copyright owner. Note that spectra A–C are similar to those reported by Smith and Steuernagel (1992) at several other field strengths).

depends on the particular ion, its oxidation state and its coordination number. To overcome broadening problems it may be necessary to ensure that the sample contains only low concentrations of unpaired electrons and to record the spectra at low temperatures. EPR spectra are usually presented as the first derivative of the absorption spectrum, and their unambiguous interpretation can be difficult.

EPR studies of Cr^{3+} substitution into the mullite structure (Fig. 2.6.15a) agree that this ion preferentially enters the octahedral sites. As indicated by an EPR study by Rager et al. (1990), Cr^{3+} can also enter octahedral interstitial sites. All the Cr-doped mullites investigated in that study showed two sharp signals at about $g_{eff} = 5$, assigned to Cr^{3+} in a strong crystal field of orthorhombic character (Cr in the normal octahedral sites), and a broad, slightly asymmetric EPR signal showing fine structure near $g_{eff} = 2.2$ arising from coupling between localized magnetic moments (interstitial Cr^{3+}). These EPR spectra were interpreted as evidence for the presence of Cr pairs in both regular octahedral sites and adjacent interstitial sites (Rager et al. 1990). The entry of Cr^{3+} only into the octahedral sites of mullite was confirmed by an EPR study of Chaudhuri and Patra (2000). EPR spectroscopy has

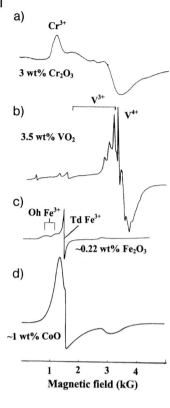

Fig. 2.6.15 A selection of typical EPR spectra of mullite containing substituted transition metal ions.
(a) Chromium-doped mullite gel precursor fired at 1650 °C showing octahedral Cr^{3+} (from Schneider et al. (1994b)). (b) Vanadium-doped mullite fired under strongly reducing conditions [from Schneider (1990)]. (c) Iron-doped 3:2 sinter-mullite showing slight evidence of octahedral substitution (from Schneider and Rager (1986)). (d) Cobalt-doped mullite showing signals attributed to octahedral Co^{2+} (from Schneider (1990)).

been used to monitor the reactions by which Cr is incorporated into mullite during the thermal treatment of gel precursors (Schneider et al. 1996). Samples heated at 400–800 °C show an EPR signal at g_{eff} = 1.96, corresponding to Cr^{5+}. At 800–1200 °C, the EPR spectrum is typical of chromium-containing glasses, but changes to the typical signal of Cr-doped mullite at about 1250 °C (Schneider et al. 1996). EPR spectroscopy has also proved useful in studies of the behavior of Cr in Cr-doped mullite glass-ceramics developed for luminescent solar converter applications and as solid-state laser materials (Reisfield et al. 1986, Knutson et al. 1989). These studies indicate that the Cr^{3+} is located at disordered low-field sites in regions of disorder within the mullite ceramic host.

The entry of other transition metal ions into the structure of mullite has also been studied by EPR spectroscopy. Klochkova et al. (2001) have shown that Sc, Ti, V, Mn, Cr and Fe all enter the octahedral sites of mullite preferentially when added as the sesquioxides. The EPR studies of Chaudhuri and Patra (2000) indicate that Mn is present in doped mullite as both Mn^{2+} (as clusters) and Mn^{3+} (in the octahedral sites).

EPR studies of Ti-doped mullite (Schneider and Rager, 1984) indicate that both Ti^{4+} and Ti^{3+} can enter the mullite structure of fusion-cast refractories. The Q-band spectra of Ti^{3+}-doped samples show two components near g_{eff} = 1.929, attributed to two inequivalent centers associated with ordering patterns of the octahedra adjacent to two different tetrahedral double chains (Rager et al. 1993).

An EPR study of vanadium-doped mullite synthesized under reducing conditions (Schneider, 1990) showed a group of signals near g_{eff} = 4.1, attributed to octahedral V^{3+}, and a weak band near g_{eff} = 1.9, attributed to a small amount of V^{4+} (Fig. 2.6.15b). Vanadium-containing mullites prepared in less reducing conditions showed signals near g_{eff} = 2.0 arising from localized octahedral V^{4+} (Schneider, 1990).

EPR spectroscopy has been used to augment results from Mössbauer spectroscopy regarding the cooperative incorporation of Fe and Ti into the mullite structure (Schneider and Rager, 1984, 1986). EPR signals near g_{eff} = 6.8 and 5.1 were attributed to Fe^{3+} in octahedral sites on the basis of their axially symmetric crystal field, while signals near g_{eff} = 4.2 were assigned to Fe^{3+} in tetrahedral sites in which the crystal field is completely orthorhombic (Fig. 2.6.15c; Schneider and Rager, 1986).

EPR studies have also been made of Co-substituted mullite (Fig. 2.6.15d; Schneider, 1990), suggesting that Co^{2+} occupies octahedral sites, and of Eu^{2+} in 3:2 and 2:1 mullite (Kutty and Nayak, 2000) in which non-emissive defect centers were identified.

2.6.3
Infrared (IR), Fourier-transform Infrared (FTIR) and Raman Spectroscopy

In the IR technique, the frequency of the IR radiation incident on the sample is varied and the quantity of radiation adsorbed or transmitted is recorded as a function of frequency. The sample may be suspended in a disk of KBr or made into a dilute paste with an organic agent such as Nujol. The IR spectra of inorganic materials such as mullite tend to be broad, but their quality and signal/noise ratio can be improved by the technique of FTIR, in which a Fourier transformation algorithm is used to extract the frequency information from the signal. For a vibrational mode to be IR-active, the associated dipole moment must vary during the vibration.

In the closely-related Raman technique, the sample is illuminated with monochromatic light, usually generated by a laser. The intensity of the Raman scattered radiation is detected in a direction perpendicular to the incident beam and plotted as a function of the frequency. For a vibration to be Raman-active, it must involve a change in polarizability.

2.6.3.1   The IR Spectrum of Mullite
The IR spectrum of 3:2 mullite (Fig. 2.6.16a) shows a number of broad bands and shoulders, typical of an aluminosilicate mineral spectrum. Two approaches have been used to identify the lattice vibrations responsible for these absorption bands. In an early attempt, calculations were made based on site-group analysis of a simplified model of the mullite structure by assuming independently vibrating

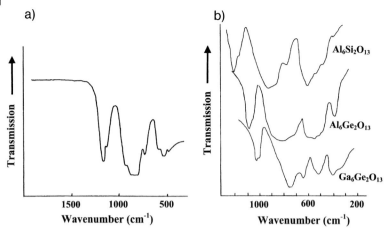

Fig. 2.6.16 (a) IR transmission spectrum of crystalline 3:2 mullite suspended in a KBr disk (from Mackenzie (1972) by permission of the copyright owner). (b) IR transmission spectra of 3:2 mullite and its Ga- and Ge-analogues, showing the resulting shifts in the position of the various bands (from Schneider (1981) by permission of the copyright owner).

structural units (MacKenzie, 1972). Since this approach does not take into account the structural disorder and the incommensurate character of the mullite structure, an alternative approach has been used, involving analysis of the experimentally observed frequency shifts in the IR spectrum resulting from the substitution of Ga for Al or Ge for Si in the mullite structure (Fig. 2.6.16b; Schneider 1981, Voll et al. 2001). This approach has led to the identification of IR bands in the region 1125–1165 cm^{-1} as Si–O stretching frequencies, the band at 950–988 cm^{-1} as an Si–O stretch, the band at 830–909 cm^{-1} as a tetrahedral Al–O stretch, the band at 730–737 as a tetrahedral (Al or Si) bend, the band at 620 cm^{-1} as a tetrahedral Al–O bend, the band at 548–578 cm^{-1} as an octahedral Al–O stretch and the band at 482–498 cm^{-1} as an Si–O bend. Similarly, comparisons between the IR spectra of Al-Si and the Ga and Ge- substituted analogues have been used to assign the IR bands in andalusite-type structures (Schneider, 1981).

2.6.3.2 IR Spectroscopic Studies of Mullite Formation

Irrespective of the details of any conflicting details in the band assignments, IR spectroscopy has proved useful in studies of mullite formation from heated kaolinite, making use of an empirical observation that the center-of-gravity of the band structure at 790 to 870 cm^{-1} progressively shifts to higher wavenumbers as the reaction proceeds (Fig. 2.6.17a), and further, that this shift bears a linear relationship to the amount of mullite formed, determined by quantitative X-ray diffraction as shown in Fig. 2.6.17b, MacKenzie 1969a). This IR method has been utilized in a kinetic study of mullite formation from synthetic kaolinite, where the amount

a)
b)

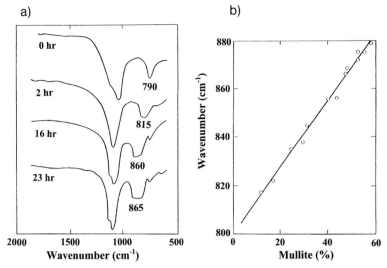

Fig. 2.6.17 (a) Changes in the infrared (IR) transmission spectra during the course of mullite development in kaolinite heated at 1100 °C for various times. Note the progressive shift of the COG of the mullite band centred at 865 cm^{-1}. (b) Relationship between the position of the 850 cm^{-1} IR band and the mullite content determined by quantitative XRD in kaolinite heated at 1100 °C for varying times (from MacKenzie, 1969, by permission of the copyright owner).

of sample available was too small for analysis by other means (Mackenzie, 1969b). More recently, the ratio of IR intensities measured at 1130 and 1170 cm^{-1} (Fig. 2.6.18a) has been shown to bear a linear relationship to the the composition of the mullite, as defined by the value of x in the structural formula $Al_2(Al_{2+2x}Si_{2-2x})O_{10-x}$ (Fig. 2.6.18b), suggesting an alternative to X-ray lattice parameter measurements for determining mullite compositions (Rüscher et al. 1996).

IR and FTIR spectroscopy have been used to study mullite formation from gel precursors (Dimitriev et al. 1998, Satoshi et al. 2001, Padmaja et al. 2001) and an emission IR technique using a heated IR cell has been used in an in situ study of mullite formation from kaolinite (Vassallo et al. 1992). The IR and Raman spectra of a number of mullite gels and their Ge-substituted analogues have been reported, and changes observed in the SiO$_4$ stretching region at 970 cm^{-1} during thermal crystallization have been explained as a progressive ordering of the silicate environment (Colomban, 1989). IR and micro-Raman spectroscopy have been used to study the thermal formation of mullite from the kaolinite polytype mineral dickite (Shoval et al. 2002). IR spectroscopy has also been used to study the formation of mullite from mechanochemically treated kaolinite (Mako Eva and Zoltan, 1997), from ultrasonically treated gel precursors (Woo et al. 1996), from co-precipitated gels doped with 2% of the transition metal ions Mn, Fe, Cr and Ti (Chaud-

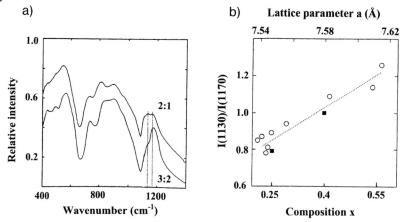

Fig. 2.6.18 (a) IR spectra of 2:1 and 3:2 mullites showing the positions of the 1130 and 1170 cm^{-1} regions (dotted lines) which can be used to determine the mullite composition. (b) Relationship between the ratio of the IR intensities at 1130 and 1170 cm^{-1} and the composition parameter x in the mullite formula $Al_2(Al_{2+2x}Si_{2-2x})O_{10-x}$ determined from the mullite a-parameter (from Rüscher et al. (1996) by permission of the copyright owner).

huri and Patra 1997) and from the 2:1 layer lattice aluminosilicate mineral pyrophyllite (Chen et al. 1991). The technique has also proved useful for studying the formation of mullite diffusion barrier coatings on SiC/C composites by pulsed laser deposition (Fritze et al. 1999) and the synthesis of nanostructured spherical mullite powders by spray pyrolysis (Janackovic et al. 1998). IR spectroscopy has been proposed as a rapid method of quality control for glassy aluminosilicate fibers, using the technique to determine the degree of mullite formation and hence the degree of fiber crystallization (Kutzendorfer and Zahradkova, 1984).

2.6.4
Mössbauer Spectroscopy

The γ-rays used in Mössbauer spectroscopy occur in the electromagnetic spectrum (Fig. 2.6.1) on the high-energy side of X-rays. Their emission from a radioactive nucleus is associated with a change in the population of the energy levels rather than a change in the atomic mass or number. Under certain conditions, called "recoilless emission", all of this energy is transmitted to the γ-rays. The resulting highly monochromatic γ-ray beam may be absorbed by a sample containing similar atoms to those responsible for the emission. Although a number of radioactive nuclides suitable for Mössbauer spectroscopy are known, ^{57}Fe and ^{119}Sn are the two of greatest chemical interest; the majority of Mössbauer studies are therefore of iron and, to a lesser extent, tin compounds. Since the nuclear energy levels of the absorber may vary somewhat, depending on the oxidation state, spin state and coordination number in the compound, the energy of the incident γ-ray beam must be modulated sufficiently to detect resonance. This is accomplished by use of the Doppler effect, in which the source or absorber is moved at constant velocity towards or away from each other. The resonance absorption is recorded as a function of source or absorber velocity and is reported as a chemical shift (CS) with respect to the resonance position of a standard substance (natural Fe metal is typically used in ^{57}Fe Mössbauer studies). Another useful diagnostic parameter is the quadrupole splitting (QS) or separation of doublet peaks arising from the splitting of the nuclear energy levels due to non-spherical charge distribution in nuclei with nuclear spin quantum numbers $I > \frac{1}{2}$. Since both the CS and QS values are sensitive to local structure and oxidation state of the target nucleus, they are used in conjunction to provide chemical and structural information.

Mössbauer spectroscopy can also provide information about the internal magnetic field from measurements of the six-line spectra resulting from magnetic hyperfine Zeeman splitting in magnetic samples. The temperature dependence of the appearance of a magnetic sextet can also provide information about magnetic ordering in the sample. For further information the reader should refer to texts such as Greenwood and Gibb (1971) and Bancroft (1973).

The principal application of Mössbauer spectroscopy in mullite studies has been to determine the location of structurally substituted iron cations. Studies on several Fe-containing mullites of various origins have been reported, including mullites from fired Fe-containing clays (Cardile et al. 1987), mullites prepared by sol-gel synthesis (Parmentier et al. 1999, Chaudhuri and Patra, 2000) and materials formed by high-temperature solid-state reaction of the oxides (Schneider and Rager, 1986, McGavin and MacKenzie, 1994, unpublished, Mack et al. 2004). A general conclusion of all these studies is that although up to three Fe^{3+} sites can be distinguished, their Mössbauer parameters (Table 2.6.1) are too similar to allow unambiguous identification of octahedral and tetrahedral sites; other factors such as relative site occupancies, Rietveld structure refinements or the temperature dependence of the spectra have therefore been invoked to interpret the Mössbauer spectra. Although there is general agreement regarding the number of sites to be fitted, the procedures adopted for assigning these sites from other indirect evidence have led to discordant interpretations of the spectra.

Since many ceramic clays contain significant iron contents that on thermal decomposition of the clay can potentially enter the resulting mullite, a Mössbauer study was made of ten clay-derived mullites (Cardile et al. 1987). Some of the iron in clays occurs as fine hydrous oxides which were removed by treatment with dithionate-citrate-bicarbonate before thermal decomposition. The Mössbauer spectra of all these mullites could best be fitted by three pairs of quadrupole doublets (Fig. 2.6.19a, Table 2.6.1). The resonance (a) of highest occupancy (IS = 0.28–0.37 mm s^{-1} w.r.t. natural Fe, QS = 1.17–1.30 mm s^{-1}) has parameters similar to those found in other fired clays (Janot and Delcroix, 1974) and were assigned to iron in the octahedral sites. The next most populated site (b) (IS = 0.28–0.38 mm s^{-1}, QS = 0.63–0.92 mm s^{-1}) was assigned to the iron in the normal tetrahedral sites, while site (c) of lowest population (IS = 0.42–

Table 2.6.1 Mössbauer parameters of Fe-containing mullites.

Sample origin	Site	IS (mm s^{-1})[a]	QS (mm s^{-1})	Relative site occupancy	Assignment	Temperature	Reference
Fired clay	a	0.28–0.37	1.17–1.30	Highest	Fe^{3+} Oh	RT	Cardile et al. (1987)
	b	0.28–0.38	0.63–0.92	Next highest	Fe^{3+} Td	RT	
	c	0.42–0.76	2.14–2.72	Lowest	Fe^{3+} Td distorted	RT	
Synthetic from sol-gel	a	0.31	1.18–1.32	High	Fe^{3+} Oh	RT	Parmentier et al. (1999)
	b	0.31	0.76–0.87	High	Fe^{3+} Td	RT	
	c	0.29–0.32	1.58–1.92	Low	Fe^{3+} Td distorted	RT	
Synthetic from sol-gel		0.28	1.14	High	Fe^{3+} Oh	77 K	Chaudhuri and Patra (2000)
Synthetic from oxides	a	0.30–0.31	1.27–1.28	Highest	Fe^{3+} Oh	RT	McGavin and MacKenzie (1994)
	b	0.31	0.80–0.82	Next highest	Fe^{3+} Td	RT	
	c	0.78–0.79	2.73–2.80	Lowest	Fe^{3+} Td distorted	RT	
	a	0.40	1.24	Highest	Fe^{3+} Oh	77 K	McGavin and MacKenzie (1994)
	b	0.41	0.77	Next highest	Fe^{3+} Td	77 K	
	c	1.0	2.8	Lowest	Fe^{3+} Td distorted	77 K	
Synthetic from oxides	a	0.31	1.14	Highest	Fe^{3+} Oh	RT	Mack et al. (2005)
	b	1.29	0.72	Next highest	Fe^{3+} Oh	RT	
	c	0.10	1.4	Very low	Fe^{3+} Td	RT	
	a	0.40	1.08	Highest	Fe^{3+} Oh	1100 °C	Mack et al (2005)
	b	0.41	0.69	Next highest	Fe^{3+} Oh	1100 °C	
	c	0.20	1.2	Very low	Fe^{3+} Td	1100 °C	

a) Isomer shift values quoted relative to natural α-Fe

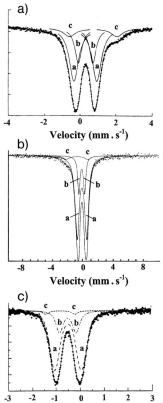

Fig. 2.6.19 (a) Typical room-temperature ^{57}Fe Mössbauer spectrum of iron-substituted mullite derived from heated kaolinite, fitted to three quadrupole doublets [from Cardile et al. (1987) by permission of copyright owner]. (b) Typical 77 K ^{57}Fe Mössbauer spectrum of 3:2 mullite containing 9 wt. % Fe$_2$O$_3$ synthesized from pure oxides. Note the extended velocity range showing an absence of magnetically split sextets (from McGavin and MacKenzie (unpublished results)). (c) Typical ^{57}Fe Mössbauer spectrum acquired at 1100 °C of 3:2 mullite containing 10.3 wt % Fe$_2$O$_3$ synthesized from pure oxides (from Mack et al. (2004) by permission of the copyright owner). The identification of all the doublet sites is given in Table 2.6.1.

0.76 mm s^{-1}, QS = 2.14–2.72 mm s^{-1}) was assigned to iron in the more distorted tetrahedral tricluster sites (Cardile et al. 1987). It was noted in this study that these site assignments were based as much on the site occupancies as on the Mössbauer parameters, which were not sufficiently different to distinguish between tetrahedral and octahedral sites. The best fit to three sites suggested that substitution of all the Al sites in mullite was occurring, but the extent of site substitution could not be determined from the Mössbauer spectra.

The Mössbauer spectra of two iron-substituted mullites of composition Al$_{4.5-y}$Fe$_y$Si$_{1.5}$O$_{9.75}$ where y = 0.25 and 0.5, prepared by a sol-gel route (Parmentier et al. 1999) were fitted to three Fe(III) sites with Mössbauer parameters: IS = 0.31 mm s^{-1}, QS = 0.76–0.87 mm s^{-1} (site a), IS = 0.31 mm s^{-1}, QS = 1.18–1.32 mm s^{-1} (site b), IS = 0.29–0.32 mm s^{-1}, QS = 1.58–1.92 mm s^{-1} (site c) (Table 2.6.1). Since these parameters are too similar to allow a distinction between octahedral and tetrahedral sites, a comparison was made of the Mössbauer site occupancies with a Rietveld X-ray analysis of these samples, suggesting that site (a) corresponds to the normal tetrahedral site, site (b) to the octahedral site and site (c) to the T* site (Parmentier et al. 1999). By contrast, another Mössbauer study of iron in hybrid gel-derived mullites adopted a single-doublet fit to the broad spectrum

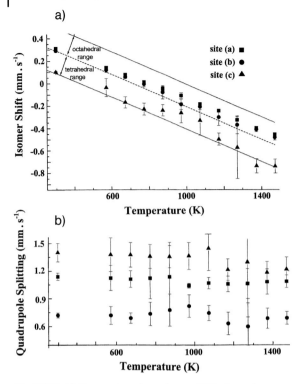

Fig. 2.6.20 (a) Temperature dependence of Mössbauer isomer shifts for 3:2 mullite containing 10.3 wt. % Fe_2O_3 synthesized from the pure oxides. (b) Temperature dependence of the quadrupole splitting for the same sample. The identification of the three sites is given in Table 2.6.1 (adapted from Mack et al. (2005) by permission of the copyright owner).

(IS = 0.28 mm s^{-1}, QS = 1.14 mm s^{-1}), assigning this to Fe substitution in the octahedral site (Chaudhuri and Patra, 2000).

The Mössbauer spectra of iron incorporated into synthetic mullites prepared by solid-state reaction of the pure oxides at 1400–1450 °C (Schneider and Rager, 1986) showed a broad doublet which could be fitted either by a single (distorted octahedral) site or by two doublets corresponding to octahedral and tetrahedral sites; it was not possible to determine the more likely assignment from these spectra (for which no Mössbauer parameters were quoted).

In another Mössbauer study of mullite synthesized from the oxides, McGavin and MacKenzie recorded the spectra of a sample containing 9 wt % Fe_2O_3 at both room temperature (RT) and 77K (unpublished results 1994). There was no sign of any magnetically-split sextets in the low-temperature spectrum (Figure 2.6.19b), and both this and the RT spectrum could be fitted best by three quadrupole doublets (Table 2.6.1). Although the IS values recorded in the 77 K spectrum were larger than at room temperature and doublet (c) was weaker, no significant differences were detected in the QS values or the site occupancies within the accuracy of the experiments.

The Mössbauer spectra of 3:2 mullite synthesized with 10.3 wt% Fe_2O_3 from the pure oxides have been determined at temperatures from RT to 1200 °C (Mack et al. 2004). The spectra at all temperatures were fitted best by three doublets with Voigt-type peak shapes (Fig. 2.6.19c). The IS values of all three sites become more negative with increasing temperature, but the QS values and peak widths showed no significant temperature dependence (Fig. 2.6.20, Table 2.6.1). On the basis of these results, the two most intense doublets were assigned as octahedral Fe^{3+} in two different structural environments, site (a) being in close proximity to TO_4 tetrahedra only and site (b) having both TO_4 and T^*O_4 nearest neighbours. The much less intense doublet (c) was assigned to tetrahedral Fe^{3+} in strongly distorted sites, possibly reflecting the preferential entry of Fe^{3+} into the more commodious T^* sites (Mack et al. 2004). The high-temperature spectra showed no evidence for any temperature-induced redistribution of Fe^{3+} in the mullite structure.

The behavior of more dilute concentrations of ferric iron in mullite has been studied by introducing isotopically-enriched ^{57}Fe as the nitrate into crystalline mullite prepared from the metal alkoxides and heated at 1500 °C (Dickson and Srivastava, 1976). Analysis of the hyperfine spectra, recorded at low temperatures (4 K to 295 K) and with external applied magnetic fields, allowed the Mössbauer parameters, including the zero-field splitting, to be determined for both the octahedral and tetrahedral Fe^{3+} sites.

References

Aksay, I. A., Dabbs, D. M. and Sarikaya, M. (1991). Mullite for structural, electronic, and optical applications. J. Am. Ceram. Soc. **74**, 2343–2358.

Alemany, L. B., Massiot, D., Sherriff, B. L., Smith, M. E. and Taulelle, F. (1991). Observation and accurate quantification of ^{27}Al MAS NMR spectra of some Al_2SiO_5 polymorphs containing sites with large quadrupole interactions. Chem. Phys. Lett. **177**, 302–306.

Alemany, L. B., Steuernagel, S., Amoureaux, J-P., Callender, R. L. and Barron, A. R. (1999). Very fast MAS and MQMAS NMR studies of the spectroscopically challenging minerals kyanite and andalusite on 400, 500 and 800 MHz spectrometers. Solid State Nucl. Magn. Res. **14**, 1–18.

Ashbrook, S. E., McManus, J., MacKenzie, K. J. D. and Wimmperis, S. (2000). Multiple-quantum and cross-polarised ^{27}Al MAS NMR of mechanically treated mixtures of kaolinite and gibbsite. J. Phys. Chem. **B 104**, 6048–6016.

Bancroft, G. M., Mössbauer Spectroscopy: An Introduction for Inorganic Chemists and Geochemists. McGraw-Hill, London, 1973.

Barin, I. and Knacke, D. (1973). Thermochemical Properties of Inorganic Substances. Springer, Berlin, Vols. 1, 2.

Barin, I., Sauert, F., Schultze-Rhonhof, E. and Sheng, W. C. (1993). Thermochemical Data of Pure Substances, Verlag Chemie, Weinheim, Vols. 1, 2.

Bauer, W. H., Gordon, I. and Moore, C. H. (1950). Flame fusion synthesis of mullite single crystals. J. Am. Ceram. Soc. **33**, 140–143.

Bauer, W. H. and Gordon, I. (1951). Flame fusion synthesis of several types of silicate structures. J. Am. Ceram. Soc. **34**, 250–254.

Beger, R. M., Burnham, C. W. and Hays, J. F. (1970). Structural changes in sillimanite at high temperature. In: Abstr. Annual Meetg. Amer. Geol. Soc., Milwankee, Nov. 11–13, 1970, pp. 490–491.

Berezhnoi (1970). Cited in: Shornikov und Archakov (2002), see below.

Bodart, P. R., Parmentier, J., Harris, R. K. and Thompson, D. P. (1999). Aluminium environments in mullite and amorphous sol-gel precursor examined by ^{27}Al triple-quantum MAS NMR. J. Phys. Chem. Solids **60**, 223–228.

Böhm, H. and Schneider, H. (unpublished). Electrical conductivity of mullite up to 1000 °C.

Braue, W., Hildmann, B. and Schneider, H. (2005a). Structural response of mullite upon shock loading. To be published.

Braue, W., Hildmann, B., Schneider, H., El-dred, B. T. and Ownby, P. D. (2005b). Reactive wetting of mullite Al$_2$ [Al$_{2+2x}$Si$_{2-2x}$] O$_{10-x}$ single crystals by yttrium-aluminosilicate glasses. To be published.

Brindley, G. W. and McKinstry, H. A. (1961). The kaolinite-mullite reaction series: IV. The coordination number of aluminium. J. Am. Ceram. Soc. **44**, 506–507.

Brindley, G. W. and Nakahira, M. (1959a). The kaolinite-mullite reaction series: I. A survey of outstanding problems. J. Am. Ceram. Soc. **42**, 311–314.

Brindley, G. W. and Nakahira, M. (1959b). The kaolinite-mullite reaction series: II. Metakaolin. J. Am. Ceram. Soc. **42**, 314–318.

Brindley, G. W. and Nakahira, M. (1959c). The kaolinite-mullite reaction series: III. The high-temperature phases. J. Am. Ceram. Soc. **42**, 319–324.

Brown, I. W. M., MacKenzie, K. J. D., Bowden, M. E. and Meinhold, R. H. (1985). Outstanding problems in the kaolinite-mullite reaction sequence investigated by ^{29}Si and ^{27}Al solid-state nuclear magnetic resonance: II. High temperature transformations of metakaolinite. J. Am. Ceram. Soc. **68**, 298–301.

Brown, I. W. M., MacKenzie, K. J. D. and Meinhold, R. H. (1987). The thermal reactions of montmorillonite studied by high-resolution solid-state ^{29}Si and ^{27}Al NMR. J. Mater. Sci. **22**, 3265–3275.

Brunauer, G., Frey, F., Boysen, H. and Schneider, H. (2001). High temperature thermal expansion of mullite: An in situ neutron diffraction study up to 1600 °C. J. Europ. Ceram. Soc. **21**, 2503–2507.

Bobkova (1992). Cited in: Shornikov and Archakov (2002), see below.

Bulens, M., Leonard, A. and Delmon, B. (1978). Spectroscopic investigations of the kaolinite-mullite reaction sequence. J. Am. Ceram. Soc. **61**, 81–84.

Cameron, W. E. (1976a). Coexisting sillimanite and mullite. Geol. Mag. **113**, 497–514.

Cameron, W. E. (1976b). Exsolution in 'stoichiometric' mullite. Nature, London **264**, 736–738.

Campos, T. W., de Souza Santos, T. and Souza Santos, P. (1976). Mullite development from fibrous kaolin mineral. J. Am. Ceram. Soc. **59**, 357–360.

Cardile, C. M., Brown, I. W. M. and MacKenzie, K. J. D. (1987). Mossbauer spectra and lattice parameters of iron-substituted mullites. J. Mater. Sci. Lett. **6**, 357–362.

Cesare, B., Gomez-Pugnaire, M. T., Sanchez-Navas, A. and Grobety, B. (2002). Andalusite-sillimanite replacement (Mazarron, SE Spain): A microstructural and TEM study. Amer. Min. **87**, 433–444.

Chakraborty, A. K. and Gosh, D. K. (1977). Re-examination of the decomposition of kaolinite. J. Am. Ceram. Soc. **60**, 165–166.

Chakraborty, A. K. and Ghosh, D. K. (1978a). Reexamination of kaolinite-to-mullite reaction series. J. Am. Ceram. Soc. **61**, 170–173.

Chakraborty, A. K. and Ghosh, D. K. (1978b). Comment on 'Interpretation of the kaolinite-mullite reaction sequence from infrared absorption spectra. J. Am. Ceram. Soc. **61**, 90–91.

Chase, N. W. (1998). NIST – JANAF Thermochemical Tables. J. Phys. Chem. Ref. Data **9**, 1–1951.

Chatterjee, N. D. and Schreyer, W. (1972). The reaction enstatite$_{ss}$ + sillimanite = sappharine$_{ss}$ + quartz in the system MgO-Al$_2$O$_3$–SiO$_2$. Contrib. Mineral. Petrol. **36**, 49–62.

Chaudhuri, S. P. and Patra, S. K. (1997). Preparation and characterisation of transition metal ion doped mullite. Brit. Ceram. Trans. **96**, 105–111.

Chaudhuri, S. P., Patra, S. K. and Chakraborty, A. K. (1999). Electrical resistivity of transition metal ion doped mullite. J. Europ. Ceram. Soc. **19**, 2941–2950.

Chaudhuri, S. P. and Patra, S. K. (2000). Electron paramagnetic resonance and Mossbauer spectroscopy of transition metal ion doped mullite. J. Mater. Sci. **35**, 4735–4741.

Chen, D., Jiang, Z. and Zhang, H. (1991). Differential thermal analysis and infrared absorption spectroscopy of pyrophyllite from China. Kuangwu Xuebao, **11**, 92–96.

Colomban, P. (1989). Structure of oxide gels and glasses by infrared and Raman scattering. J. Mater. Sci., **24**, 2011–2020.

Comer, J. J. (1960). Electron microscope studies of mullite development in fired kaolinites. J. Am. Ceram. Soc. **43**, 378–384.

Comer, J. J. (1961). New electron-optical data on the kaolinite-mullite transformation. J. Am. Ceram. Soc. **44**, 561–563.

Comoforo, J. E., Fischer, R. B. and Bradley, W. F. (1948). Mullitization of kaolinite. J. Am. Ceram. Soc. **31**, 254–259.

Crank, J. (1975). The Mathematics of Diffusion. 2nd edn., Oxford University Press, Oxford.

Dabbs, D. M., Yao, N. and Aksay, I. A. (1999). Nanocomposite mullite/mullite powders by spray pyrolysis. J. Nanopart. Res. **1**, 127–130.

Davis, R. F. and Pask, J. A. (1972). Diffusion and reaction studies in the system Al_2O_3–SiO_2. J. Am. Ceram. Soc. **55**, 524–531.

Day, R. A., Vance, E. R., Cassidy, D. J. and Hartman, J. S. (1995). The topaz to mullite transformation on heating. J. Mater. Res. **10**, 2693–2699.

Dec, S. F., Fitzgerald, J. J., Frye, J. S., Shatlock, M. P. and Maciel, G. E. (1991). Observation of five-coordinate aluminium in andalusite by solid-state ^{27}Al MAS NMR. J. Magn. Res. **93**, 403–406.

Dickson, B. L. and Srivastava, K. K. P. (1976). A Mossbauer study of the relaxation behaviour of dilute Fe^{3+} in $LiScO_2$ and mullite. J. Solid State Chem. **19**, 117–123.

Dimitriev, Y., Samuneva, B., Kashchieva, E. and Gattef, E. (1998). Microstructure of glasses and glass-ceramics in the system Al_2O_3-SiO_2 obtained by sol-gel technology. Proc. 18th Intern. Conf Glass, San Francisco, U. S. A., July 1998, Amer. Ceram. Soc., Westerville, pp. 2873–2878.

Dreyer, W. (1974). Materialverhalten Anisotroper Festkörper. Springer, Vienna, New York.

Dokko, P. C., Pask, J. A. and Mazdiyasni, K. S.

(1977). High-temperature mechanical properties of mullite under compression. J. Am. Ceram. Soc. **60**, 150–155.

Eils, N., Rüscher, C. H. and Schneider, H. (2005). Hydrothermal transformation of mullite to sillimanite plus α-alumina. To be published.

Faber, K. T. and Evans, A. G. (1983). Crack deflection process – II. Experiment. Acta Metall. **31**, 577–584.

Fielitz, P., Borchart, G., Schmücker, M., Schneider, H., Wiedenbeck, M., Rhede, D., Weber, S. and Scherrer, S. (2001a). Secondary ion mass spectroscopy study of oxygen-18 tracer diffusion in 2/1-mullite single crystals. J. Am. Ceram. Soc. **84**, 2845–2848.

Fielitz, P., Borchert, G., Schneider, H., Schmücker, M., Wiedenbeck, M. and Rhede, D. (2001b). Self diffusion of oxygen in mullite. J. Europ. Ceram. Soc. **21**, 2577–2582.

Fielitz, P., Borchert, G., Schmücker, M., Schneider, H. and Willich, P. (2003a). Measurement of oxygen grain boundary diffusion in mullite ceramics by SIMS depth profiling. Appl. Surf. Sci. **203–204**, 639–643.

Fielitz, P., Borchert, G., Schmücker, M. and Schneider, H. (2003b). Silicon tracer diffusion in single crystalline 2/1-mullite measured by SIMS depth profiling. Phys. Chem. Chem. Phys. **5**, 2279–2282.

Fischer, R. X., Schmücker, M., Angerer, H. and Schneider, H. (2001). Crystal structures of Na and K aluminate mullites. Amer. Min. **86**, 1513–1518.

Fischer, R. X., Schneider, H. and Schmücker, M. (1994). Crystal structure of Al-rich mullite. Amer. Min. **79**, 983–990.

Fischer, R. X., Schneider, H. and Voll, D. (1996). Formation of aluminium rich 9:1 mullite and its transformation to low alumina mullite upon heating. J. Europ. Ceram. Soc. **16**, 109–113.

Fischer, W. E. and Janke, D. (1969). Festelektrolytzellen mit gasförmigen Vergleichselektronen zur Messung der Sauerstoffaktivität in Eisenschmelzen. Arch. Eisenhüttenwesen **40**, 1027–1033.

Fitzgerald, J. J., Dec, S. F. and Hamza, A. I. (1989). Observation of five-coordinated Al in pyrophyllite dehydroxylate by solid-state 27-Al NMR spectroscopy. Amer. Min. **74**, 1405–1408.

Freund, F. (1967). Kaolinit-Metakaolinit, Modellfall eines Festkörpers mit extrem hohen Störstellenkonzentrationen. Ber. Dtsch. Keram. Ges. **44**, 5–13.

Fritze, H., Schnittker, A., Witke, T., Rüscher, C., Weber, S., Scherrer, S., Schultrich, B. and Borchardt, G. (1999). Mullite diffusion barriers for SiC-C/C composites produced by pulsed laser deposition. Mater. Res. Soc. Symp. Proc. **555**. Properties and Processing of Vapour-Deposited Coatings. pp. 79–84.

Frost, R. L. and Barron, P. F. (1984). Solid-state silicon-29 and aluminium-27 nuclear magnetic resonance investigation of the dehydroxylation of pyrophyllite. J. Phys. Chem. **88**, 6206–6209.

Gehlen von, K. (1962). Die orientierte Bildung von Mullit aus Al–Si-Spinell in der Umwandlungsreihe Kaolinit-Mullit. Ber. Dtsch. Keram. Ges. **39**, 315–320.

Gerardin, C., Sundaresan, S., Benziger, J., Navrotsky, A. (1994). Structural unvestigation and energetics of mullite formation from solgel precursors. Chem. Mater. **6**, 160–170.

Ghate, B. B., Hasselman, D. P. H. and Spriggs, R. M. (1975). Kinetics of pressure-sintering and grain growth of ultra fine mullite powders. Ceram. Intern. **1**, 105–110.

Gilson, J.-P., Edwards, G. C., Peters, A. W., Rajagopalan, K., Wormsbecher, R. F., Roberie, T. G. and Shatlock, M. P. (1987). Penta-coordinated aluminium in zeolites and aluminosilicates. J. Chem. Soc. Chem. Comm. 91–92.

Greenwood, N. N. and Gibb, T. C. (1971). Mossbauer Spectroscopy, Chapman and Hall, London.

Grimshaw, R. W., Heaton, E. and Roberts, A. L. (1945). Refractory clays, II. Trans. Brit. Ceram. Soc. **44**, 76–92.

Guo, J., He, H., Wang, F., Wang, D., Zhang, H. and Hu, C. (1997). Kaolinite-mullite reaction series: ^{27}Al and ^{29}Si MAS NMR study. Kuangwu Xuebao, **17**, 250–259.

Guse, W., Saalfeld, H. and Tjandra, J. (1979). Thermal transformation of sillimanite single crystals. N. Jb. Min. Mh. 175–185.

Gypesova, D. and Durovic, S. (1977). Single-crystal study of thermal decomposition of sillimanite. Silikaty **21**, 147–149.

Hammonds, K. D., Bosenick, A., Dove, M. T. and Heine, V. (1998). Rigid unit modes in crystal structures with octahedrally coordinated atoms. Amer. Min. **83**, 476–479.

Hariya, Y., Dollase, W. A. and Kennedy, G. C. (1969). An experimental investigation of the relationship of mullite to sillimanite. Amer. Min. **54**, 1419–1441.

Hartman, J. S. and Sherriff, B. L. (1991). ^{29}Si MAS NMR of the aluminosilicate mineral kyanite: Residual dipolar coupling to ^{27}Al and nonexponential spin-lattice relaxation. J. Phys. Chem. **95**, 7575–7579.

Heller, L. (1962). The thermal transformation of pyrophyllite to mullite. Amer. Min. **47**, 156–157.

Hildebrand, B., Ledbetter, H., Kim, S. and Schneider, H. (2001). Structural control of elastic constants of mullite in comparison to sillimanite. J. Am. Ceram. Soc. **84**, 2409–2414.

Hildmann, B. and Schneider, H. (2004). Heat capacity of mullite: New data for a high temperature phase transformation. J. Am. Ceram. Soc. **87**, 227–234.

Hildmann, B. and Schneider, H. (2005). Thermal conductivity of mullite. J. Am. Ceram. Soc., in press.

Holm, J. L. and Kleppa, O. J. (1966). The thermodynamic properties of the aluminium silicates. Amer. Min. **51**, 1608–1622.

Huang, Y. X., Senos, A. M. R. and Baptista, J. L. (2000). Thermal and mechanical properties of aluminum titanate-mullite composites. J. Mater. Res., **15**, 357–363.

Hülsmans, A., Schmücker, M., Mader, W. and Schneider, H. (2000a). The transformation of andalusite to mullite and silica. Part I. Transformation mechanism in $[001]_A$ direction. Amer. Min. **85**, 980–986.

Hülsmans, A., Schmücker, M., Mader, W. and Schneider, H. (2000b). The transformation of andalusite to mullite and silica. Part II. Transformation mechanism in $[100]_A$ and $[010]_A$ directions. Amer. Min. **85**, 987–992.

Hyatt, M. J. and Bansal, N. P. (1990). Phase transformations in xerogels of mullite composition. J. Mater. Sci. **25**, 2815–2821.

Ikeda, Y., Yokoyama, T., Yamashita, S., Watanabe, T. and Wakita, H. (1996). Quantitative coordination analysis of Al in mullite precursors: comparison between XAFS and ^{27}Al MA NMR methods. X-sen Bunseki no Shinpo, **27**, 211–219.

Ikuma, Y., Shimada, E., Sakano, S., Oishi, M., Yokoyama, M. and Nakagawa, Z. (1999). Oxygen self diffusion in cylindrical single-crystal mullite. J. Electrochem. Soc. **146**, 4672–4675.

Insley, H. and Ewell, R. H. (1935). Thermal behaviour of the kaolin minerals. J. Res. Natl. Bur. Stand. **14**, 615–627.

Ismail, M. G. M. U., Nakai, Z. and Somiya, S. (1987). Microstructure and mechanical properties of mullite prepared by the sol-gel method. J. Am. Ceram. Soc. **70**, C-7–C-8.

Iwai, S., Tagai, H. and Shimamune, T. (1971). Procedure for dickite structure modification by dehydration. Acta Cryst. **B27**, 248–250.

Jäger, C., Rocha, J. and Klinowski, J. (1992). Highspeed satellite transition ^{27}Al MAS NMR spectroscopy. Chem. Phys. Lett., **188**, 208–212.

Janackovic, D., Jokanovic, V., Kostic-Gvozdenovic, L. and Uskokovic, D. (1998). Synthesis of mullite nanostructured spherical powder by ultrasonic spray pyrolysis. Nanostruct. Mater. **10**, 341–348.

JANAF (1965–1968). Thermochemical Tables. US Depart. Commerce, Nat. Bur. Stand., Inst. Appl. Techn., Michigan.

Janot, C. and Delcroix, P. (1974). Moessbauer study of ancient French ceramics. J. Phys. Suppl. 12, **35**, 557–561.

Jaymes, I., Douy, A., Florian, P., Massiot, D. and Coutures, J.-P. (1994). New synthesis of mullite. Structural evolution study by ^{17}O, ^{27}Al and ^{29}Si MAS NMR spectroscopy. J. Sol-Gel Sci. Techn. **2**, 367–370.

Jaymes, I., Douy, A., Massiot, D. and Coutures, J-P. (1995). Evolution of the Si environment in mullite solid solution by ^{29}Si MAS-NMR spectroscopy. J. Non-Cryst. Solids **204**, 125–134.

Johnson, S. and Pask, J. A. (1982). Role of impurities on formation of mullite from kaolinite and Al_2O_3–SiO_2 mixtures. Am. Ceram. Soc. Bull. **61**, 838–842.

Kay, D. A. R. and Taylor, J. (1960). Acitivities of silica in the lime + alumina + silica system. Trans. Faraday Soc. **56**, 1372–1386.

Kelly, W. H., Palazotto, A. N., Ruh, R., Heuer, J. K. and Zangoil, A. (1997). Thermal shock resistance of mullite and mullite–ZrO_2–SiC whisker composites. Ceram. Eng. Sci. Proc. **18**, 195–203.

Kingery, W. D., Bowen, H. K. and Uhlmann, D. R. (1976). Introduction to Ceramics. 2nd edn. Wiley, New York.

Kirschen, M., DeCapitani, C. and Millot, F. (1999). Immiscibility of silicate liquids in the system SiO_2–TiO_2–Al_2O_3. J. Europ. Min. Soc. **11**, 427–440.

Kiyono, H., Shimada, S. and MacKenzie, K. J. D. (2001). Kinetic and magic angle spinning nuclear magnetic resonance studies of wet oxidation of beta-sialon powders. J. Electrochem. Soc. **148**, B86–B91.

Kleebe, H.-J., Siegelin, F., Straubinger, T. and Ziegler, G. (1996). Conversion of Al_2O_3-SiO_2 powder mixtures to 3:2 mullite following the stable or metastable phase diagram. J. Europ. Ceram. Soc. **21**, 2521–2533.

Klochkova, I. V., Dudkin, B. N., Shveikin, G. P., Goldin, B. A. and Nazarova, L. Y., (2001). The effect of sesquioxides of 3d-transition elements on the strength of synthetic mullite and mullite-based materials. Refract. Indust. Ceramics **42**, 351–354.

Knutson, R., Liu, H. and Yen, W. M., (1989). Spectroscopy of disordered low-field sites in Cr^{3+}: Mullite glass ceramic. Phys. Rev. B **40**, 4264–4270.

Kollenberg, W. and Schneider, H. (1989). Microhardness of mullite at temperatures up to 1000 °C. J. Am. Ceram. Soc. **72**, 1739–1740.

Konopicky, K., Routschka, G. and Baum, M. (1965). Morphologie and Kristallgrößenverteilung des Mullits in Schamotte. Sprechsaal **98**, 1–15.

Kriven, W. M., Palko, J. W., Sinogeikin, S., Bass, J. D., Sayir, A., Brunauer, G., Boysen, H., Frey, F. and Schneider, J. (1999). High temperature single crystal properties of mullite. J. Europ. Ceram. Soc. **19**, 2529–2541.

Kriven, W. M., Siah, L. F., Schmücker, M. and Schneider, H. (2004). High temperature microhardness of single crystal mullite. J. Am. Ceram. Soc., **87**, 970–972.

Kunath-Fandrei, G., Rehak, P., Steurnagel S., Schneider, H. and Jäger, C. (1994). Quantitative structural analysis of mullite by ^{27}Al nuclear magnetic resonance satellite transition spectroscopy. Solid State Nucl. Magn. Res. **3**, 241–248.

Kutty, T. R. N. and Nayak, M. (2000). Photoluminescence of Eu^{2+}-doped mullite (xAl$_2$O$_3$.ySiO$_2$; x/y = 3/2 and 2/1) prepared by a hydrothermal method. J. Mater. Chem. Phys. **65**, 158–165.

Kutzendorfer, J. and Zahradkova, M. (1984). Determination of the degree of crystallization of refractory fibres. Chemie a Technologie Silikatu **L12**, 231–241.

Lacks, D., Hildmann, B. and Schneider, H. (2005). Molecular dynamic (MD) simulation of the high temperature phase transformation of mullite. To be published.

Lacy, E. D. (1963). Aluminium in glasses and melts. Phys. Chem. Glasses **4**, 234–238.

Lambert, J. F., Millman, W. F. and Fripiat, J. J. (1989). Revisiting kaolinite dehydroxylation: a ^{29}Si and ^{27}Al MAS NMR study. J. Am. Chem. Soc. **111**, 3517–3522.

Ledbetter, H., Kim, S., Balzar, D., Crudele, S. and Kriven, W. M. (1998). Elastic properties of mullite. J. Am. Ceram. Soc. **81**, 1025–1028.

Lee, J. S. and Wu, S. C. (1992). Characteristics of mullite prepared from coprecipitated 3Al$_2$O$_3$ · 2SiO$_2$ powders. J. Mater. Sci. **27**, 5203–5208.

Lee, S., Kim, Y. J. and Moon, H.-S. (1999). Phase transformation sequence from kaolinite to mullite investigated by an energy-filtering transmission electron microscope. J. Am. Ceram. Soc. **82**, 2841–2848.

Leonard, A. J. (1977). Structural analysis of the transition phases in the kaolinite-mullite thermal sequence. J. Am. Ceram. Soc. **60**, 37–43.

Low, I. M. and McPherson, R. (1989). The origins of mullite formation. J. Mater. Sci. **24**, 926–936.

Lugmair, C. G., Fujdala, K. L. and Tilley, T. D. (2002). New tris(tert-butoxy)siloxy complexes of aluminium and their transformation to homogeneous aluminosilicate minerals via low-temperature thermolytic pathways. Chem. Mater. **14**, 888–898.

Mack, D. E., Becker, K. D. and Schneider, H. (2005). High-temperature Mössbauer study of Fe-substituted mullite. Amer. Min. In press.

MacKenzie, K. J. D. (1969a). An infrared frequency shift method for the determination of the high-temperature phases of aluminosilicate minerals. J. Appl. Chem. **19**, 65–67.

MacKenzie, K. J. D. (1969b). Infrared kinetic study of high-temperature reactions of synthetic kaolinite. J. Am. Ceram. Soc. **52**, 635–637.

MacKenzie, K. J. D., (1972). Infrared frequency calculations for ideal mullite (3Al$_2$O$_3$.2SiO$_2$). J. Am. Ceram. Soc. **55**, 68–71.

MacKenzie, K. J. D., Brown, I. W. M., Meinhold, R. H. and Bowden, M. E. (1985a). Outstanding problems in the kaolinite-mullite reaction sequence investigated by ^{29}Si and ^{27}Al solid-state nuclear magnetic resonance. I. Metakaolinite. J. Am. Ceram. Soc. **68**, 293–297.

MacKenzie, K. J. D., Brown, I. W. M., Meinhold, R. H. and Bowden, M. E. (1985b). Thermal reactions of pyrophyllite studied by high-resolution solid-state ^{27}Al and ^{29}Si nuclear magnetic resonance spectroscopy. J. Am. Ceram. Soc. **68**, 266–272.

MacKenzie, K. J. D., Brown, I. W. M., Cardile, C. M. and Meinhold, R. H. (1987). The thermal reactions of muscovite studied by high-resolution solid-state 29-Si and 27-Al NMR. J. Mater. Sci. **22**, 2645–2654.

MacKenzie, K. J. D., Bowden, M. E., Brown, I. W. M. and Meinhold, R. H. (1989). Structure and thermal transformations of imogolite studied by ^{29}Si and ^{27}Al high-resolution solid state nuclear magnetic resonance. Clays Clay Minerals **37**, 317–324.

MacKenzie, K. J. D., Bowden, M. E. and Meinhold, R. H. (1991). The structure and thermal transformations of allophanes studied by ^{29}Si and ^{27}Al high resolution solid-state NMR. Clays Clay Minerals **39**, 337–346.

MacKenzie, K. J. D., Meinhold, R. H., Patterson, J. E., Schneider, H., Schmücker, M. and Voll, D. (1996a). Structural evolution in gel-derived mullite precursors. J. Europ. Ceram. Soc. **16**, 1299–1308.

MacKenzie, K. J. D., Okada, K. and Hartman, J. S. (1996b). MAS NMR evidence for the presence of silicon in the alumina spinel from thermally transformed kaolinite. J. Am. Ceram. Soc. **79**, 2980–2982.

MacKenzie, K. J. D., Meinhold, R. H., Brown, I. W. M. and White, G. V. (1996c). The formation of mullite from kaolinite under various reaction atmospheres. J. Europ. Ceram. Soc. **16**, 115–119.

MacKenzie, K. J. D. and Meinhold, R. H. (1997). MAS NMR study of pentacoordinated magnesium in grandidierite. Amer. Min. **82**, 479–482.

MacKenzie, K. J. D., Sheppard, C. M., Barris, G. C., Mills, A. M., Shimada, S. and Kiyono, H. (1998). Kinetics and mechanism of thermal oxidation of sialon ceramic powders. Thermochim. Acta **318**, 91–100.

MacKenzie, K. J. D., Sheppard, C. M. and McCammon, C. (2000). Effect of Y_2O_3, MgO and Fe_2O_3 on silicothermal synthesis and sintering of X-sialon. An XRD, multinuclear MAS NMR and 57Fe Mossbauer study. J. Europ. Ceram. Soc. **20**, 1975–1985.

MacKenzie, K. J. D., Smith, M. E., Schmücker, M., Schneider, H., Angerer, P., Gan, Z., Anupold, T., Reinhold, A. and Samosan, A. (2001). Structural aspects of mullite-type $NaAl_9O_{14}$ studied by ^{27}Al and ^{23}Na solid-state MAS and DOR NMR techniques. Phys. Chem. Chem. Phys. **3**, 2137–2142.

MacKenzie, K. J. D. and Smith, M. E. (2002). Multinuclear Solid State NMR of Inorganic Materials, Pergamon Materials Series Vol. **6**, Pergamon-Elsevier, Oxford.

Magi, M., Lippmaa, E., Samosan, A., Grimmer, A-R. (1984). Solid-state high-resolution silicon-29 chemical shifts in silicates. J. Phys. Chem. **88**, 1518–1522.

Mako E., K. and Zoltan, J. A. (1997). Mechanochemical transformations in the crystal structure of kaolinite. Epitoanyag, **49**, 2–6.

Malow, T. R. and Koch, C. C. (1997). Grain growth in nanocrystalline iron prepared by mechanical attrition. Acta Mater. **45**, 2177–2186.

Massiot, D., Dion, P., Alcover, J. F. and Bergaya, F. (1995). ^{27}Al and ^{29}Si MAS NMR study of kaolinite thermal decomposition by controlled rate thermal analysis. J. Am. Ceram. Soc. **78**, 2940–2944.

McConnell, J. D. C. and Heine, V., (1985) Incommensurate structure and stability of mullite. Phys. Rev. **B 32**, 6140–6143.

C. J. McConville (1997). Related microstructural development on firing kaolinite, illite and smectite clays. Ph. D. Thesis Univ. Sheffield.

McManus, J., Ashbrook, S. E., MacKenzie, K. J. D. and Wimperis, S. (2001). ^{27}Al multiple-quantum MAS and $^{27}Al\{^1H\}$ CPMAS NMR study of amorphous aluminosilicate. J. Non-Cryst. Solids **282**, 278–290.

Mechnich, P., Schmücker, M. and Schneider, H. (1999). Reaction sequence and microstructural development of CeO_2-doped reaction-bonded mullite. J. Am. Ceram. Soc. **82**, 2517–2522.

Meinhold, R. H., MacKenzie, K. J. D. and Brown, I. W. M. (1985). Thermal reactions of kaolinite studied by solid state 27-Al and 29-Si NMR. J. Mater. Sci. Lett. **4**, 163–166.

Meinhold, R. H. and MacKenzie, K. J. D. (2000). The system $Ga_2O_3(Al_2O_3)$–$GeO_2(SiO_2)$ studied by NMR, XRD, IR and DTA. J. Mater. Chem. **10**, 701–708.

Meinhold, R. H., Slade, R. C. T. and Davies, T. W. (1993). High-field ^{27}Al MAS NMR studies of the formation of metakaolinite by flash calcination of kaolinite. Appl. Magn. Res. **4**, 141–155.

Menard, D. and Doukhan, J. C. (1978). Defauts de reseau dans la sillimanite: Al_2O_3–SiO_2. J. Physique Lett. **39**, L-19–L-22.

Merwin, L. H., Sebald, A., Rager, H. and Schneider, H. (1991). ^{29}Si and ^{27}Al MAS NMR spectroscopy of mullite. Phys. Chem. Minerals **18**, 47–52.

Miller, J. M. and Lakshmi, L. J. (1998). Spectroscopic characterization of sol-gel-derived mixed oxides. J. Phys. Chem. **B 102**, 6465–6470.

Nass, R., Tkalcec, E. and Ivankovic, H. (1995). Single phase mullite gels doped with chromium. J. Am. Ceram. Soc. **78**, 3097–3106.

Navrotsky, A., Newton, R. C. and Kleppa, O. J. (1973). Sillimanite-disordering by enthalpy calorimetry, Geochim. Cosmochim. Acta **37**, 2497–2508.

Nowick, A. S. and Berry, B. S. (1972). Anelastic Relaxation in Crystalline Solids. Academic Press, New York.

Okada, K. and Otsuka, N. (1986). Characterization of the spinel phase from SiO_2–Al_2O_3 xerogels and the formation process of mullite. J. Am. Ceram. Soc. **69**, 652–656.

Okada, K. and Otsuka, N. (1988). Chemical composition change of mullite during formation process. Sci. Ceram. **14**, 497–502.

Okada, K., Otsuka, N. and Ossaka, J. (1986). Characterization of the spinel phase formed in the kaolin-mullite thermal sequence. J. Am. Ceram. Soc. **69**, C251–C253.

Padmaja, P., Anilkumar, G. M., Mukundan, P., Aruldhas, G. and Warrier, K. G. K. (2001). Characterisation of stoichiometric sol-gel mullite by fourier transform infrared spectroscopy. Intern. J. Inorg. Mater. **3**, 693–698.

Palko, J. W., Sayir, A., Sinogeikin, S. V., Kriven, W. M. and Bass, J. D. (2002). Complete elastic tensor for mullite (~ 2.5 $Al_2O_3 \cdot SiO_2$) to high temperatures measured from textured fibers. J. Am. Ceram. Soc. **85**, 2005–2012.

Pankratz, L. B., Weller, W. W. and Kelley, K. K. (1963). Low-temperature heat content of mullite. US Bur. Mines Rep. Invest. No. 6287, 7.

Pannhorst, W. and Schneider, H. (1978). The high-temperature transformation of andalusite (Al_2SiO_5) into 3/2-mullite ($3Al_2O_3 \cdot 2SiO_2$) and vitreous silica (SiO_2). Min. Mag. **42**, 195–198.

Parmentier, J., Vilminot, S. and Dormann, J.-L. (1999). Fe- and Cr-substituted mullites: Mossbauer spectroscopy and Rietveld structure refinement. Solid State Sci. **1**, 257–265.

Paulmann, C. (1996). Study of oxygen vacancy ordering in mullite at high temperatures. Phase Transitions **59**, 77–90.

Peeters, M. J. P. and Kentgens, A. P. M. (1997). A ^{27}Al MAS, MQMAS and off-resonance nutation NMR study of aluminium containing silica-based sol-gel materials. Solid State Nucl. Magn. Res. **9**, 203–217.

Pentry, R. A., Hasselman, D. P. H. and Spriggs, R. M. (1972), Young's modulus of high density polycrystalline mullite. J. Am. Ceram. Soc. **55**, 169–170.

Percival, H. J., Duncan, J. F. and Foster, P. K. (1974). Interpretation of the kaolinite-mullite reaction sequence from infrared absorption spectra. J. Am. Ceram. Soc. **57**, 57–61.

Pitchford, J. E., Stearn, R. J. Kelly, A. and Clegg, W. J. (2001). Effect of oxygen vacancies on the hot hardness of mullite. J. Am. Ceram. Soc. **84**, 1167–1168.

Prochazka, S. and Klug, F. J. (1983). Infrared-transparent mullite ceramic. J. Am. Ceram. Soc. **66**, 874–880.

Rager, H., Schneider, H. ans Graetsch, H. (1990). Chromium incorporation in mullite. Amer. Min. **75**, 392–397.

Rager, H., Schneider, H. and Bakhshandeh, A. (1993). Ti^{3+} centres in mullite. J. Europ. Min. Soc. **5**, 511–514.

Rahman, S., Fenstel, U. and Freimann, S. (2001). Structure description of the thermic phase transformation sillimanite-mullite. J. Europ. Ceram. Soc. **21**, 2471–2478.

Rehak, P., Kunath-Fandrei, G., Losso, P., Hildmann, B., Schneider, H. and Jäger, C. (1998). Study of the Al coordination in mullites with varying Al:Si ratio by ^{27}Al NMR spectroscopy and X-ray diffraction. Amer. Min. **83** 1266–1276.

Rein, R. H. and Chipman, J. (1965). Activities in the liquid solution SiO_2–CaO–MgO–Al_2O_3 at 1600 °C. Trans. AIME **223**, 415–425.

Reisfeld, R., Kisilev, A., Buch, A. and Ish-Shalom, M. (1986). Spectroscopy and EPR of chromium(III) in mullite transparent glass-ceramics. Chem. Phys. Lett. **129**, 446–449.

Risbud, S. H., Kirkpatrick, R. S., Tagliavore, A. P. and Montez, B. (1987). Solid-state evidence of 4-, 5-, and 6-fold aluminium sites in roller-quenched SiO_2–Al_2O_3 glasses. J. Am. Ceram. Soc., **70**, C10–C12.

Roberts, A. L. (1945). Refractory clays, I. Trans. Brit. Ceram. Soc. **44**, 69–75.

Robie, R. A. and Hemingway, B. S. (1995). Thermodynamic properties of minerals and related substances at 298.15 K and 1 bar (10^5Pa) pressure and high temperatures. US Geol. Surv. Bull. No. 2131.

Rocha, J. (1999). Single- and triple-quantum ^{27}Al MAS NMR study of the thermal transformation of kaolinite. J. Phys. Chem. **B 103**, 9801–9804.

Rocha, J. and Klinowski, J. (1990). ^{29}Si and ^{27}Al magic-angle spinning NMR studies of the thermal transformation of kaolinite. Phys. Chem. Minerals **17**, 179–186.

Rocha, J., Klinowski, J. and Adams, J. M. (1991). Solid-state NMR elucidation of the role of mineralisers in the thermal stability and phase transformations of kaolinite. J. Mater. Sci. **26**, 3009–3018.

Rodriguez-Navarro, C., Cultrone, G., Sanchez-Newas, A. and Sebastian, E. (2003). TEM study of mullite growth after muscovite breakdown. Amer. Min. **88**, 713–724.

Rommerskirchen, J., Chavez, F. and Janke, D. (1994). Ionic conduction behaviour of mullite ($3Al_2O_3 \cdot 2SiO_2$) at 1400° to 1600 °C. Solid State Ionics **74**, 179–187.

Rüscher, C. H. (2001). Thermic transformation of sillimanite single crystals to 3:2 mullite plus melt: Investigations by polarized IR-reflection micro spectroscopy. J. Europ. Ceram. Soc. **21**, 2463–2470.

Rüscher, C. H., Schrader, G. and Gotte, M. (1996). Infrared spectroscopic investigation in the mullite field of composition: $Al_2(Al_{2+2x}Si_{2-2x})O_{10-x}$ with $0.55 > x > 0.25$. J. Europ. Ceram. Soc. **16**, 169–175.

Saalfeld, H. (1977). Röntgenographische Einkristalluntersuchungen über die thermische Umwandlung von Sillimanit und Kyanit. Fortschr. Miner. **55**, 118–119.

Sadanaga, R., Tokonami, M. and Takeuchi, Y. (1962). The structure of mullite, $2Al_2O_3 \cdot SiO_2$, and relationship with the structures of sillimanite and andalusite. Acta Cryst. **15**, 65–68.

Sanchez-Soto, P. J., Sobrados, I., Sanz, J. and Perez-Rodriguez, J. L. (1993). 29-Si and 27-Al magic-angle spinning nuclear magnetic resonance study of the thermal transformations of pyrophyllite. J. Am. Ceram. Soc. **76**, 3024–3028.

Sánchez-Soto, P. J., Pérez-Rodríguez, J. L., Sobrados, I. and Sanz, J. (1997). Influence of grinding in pyrophyllite-mullite thermal transformation assessed by ^{29}Si and ^{27}Al MAS NMR spectroscopies. J. Chem. Mater. **9**, 677–684.

Sanz, J., Madani, A., Serratosa, J. M., Moya, J. M. and de Aza, S. (1988). Aluminium-27 and silicon-29 magic-angle spinning nuclear magnetic resonance study of the kaolinite-mullite transformation. J. Am. Ceram. Soc. **71**, C418–C421.

Sato, R. K., McMillan, P. F., Dennison, P. and Dupree, R. (1991). High-resolution ^{27}Al and ^{29}Si MAS NMR investigation of SiO_2–Al_2O_3 glasses. J. Phys. Chem. **95**, 4483–4489.

Satoshi, S., Contreras, C., Juarez, H., Aguilera, A. and Serrato, J. (2001). Homogeneous precipitation and thermal phase transformation of mullite ceramic precursor. Intern. J. Materials, **3**, 625–632.

Schmücker, M., Schneider, H., Poorteman, M., Cambier, F. and Meinhold, R., (1995). Constitution of mullite glasses produced by ultra-rapid quenching of plasma-sprayed melts. J. Europ. Ceram. Soc. **15**, 1201–5.

Schmücker, M., Flucht, F. and Schneider, H. (1996). High temperature behaviour of polycrystalline aluminiumsilicate fibers with mullite bulk composition. I. Micro-

structure and strength properties. J. Europ. Ceram. Soc. **16**, 281–285.

Schmücker, M. and Schneider, H. (1996). A new approach on the coordination of Al in non-crystalline gels and glasses of the system Al_2O_3–SiO_2. Ber. Bunsen- Ges. Physik. Chemie **100**, 1550–1553.

Schmücker, M., MacKenzie, K. J. D., Schneider, H. and Meinhold, R. H. (1997). NMR studies on rapidly solidified SiO_2-Al_2O_3 and SiO_2-Al_2O_3-Na_2O glasses. J. Non-Cryst. Solids **217**, 99–105.

Schmücker, M., Schneider, H. and MacKenzie, K. J. D. (1998). Mechanical amorphization of mullite and thermal recrystallization. J. Non-Cryst. Solids **226**, 99–104.

Schmücker, M., Schneider, H., MacKenzie, K. J. D. and Okuno, M. (1999). Comparative ^{27}Al NMR and LAXS studies on rapidly quenched aluminosilicate glasses. J. Europ. Ceram. Soc. **19**, 99–103.

Schmücker, M. and Schneider, H. (1999). Transformation of X-phase-SiAlON to mullite. J. Am. Ceram. Soc. **82**, 1934–1936.

Schmücker, M., Flucht, F. and Schneider, H. (2001). Temperature stability of 3M Nextel 610, 650, and 720 fibers – A microstructural study. In: High Temperature Ceramic Matrix Composites, Eds.: Krenkel, W., Naslain, R. and Schneider, H. Wiley–VCH, Weinheim, pp. 74–78.

Schmücker, M., Hildmann, B. and Schneider, H. (2002). Mechanism of 2/1- to 3/2-mullite transformation at 1650 °C. Amer. Min. **87**, 1190–1193.

Schmücker, M. and Schneider, H. (2002). New evidence for tetrahedral triclusters in aluminosilicate glasses. J. Non-Cryst. Solids **311**, 211–215.

Schmücker, M., Schneider, H. and Kriven, W. M. (2003). Indentation-induced amorphization in mullite single crystals. J. Am. Ceram. Soc. **86**, 1821–1822.

Schmücker, M., Schneider, H., Mauer, T. and Clauß, B. (2005a). Kinetics of mullite grain growth in alumino silicate fibers. J. Am. Ceram. Soc., in press.

Schmücker, M., MacKenzie, K. J. D., Smith, M. E., Carroll, D. and Schneider, H. (2005b). AlO_4/SiO_4 distribution in tetrahedral double chains in mullite. J. Am. Ceram. Soc., in press.

Schneider, H. and Homemann, U. (1977). The disproportionation of andalusite

(Al$_2$SiO$_5$) to Al$_2$O$_3$ and SiO$_2$ under shock compression. Phys. Chem. Minerals **1**, 257–264.

Schneider, H. and Majdic, A. (1979). Kinetics and mechanism of the solid-state high-temperature transformation of andalusite (Al$_2$SiO$_5$) into 3/2-mullite (3Al$_2$O$_3$ · 2SiO$_2$) and silica (SiO$_2$). Ceramurgia Int. **5**, 31–36.

Schneider, H. and Majdic, A. (1980). Kinetics of the thermal decomposition of kyanite. Ceramurgia Int. **6**, 32–37.

Schneider, H. (1981). Infrared spectroscopic investigations on andalusite- and mullite-type structures in the systems Al$_2$GeO$_5$–Fe$_2$GeO$_5$ and Al$_2$GeO$_5$–Ga$_2$GeO$_5$. N. Jb. Min. Abh. **142**, 111–123.

Schneider, H. and Hornemann, U. (1981). Shock-induced transformation of sillimanite powders. J. Mater. Sci. **16**, 45–49.

Schneider, H. and Majdic, A. (1981). Preliminary investigations on the kinetics of the high-temperature transformation of sillimanite to 3/2-mullite plus silica and comparison with the behaviour of andalusite and kyanite. Sci. Ceram. **11**, 191–196.

Schneider, H. and Rager, H. (1984). Occurrence of Ti^{3+} and Fe^{2+} in mullite. J. Am. Ceram. Soc. 67, C248–C250.

Schneider, H. and Rager, H. (1986). Iron incorporation in mullite. Ceram. Intern. **12**, 117–125.

Schneider, H. (1990). Transition metal distribution in mullite. Ceramic Trans. **6**, 135–158.

Schneider, H. and Eberhard, E. (1990). Thermal expansion of mullite. J. Am. Ceram. Soc. **73**, 2073–2076.

Schneider, H., Merwin, L. and Sebald, A. (1992). Mullite formation from non-crystalline precursors. J. Mater. Sci. **27**, 805–812.

Schneider, H. and Pleger, R. (1993). The reconstructive 2/1-to 3/2-mullite transformation in the presence of Fe$_2$O$_3$-richglass at 1570 °C. J. Europ. Min. Soc. **5**, 515–521.

Schneider, H., Schmücker, M., Ikeda, K. and Kaysser, W. A. (1993a). Optically translucent mullite ceramics. J. Am. Ceram. Soc. **76**, 2912–2914.

Schneider, H., Fischer, R. X. and Voll, D. (1993b). A mullite with lattice constants a > b. J. Am. Ceram. Soc. **76**, 1879–1881.

Schneider, H., Rodewald, K. and Eberhard, E. (1993c). Thermal expansion discontinuities

of mullite. J. Am. Ceram. Soc. **76**, 2896–2898.

Schneider, H., Voll, D., Schmücker, M., Saruhan, B., Schaller, T. and Sebald, A. (1994). Constitution of γ-alumina spinel in mullite precursors. J. Europ. Ceram. Soc. **13**, 441–448.

Schneider, H., Voll, D., Saruhan, B., Sanz, J., Schrader, G., Rüscher, C. and Mosset, A. (1994a). Synthesis and structural characterization of non-crystalline mullite precursors. J. Non-Cryst. Solids **178**, 262–271.

Schneider, H., Okada, K., Pask, J. A. (1994b). Mullite and Mullite Ceramics. Wiley, Chichester.

Schneider, H., Ikeda, K., Saruhan, B. and Rager, H. (1996). Electron paramagnetic resonance and optical absorption studies on Cr-doped mullite precursors. J. Europ. Ceram. Soc. **16**, 211–215.

Schneider, H., Göring, J., Schmücker, M. and Flucht, F. (1998). Thermal stability of Nextel 720 alumino silicate fibers. In: Ceramic Microstructures: Control at Atomic Level Eds.: Tomsia, A. P. and Glaser, A. M., Plenum Press, New York, pp. 721–730.

Schneider, J. and Schneider, H. (2005). Structural refinement of mullite up to 1400 °C. To be published.

Schreuer, J., Hildmann, B. and Schneider, H. (2005a). Elastic properties of mullite up to 1400 °C. To be published.

Schreuer, J., Hildmann, B. and Schneider, H. (2005b). Thermal expansion of mullite from single crystal dilatometry up to 1400 °C. To be published.

Sheppard, C. M., MacKenzie, K. J. D., Barris, G. C. and Meinhold, R. H. (1997). A new silicothermal route to the formation of X-phase sialon: the reaction sequence in the presence and absence of Y$_2$O$_3$. J. Europ. Ceram. Soc. **17**, 667–673.

Sheppard, C. M. and MacKenzie, K. J. D. (1999). Silicothermal synthesis and densification of X-sialon in the presence of metal oxide additives. J. Europ. Ceram. Soc. **19**, 534–541.

Sherriff, B. L. and Grundy, H. D. (1988) Nature **332**, 819–822.

Shornikov, S. I., Archakov, I. Yu and Shul'ts, M. M. (2003). A mass spectrometric study of the thermodynamic properties of mullite. Russ. J. Phys. Chem. **77**, 1044–1050.

Shornikov, S. I. and Archakov, I. Yu (2002). A

mass spectrometric determination of the enthalpies and entropies of Al_2O_3–SiO_2 melts. Russ. J. Phys. Chem. **76**, 1054–1060.

Shoval, S., Michaelian, K. H., Boudeulle, M., Panczer, G., Lapides, I. and Yariv, S. (2002). Study of thermally treated dickite by infrared and micro-Raman spectroscopy using curve-fitting technique. J. Therm. Anal. Calorim. **69**, 205–225.

Slade, R. C. T. and Davies, T. W. (1991). Evolution of structural changes during flash calcination of kaolinite. J. Mater. Chem. **1**, 361–364.

Smith, M. E. and Steuernagel, S. (1992). A multinuclear magnetic resonance examination of the mineral grandidierite. Identification of a ^{27}Al resonance from a well-defined AlO_5 site. Solid State Nucl. Magn. Res. **1**, 175–183.

Smith, M. E. (1994). ^{29}Si and ^{27}Al magic-angle spinning nuclear magnetic resonance of sialon X-phase. Solid State Nucl. Magn. Res. **3**, 111–114.

Sonuparlak, B., Sarikaya, M. and Aksay, I. A. (1987). Spinel phase formation during the 980 °C exothermic reaction in the kaolinite-to-mullite reaction series. J. Am. Ceram. Soc. **70**, 837–842.

Srikrishna, K., Thomas, G., Martinez, R., Corral, M. P., De Aza, S. and Moya, J. S. (1990). Kaolinite-mullite reaction series: a TEM study. J. Mater. Sci. **25**, 607–612.

Stöffler, D. (1970). Shock deformation of sillimanite from the Ries Crater, Germany. Earth Planet. Sci. Lett. **10**, 115–120.

Temuujin, J., MacKenzie, K. J. D., Okada, K. and Jadambaa, T. (1999). The effect of water vapour atmospheres on the thermal transformation of kaolinite investigated by XRD, FTIR and solid state MAS NMR. J. Europ. Ceram. Soc. **19**, 105–112.

Temuujin, J., MacKenzie, K. J. D., Schmücker, M., Schneider, H., McManus, J. and Wimperis, S. (2000). Phase evolution in mechanically treated mixtures of kaolinite and alumina hydrates (gibbsite and boehmite). J. Europ. Ceram. Soc. **20**, 413–421.

Tröger, W. E. (1982). Optische Bestimmung der gesteinsbildenden Minerale. Teil 1: Bestimmungstabellen. 5. Aufl. Schweizerbarth, Stuttgart.

Turner, G. L., Kirkpatrick, R. J., Risbud, S. H. and Oldfield, E. (1987). Multinuclear magic-angle sample spinning nuclear magnetic resonance spectroscopic studies of crystalline and amorphous ceramic materials. Amer. Ceram. Soc. Bull. **66**, 656–663.

Vassallo, A. M., Cole-Clarke, P. A., Pang, L. S. K. and Palmisano, A. J. (1992). Infrared emission spectroscopy of coal minerals and their thermal transformations. Appl. Spectr. **46**, 73–78.

Vaughan, M. T. and Weidner, D. J. (1978). The relationship of elasticity and crystal structure in andalusite and sillimanite. Phys. Chem. Minerals **3**, 133–144.

Vernon, R. H. (1987). Oriented growth of sillimanite in andalusite, Placitas-Juan Tabo aera, New Mexico, U. S. A. Canad. J. Earth Sci. **24**, 580–590.

Voll, D., Lengauer, C., Beran, A. and Schneider, H. (2001). Infrared band assignment and structural refinement of Al–Si, Al–Ge and Ga–Ge mullites. J. Europ. Min. Soc. **13**, 591–604.

Voll, D., Schmücker, M. and Schneider, H. (2005). Structural evolution of mullite from type I single phase precursors. To be published.

Wei, W. and Halloran, J. W. (1988). Phase transformation of diphasic aluminosilicate gels. J. Am. Ceram. Soc. **71**, 166–172.

Weller, M. (1996). Anelastic relaxation of point defects in cubic crystals. J. Phys. IV. (Suppl. J. Phys. III 6), C8–83–C8–72.

Wilson, H. H. (1969). Mullite formation from the sillimanite group minerals. Am. Ceram. Soc. Bull. **48**, 796–797.

Winter, J. K. and Ghose, S. (1979). Thermal expansion and high-temperature crystal chemistry of the Al_2SiO_5 polymorphs. Amer. Min. **64**, 573–586.

Woo, J., Lee, C., Hong, Y., Yoon, Y., Hanm, Y. and Chang, Y. (1996). Effects of ultrasound on synthesis of mullite by sol-gel process. Hwahak Konghak, **34**, 208–214.

Yang, H., Hazen, R. M., Finger, L. W., Prewitt, C. T. and Downs, R. T. (1997). Compressibility and crystal structure of sillimanite, Al_2SiO_5, at high pressure. Phys. Chem. Minerals **25**, 39–47.

Yao, J., Gao, Z. and Hu, C. (2001). ^{29}Si and ^{27}Al MAS/NMR study of the thermal transformations of kaolinite. Kuangwu Xuebao., **21**, 448–452.

Zhao, H., Hiragushi, K. and Mizota, Y. (2002). Phase segregation of non-stoichiometric aluminosilicate gels characterised by ^{27}Al and ^{29}Si MAS NMR. J. Non-Cryst. Solids **311**, 1990–1996.

3
Phase Equilibria and Stability of Mullite

An important requirement for the fabrication and technical application of mullite-based ceramic compounds and structures is a knowledge and understanding of the Al_2O_3–SiO_2 phase diagram. Mullite, nominally $3Al_2O_3 \cdot 2SiO_2$ (i.e. 3/2-mullite), is the only stable crystalline phase of the binary system Al_2O_3–SiO_2 at atmospheric pressure above about 700 °C (see Section 2.2.1.1). Besides Al_2O_3–SiO_2, ternary or multi-component metal oxide-aluminosilicate systems are of scientific and application-related interest. This is particularly true for alkali oxide-, alkaline earth oxide- and iron oxide–Al_2O_3–SiO_2 systems. Because these systems often form low temperature eutectics they melt at relatively low temperatures, and thus often dominate the thermo-mechanical properties of ceramics. Recently the high-temperature stability of mullite and mullite ceramics under corrosive chemical environments, especially in the presence of water vapor, has also gained major importance.

While most of the contents of this chapter are based on the monograph of Schneider et al. (1994), the section dealing with the behavior of mullite in agressive high-temperature environments is completely new.

3.1
The Al_2O_3–SiO_2 Phase Diagram
J. A. Pask and H. Schneider

3.1.1
Experimental Observations

Major disagreement exists in the literature whether mullite melts congruently or incongruently, and on the extent and range of the solid-solution region of mullite. Shephard et al. (1909) published a diagram with sillimanite ($Al_2O_3 \cdot SiO_2$) as the only binary compound in the Al_2O_3–SiO_2 system. However, this compound was shown to be metastable under standard conditions and stable only at higher pressures. Bowen and Greig (1924) were the first to publish an equilibrium phase diagram with mullite as a stable compound at room temperature melting incongruently at 1828 °C (but see Section 2.2.1). They investigated mixes of different

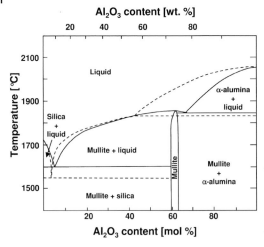

Fig. 3.1.1 The Al$_2$O$_3$–SiO$_2$ phase equilibrium diagram. The results of Bowen and Greig (1924, dashed lines), and of Amaraki and Roy (1962, full lines), show incongruent and congruent melting, respectively.

compositions prepared from synthetic α-alumina and silica, which were annealed in air atmosphere at selected temperatures, and quenched in a suitable cooling agent. Their mullite displays no solid-solution range (Fig. 3.1.1).

Bauer et al. (1950) and Bauer and Gordon (1951) grew mullite single crystals containing 83 wt. % Al$_2$O$_3$ (molar ratio: 3Al$_2$O$_3$: 1SiO$_2$ i.e. 3/1-mullite) by flame fusion. Their experiments raised doubts as to the incongruent melting of mullite. In 1951, Toropov and Galakhov (1951) heated mixtures of aluminum oxide gel and quartz. These mullites display congruent melting at about 1900 °C. Shears and Archibald (1954) described mullites with a solid solution range from 3Al$_2$O$_3$ · 2SiO$_2$ (about 72 wt. % Al$_2$O$_3$ i.e. 3/2-mullite) to 2Al$_2$O$_3$ · SiO$_2$ (about 78 wt. % Al$_2$O$_3$ i.e. 2/1-mullite) which melt incongruently at approximately 1810 °C. Welch (1960) supported the suggestion of a solid solution between 3/2- and 2/1-mullite and the incongruent mullite melting approach. Trömel et al. (1957) showed that in short duration synthesis runs no α-alumina is obtained, although it should have been found if mullite melts incongruently. On the other hand, Trömel et al. stated that α-alumina appears in longer lasting experiments, which supports the idea of incongruent melting of mullite.

Aramaki and Roy (1962) described congruent melting of mullite with a solid solution ranging from 71.8 to 74.3 wt. % Al$_2$O$_3$ (Fig. 3.1.1), which extends to 77.3 wt. % under metastable conditions. The samples, which were prepared from dry mixtures of α-alumina and powder silica glass were held at temperature and quenched in mercury or water. Davis and Pask (1972) used semi-infinite diffusion couples of sapphire (α-alumina) and fused silica at temperatures up to 1750 °C and determined a solid solution range of mullite from 71.0 to 74.0 wt. % Al$_2$O$_3$. Aksay and Pask (1975) extended these experiments to higher temperatures and reported an α-alumina liquidus profile (Fig. 3.1.3) similar to that of Bowen and Greig (1924) with a peritectic at about 55 wt. % Al$_2$O$_3$, supporting incongruent melting of mullite.

Mullite single crystals have been grown by Guse (1974) and Guse and Mateika

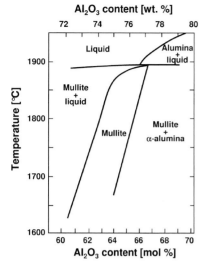

Fig. 3.1.2 The Al$_2$O$_3$–SiO$_2$ phase diagram according to Klug et al. (1987) in the Al$_2$O$_3$-rich region. The diagram shows incongruent melting behavior of mullite. Note the change of the solid solution with temperature.

(1974) using the Czochralski technique. They obtained a 2/1- instead of the 3/1-composition typical for the flame fusion process (see above). Shindo (1980) also grew mullite single crystals by the "Slow cooling float zone (SCFZ) method" with 2/1-composition and incongruent melting (for further work see, for example, Schneider et al. 1994). The mullite phase diagram of Prochazka and Klug (1983) proposed congruent melting of mullite and a solid-solution range that shifts to higher Al$_2$O$_3$ contents equivalent to the 2/1-type mullite at temperatures from 1600 °C up to melting. Klug et al. (1987) confirmed the results of Prochazka and Klug but modified the position of the Al$_2$O$_3$ liquidus assuming incongruent melting of mullite (Fig. 3.1.2). Their phase boundaries join at 1890 °C at a mullite composition of 77.2 wt. % Al$_2$O$_3$, and the peritectic composition was between 76.5 and 77.0 wt. % Al$_2$O$_3$. The solid solution range of mullite bends towards higher Al$_2$O$_3$ contents at temperatures above 1600 °C. More information on phase equilibria, solid state reactions and on whether mullite melts congruently or incongruently has been intensively discussed by Pask in several review papers (e.g. Davis and Pask, 1971, Pask, 1990, 1998). Pask believed that it is unhelpful to assume that one concept is correct and the other incorrect. It seems to be necessary to review the nature of the experiments with a view to identifying stable and metastable reaction products which can explain the different findings. A feature of the phase diagram that can be used for deductive analysis is the position of the α-alumina liquidus. When the Al$_2$O$_3$ content of the liquidus peritectic is less than that of mullite at the melting temperature, mullite melts incongruently. When the liquidus peritectic has a higher Al$_2$O$_3$ content, then mullite melts congruently.

Shornikov and coworkers (e.g. Shornikov et al. 2002) re-examined the thermodynamic properties of mullite. The dependencies identified in the temperature-concentration functions of the dominant components in the gas phase over the subsolidus regions in the system Al$_2$O$_3$–SiO$_2$ yield stoichiometric mullite. The approach of Shornikov et al. provides no evidence for a (stable) mullite solid solution and suggests congruent melting of mullite. However, Shornikov et al. do not discuss the discrepancy between their results and the experimental findings of other researchers.

3.1.2
Processing Parameters and Reaction Mechanisms

In evaluations of stable or unstable phase diagrams with associated congruent or incongruent melting of mullite, significant differences in synthesis procedures and starting materials can be noted: α-alumina and silica are starting materials in the Al_2O_3–SiO_2 system investigated by Bowen and Greig (1924), Aksay and Pask (1975) and Aramaki and Roy (1962). In the respective phase diagrams, the mullite solid-solution region is constant with increase in temperature. The starting materials used by Klug et al. (1987) are homogeneous aluminosilicate powders formed by sol-gel processing that are free of α-alumina particles. Their phase diagram shows mullite solid solutions with the mean Al_2O_3 content increasing with temperature. It has therefore been postulated that reactions between α-alumina and silica are critical and that reactions in the presence or absence of α-alumina can follow different reaction paths.

The basic nature of reactions in the presence of α-alumina can be explored by reviewing the procedures followed by Davis and Pask (1972), and Aksay and Pask (1975). Diffusion couples of single crystal sapphire (α-alumina) and fused silica were held at a selected temperature and subsequently quenched. The analyses provided data for calculating concentration profiles for Al_2O_3 and SiO_2. Repeating this experiment at constant temperature for a number of different times shows that the composition at the interface remains constant. Also, the distance from the Boltzmann-Matano interface for a given concentration is directly proportional to the square root of time. These results indicate that the reaction rates at the interface are faster than the diffusion rates and that diffusion is the process-controlling step. The compositions at the interfaces at a given temperature are thus believed to be at equilibrium, which means that the aluminosilicate composition is a point on the α-alumina liquidus. Concentration profiles for Al_2O_3 at 1950 and 1750 °C are shown for phase equilibrium in Figs. 3.1.3a,b. At temperatures above 1828 °C, no mullite appears at the interface; below 1828 °C mullite occurs together with α-alumina and liquid. The data for the liquid/mullite interface determines a point on the mullite liquidus at 1750 °C. With a sufficient number of such composition points on the liquidi, the peritectic temperature and composition can be determined. The composition profile also indicates the composition of mullite in equilibrium with α-alumina. The SiO_2 profile is not shown because silica solubility in α-alumina was not discernable.

It is noteworthy that the peritectic Al_2O_3 composition determined by Klug et al. (1987) is at about 76 wt. %, while that of Aksay and Pask (1975) is at about 55 wt. % Al_2O_3. The reactions occurring at the interface of the α-alumina/silica diffusion couple may help to account for this large discrepancy. At temperatures above 1828 °C, silica dissolves alumina as indicated by the Al_2O_3 composition profile

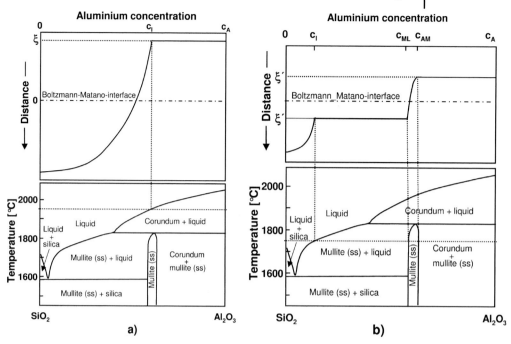

Fig. 3.1.3 Relationships between the concentration profile for alumina of semi-infinite Al$_2$O$_3$–SiO$_2$ diffusion couples and the phase equilibrium diagram. The Boltzmann-Matano interface corresponds to the boundary at $t = 0$. (a) 1950 °C, (b) 1750 °C (after Aksay and Pask, 1975).

(Fig. 3.1.3a). The driving force of the process is a reduction of the internal free energy of the fused silica as alumina is incorporated into the liquid. "Free" alumina is obtained by dissociation from crystalline α-alumina, and this requires energy. This energy is provided by the internal free energy released by the siliceous liquid as the alumina dissolves. Summation of these step reactions results in the observed net reaction:

Step I: Dissociation of α-alumina
$$\alpha\text{–Al}_2\text{O}_3 = x\,\text{Al}_2\text{O}_3 + (1-x)\,\alpha\text{–Al}_2\text{O}_3 \qquad \Delta G_\text{I}\ (\text{"Energy loss"})$$
Step II: Al$_2$O$_3$ incorporation into liquid
$$\text{SiO}_2 + x\,\text{Al}_2\text{O}_3 = \text{SiO}_2 \cdot x\text{Al}_2\text{O}_3 \qquad \Delta G_\text{II}\ (\text{"Energy gain"})$$
Net reaction:
$$\text{SiO}_2 + \text{Al}_2\text{O}_3 = \text{SiO}_2 \cdot x\text{Al}_2\text{O}_3 + (1-x)\,\text{Al}_2\text{O}_3 \qquad \Delta G^* \qquad (1)$$

If $\Delta G_\text{I} < \Delta G_\text{II}$, then the net ΔG^* is negative and the reaction will proceed with evolution of thermal energy. This means that the energy released as alumina is absorbed into the liquid structure is sufficient to dissociate the necessary alumina molecules from α-alumina. If, $\Delta G_\text{I} > \Delta G_\text{II}$, then ΔG^* is positive and the reaction

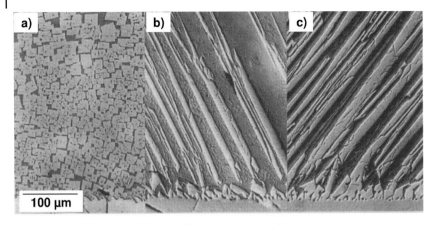

Fig. 3.1.4 Optical micrograph showing the microstructures of diffusion zones in reaction couples of sapphire (α-alumina, bottom of pictures) and silica glass, annealed at 1900 °C (15 min) submitted to different cooling procedures. (a) Rapidly quenched; (b) Cooled at a moderate rate, (c) Cooled slowly. Precipitates are: in (a) mullite (light gray); in (b) alumina (light gray needles) and mullite (fine precipitates between alumina needles); in (c) alumina (light gray needles). Precipitates along the interface in (b) and (c) are also alumina (from Schneider et al. 1994).

will not proceed. Therefore, when $\Delta G_I = \Delta G_{II}$, or $\Delta G^* = 0$, the total alumina dissolved in the silica liquid is the equilibrium amount and it becomes a point on the α-alumina liquidus for that temperature. This equilibrium relative to α-alumina, however, does not necessarily mean that the aluminosilicate liquid itself is saturated with alumina molecules. If instead of α-alumina another source of alumina is used whose bonds are not as strong, making ΔG_I less positive, then the liquid can dissolve and incorporate more alumina molecules before ΔG^* becomes zero.

Aksay and Pask (1975) investigated the influence of cooling rates from high temperatures on the reaction processes. For that purpose three diffusion couples of sapphire and fused silica were heated to 1900 °C and then cooled at different rates. Fig. 3.1.4 shows cross-sections perpendicular to the interfaces. The specimen, with the fastest cooling rate shows precipitates of mullite in the liquid (Fig. 3.1.4a), while that with the slowest cooling rate shows precipitates of α-alumina (Fig. 3.1.4c). The reaction couple that was cooled at an intermediate rate shows large mullite precipitates with smaller crystals of α-alumina in the in-between glass phase (Fig. 3.1.4b). At the interface of samples with slow and intermediate cooling rates a layer of α-alumina crystals grows since sapphire acts as a heterogeneous nucleation site (Figs. 3.1.4b and c). With fast cooling, as represented by Fig. 3.1.4a, sufficient time to form α-alumina is not available, and mullite crystallizes at lower temperatures instead. The conclusion is that α-alumina nucleation is difficult from a liquid that is not saturated with alumina or in the absence of α-alumina nucleation sites. It should be noted that α-alumina crystals,

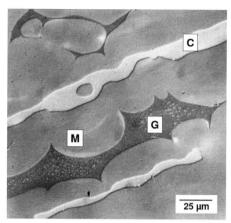

Fig. 3.1.5 Optical micrograph showing the peritectic crystallization of mullite. The micrograph shows the microstructure of a specimen containing 71.8 wt. % Al_2O_3 (stoichiometric mullite composition) held at 1950 °C (7.7 h), cooled to 1750 °C in 30 min, annealed at 1750°C (29.4 days), and then quenched to room temperature. Light gray precipitates are α-alumina (C) completely surrounded by layers of mullite (M). Dark gray portions between mullite layers are glass (G) with mullite precipitations, formed on quenching. The microstructure of the sample indicates that equilibrium has not been achieved (from Schneider et al. 1994).

as observed in Fig. 3.1.4c, would not have formed if congruent mullite melting was thermodynamically stable. The observations thus provide strong evidence for an incongruent melting of mullite.

Another experiment performed by Aksay and Pask (1975) showed mullite formation by a peritectic reaction. For that purpose a mixture containing 71.8 wt. % Al_2O_3 was first homogenized at 1950 °C, slowly cooled to 1750 °C, then long-term annealed at 1750 °C and subsequently quenched. The micrograph of a cross-section shows crystals of α-alumina surrounded by mullite with an aluminosilicate glass occurring between the mullite crystals (Fig. 3.1.5). Obviously the homogenized liquid was cooled to 1828 °C sufficiently slowly to allow precipitation of α-alumina crystals. Annealing at 1750 °C then caused a peritectic reaction between α-alumina and silica in the liquid to form mullite. This reaction is slow because the rate-determining step is aluminum diffusion from α-alumina through the newly formed mullite to the silica liquid. The important point of this experiment is the formation of α-alumina, which should not precipitate from the melt under congruent melting conditions. Thus Pask and coworkers favor incongruent melting.

Kleebe et al. (2001) studied the formation of 3/2-mullite from amorphous silica spheres (500 nm in diameter), and two different alumina powders (300 nm and 2 μm in diameter). When the nanosized (300 nm) alumina powders were used fast dissolution into the coalesced silica glass was observed, followed by homogeneous nucleation and growth of mullite in the glass ("stable crystallization path"). Using the microsized (2 μm) alumina grain fraction resulted in the formation of two glasses, one being Al_2O_3- and the other SiO_2-rich, where mullite crystallized from the former ("metastable crystallization path"). Kleebe et al. pointed out that in practice both stable and metastable mullite formation can occur simultaneously. From theoretical criteria, and using data along the liquidus curves of mullite and α-alumina, Djuric et al. (1996) suggested immiscibility regions at both ends of the phase diagram.

3.1.3
Solid-solution Range of Mullite

According to the experiments of Klug et al. (1987) mullite with maximally 77 wt. % of Al_2O_3 (i.e. approximately 2/1-composition: $2Al_2O_3 \cdot SiO_2$) forms under equilibrium conditions (Fig. 3.1.2). Bauer et al. (1950), Bauer and Gordon (1951), and Kriven and Pask (1983) suggested that melt-grown mullites can achieve up to 83 wt. % Al_2O_3, which corresponds to a 3/1-composition ($3Al_2O_3 \cdot SiO_2$). In this context it is interesting that the Al_2O_3 content of sol-gel-derived mullite can be as high as 92 wt. %, corresponding to a 9/1-composition ($9Al_2O_3 \cdot SiO_2$, see Fischer et al. 1996 and Sections 1.1.3.7 and 2.5.1). Both Al_2O_3-rich compositions are, without doubt, metastable mullite forms. Important factors for the formation of these Al_2O_3-rich mullites are the (super)saturation of the melts or gels with Al_2O_3, and the lack of effective aluminum sinks (e.g. α-alumina or γ-alumina). The occurrence of 3/2-, 2/1-, 3/1- and 9/1-mullites raises the question of whether some mullite compositions are more stable than others. If so, specifically "stable" oxygen-vacancy distribution schemes may be taken into account for these mullites (see also Sections 1.1.3.7 and 2.5.1).

3.1.4
Melting Behavior of Mullite

As shown in Sections 3.1.1 and 3.1.2 mullite melts incongruently. This conclusion is based on the appearance of mullite at the α-alumina/silica interface at temperatures below, and its absence at temperatures above the peritectic line using the diffusion couple technique. It also refers to the Al_2O_3 content of the peritectic, which is lower than that of mullite. Bowen and Greig (1924) used mixtures of α-alumina and silica as starting materials. They suggested that above 1828 °C and in mixtures with more than 50% Al_2O_3 the solution reaction is complete with some α-alumina still remaining (see Fig. 3.1.1). The compositions prepared by Klug et al. (1987) were homogeneous on an atomic scale. On heating to 1890 °C, Al_2O_3-rich mullite appears (with an approximately 2/1-composition, i.e. $2Al_2O_3 \cdot SiO_2$), and silica is expelled as a liquid. This mullite melts at 1890 °C and, if α-alumina seeds are present, α-alumina growth will proceed with continued heating.

3.1.5
Simulations of the Al_2O_3–SiO_2 Phase Diagram

Eriksson and Pelton (1993) presented a calculated Al_2O_3–SiO_2 phase diagram at 1 bar pressure from 25 °C up to above liquidus temperature. Their simulations relate to the experimental work of Klug et al. (1987), who studied the region of the Al_2O_3–SiO_2 phase diagram close to the melting point of mullite, and the non-stoichiometry of mullite above 1600 °C (see Sections 3.1.1 and 3.1.2). The simulated phase diagram follows well the experimentally determined liquidus curve and the non-stoichiometry of mullite of Klug et al. (1987). According to the calculated phase diagram, mullite with 79.6 mol% Al_2O_3 melts congruently at about 1890 °C. The disagreement of simulations with experimental observations may be explained with the fact that congruent and incongruent melting of mullite are only few degrees Celsius apart.

3.1.6
General Remarks

Studies in the Al_2O_3-rich region of the Al_2O_3–SiO_2 system provide evidence for both thermodynamically stable incongruent and metastable congruent melting of mullite at atmospheric pressure. The controlling feature seems to be the position of the α-alumina liquidus which determines whether incongruent or congruent melting is observed. If the α-alumina plus silica liquidus intersects the mullite melting temperature line at Al_2O_3 contents below that of mullite then mullite melts incongruently. If this intersection point is at higher Al_2O_3 contents mullite will melt congruently. According to Pask and coworkers the position of the α-alumina plus silica liquidus depends on the nature of the starting materials and on experimental methods. Thus bond strengths, free energies and crystal structures of the participating phases can influence the position of the equilibria. The driving force ("energy gain") for alumina dissolution is the energy released on incorporating alumina into the liquid, caused by the reduction of its free energy. The resisting force ("energy loss") is the energy necessary to dissolve Al_2O_3 from α-alumina. Equilibrium solution occurs if both forces are equal. If different sources of alumina are used with different energies required to release alumina, varying amounts of alumina may be dissolved. If the liquid phase becomes Al_2O_3-saturated on cooling, it will expel α-alumina. If Al_2O_3-saturation is not reached on cooling, then the formation of α-alumina becomes unlikely.

The works of Pask and coworkers and the careful experimental study of Klug et al. (1987) provide strong evidence that mullite under equilibrium conditions and in the presence of α-alumina melts incongruently, whereas in the absence of α-alumina mullite shows (non-equilibrium) congruent melting. In spite of the experimental and theoretical evidence for stable incongruent and metastable congruent melting, a final decision upon this point is difficult. A possible reason for this confusing situation is probably that incongruency is only a few degrees Celsius away from congruency (Klug et al. 1987, Figs. 3.1.1 and 3.1.2). Thus both congruent and incongruent melting models remain the subject of further discussion.

3.2
Influence of Environmental Conditions on the Stability of Mullite
H. Schneider

This section provides information on the high-temperature stability of mullite under various environmental conditions. Discussions focus on reducing and water-vapor-rich atmospheres and on attack by molten alkaline salts and fluorine-containing compounds. These data are crucial for the prediction of the stability of mullite ceramics in severe environments.

3.2.1
Interactions with Reducing Environments

Reducing atmospheres cause decomposition of mullite at high temperature. Mullite specimen surfaces heat-treated in helium between 1650 and 1800 °C degrade with formation of α-alumina and volatile silicon species, the recession progressing from the surface towards the bulk of specimens (Davis et al. 1972). The following reactions have been considered:

$$3\ Al_2O_3 \cdot 2\ SiO_2(s) \rightarrow 3\ Al_2O_3(s) + 2\ SiO(g) + O_2(g) \qquad (2)$$
$$\text{mullite} \qquad\qquad\qquad \text{α-alumina}$$

and

$$2\ SiO(g) + O_2(g) \rightarrow 2\ SiO_2(g) \qquad (3)$$

Net reaction:

$$3\ Al_2O_3 \cdot 2\ SiO_2(s) \rightarrow 3\ Al_2O_3(s) + 2\ SiO_2(g) \qquad (4)$$
$$\text{mullite} \qquad\qquad\qquad \text{α-alumina}$$
$$s = \text{solid, } g = \text{gaseous}$$

Krönert and Buhl (1978a,b) studied the influence of a carbon monoxide atmosphere on the high-temperature behavior of mullite. They identified a reaction of the type:

$$3Al_2O_3 \cdot 2SiO_2(s) + 2\ CO(g) \rightarrow 3\ Al_2O_3(s) + 2\ SiO(g) + 2\ CO_2(g) \qquad (5)$$
$$\text{mullite} \qquad\qquad\qquad\qquad \text{α-alumina}$$
$$s = \text{solid, } g = \text{gaseous}$$

The recession of mullite is a heterogeneous process with transport of carbon monoxide to the ceramic surface, reaction with mullite at phase boundaries and subsequent removal of the gaseous compounds. As a result of mullite decomposition, a porous α-alumina reaction layer forms (Fig. 3.2.1), whose thickness is a function of reaction time and through which the gaseous reactants have to migrate. Krönert and Buhl suggested that the decomposition rate of mullite is essentially controlled by the gas diffusion velocity, and only to a lower degree by phase boundary reactions. The recession of mullite with carbon monoxide in powders is much higher than in pore-free, smooth-surfaced specimens. This is explained by the high free surface area in powders, which enhances diffusion and reaction processes.

A more recent study of mullite decomposition under reducing conditions in the

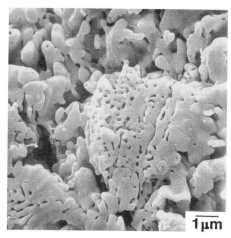

Fig. 3.2.1 Scanning electron micrograph of a mullite ceramic after heat treatment in carbon monoxide (CO) atmosphere. Note the porous surface of the grains and the intense corrosion, especially of the bonding phase at the grain boundaries (sample heat-treated at 1600 °C, 1 h, reaction atmosphere 10 l CO h^{-1}; from Schneider et al. 1994).

1 μm

presence of carbon has been performed by Xiao and Mitchell (2000) between 1900 and 2000 °C. Although their experimental temperature range was beyond the melting point of mullite (about 1830 °C: Pask, 1994; about 1890 °C: Klug et al. 1987) a phase mixture consisting of α-alumina, aluminosilicate glass and mullite was found, probably due to non-equilibrium reactions. Xiao and Mitchell believe that mullite decomposition is a stepwise process:

Step 1: $3\,Al_2O_3 \cdot 2\,SiO_2(s) \rightarrow 3\,Al_2O_3(s) + 2\,SiO_2(g)$ (6)
       mullite   α-alumina

Step 2: $SiO_2(g) + C(s) \rightarrow SiO(g) + CO(g)$ (7)

Net reaction:

$3\,Al_2O_3 \cdot 2\,SiO_2(s) + 2\,C(s) \rightarrow 3\,Al_2O_3(s) + 2\,SiO(g) + 2\,CO(g)$ (8)
mullite   α-alumina

s = solid, g = gaseous

Owing to the release of volatile SiO, the residual material becomes enriched in Al$_2$O$_3$ leading to α-alumina formation. Over time large α-alumina crystals grow by coalescence of smaller individuals. In a final stage condensed α-alumina sample surface layers can be observed on the mullite substrates (Fig. 3.2.2). The addition of SiO$_2$ slows down the mullite decomposition, since the equilibrium of reaction step 1 is pushed to the left-hand side of the equation.

Shornikov et al. (2002) described the decomposition of mullite to the gaseous species silicon monoxide (SiO(g)), aluminum (Al(g)) and Oxygen (O(g)) on the basis of Knudsen mass spectroscopy. Partial vapor pressures p_{SiO}, p_{Al} and p_O are shown in Fig. 3.2.3 in the sub-solidus of the system Al$_2$O$_3$–SiO$_2$. Partial pressures are constant between silica and mullite, and between mullite and alumina, respectively, with p_{SiO} being several orders of magnitude higher than p_{Al} and p_O. The

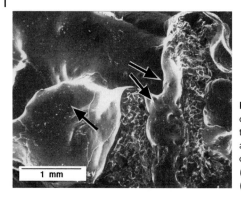

Fig. 3.2.2 Scanning electron micrograph of a mullite sinter body after heat treatment at 1950 °C in a reducing atmosphere. Note the occurrence of condensed α-alumina and of silica glass (arrows) at the surface of samples (Courtesy B. Mitchell).

p_{Al} value is relatively low in the silica-mullite and higher in the mullite-alumina field, while p_{SiO} and p_O are similar in both fields. The data confirm that decomposition of mullite under reducing conditions is primarily caused by volatilization of silica in the form of silicon oxide (SiO).

The decomposition of mullite under exposure to dry hydrogen gas at high temperature was investigated by Tso and Pask (1982) and Herbell et al. (1990, 1998). Herbell et al. found that the recession of mullite ceramics started at the silica-rich glass grain boundary phase at about 1050 °C. Above about 1250 °C decomposition of mullite is observed according to the equation:

$$3\ Al_2O_3 \cdot 2\ SiO_2(s) + 2\ H_2(g) \rightarrow 3\ Al_2O_3(s) + 2\ SiO(g) + 2\ H_2O(g) \tag{9}$$

mullite   α-alumina

s = solid, g = gaseous

During the decomposition process gaseous SiO and H_2O compounds are removed from the reaction zone towards the cooler parts of the furnace, with re-oxidation of SiO to SiO_2 and precipitation of solid SiO_2. The scrubbing of the aluminosilicate grain boundary glass, and the degradation of mullite finally produces porous α-alumina layers on the mullite specimens. Herbell et al. suggest that the removal of gaseous products from the reaction zone is rate controlling rather than their diffusion through the porous reaction layer. High temperature hydrogen gas-exposed mullite ceramics display reduced room temperature strength. This can be explained by the preferential attack of the grain boundary glass phase, enveloping the mullite crystals. The strength enhancement after short-time hydrogen gas exposure has been associated with the healing of surface flaws triggered by the viscous flow of the aluminosilicate melts towards the ceramic surface. Zaykoski et al. (1991) studied the stability of mullite whisker compacts in vacuum and helium atmosphere between 1350 and 1550 °C, and observed recession of the mullite whiskers caused by evaporation of SiO. The reactions are temperature and gas pressure-controlled and start at the periphery of compacts.

The experimental results described above agree well with the poor resistance of mullite refractories against reducing environments. In these materials attack takes

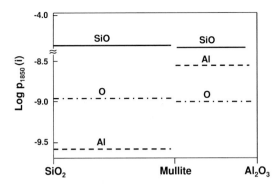

Fig. 3.2.3 Partial vapor pressures of silicon monoxide (SiO), aluminum (Al) and oxygen (O) over compounds in the system $Al_2O_3–SiO_2$ at 1850 K (1577 °C; after Shornikov et al. 2002).

place preferentially on the fine-grained mullite bonding phase. However, a beneficial influence of reducing conditions on the creep behavior of mullite refractories has been reported as well (Vishnevskii and Tal'yanskaya, 1979), who worked with argon and hydrogen atmospheres, and under vacuum. They suggested that if the reducing character is relatively strong, a fine grained α-alumina matrix is formed by decomposition of mullite, which strengthens the material. In the case of lower reduction power, re-oxidation of volatile SiO to SiO_2 and subsequent precipitation to an SiO_2 melt takes place (see above). With increasing amount of melt, however, the mechanical strength of mullite ceramics decreases at high temperature. Provost and Le Doussal (1974a,b) pointed out that creep of mullite ceramics can be three times more intensive in reducing than in oxidizing atmospheres. However, they ascribed this observation to the lower viscosity, not to the increased amount of the coexisting glass phase. Another explanation for decreased strength is that sublimation of gaseous silicon monoxide produces porosity, primarily along grain boundaries (Pitak and Ansimova, 1974).

3.2.2
Interaction with Water Vapor-rich Environments

The high temperature stability of materials under water vapor-rich conditions is crucial for many technical applications (e. g. thermal protection systems in combustor chambers of gas turbine engines). Silicon-based non-oxide ceramics (silicides, nitrides, carbides, borides) are not long-term stable in these environments, owing to decomposition of their silica surface layers to volatile silicon compounds. Both constituents of mullite, Al_2O_3 and SiO_2, can also form volatile hydroxides and oxyhydroxides. The high chemical activity of silica compared with alumina (at 1879 K, i.e. 1606 °C, the activity of silica has been given as 1.00 and that of alumina as 0.37, see Shornikov and Archakov, 2002) explains the preferred release of Si–O–H species from mullite, and correlated high gas pressures of volatile Si–O–H species:[1] The dominating gaseous silicon species is $Si(OH)_4$. Additionally, vola-

---

1) Under oxidizing conditions and a total gas pressure of 1 atm at 850 °C the partial gas pressure of $Si(OH)_4$ is 0.5×10^{-8} and that of $Al(OH)_3$ is 0.5×10^{-12} (Crossland et al. 1997).

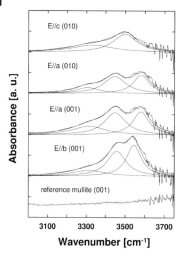

Fig. 3.2.4 Infrared absorption spectra of mullite (010) and (001) single crystal plates after treatment in a water vapor-rich environment (Ar/H$_2$O = 90/10, 10 kPa, 1600 °C). Note the incorporation of hydroxyl (OH) groups into the mullite structure, while the untreated reference sample displays no protonization of mullite (Courtesy C. Rüscher).

tile SiO(OH)$^+$, SiO(OH)$_2$ and SiO compounds have been described (Opila et al. 1999).

Rüscher et al. (2002) provided a model explaining mullite decomposition due to water vapor attack in several steps:

- Dissociation of H$_2$O into protons, which diffuse into mullite, and form OH groups: A maximum amount of OH corresponding to 0.02 wt. % H$_2$O has been estimated (Fig. 3.2.4). Charge compensation is believed to be effected by "exsolution" of Na$^+$ ions found as trace impurity in mullite. The proton uptake may cause hydroxyl weakening of the mullite structure and in ceramics damage of grain boundaries.

- When the maximum level of proton incorporation in mullite is exceeded, Si^{4+} ions are set free, migrate to the mullite crystal surfaces and react with water vapor to form Si–O–H species. The substrate material thus gradually becomes enriched in α-alumina. This process proceeds from the surface towards the bulk.

The water vapor-induced decomposition of mullite in stationary systems is diffusion controlled, since the liberated silicon species must diffuse through a Si–O–H gas boundary layer between mullite and the water-rich gas (Fig. 3.2.5). Depending on temperature and gas pressure, these boundary layers act more or less as barriers against further mullite decomposition. This explains the high stability of dense and glass-free mullite coatings in stationary (low gas pressure, no streaming gas) high temperature water vapor-rich environments (Haynes et al. 1999, 2000). In open high temperature systems (with high gas pressure and rapidly flowing gas) the boundary diffusion barriers become thin or break down totally, which results in rapid recession of mullite. The basic reaction of the mullite decomposition in the presence of water vapor at high temperature can thus be expressed as:

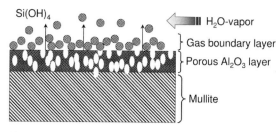

Fig. 3.2.5 Suggested mechanism of mullite recession in water vapor-rich, rapidly flowing gases at high temperature and gas pressure. The silicon compound formed by mullite decomposition is primarily Si(OH)$_4$. The volatile reaction products have to diffuse through the newly formed porous α-alumina and the gas boundary layer of Si(OH)$_4$ at the mullite surface.

$$3 \, Al_2O_3 \cdot 2 \, SiO_2(s) + 2 \, H_2O(g) \rightarrow 3 \, \alpha\text{-}Al_2O_3(s) + 2 \, Si(OH)_4(g) \qquad (10)$$

mullite $\qquad\qquad\qquad$ α-alumina

s = solid, g = gaseous

Mullite decomposition is enhanced by temperature, time, gas pressure and gas stream velocity (see Robinson and Smialek, 1999).

The alteration of Nextel 720 aluminosilicate fiber-reinforced alumina matrix composites in high temperature water vapor environments has been studied by Wannaparhun and Seal (2003) by X-ray photoelectron spectroscopy (XPS) and thermodynamic calculations. Their results confirmed the mullite decomposition and the formation of volatile silicon hydroxide, Si(OH)$_4$. The authors observed recondensation of the gaseous silicon species at the composite's surface with formation of aluminosilicate phases.

The instability of mullite in high temperature water vapor-rich environments can become a serious problem for the application of these materials. A suitable way to overcome this instability is by deposition of environmental barrier coatings (EBCs).

3.2.3
Interactions with Molten Sodium Salts

Jacobson et al. (1996) studied the interactions between mullite and molten sodium salts in a temperature range from 1000 to 1089 °C. From the phase diagram Na$_2$O–Al$_2$O$_3$–SiO$_2$ (Fig. 3.2.6) they deduced that low Na$_2$O activities give rise to the formation of α-alumina (Al$_2$O$_3$), mullite (3Al$_2$O$_3$ · 2SiO$_2$) and albite (Na$_2$O · Al$_2$O$_3$ · 6SiO$_2$), whereas at higher Na$_2$O contents α-alumina, nepheline (Na$_2$O · Al$_2$O$_3$ · 2SiO$_2$) and albite or α-alumina, nepheline and β-alumina (Na$_2$O · 11Al$_2$O$_3$), are observed. The experiments of Takahashi et al. (2002) confirmed this view by means of hot corrosion tests on cordierite-mullite materials with sodium chloride (NaCl) and sodium sulfate (Na$_2$SO$_4$) at moderate Na$_2$O activities and a temperature of

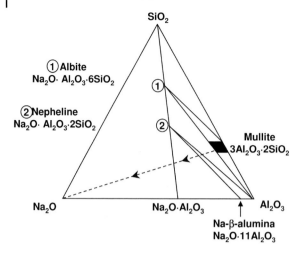

Fig. 3.2.6 Phase equilibrium diagram of the ternary system $Na_2O–Al_2O_3–SiO_2$ (after Schairer and Bowen, 1956). The dashed line and arrows represent the phase development with the increase of the amount of sodia (Na_2O) added to the mullite composition (after Jacobsen et al. 1996).

1000 °C. Under these conditions mullite partially decomposes to carnegieite ($Na_2O \cdot Al_2O_3 \cdot 2SiO_2$, the high-temperature polymorph of nepheline). The sodium corrosion of mullite starts at sample surfaces and at mullite grain boundaries and proceeds into the bulk of the ceramics. Owing to the highly refractory nature of the decomposition products, mullite coatings are expected to be suitable against sodium corrosion for silicon carbide or silicon nitride substrates.

3.2.4
Interactions with Fluorine Salt Environments

Moyer (1995) investigating the phase diagram mullite-silicon tetrafluoride (SiF_4, Fig. 3.2.7) showed that mullite is stable in the presence of gaseous silicon tetrafluoride above 1060 °C. Below 1060 °C and depending on the SiF_4 concentration mullite partially or completely decomposes to fluorotopaz:

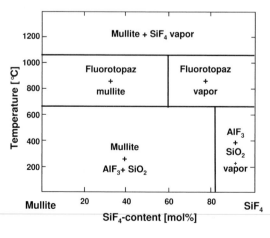

Fig. 3.2.7 Phase equilibrium diagram for mullite-SiF_4. The system is pseudo-binary below 1056 °C with four compounds (mullite, AlF_3, SiO_2, SiF_4; after Moyer, 1995).

3 Al$_2$O$_3$ · 2 SiO$_2$(s) + 1.5 SiF$_4$(g) → 3 Al$_2$O$_3$ · 2 SiO$_2$ · 1.5 SiF$_4$(s)   (11)
mullite   fluorotopaz

s = solid, g = gaseous

The reverse reaction, i.e. pyrolysis of fluorotopaz above 1060 °C to form mullite and volatile silicon tetrafluoride (SiF$_4$), has been used for the synthesis of mullite whiskers by a gas transport reaction (see also Section 6.1.2). At temperatures below 660 °C the stability field of mullite plus aluminum fluoride (AlF$_3$) and silica (SiO$_2$) is extended towards SiF$_4$ contents above 80%. Moyer (1995) showed that both fluorotopaz and mullite can grow from a metastable melt in the presence of silicon tetrafluoride.

3.3
Ternary X–Al$_2$O$_3$–SiO$_2$ Phase-equilibrium Diagrams
H. Schneider

Ternary phase diagrams are of basic interest, providing information on possible reactions of specific oxide compounds with the Al$_2$O$_3$–SiO$_2$ system. Three component systems have been widely used to predict the thermal stability of aluminosilicate materials, since refractories can only seldomly be regarded as part of the binary system Al$_2$O$_3$–SiO$_2$. Almost every natural ceramic raw material contains iron oxide (Fe$_2$O$_3$), magnesia (MgO), titania (TiO$_2$) and alkali and alkaline earth oxides in higher or lower amounts. Moreover, mullite may react with dust, slags and metal melts during the industrial use of ceramics. Thus impurities can have a strong influence on the high temperature properties of ceramics. Selected ternary phase diagrams will illustrate relationships.

3.3.1
Alkaline Oxide–Al$_2$O$_3$–SiO$_2$

Phase relations at liquidus temperatures in the system K$_2$O–Al$_2$O$_3$–SiO$_2$ were determined by Schairer and Bowen (1947, 1955). Potash (K$_2$O) has a strong fluxing effect on Al$_2$O$_3$–SiO$_2$ mixtures with high SiO$_2$ contents. Liquidus temperatures decrease to values as low as 958 °C (ternary eutectic) if K$_2$O in the order of 10 wt. % is added to a mixture containing 87 wt. % SiO$_2$ and 13 wt. % Al$_2$O$_3$. The system Na$_2$O–Al$_2$O$_3$–SiO$_2$ (Schairer and Bowen, 1956) exhibits similar phase relationships, as does the system K$_2$O–Al$_2$O$_3$–SiO$_2$ (Fig. 3.3.1). The phase relationships show that materials with low alkali content have to be selected if high refractoriness is required.

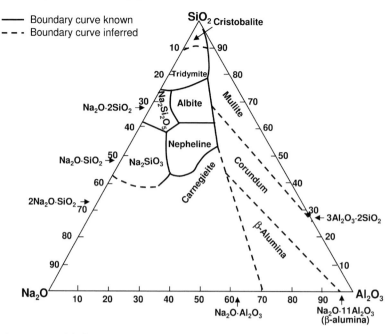

Fig. 3.3.1 Simplified ternary phase equilibrium diagram
Na$_2$O–Al$_2$O$_3$–SiO$_2$ (after Schairer and Bowen, 1956).

3.3.2
Iron Oxide–Al$_2$O$_3$–SiO$_2$

The effect of iron oxide on mullite stability depends on the oxygen partial pressure,
i.e. whether Fe^{3+} or Fe^{2+} is the major iron species (Schairer and Yagi, 1952, Muan,
1957):

- Reducing conditions in a ceramic body consisting of mullite plus silica can give
  rise to the development of liquids at temperatures as low as 1210 °C if small
  amounts of iron oxide are present in the refractory (Muan and Osborn, 1965). If
  the composition of the ceramic body is richer in Al$_2$O$_3$ and consists of α-alumina
  plus mullite, a liquid will form at about 1380 °C.
- Oxidizing conditions cause higher liquidus temperatures (see Muan and Osborn, 1965): If the refractory consists of mullite plus silica, a liquid develops
  above 1380 °C. A mullite plus α-alumina material withstands temperatures of
  more than 1460 °C in the presence of iron oxide. The different fluxing behaviors
  can be related to the incorporation of Fe^{3+} (oxidizing conditions), but not of Fe^{2+}
  (reducing conditions) into the crystalline phases α-alumina, mullite and tieillite
  (Al$_2$TiO$_5$). Therefore, small amounts of Fe^{2+} oxide will cause the formation of
  liquids, whereas the presence of Fe^{3+} does not have this effect. Iron oxide can
  also act as a catalyst causing the deposition of carbon produced by decomposition of carbon monoxide. This can lead to a disintegration of refractory bodies.

This undesirable effect is prevented by high temperature firing by which the iron oxide is brought into the liquid phase.

3.3.3
Alkaline Earth Oxide–Al$_2$O$_3$–SiO$_2$

Although magnesia (MgO) itself is an excellent refractory compound, the addition of small amounts of magnesia to the system Al$_2$O$_3$–SiO$_2$ causes the formation of large quantities of liquid at high temperature. Phase relationships in the system MgO–Al$_2$O$_3$–SiO$_2$ were first determined by Rankin and Merwin (1918) and Greig (1927) and have been modified many times (see Muan and Osborn, 1965).

The phase diagram CaO–Al$_2$O$_3$–SiO$_2$ was first proposed by Rankin and Wright (1915) and Greig (1927) and has often been reinvestigated since that time (e.g. Osborn and Muan 1960, Filonenko and Lavrov 1953, Langenberg and Chipman 1956, Aramaki and Roy 1959, Gentile and Foster, 1963). A large number of crystalline phases occur in this system. In spite of the high melting temperatures of the three end members calcia (CaO), α-alumina (Al$_2$O$_3$) and silica (SiO$_2$), liquidus temperatures lower than 1300 °C can be observed. There is only little mutual solid solution between the end members calcia, alumina and silica, because of the different sizes and charges of the cations Ca^{2+}, Al^{3+} and Si^{4+}. The same is true for the system MgO–Al$_2$O$_3$–SiO$_2$.

3.3.4
MnO–Al$_2$O$_3$–SiO$_2$

Knowledge of the stability of aluminosilicate ceramics in the presence of manganese oxide (MnO) is important for the evaluation of possible reactions with steel plant slags. Since these processes take place in reducing atmosphere, they can be described by means of the ternary system MnO–Al$_2$O$_3$–SiO$_2$. Snow (1943) found that liquidus temperatures are as low as 1100 to 1200 °C in major parts of the system, documenting the strong fluxing effect of MnO on aluminosilicate bodies.

3.3.5
TiO$_2$–Al$_2$O$_3$–SiO$_2$

The system TiO$_2$–Al$_2$O$_3$–SiO$_2$ is important for the understanding of the thermal behavior of titanium-rich aluminosilicate refractories such as refractory grade bauxites. Phase relations were investigated by Agamawi and White (1952), and De Vries et al. (1954). Mullite, α-alumina (Al$_2$O$_3$), tridymite (or cristobalite, both SiO$_2$), rutile (TiO$_2$), and tieillite (Al$_2$TiO$_5$) are stable phases. The fluxing effect of titania (TiO$_2$) on aluminosilicate ceramic bodies is relatively low. Relationships change drastically if other compounds like Fe$_2$O$_3$ are also involved, since the melting point of tieillite mixed crystal containing Fe$_2$O$_3$ then decreases by more than 400 °C (see Caldwell et al. 1967).

3.4
Multicomponent Systems
H. Schneider

For the understanding of reaction processes during industrial uses of aluminosilicate refractories, a six-component system often is suitable: Al_2O_3–SiO_2–TiO_2–CaO–Fe_2O_3–FeO. Crystalline phases occurring in this system are mullite, α-alumina (Al_2O_3), tieillite mixed crystals (Al_2TiO_5–Fe_2TiO_5), anorthite ($CaAl_2Si_2O_8$), aluminate mixed crystals ($CaO \cdot 6\,Al_2O_3$–$CaO \cdot 6\,Fe_2O_3$), hematite mixed crystals (Fe_2O_3–Al_2O_3), spinel mixed crystals (Fe_3O_4–$FeAl_2O_4$–Fe_2TiO_4) and $CaTiO_3$. Assuming constant pressure in a six-component system, the maximum number of phases coexisting at equilibrium over a temperature range is six. In practice, the number of phases is smaller owing to the formation of solid solutions between individual constituents. Restrictions also arise from the incompatibility of phases. For example mullite cannot exist together with $CaO \cdot 6\,Al_2O_3$ since $CaO \cdot Al_2O_3 \cdot 2SiO_2$ (anorthite) is formed instead.

References

Agamawi, Y. M. and White, I. (1952). The system Al_2O_3–TiO_2–SiO_2. Trans. Brit. Ceram. Soc. **51**, 293–325.

Aksay, I. A. and Pask, J. A. (1975). Stable and metastable equilibria in the system SiO_2–Al_2O_3. J. Am. Ceram. Soc. **58**, 507–512.

Aramaki, S. and Roy, R. (1959). The mullite-corundum boundary in the system MgO–Al_2O_3–SiO_2 and CaO–Al_2O_3–SiO_2. J. Am. Ceram. Soc. **42**, 644–645.

Aramaki, S. and Roy, R. (1962). Revised phase diagram for the system Al_2O_3–SiO_2. J. Am. Ceram. Soc. **45**, 229–242.

Bauer, W. H. and Gordon, I. (1951). Flame fusion synthesis of several types of silicate structures. J. Am. Ceram. Soc. **34**, 250–254.

Bauer, W. H., Gordon, I. and Moore, C. H. (1950). Flame fusion synthesis of mullite single crystals. J. Am. Ceram. Soc. **33**, 140–143.

Bowen, N. L. and Greig, J. W. (1924). The system Al_2O_3–SiO_2. J. Am. Ceram. Soc. **7**, 238–254.

Caldwell, M., Hayhurst, A. and Webster, R. (1967). Role of titania in bauxite refractories. Trans. Brit. Ceram. Soc. **66**, 107–119.

Crossland, C. E., Shelleman, D. L., Spear, K. E. and Tressler, R. E. (1997). Thermochemistry of corrosion of ceramic hot gas filters in service. Mater. High Temp. **14**, 365–370.

Davis, R. F. and Pask, J. A. (1971). Mullite. In: Alper, A. M. (ed.) Refractory materials. High temperature oxides: Part IV, Refractory glasses, glass-ceramics, and ceramics. Academic Press, New York and London, pp. 37–76.

Davis, R. F. and Pask, J. A. (1972). Diffusion and reaction studies in the system Al_2O_3–SiO_2. J. Am. Ceram. Soc. **55**, 525–531.

Davis, R. F., Aksay, I. A. and Pask, J. A. (1972). Decomposition of mullite. J. Am. Ceram. Soc. **55**, 98–101.

De Vries, R. C., Roy, R. and Osborn, E. F. (1954). The system TiO_2–SiO_2. Trans. Brit. Ceram. Soc. **53**, 525–540.

Djuric, M., Mihajlov, A., Petrasinovic-Stojkanovic, L. and Zivanovic, B. (1996). Thermodynamic analysis of the metastable regions for the Al_2O_3–SiO_2 system. J. Am. Ceram. Soc. **79**, 1252–1256.

Eriksson, G. and Pelton, A. D. (1993). Critical evaluation and optimization on the thermodynamic properties and phase diagrams of the CaO–Al_2O_3, Al_2O_3–SiO_2, and CaO–Al_2O_3–SiO_2 systems. Metal. Trans. **24B**, 807–816.

Filonenko, N. E. and Lavrov, I. V. (1953). Fusion of mullite. Quoted in: Davis, R.F. and Pask, J.A.: Mullite. In: Alper, A.M. (ed.) Refractory materials. High temperature oxides: Part IV, Refractory glasses, glass-ceramics, and ceramics. Academic Press, New York and London, pp. 37–76.

Fischer, R.X., Schneider, H. and Voll, D. (1996). Formation of aluminium rich 9:1 mullite and its transformation to low alumina mullite upon heating. J. Europ. Ceram. Soc. **16**, 109–113.

Gentile, A.L. and Foster, W.R. (1963). Calcium hexaaluminate and stability relations in the system CaO–Al$_2$O$_3$–SiO$_2$. J. Am. Ceram. Soc. **46**, 74–76.

Greig, J.W. (1927). Immiscibility in silicate melts. Amer. J. Sci. **13**, 1–44.

Guse, W. (1978). Compositional analysis of Czochralski grown mullite single crystals. J. Crystal Growth **26**, 151–152.

Guse, W. and Mateika, D. (1974). Growth of mullite single crystals (2Al$_2$O$_3$ · SiO$_2$) by the Czochralski method. J. Crystal Growth **22**, 237–240.

Haynes, J.A., Cooley, K.M., Stinton, D.P., Lowden, R.A. and Lee, W.Y. (1999). Corrosion-resistant CVD mullite coatings for Si$_3$N$_4$. Ceram. Eng. Sci. Proc. **20**, 355–362.

Haynes, J.A., Lance, M.J., Cooley, K.M., Ferber, M.K., Lowden, R.A. and Stinton, D.P. (2000). CVD mullite coatings in high-temperature, high-pressure air-H$_2$O. J. Am. Ceram. Soc. **83**, 657–659.

Herbell, T.P., Hull, R.D. and Hallum, G.W. (1990). Effect of hight temperature hydrogen exposure on the strength and microstructure of mullite. In: Moody, N.R. and Thompson, A.W. (eds.) Hydrogen Effects on Material Behaviour. The Minerals, Metals & Materials Society, pp. 351–359.

Herbell, T.P., Hull, R.D. and Garg, A. (1998).Hot hydrogen exposure degradation of the strength of mullite. J. Am. Ceram. Soc. **81**, 910–916.

Jacobson, N.S., Lee, K.N. and Yoshio, T. (1996). Corrosion of mullite by molten salts. J. Am. Ceram. Soc. **79**, 2161–2167.

Kleebe, H.-J., Siegelin, F., Straubinger, T. and Ziegler, G. (2001). Conversion of Al$_2$O$_3$–SiO$_2$ powder mixtures to 3:2 mullite following the stable or metastable phase diagram. J. Europ. Ceram. Soc. **21**, 2521–2533.

Klug, F.J., Prochazka, S. and Doremus, R.H. (1987). Alumina-silica phase diagram in the mullite region. J. Am. Ceram. Soc. **70**, 750–759.

Kriven, W.M. and Pask, J.A. (1983). Solid solution range and microstructures of melt-grown mullite. J. Am. Ceram. Soc. **66**, 649–654.

Krönert, W. and Buhl, H. (1978a). The influence of various gaseous atmospheres on the melting behaviour of mullite (Part I). Interceram **1**, 68–72.

Krönert, W. and Buhl, H. (1978b). The influence of various gaseous atmospheres on refractories of the system Al$_2$O$_3$–SiO$_2$ (Part II). Interceram **2**, 140–146.

Langenberg, F.C. and Chipman, J. (1956). Determination of 1600 °C and 1700 °C liquidus lines in CaO–2Al$_2$O$_3$ and Al$_2$O$_3$ stability fields of the system CaO–Al$_2$O$_3$–SiO$_2$. J. Am. Ceram. Soc. **39**, 423–433.

Moyer, J.R. (1995). Phase diagram for mullite–SiF$_4$. J. Am. Ceram. Soc. **78**, 3253–3258.

Muan, A. (1957). Phase equilibria at liquidus temperatures in the system iron oxide-Al$_2$O$_3$–SiO$_2$ in air atmosphere. J. Am. Ceram. Soc. **40**, 121–133.

Muan, A. and Osborn, E.F. (1965). Phase Equilibria among Oxides in Steel Making. Addison-Wesley, Reading.

Opila, E.J., Smialek, J.L, Robinson, R.C., Fox, D.S. and Jacobsen, N.S. (1999). SiC recession caused by SiO$_2$ scale volatility under combustion conditions: II, Thermodynamics and gaseous diffusion model. J. Am. Ceram. Soc. **82**, 1826–1834.

Osborn, E.F. and Muan, A. (1960). Phase equilibrium diagrams in oxide systems. The American Ceramic Society, Columbus.

Pask, J.A. (1990). Critical Review of phase equilibira in the Al$_2$O$_3$–SiO$_2$ system. Ceramic Trans. **6**, 1–13.

Pask, J.A. (1994). Solid state reactions and phase equilibria in the Al$_2$O$_3$–SiO$_2$ system. (unpublished results).

Pask, J.A. (1998). The Al$_2$O$_3$–SiO$_2$ system: Logical analysis of phenomenological experimental data. In: Tomsia, A.P. and Glaeser, A. (eds.) Ceramic microstructure: Control at the atomic level. Plenum Press, New York, pp. 255–262.

Pitak, N.V. and Ansimova, T.A. (1974). Mechanism of destruction of mullite-corundum products in a variable redox medium. Refractories **15**, 38–41.

Prochazka, S. and Klug, F. J. (1983). Infrared transparent mullite ceramic. J. Am. Ceram. Soc. **66**, 874–880.

Provost, G. and Le Doussal, H. (1974a). Untersuchung des Fließverhaltens feuerfester Werkstoffe bei unterschiedlichen Ofenatmosphären (Teil 1). Tonind. Ztg. **98**, 103–109.

Provost, G. and Le Doussal, H. (1974b). Untersuchung des Fließverhaltens feuerfester Werkstoffe bei unterschiedlichen Ofenatmosphären (Teil 2). Tonind. Ztg. **98**, 168–172.

Rankin, G. A. and Merwin, H. E. (1918). The ternary system MgO–Al$_2$O$_3$–SiO$_2$. Amer. J. Sci. (4th series) **45**, 301–325.

Rankin, G. A. and Wright, F. E. (1915). The ternary system CaO–Al$_2$O$_3$–SiO$_2$. Amer. J. Sci. (4th series) **39**, 1–79.

Robinson, R. C. and Smialek, J. L. (1999). SiC recession caused by SiO$_2$ scale volatility under combustion conditions: I, Experimental results and empirical model. J. Am. Ceram. Soc. **82**, 1817–1825.

Rüscher, C. H., Shimada, S. and Schneider, H. (2002). High-temperature hydroxylation of mullite. J. Am. Ceram. Soc. **85**, 1616–1618.

Schairer, J. F. and Bowen, N. L. (1947). Melting relations in the systems Na$_2$O–Al$_2$O$_3$–SiO$_2$ and K$_2$O–Al$_2$O$_3$–SiO$_2$. Amer. J. Sci. **245**, 193–204.

Schairer, J. F. and Bowen, N. L. (1955). The system K$_2$O–Al$_2$O$_3$–SiO$_2$. Amer. J. Sci. **253**, 681–746.

Schairer, J. F. and Bowen, N. L. (1956). The system Na$_2$O–Al$_2$O$_3$–SiO$_2$. Amer. J. Sci. **254**, 129–195.

Schairer, J. F. and Yagi, K. (1952). The system FeO–Al$_2$O$_3$–SiO$_2$. Amer. J. Sci. Bowen Volume, pp. 471–512.

Schneider, H., Okada, K. and Pask, J. A. (1994). Mullite and Mullite Ceramics. Wiley, Chichester, pp. 83–104.

Shears, E. C. and Archibald, W. A. (1954). Aluminosilicate refractories. Iron and Steel **27**, 26–30, 61–65.

Shephard, E. S., Rankin, G. A. and Wright, W. (1909). The binary systems of alumina and silica, lime and magnesia. Amer. J. Sci. **28**, 301.

Shindo, I. (1980). Applications of the floating zone technique in phase equilibria study and in single crystal growth, D. Sc. Thesis, Tohoku Univ., Tokyo.

Shornikov, S. I. (2002). Thermodynamic study of mullite solid solution region in the Al$_2$O$_3$–SiO$_2$ system by mass spectrometric techniques. Geochem. Intern. **40**, 46–60.

Shornikov, S. I. and Archakov, I. Yu (2002). A mass spectrometric determination of the enthalpies and entropies of Al$_2$O$_3$–SiO$_2$ melts. Russ. J. Phys. Chem. **76**, 1054–1060.

Shornikov, S. I., Archakov, I. Yu, Schultz, M. M. and Vorisova, N. O. (2002). Mass spectrometric study of evaporation and the thermodynamic properties of solid phases in the Al$_2$O$_3$–SiO$_2$ system. Doklady Chemistry **383**, 82–85.

Snow, R. B. (1943). Equilibrium relationships on the liquidus surface in part of the MnO–Al$_2$O$_3$–SiO$_2$ system. J. Am. Ceram. Soc. **26**, 11–20.

Takahashi, J., Kawai, Y. and Shimada, S. (2002). Hot corrosion of cordierite/mullite composites by Na-salts. J. Europ. Ceram. Soc. **22**, 1959–1969.

Toropov, N. A. and Galakhov, F. Y. (1951). New data for the system Al$_2$O$_3$–SiO$_2$. Quoted in: Davis, R. F. and Pask, J. A.: Mullite. In: Alper, A. M. (ed.) Refractory materials. High temperature oxides: Part IV, Refractory glasses, glass-ceramics, and ceramics. Academic Press, New York and London, pp. 37–76.

Trömel, S., Obst, K. H., Konopicky, K., Bauer, H. and Patzak, I. (1957). Untersuchungen im System SiO$_2$–Al$_2$O$_3$. Ber. Dtsch. Keram. Ges. **34**, 397–402.

Tso, S. T. and Pask, J. A. (1982). Reaction of silicate glasses and mullite with hydrogen gas. J. Am. Ceram. Soc. **65**, 383–387.

Vishnevskii, I. I. and Tal'yanskaya, N. D. (1979). Creep of a mullite refractory in neutral and reducing media. Refractories **20**, 707–711.

Wannaparhun, S. and Seal, S. (2003). Combined spectroscopic and thermodynamic investigation of Nextel-720 fiber/alumina ceramic-matrix composite in air and water vapor at 1100 °C. J. Am. Ceram. Soc. **86**, 1628–1630.

Welch, J. H. (1960). New interpretation of the mullite problem. Nature **186**, 545–546.

Xiao, Z. and Mitchell, B. S. (2000). Mullite decomposition kinetics and melt stabilization in the temperature range 1900–2000 °C. J. Am. Ceram. Soc. **83**, 761–767.

Zaykoski, J., Talmy, I., Norr, M. Wuttig, M. (1991). Disiliconisation of mullite felt. J. Am. Ceram. Soc. **74**, 2419–2427.

4
Mullite Synthesis and Processing

4.1
Mullite Synthesis
S. Komarneni, H. Schneider and K. Okada

The synthesis of mullite can be classified into solid-state processes, including the conventional and solution-sol-gel (SSG) processes, the liquid-state or hydrothermal process and the vapor-state process. Mullites synthesized by solid- or liquid-state processes have been designated as sinter-mullite and fused-mullite, respectively, depending upon the heat-treatment temperature of silica (SiO_2) and alumina (Al_2O_3) compounds: The term sinter-mullite refers to mullite synthesized by heating to a temperature below the melting point to crystallize and densify mullite, while fused-mullite is prepared by heating alumina and silica mixtures to a temperature above the melting point followed by cooling to crystallize mullite. Solution-sol-gel-derived mullites have been designated as chemical-mullites. They have been synthesized by chemical reaction, pyrolysis and mullitization. The purity, homogeneity, crystallization temperature, densification, and properties of mullite are highly dependent on the synthesis method. This section gives a summary of the characteristics of the various preparation methods and also their mullitization routes.

4.1.1
Solid-state-derived Mullite

Mullite prepared by the solid-state process involves natural aluminosilicate minerals such as kaolinite, the sillimanite group (sillimanite, andalusite, kyamite) and many types of oxides, oxyhydroxides, hydroxides, inorganic salts and metal organics as alumina and silica precursors.

4.1.1.1 Formation from Kaolinite and Related Phases

Perhaps the earliest synthesis of mullite was done unwittingly by firing kaolinite clay to make bricks, refractories, whitewares and porcelains. Other natural aluminosilicate sheet silicates such as pyrophyllite ($Si_4Al_2O_{10}(OH)_2$) and muscovite ($KAl_3Si_3O_{10}(OH)_2$) can also be used for the synthesis of mullite. However, the use of these naturally occurring aluminosilicate minerals leads to mullite and to silica or silicate phases, which are intimately mixed in this conventional process. When kaolinite is heated to temperatures above 1000 °C it decomposes according to the following reaction:

$$2\ Al_2Si_2O_5(OH)_4 \rightarrow 0.66\ (3\ Al_2O_3 \cdot 2\ SiO_2) + 2.68\ SiO_2 \qquad (1)$$

    kaolinite   3/2-mullite   amorphous silica

To avoid the formation of silica along with mullite, some alumina must be admixed with the starting material. Kaolinite clay has been mixed with refractory-grade bauxite (Hawkes, 1962, Chatterjee and Panti, 1965), aluminum fluoride (AlF_3, Locsei 1968), aluminum hydroxide ($Al(OH)_3$, Rossini et al. 1970) or α-alumina (Al_2O_3, Hawkes, 1962, Moya et al. 1982, Sacks, 1977). A more complicated approach to obtaining single-phase mullite from kaolinite-type clays is by removing excess silica through sodium hydroxide (NaOH) leaching. Moya et al. (1982) first heated halloysite (a hydrated form of kaolinite) to 1000 °C and leached out the excess silica with sodium hydroxide to prepare more or less stoichiometric mullite precursors. A detailed description of the structural mechanisms of the kaolinite to mullite transformation is given in Section 2.4.1.1.

4.1.1.2 Formation from Kyanite, Andalusite and Sillimanite

The polymorphic aluminosilicate minerals kyanite, andalusite and sillimanite ($Al_2O_3 \cdot SiO_2 = Al_2SiO_5$, in the refractory literature sometimes designated as the sillimanite group), transform to 3/2-mullite plus silica (SiO_2) on heating at high temperature under oxidizing conditions:

$$3\ Al_2SiO_5 \quad \rightarrow \quad 3\ Al_2O_3 \cdot 2SiO2 \quad + \quad SiO_2 \qquad (2)$$

    kyanite, andalusite,   3/2-mullite   amorphous silica,
    sillimanite   cristobalite

The transformation temperature and the width of the transformation interval increase in the sequence sillimanite, andalusite and kyanite. The structural mechanisms and the kinetics of the thermal decomposition of sillimanite and andalusite to mullite plus silica are discussed in detail in Section 2.4.1.2.

4.1.1.3 Formation from Staurolite and Topaz

Staurolite, with an idealized stoichiometry of $Fe_4^{2+}Al_{18}Si_8O_{46}(OH)_2$ (Ganguly, 1972) can be imagined to be built of layers parallel to (010) composed alternatively of kyanite (Al_2SiO_5) and of $AlOOH \cdot 2FeO$. On dry heating staurolite decomposes to mullite plus silica and iron oxide:

$$Fe_4^{2+}Al_{18}Si_8O_{46}(OH)_2(s) \rightarrow 3(3\ Al_2O_3 \cdot 2\ SiO_2)(s) + \text{``4 FeO''}^{1)}(s) + 2\ SiO_2(s) + H_2O(g)$$

| staurolite | 3/2-mullite | iron oxide | amorphous silica | (3) |

Heat treatment above 1400 °C makes topaz ($Al_2(F, OH)_2[SiO_4]$) transform to mullite plus silica:

$$3(Al_2(F,OH)_2[SiO_4])(s) \rightarrow 3\ Al_2O_3 \cdot 2\ SiO_2(s) + SiO_2(s) + 3\ H_2O/3\ F_2(g) \qquad (4)$$

| topaz | 3/2-mullite | amorphous silica |

s = solid, g = gaseous

The topaz-derived mullites are whisker shaped and are suggested to grow via a gas-phase transport reaction (see Section 6.1.2). X-ray single crystal studies on heat-treated topaz, partially transformed to mullite, showed no coherence between the two phases.

4.1.1.4 Reaction Sintering of Alumina and Silica

The synthesis of mullite from the basic compounds alumina (Al_2O_3) and silica (SiO_2) started with the work of Bowen and Greig (1924), whose reactant charges were made up from specially purified precipitated α-alumina and quartz (i.e. silica). The raw materials were homogenized together in the desired proportion and heated to a temperature close to the melting point of platinum. The charge was then ground and reheated in the furnace. This process was carried out five times before controlled runs were performed to investigate the phase relationships (see also Section 3.1). Since then numerous studies have been carried out with many combinations of silica sources such as quartz, cristobalite, silicic acid, fumed silica and fused silica with alumina sources such as α-alumina, γ-alumina, diaspore (AlOOH), gibbsite (Al(OH)$_3$) and boehmite (AlOOH). In their investigation of the system Al_2O_3–SiO_2–H_2O, Roy and Osborn (1954) started with mechanical mixtures of γ-alumina and silica gel but abandoned these in favor of reactive alumina-silica gels because of slow reactivity in the former. Aluminum hydroxide and silicic acid were used in the synthesis of mullite and these precursors led to complete mullitization at 1700 °C after 8 h of heat treatment (Murthy and Hummel, 1960). Aluminum oxide and potters' flint were reacted between 1650 and 1710 °C to synthesize mullite after ball milling this mixture for 48 h (Fenstermacher and Hummel, 1961).

Wahl et al. (1961) used an extensive range of alumina and silica precursors in the synthesis of mullite. They examined the phase transformations including the formation of mullite by continuous X-ray diffractometry using mixtures of quartz, cristobalite or silicic acid with α-alumina, gibbsite or diaspore. Wahl and coworkers

---

1) Under oxidizing transformation conditions iron may occur as Fe_2O_3.

heated these mixtures to 1450 °C to follow their reactivity in the formation of mullite. Their results show that, in the formation of mullite, diaspore is the most reactive among the alumina sources mentioned while cristobalite is the most reactive among the silica sources. The influence of particle size on the reactivity was, however, not determined by these authors. Dry mixtures of α-alumina and powdered silica glass were used for the synthesis of mullite by Aramaki and Roy (1962) in their investigation to revise the Al_2O_3–SiO_2 phase diagram (see also Section 3.1). Pankratz et al. (1963) showed that unreacted aluminum oxide remained after repeated grinding and heat treatments between 1500 and 1540 °C using a mixture of high-purity aluminum oxide and quartz. Rana et al. (1982) and Nurishi and Pask (1982) used α-alumina and cristobalite, quartz or fused silica and showed that mullitization started at 1415 °C after 24 h with quartz and at 1480 °C with fused silica. These results confirm the higher reactivity of cristobalite, which was found by Wahl et al. (1961) and which has been attributed to the formation of a metastable liquid phase (Rana et al. 1982). Quartz and α-alumina mixtures were also used by Sacks and Pask (1978) who showed that mullitization was complete only after 8 h of treatment at 1700 °C. Temuujin et al. (1998) prepared aluminosilicate precursors by mechano-chemical treatment of gibbsite-silica-gel mixtures and characterized them by ^{29}Si and ^{27}Al magic angle spinning nuclear magnetic resonance (MAS NMR) spectroscopy to follow the molecular level of mixing. Mullite crystallized at 1200 °C from samples ball-milled for 8 to 20 h, which showed intimate mixing. Kong et al. (2003) reported mullite phase formation and reaction sequences with the presence of vanadium oxide (V_2O_5), niobium oxide (Nb_2O_5) and tantalum oxide (Ta_2O_5). They found that vanadium oxide accelerated the mullite phase formation while niobium and tantalum oxide inhibited the mullitization with precipitated silica and alumina as mullite precursors. The formation temperature of mullite is thus dependent on the stability, crystallinity, particle size and impurities of the starting alumina and silica precursors. The readers are referred to two excellent reviews on the synthesis of mullite, one by Rodrigo and Boch (1985) and another by Sacks et al. (1990).

4.1.1.5 **Effects of Mineralizers, Reaction Atmosphere and Structural Defects**

Effect of mineralizers   It has long been known that the addition of even small amounts of impurities can strongly influence the formation of mullite, though the experimental results are contradictory, even to the point of diametrically opposed effects on the same material with the same mineralizer.

Chaudhuri (1969) found that sodium oxide (sodia, Na_2O), potassium oxide (potassia, K_2O), and calcium oxide (calcia, CaO) in concentrations less than 1 wt. %, and the addition of titania (TiO_2) and iron oxide (Fe_2O_3), enhanced the transformation of kaolinite to mullite. Bulens and Delmon (1977), and Bulens et al. (1978) found that pure and magnesia (MgO)-doped kaolinites which are heat-treated at 900 °C produce γ-alumina and silica, whereas calcia-mineralized samples seem to decompose directly to mullite (Fig. 4.1.1). Johnson and Pask (1982) published a

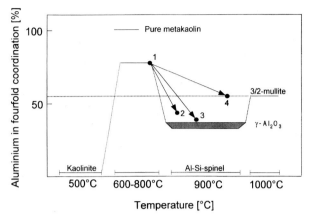

Fig. 4.1.1 Reaction processes and variation of the coordination number of aluminum in the kaolinite to mullite reaction sequence depending on the occurrence of impurity oxides. Experimental procedures: **1** Pure kaolinite heated at 800 °C (24 h); **2** pure kaolinite heated at 900 °C (124 h); **3** kaolinite + MgO heated at 900 °C (124 h); **4** kaolinite + CaO heated at 900 °C (124 h), (after Bulens et al. 1978, from Schneider et al. 1994).

paper on the role of impurities in formation of mullite from kaolinite and mixtures of alumina and silica. They stated that addition of titania, iron oxide and calcia to alumina and silica mixtures enhanced mullite crystal growth, while the addition of alkalis astonishingly does not. The former was explained by a decrease of the viscosity of the coexisting liquid and enhanced migration of diffusing species by the impurities. The effect was less significant for kaolinite than for the oxide mixtures. The observation that (larger amounts of) alkalis do not have a stimulating influence on the growth of mullite, in spite of their strong fluxing effect, may be caused by the fact that mullite becomes unstable at the expense of α-alumina plus silica (see also Section 3.3.1).

Data concerning the influence of iron oxide and titania addition on the transformation of kyanite to 3/2-mullite plus silica were published by Schneider and Majdic (1980). They mentioned that addition of the oxides causes a decrease in the reaction temperatures. The addition of other metal oxides may probably be similarly effective: Wilson (1969) stated that mullite formation increases about 20% if 0.5 wt. % magnesia (MgO) is added to kyanite at 1250 °C.

Chaudhuri (1982) investigated the influence of mineralizers on the crystallization of mullite from alkali oxide–Al_2O_3–SiO_2 glasses in whitewares. The formation of mullite was reported to depend on such factors as the radius, charge, field strength and bond strength of the cations of the mineralizers. The reaction leading to mullite was believed to take place in separate steps: Replacement of aluminum ions in the glass by the mineralizer cations, diffusion of aluminum to the SiO_2-rich glass, and mullite nucleation and growth.

Yamuna et al. (2002) reported that kaolinite in the presence of potassium carbonate (K_2CO_3) as a mineralizer acted as the basic precursor material for the production of mullite. Addition of 1 to 10 wt. % of potassium carbonate conclusively proved this result. The presence of other carbonate mineralizers such as sodium (Na_2CO_3) and calcium ($CaCO_3$) gave rise to cristobalite formation. Soro et al. (2003) investigated the role of iron in kaolinites and found that poorly crystallized kaolinite favors rapid mullite nucleation and a simultaneous incorporation of iron in the mullite structure, whereas a highly crystallized kaolinite with lower kinetics of nucleation and mullite growth favors larger quantities of mullite structural iron.

Effect of reaction atmosphere   A reducing atmosphere, especially carbon monoxide, has a stimulating influence on the high temperature transformation of kyanite and sillimanite and possibly of andalusite. Lysak and Drizheruk (1977) explained this phenomenon in terms of a reaction of the carbon monoxide with the surface of the aluminosilicate. Structural defects so produced reduce the activation energy of the phase transition. A reducing atmosphere also favors oxygen diffusion and thus increases the transformation velocity. The influence of the reaction atmosphere on the sillimanite transformation is stronger than in the case of kyanite, because the higher reaction temperatures of sillimanite decomposition cause forced oxygen diffusion. An increased possibility of silica reduction liberating $SiO(g) + O_2$ (g = gaseous) in spite of the formation of vitreous SiO_2, may also accelerate the reaction process (see also Section 3.2.1).

Effect of structural defects   Heat treatment of experimentally shock-loaded kyanite, andalusite and sillimanite powders shift the transformation curves towards lower temperatures (Schneider et al. 1980, Schneider and Hornemann, 1981). This has been explained by the production of structural defects in the shock-loaded minerals, which in turn decrease the activation energies of the transformation process to mullite plus silica.

4.1.1.6   Commercial Production (Sinter-mullite)

Commercial sinter-mullite is usually synthesized from mixtures of clay minerals and bauxite, aluminum hydroxide or alumina. Kaolinite is the clay mineral most commonly used. Mullite synthesized by this process is mainly used for refractory purposes and furnace materials such as crucibles and tubes. The term "old mullite" has sometimes been taken for this type of mullite in comparison with "new mullite" which represents high-purity mullite (chemical-mullite). Table 4.1.1 shows the chemical compositions and some characteristics of various mullite raw materials commercially available in Japan (Okada, 1991).

Sinter-mullite normally contains considerable amount of impurities. These impurities and the relatively low reactivity of the starting materials limit the quality of sinter-mullite ceramics. Fig. 4.1.2 shows a typical microstructure of sinter-mullite ceramics with the occurrence of prismatic euhedral mullite. A liquid phase-as-

Table 4.1.1 Composition and properties of commercial mullite powders and production in Japan in 1990.

	Solid mix	Electro-fused	Sol-mix	Sol-mix	Sol-mix	Alkoxide
Al_2O_3(%)	71.37	77.0	71.8	71.8	71.41	71.8
SiO_2(%)	26.39	22.5	28.2	28.2	27.90	28.2
Fe_2O_3(%)	0.72	0.06	<0.01	0.01	0.03	–
TiO_2(%)	0.07	–	–	0.01	–	–
CaO(%)	0.25	–	–	<0.01	0.01	–
MgO(%)	0.02	–	–	<0.01	0.01	–
K_2O(%)	0.18	–	–	<0.01	0.01	0.02
Na_2O(%)	0.36	0.20	<0.01	<0.01	0.03	0.02
Particle size (μm)	–	–	1.5	1.2	–	–
Surface area ($m^2\,g^{-1}$)	–	–	10	12	–	28
Production (kg per month)	–	–	5000	300	500	500
Company[a]	NT	SD	CC	SD	KC	HC

a) NT = Naigai; SD = Shown Denko; CC = Chichibu Cement; KC = Kyoritsu Ceramic Materials; HC = Hokko Chemicals. From Schneider et al. (1994).

sisted crystal growth occurred apparently by the presence of impurities. Hamano et al. (1988) tried to overcome these issues by using high purity kaolinite, in which the impurity level is several thousands ppm only, and also by a prolonged milling of the starting materials. Mullite ceramics produced in that way can have bending strengths of about 300 to 400 MPa due to optimized processing. These values are comparable to those of high-purity mullites, and much higher than those reported for conventional sinter-mullites, which display strengths of 100 MPa or less. Low impurity contents of the starting materials correspond to little glass contents at the grain boundaries of mullite ceramics, whereas prolonged grinding develops uniform and fine grained microstructures. These two points are considered to be the reasons of the enhancement of bending strength. Mechano-chemical effects caused by prolonged grinding promote mullitization (Kawai et al. 1990), and therefore may also contribute to the enhancement of strength.

50 nm

Fig. 4.1.2 Thin section photograph of commercial 3/2-mullite ($3Al_2O_3 \cdot 2SiO_2$) showing the typical fine-grained microstructure of these materials. The sinter-mullite was produced by calcination at 1700 °C in a rotary kiln (from Schneider et al. 1994).

Table 4.1.2 Firing procedure for sinter-mullite raw materials.

	Mixes of Bayer alumina with clay			Bauxite
	1	2	3	
Setting	Pellets	Dobies	Dobies	Dobies
Furnace type	Rotary kiln	Recuperative kiln	Tunnel kiln	Tunnel kiln
Firing temperature (°C)	1760	1790	1720–1730	1720–1730

Data are from Hawkes (1962).
From Schneider et al. (1994).

Details of the processing of commercial sinter-mullite have been given by Hawkes (1962): Mixtures of Bayer alumina and clay, and refractory bauxite in powdered and lump form, are used as starting materials. The raw material is crushed, if necessary, passed through a rotary drier and subsequently milled. The blended products are fed into a tube mill with steel balls of different diameter and are ground to the required grain size. The milled material is pressed into briquettes having a moisture content of 20 to 22 wt. %, and passed through a tunnel dryer operating with the waste heat from the calcination kiln's cooling zone. As they leave the dryer, the briquettes' moisture content is about 2%. Tunnel or rotary kilns are used for the main calcination process. Calcination temperatures are usually above 1750 °C (Table 4.1.2). The kilns are usually short, since it is their function to raise the material as quickly as possible to the sintering temperature and then to stabilize it. Any cracking of the material taking place during the firing process is positive, since it aids subsequent crushing. After leaving the tunnel kiln at about 200 °C and further cooling, the very hard briquettes are crushed in a jaw crusher. The crushed material then passes to a perforated bottom pan mill, producing rounded particles and countering the tendency of the material to splinter: rounded particles give refractory bricks with better strength than splintered grains. After pan milling, the material passes over rotary magnets and is screened, thus removing about 95% of the free iron that was taken up during processing. Before being weighed and bagged, the grog passes through a magnetic filter to remove the remaining free iron.

Some mullite qualities are produced in sinter ladles by alternate rapid heating and cooling. These mullites are characterized by a high porosity. Owing to their high Al_2O_3-rich glass content, these mullites are more slag and vapor resistant than other synthetic mullites. During use of the bricks, secondary mullitization does occur, and the slag and vapor resistance of the material decreases with the duration of use in the high temperature aggregate. Characteristic properties of commercial synthetic mullites are compiled in Table 4.1.1.

4.1.2
Liquid-state-derived Mullite

4.1.2.1 Crystal Growth Techniques

Guse and Mateika (1974) and Guse and Saalfeld (1990) were successful in growing mullite single crystals of 2/1-composition ($2Al_2O_3 \cdot SiO_2$) up to about 20 to 50 mm in size with the Czochralski method. A schematic drawing of the pulling chamber is given in Fig. 4.1.3. An iridium tube containing the seed crystal is mounted on the pulling rod. Growth experiments are carried out with mixtures of alumina and silica powders. Covered iridium crucibles are taken in order to prevent heat radiation. The crystal growth crucible is slowly heated to 1850 °C in nitrogen atmosphere. Growth directions were [100] and [001], respectively. Inclusion-free single crystals up to 50×10 mm have been grown with the Czochralski method (Fig. 4.1.4).

The slow cooling float zone (SCFZ) method with a quartz halogen lamp furnace was used by Shindo (1990) for the production of mullite single crystals (up to 5 to 70 mm in size) from alumina and silica powders (Fig. 4.1.5). The SCFZ method requires that the melt is completely homogeneous and that solid and liquid are in

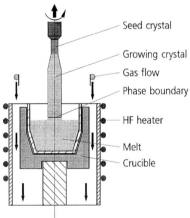

— Seed crystal

— Growing crystal

— Gas flow

— Phase boundary

— HF heater

— Melt

— Crucible

Fig. 4.1.3 Schematic drawing of the pulling chamber for mullite single crystal growth using the Czochralski technique (from Schneider et al. 1994).

Fig. 4.1.4 Inclusion-free mullite single crystals grown by the Czochralski process. The crystals were synthesized by F. Wallrafen (University of Bonn, Germany) following the route described by Guse and Mateika (1974).

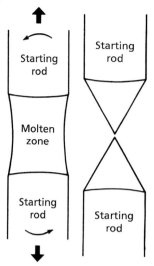

Fig. 4.1.5 Schematic drawing of the floating melt technique used for single crystal growth of mullite. The situation before the cooling in a slow cooling float zone experiment (SCFZ) is shown (left), and the resulting specimen after cooling (right). The solidified specimen (right) shows a zonal structure (after Shindo, 1990, from Schneider et al. 1994).

equilibrium at their interface during crystallization. The former is achieved by continuous counter-rotation of the upper and lower rods, while the latter is obtained by controlling the rate of advance of the solidification front. Good quality single crystals can also be grown by the traveling solvent float zone (TSFZ) technique, which is very similar to the SCFZ procedure but with the melt zone being gradually moved along the rod (see Shindo, 1990). More information on mullite single-crystal growth is provided in Section 6.3 dealing with single crystal mullite fibers.

4.1.2.2 **Commercial Production (Fused-mullite)**

Commercial fused-mullite is produced by melting the starting materials in an electric arc furnace above about 2000 °C. The obtained aluminum silicate melt is then cast into ingot molds and cooled to room temperature (Fig. 4.1.6). Raw materials for fused-mullite ceramics and refractories are Bayer alumina, quartz sand, rock crystals and fused silica. The impurity level of the raw materials is relatively low as shown in Table 4.1.2. For fused-mullite of lower quality, bauxite, or mixtures of α-alumina and kaolinite, have also been used.

The chemical composition of fused-mullite depends largely on the crystallization temperature and the cooling speed and only to a minor degree on the initial composition of the starting materials. This can be understood from the metastable phase diagram of silica and mullite reported by Aksay and Pask (1975). Mullites with chemical composition up to around 83 wt. % (73 mol%) Al_2O_3 have been reported (Bauer et al. 1950, Bauer and Gordon, 1951, Kriven and Pask, 1983), which is much richer in Al_2O_3 than sinter-mullite (about 78 wt. % or 60 mol%) obtained by solid-state reaction. Commercial fused-mullites usually have Al_2O_3 compositions in the range of 2/1-mullite (about 66 mol%, Table 4.1.1). They frequently contain low amounts of glass phase owing to incomplete crystallization.

Fig. 4.1.6 Casting of mullite melt in an electrical arc furnace for commercial production of fused-mullite. (Courtesy Hüls Comp. Troisdorf, from Schneider et al. 1994).

While the crystal size of sinter-mullite is essentially controlled by the synthesis process (firing temperature, duration of firing, microstructure and composition of the raw material) that of fused-mullite is a function of the cooling conditions of the melt. This implies that the microstructural features of fused-mullites can vary widely (Fig. 4.1.7), whereas sinter-mullites usually exhibit similar microstructures. The cooling rate of the melt may also influence the α-alumina/mullite ratio of the product. The cooling of the melt for technical products should be controlled in a way that coarse interlocking mullite crystals develop and a minimum of glass phase occurs. Common phase compositions of commercial fused-mullites are >80 wt. % mullite, <10 wt. % α-alumina and <10 wt. % glass phase. Good bulk density values lie near 3.3 g cm^{-3} and the apparent porosity is near 1% (Grofczik and Tamas, 1961). The low porosity and the large crystal size of fused-mullites explain their high slag resistance.

Raw materials with relatively high alkali or alkaline earth oxide contents cannot be tolerated for the production of commercial mullite since the stability fields of α-alumina and vitreous silica are extended at the expense of mullite (see Section 3.3.1). The phase composition of commercial fused-mullite refractory products depend on the Al_2O_3/SiO_2-ratios and on the addition of foreign cation oxides:

- Mullite specimens with Al_2O_3/SiO_2-ratios between 2.2 and 3.2 consist mainly of mullite and glass phase. Initially, mullite is acicular in morphology, but becomes more equiaxed as the Al_2O_3/SiO_2-ratio exceeds 2.7. α-alumina crystals begin to crystallize at an Al_2O_3/SiO_2-ratio above 3.3.
- Addition of calcium oxide, sodium oxide, magnesium oxide, manganese oxide

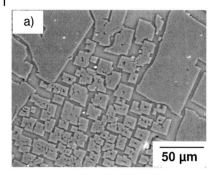 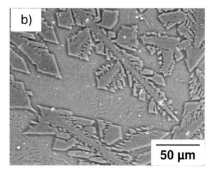

Fig. 4.1.7 Electron microprobe photographs of commercially produced fused-mullite. The material consists of mullite plus coexisting glass phase. (a) Mullite with pseudo-orthorhombic contours in a view down the crystallographic **c** axis from the inner relatively slowly cooled parts of the brick; (b) Mullite with dentritic crystal growth from the outer rapidly cooled parts of the mullite brick.

(MnO), or titanium oxide (titania = TiO_2) produces larger amounts of uniformly distributed glass phase. Mullite refractories prepared with addition of zirconia (ZrO_2) to the melt were successfully used in furnaces operating under vacuum conditions at high temperatures. The most advantageous composition range of such refractories was found to lie between 50 and 55 wt. % Al_2O_3, 21 to 36 wt. % ZrO_2, and 14 to 18 wt. % SiO_2 (see Fig. 4.6.2b below and Section 1.3.2, Fig. 1.3.12).

4.1.3
Solution-sol-gel-derived Mullite

The solution-sol-gel (SSG) process utilizes atomic, molecular or nanoscale mixing of components to prepare mullite powder at low temperatures and/or in short duration. Mixing of components on these scales increases the reaction rate by shortening the diffusion distances, unlike the processes described above where natural minerals or micrometer-sized silica and alumina particles are used. Mixing of components in solution can be accomplished by using metal organics, metal salts, oxides, oxyhydroxides or hydroxide sols or any combinations thereof. The solution-sol-gel process can be subdivided into three types: (a) Solution plus solution, (b) Solution plus sol, and (c) Sol plus sol. The solution-sol-gel process of mullite preparation comes under the category of advanced processing. The importance of the solution-sol-gel methods has increased since the 1980s, and many preparation methods have been developed although industrial fabrication has been restricted to a small number of methods.

4.1.3.1 Solution-plus-solution Process

The earliest use of the solution-sol-gel process for mullite synthesis was perhaps the one of Ewell and Insley (1935) who used gels prepared from two inorganic salts to follow the high temperature reactions of clay compositions. Sodium silicate solution was added dropwise to aluminum sulfate or chloride solution under vigorous stirring, followed by the addition of sodium hydroxide (NaOH) to precipitate the two components. The gelatinous precipitate was then filtered, washed and dried, then crushed and re-washed with hot water. The main problems with this technique were sodium impurity and perhaps inhomogeneity and non-stoichiometry, the latter due to washing. However, they detected mullite formation through the 980 °C exotherm in the differential thermal analysis (DTA) and by X-ray diffraction (XRD) analysis of samples heated near this temperature.

A slight variation of this process was used by Ossaka (1961), who used sodium silicate and potassium aluminum sulfate as the precursors. These two precursors were dissolved in dilute sulfuric acid solution, coprecipitated with hexamethylenetetramine, filtered and washed to remove sodium, potassium and the amine. Powder X-ray diffraction revealed a pseudo-tetragonal mullite (see Sections 1.1.3.7 and 2.5.1 and Schneider and Rymon-Lipinski, 1988) after heat treatment of these coprecipitated gels in a temperature range of 910 to 1100 °C and well crystallized mullite was observed at 1250 °C after heat treatment for 5 h. This procedure has the same drawbacks as the procedure used by Ewell and Insley (1935), and the extent of segregation of alumina and silica components is difficult to judge. Judging from the mullitization routes, silica and alumina components can be considered to be in a molecularly mixed state. However, this preparation method is not without problems, since the yield of the precipitates is very small and the chemical composition of the precipitates may deviate considerably from the starting compositions because the solubility of the aluminum ion can be ignored under the precipitation conditions but not that of the silicon ion. Another point is that a considerable amount of alkali components may remain in the precipitates in spite of careful washing.

Preparation using chlorides for both components has been examined by Horte and Wiegmann (1956) who used ammoniated water to precipitate the components in order to avoid alkali ions in the system. Mullitization was reported in a temperature range of 1000 to 1200 °C. Powders showed γ-alumina and cristobalite after calcination at 800 °C and 1200 °C, respectively. The above results suggest that there is segregation of components during mixing and/or calcination. The use of silicon chloride is unsuitable because it is unstable and it is very difficult to control the stoichiometric chemical composition of precipitates. McGee and Wirkus (1972) also used chlorides for both the components, and these were initially dissolved in absolute methanol followed by precipitation with ammonium hydroxide. They reported powders are amorphous up to a temperature of 1050 °C and mullite crystallization occurs at 1100 °C.

Jaymes and Douy (1995) synthesized mullite powders or gels from an aqueous precursor solution by hydrolyzing tetraethyloxysilane (TEOS) and aluminum nitrate in different ways. When this precursor solution is sprayed into ammonia isopropanol solution, a monophasic precursor is obtained whereas when it is sprayed into an aqueous ammonia solution, a diphasic precipitate forms. When urea is used for *in-situ* hydrolysis, monophasic gels are obtained using the same precursor solution.

Preparation of precipitates from a clear solution using a precipitant was the common theme in all of the above syntheses. In general, there are two kinds of precipitation methods, i.e. coprecipitation and homogeneous precipitation. The former involves the addition of a precipitant (usually ammonium hydroxide or alkali hydroxide) to a solution to form precipitates. The latter works by first dissolving a precipitant such as urea or hexamethylenetetramine in the solution and then, by a change in the pH of the solution, decomposing the precipitant to form precipitates.

The preparation method using a combination of silicon alkoxide and aluminum alkoxide or another salt has been examined by many workers. The starting components of this technique are dissolved in ethanol or propanol and precipitates are obtained by addition of ammonium hydroxide or ammoniated water. Mazdiyasni and Brown (1972) gave a detailed synthesis of aluminum tris-isopropoxide ($Al(OC_3H_7)_3$) and silicon tetrakis-isopropoxide ($Si(OC_3H_7)_4$) which were then mixed and refluxed in excess isopropyl alcohol to get a thorough mixing of the two components. The hydroxyl aluminosilicate was prepared by slowly adding the mixed alkoxide solution to ammoniated distilled water. Thus this method also involves precipitation. This precursor yields mullite at 1185 to 1200 °C and the mullite content increases with temperature in the range of 1200 to 1700 °C. West and Gray (1958) made a series of alumina and silica mixtures from very pure metal organic compounds. These mixtures, containing less than five parts per million (ppm) of residual metallic impurities and less than ten ppm of inorganic acid anions, were vacuum dehydrated at 450 °C before crystallization studies. The authors, however, did not indicate the type of metal organics used or the type of procedures that were followed in making the silica alumina mixtures. Although trace amounts of mullite were detected in some mixtures at about 1080 °C, almost complete mullitization occurred only at 1500 °C and above.

Yamada and Kimura (1962) used tetraethyloxysilane (TEOS) and aluminum triethoxide ($Al(OEt)_3$; Et = C_2H_5) as starting materials, and dissolved them in an ethanol-water solution with urea. Precipitates are obtained by changing the pH of the solution from around 5 to 8 by decomposing the urea by heating at 70 °C. A small amount of mullitization was observed at above 1000 °C along with the formation of spinel phase (this is transitional alumina with varying amounts of silicon, designated in the following as γ-alumina), and extensive mullitization occurred above 1200 °C. Hirata et al. (1985) mixed TEOS and aluminum isopropoxide ($Al(O^iPr)_3$) solutions vigorously and added ammoniated water, adjusted to a pH ranging between about 7 to 12, drop by drop. The solution was refluxed and centrifuged to separate the precipitates. Mullitization of these powders was observed by firing at 1200 °C for 2 h, and extensive mullitization was detected above 1300 °C. Similar results were also reported by Hamano et al. (1986), who performed syntheses by using tetramethyloxysilane (TMOS) and aluminum isopropoxide (Al(O^{i}Pr)$_3$). Hamano et al. (1986) prepared two sets of gels, one with ammonium hydroxide precipitant and another set without any precipitant but hydrolyzed with an isopropanol/water solution. In the latter gels, mullite is detected at about 970 °C indicating more intimate mixing while the former gels yielded aluminum silicon spinel (γ-phase) at this temperature. These gels crystallize to mullite above 1250 °C because of less intimate mixing.

Mitachi et al. (1990) reported the preparation of aluminum silicate powders by hydrolysis of mixed benzene solutions of aluminum isopropoxide and tetramethyloxysilane (TMOS) with water (A-series) and aqueous ammonia (B-series). The starting amorphous powder of the A-series transforms directly to crystalline mullite above 980 °C (corresponding to type I mullite precursors), while the B-series powder converts to mullite above 1250 °C as a result of segregation of components in the latter series of gels (corresponding to type III mullite precursors, see Sections 1.4.1 and 1.4.3). Imose et al. (1998) used hydrous aluminum chloride, TEOS and hydrazine to prepare mullite precursor at pH 9. This precursor heated below 1200 °C yielded very fine needle-like particles whereas the mullite powder synthesized at 1300 °C showed prismatic morphology. Thim et al. (2001) synthesized mullite precursors using aqueous solutions of silicic acid and aluminum nitrate in urea and found that this precursor crystallized to pseudo-tetragonal mullite at about 1050 °C (see Sections 1.1.3.7 and 2.5.1).

Okada and Otsuka (1986) dissolved TEOS and aluminum nitrate in ethanol, the same precursors as used previously for hydrolysis by Roy and associates without any precipitants (Roy and Osborn, 1952, 1954, Roy and Roy, 1955, Aramaki and Roy, 1962, 1963, Hoffman et al. 1984) but precipitates were obtained in this study by adding ammonium hydroxide while vigorously stirring the solution. Mullitization occurred extensively by firing at 1150 °C for 24 h and γ-phase formation was observed before mullitization at 980 °C. Similar mullitization reactions have commonly been reported for these preparation methods. Since the pH of the solution was changed from acidic to neutral by the addition of ammonium hydroxide, the precipitation was considered to be initiated by the formation of nuclei of the aluminum component with the accompanying silicon component to form a co-precipitate. The reason for this is that the solubility of aluminum ion drops quickly between the different pH conditions while that of silicon ion does not (Loughnan, 1969). The particles obtained are very small but some inhomogeneity of the chemical composition exists because of the above mentioned co-precipitation mechanism. The inner part of the particles is considered to be of Al$_2$O$_3$-rich composition whereas the surface parts are silica-rich. Schematic microstructural models for these particles are shown in Fig. 4.1.8.

Roy and associates (Roy and Osborn, 1952, 1954, Roy and Roy, 1955, Aramaki and Roy, 1962, 1963) popularized the use of gels for mullite crystallization in the liquid and solid states. Perhaps the above reports were the first to use a silicon metal organic and a soluble aluminum salt to prepare mullite gels that did not involve alkalis or precipitants. Most of their gels were prepared from aluminum nitrate as the alumina source and TEOS or ammonium-stabilized silica sol as the silica source. Pre-standardized TEOS is dissolved in absolute ethanol, and to it is

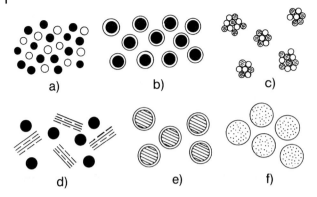

Fig. 4.1.8 Microstructural schematics of mullite starting materials prepared by: (a) Sol mixing; (b) Composite particles; (c) Hydrolysis of alkoxides; (d) Conventional mixing of clay minerals and alumina; (e) Coprecipitation; (f) Spray pyrolysis (from Schneider et al. 1994).

added the alcoholic (or aqueous) aluminum nitrate solution. This solution is mixed, water is added and the mixture is heated at 60 °C to accelerate the hydrolysis, condensation and polymerization of the two components to form a gel (Roy, 1956). Preparation using a combination of silicon alkoxides and aluminum salts as described above, and admixtures of aluminum and silicon alkoxides have been examined by many investigators. Hydrolysis occurs by addition of water, and acid or base is sometimes added as catalyst for hydrolysis of the alkoxides.

Hydrolytic and polymeric reactions of silicon alkoxide are influenced by many factors, as shown by Brinker and Scherer (1990). Very important factors among them are considered to be the solution pH and the amount of water. There are many publications (e.g. Iler, 1979, Keefer, 1984) that clarify the relations between the reaction and the hydrolysis conditions. The hydrolytic and polymerization reactions can be represented schematically by the following equations:

$$Si(OR)_4 + H_2O \rightarrow Si(OR)_3OH + ROH \tag{5}$$

$$Si(OR)_3OH + Si(OR)_3OH \rightarrow (OR)_3Si-O-Si(OR)_3 \rightarrow ... \rightarrow [-Si-O-Si-]_n \tag{6}$$

$$Si(OR)_4 + 4 H_2O \rightarrow Si(OH)_4 + 4 ROH \tag{7}$$

where R represents the alkyl groups (see also Section 6.2.1).

If the solution is prepared under basic conditions, a nucleophilic reaction of the silicon atom with the OH hydroxyl radical is expected. $Si(OR)_3OH$ rather than $Si(OR)_4$ is suceptible to be attacked and promotes the reaction. Therefore the reaction leading to $[-Si-O-Si-]_n$ predominates over that leading to $Si(OH)_4$ and polymerization to form a three-dimensional framework structure occurs as a result. If

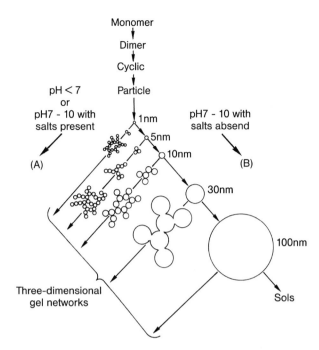

Fig. 4.1.9 Schematic drawing of the polymerization behavior of silica. In basic solution (B) particles in the sol grow in size but decrease in number; in acid solution or in the presence of flocculating salts (A), particles aggregate into three dimensional networks and form gels (Iler, 1979). The polymerization of mullite sols follows similar reaction paths (see also Section 6.2, from Schneider et al. 1994).

the solution is acidic, a nucleophilic reaction corresponding to the proton is expected to occur. Since $Si(OR)_4$ has a higher electron density around the silicon atom than $Si(OR)_3OH$ the former is much more susceptible to attack by the OH radical. The main reaction proceeds from $Si(OR)_4$ to $Si(OR)_3OH$ to $Si(OR)_2(OH)_2$ to $Si(OR)(OH)_3$ to finally $Si(OH)_4$. Therefore, the reaction producing $Si(OH)_4$ is dominant to that producing $[-Si-O-Si-]_n$, and a hydrolytic reaction to form dendritic structure occurs as a result. A schematic model for hydrolysis and polymerization of $Si(OR)_4$ is shown in Fig. 4.1.9 (Iler, 1979).

Hoffman et al. (1984) prepared xerogels by dissolving TEOS and aluminum nitrate nonahydrate $(Al(NO_3)_3 \cdot 9H_2O)$ in absolute ethanol and gelled the solution in a water bath by heating at 60 °C. Slow evaporation of solvent and the hydrolysis-polymerization reaction converted it to wet gels after several days. The transparent gels were then dried at 60 °C in air to form the powders. Differential thermal analyses (DTA) of these powders exhibited an exothermic peak at about 960 °C and the workers reported the presence of poorly crystalline mullite at about 1015 °C, although a closer observation of their X-ray diffraction data reveals the presence of γ-phase as well as mullite. Hoffman and coworkers used the term "single phase" to identify these xerogels. ^{29}Si MAS NMR spectroscopy showed the incorporation of aluminum in the silicon nearest neighbor environment (Komarneni et al. 1986a), corroborating the single phase nature of the above gels (see Sections 1.4 and 2.6). Thus the advent of solid-state NMR spectroscopy has revolutionized the characterization of gel solids with respect to the nearest neighbor environment. Solid-state ^{29}Si and ^{27}Al NMR spectroscopy clearly gives information on the level of mixing (Komarneni et al. 1986b, see also Sections 1.4 and 2.6). Okada et al. (1986a) used a process similar to that of Hoffman et al. (1984) and examined the mullitization route. Mullite was observed at 900 to 950 °C and directly crystallized from the amorphous state. From this result, the chemical homogeneity of the sample was considered to be high and attributed to molecular mixing. Schematic microstructural models of various starting materials are shown in Fig. 4.1.8 (Okada et al. 1991). Differential thermal analysis (DTA) yields two-step patterns with plateaus between 1000 and 1100 °C and above 1200 °C. The second step is similar with the formation curve of materials synthesized by the precipitation method (Okada and Otsuka, 1986). Okada and Otsuka (1990b) prepared mullite thin films by the dip coating method using a solution prepared by the same method as published by Okada et al. (1986). The crystalline phase in the thin film formed by calcination at 1000 °C was mullite when the aging time of the dip solution was shorter than one month at room temperature but it changed to γ-phase if the aging time of the dip solution was prolonged. The gelation time of the solution is extremely long, exceeding six months at room temperature because the pH is very low and the solution contains salt. Using ^{29}Si and ^{27}Al NMR spectroscopy, Aoki et al. (1992) examined the local structures around silicon and aluminum atoms in solutions aged for various times. They recognized that the degree of polymerization of SiO_4 tetrahedra increased strongly with the aging time, but little change was detected in the local structure of sixfold coordinated aluminum atoms. A conclusion is that the polymerization of SiO_4 tetrahedra inhibits mullitization. Huling and Messing (1991) used the same precursors as Hoffman et al. (1984) but with two different aging treatments of the gels. When alumina phase separation was avoided in these gels, orthorhombic mullite crystallized at a temperature as low as 700 °C. According to the findings of Schneider et al. (1993c) and Fischer et al. (1996) these mullites are extremely Al_2O_3-rich with compositions beyond the pseudo-tetragonal point ($x > 0.67$, see Sections 1.1.3.7 and 2.5.1).

Al-Jarsha et al. (1985) used a combination of TEOS and aluminum chloride instead of aluminum nitrate and observed mullitization above 1200 °C. Therefore it can be stated that the mullitization temperature is controlled by the kind of aluminum salt used. Nishio et al. (1994) studied the mullite crystallization behavior using two kinds of gel obtained from aluminum isopropoxide and aluminum nitrate and tetraethyloxysilane (TEOS). One kind of gel was obtained by adding aluminum isopropoxide and TEOS simultaneously to aqueous aluminum nitrate solution and the second kind was obtained by first adding aluminum propoxide to aqueous aluminum nitrate solution followed by the addition of TEOS. The first type of gel was found to be more homogeneous than the second kind through mullitization studies. Mullite crystals were found in the former gel at 600 °C but at

800 °C in the latter type, reflecting their scale of mixing and probably their very high Al_2O_3-contents (see above). Simendic and Radonjic (1998) used a similar procedure as that of Hoffman et al. (1984) and Komarneni et al. (1986b) in preparing polymeric aluminosilicate gels using metal salts of aluminum and silicon alkoxide and found that the temperature of mullite formation depended on the structure of the gel. Imose et al. (1998) used hydrous aluminum chloride and TEOS along with hydrazine to prepare gels at pH 9 and found that boehmite precipitates in these gels. The mullite powder prepared by this method at 1300 °C was found to have a surface area of 87 $m^2 g^{-1}$.

Prochazka and Klug (1983) used aluminum isopropoxide and TEOS to prepare mullite powders by separately dissolving the alkoxides in cyclohexane and mixing them in the appropriate ratio followed by hydrolysis and condensation with a water plus butyl alcohol solution using as blender. The obtained gels were dried and calcined at different temperatures up to 1200 °C. Powders crystallized to γ-alumina or aluminum silicon spinel (γ-phase) upon prolonged heating at 850 °C but were amorphous below this temperature. Heating at 1200 °C and above led to sharp X-ray diffraction peaks of mullite. Thus Prochazka and Klug used two alkoxides but used no precipitants unlike Mazdiyasni and Brown (1972). Hirata et al. (1989) used the same combination of alkoxides as Prochazka and Klug (1983) but dissolved them in benzene. The solution was hydrolyzed by slowly adding water diluted by isopropanol and refluxing at around 80 °C. The amount of water used was sufficient for complete hydrolysis of the alkoxides. The precursors, recovered by evaporation of the solvent at 188 °C, show mullitization as well as γ-phase formation at 1000 °C. Therefore the production of chemically homogeneous materials is difficult using these methods.

It is well known that the hydrolysis rate of aluminum alkoxide is much greater than that of silicon alkoxide. This difference cannot be adjusted by only replacing their alkoxyl groups. Taylor and Holland (1993) produced stoichiometric mullite powders using a water-free solution-sol-gel (SSG) approach with TEOS and aluminum isopropoxide as the starting chemicals. Materials with varying degrees of homogeneity were produced and characterized by solid-state ^{27}Al MAS NMR spectroscopy. They correlated the level of homogeneity with the amount of penta-co-ordinated aluminum, and developed a crystallization model for mullite. Sales and Alarcon (1996) prepared mullite gels using ethoxides ($Al(OC_2H_5)_3$) of aluminum and silicon. The aluminum ethoxide was dissolved in ethanol by refluxing and added to a pre-hydrolyzed TEOS solution. The formation of mullite started around 1000 °C with full crystallization at 1400 °C.

Osendi and Miranzo (1997) studied the thermal evolution of a mullite gel of composition $2Al_2O_3 \cdot SiO_2$ prepared from a commercially produced sol-gel precursor of silicon and aluminum. This gel crystallized into an Al_2O_3-rich mullite and α-alumina, instead of single-phase $2Al_2O_3 \cdot SiO_2$ mullite at 1300 °C.

Kansal et al. (1997) used an entirely new approach to synthesize a processable mullite precursor by reacting silica and aluminum hydroxide with triethanolamine in ethylene glycol, which yields a single phase. This single phase precursor was characterized by ^{27}Al and ^{29}Si NMR spectroscopy and found to crystallize to pseudo-tetragonal mullite above 950 °C (see Sections 1.1.3.7 and 2.5.1). Suzuki et al. (1984) treated admixtures of alkoxides or sols hydrothermally, and examined the thermal phase changes of the obtained powders. A pseudo-boehmite-like phase containing some silica besides alumina is obtained when the mixtures are treated under 50 MPa at 300 to 400 °C for 2 h. Dehydration occurs by firing of these hydrothermally treated powders at 600 °C, and this phase transforms to γ-phase. Mullitization is observed by firing at 1300 °C for 1 h. On the other hand, some aluminum silicates as well as a small amount of mullite are formed by the hydrothermal treatment of admixtures of sols at 500 to 600 °C and 50 MPa for 2 h. Mullitization of these hydrothermally treated powders is observed at temperatures as low as 800 °C with the newly formed mullites being probably very Al_2O_3-rich (see above). Formation of aluminosilicates in the samples is considered to promote the mullitization.

Suzuki et al. (1988) adopted the prehydrolysis method for mullite synthesis in order to adjust the differences of hydrolysis rates of silicon and aluminum alkoxides. Tetraethyloxysilane (TEOS) was dissolved in a solution of ethanol, water and catalyst (HCl), and refluxed at 70 °C for 50 h. It was mixed and reacted with aluminium propoxide, $Al(O^iPr)_3$, which was dissolved and refluxed in isobutanol. The above mullite precursor solution was hydrolyzed by adding water and refluxing. Suzuki et al. (1990) first partially hydrolyzed TEOS with acidic water and then mixed it with aluminum alkoxide that had been previously refluxed in isobutanol. This mixed solution was reacted at room temperature for 24 h and subsequently hydrolyzed with drop-wise addition of excess distilled water. This precursor showed a pseudo-boehmite-like structure or amorphous nature below 800 °C but crystallized to ultrafine mullite powder above 900 °C. Kamiya et al. (1998) prepared ultrafine mullite powders by chemical polymerization between partially hydrolyzed aluminum and silicon alkoxide species, which led to high compositional homogeneity using the same procedure as that of Suzuki et al. (1988). Osendi and Miranzo (1997) used a gel prepared from a precursor with an aluminum to silicon ratio of 2:1, which was commercially obtained. However, they did not give the gel preparation procedure from this precursor. The gel crystallized into an Al_2O_3-rich mullite along with α-alumina at 1300 °C, as was to be expected. Paulick et al. (1987) used aluminum isopropoxide and TEOS as precursors and studied the role of pH, reaction time and temperature in preparing the mullite gels. Both acidic and basic conditions have been used for gel preparation with nitric acid and ammonium hydroxide in the pH range of 2 to 10, or alternatively with no acid or base. These investigators found that the most intimately mixed gels were obtained at low pH where hydrolysis predominates over condensation (Keefer, 1984, Brinker and Scherer, 1990). Suzuki et al. (1988) detected the occurrence of Al–O–Si bonding in the mullite precursor solution by infrared spectroscopy and proposed the formation of Al–O–Si double alkoxide complexes with mullite composition ($Al_6Si_2O_6(OR)_{18}$). However, instead of mullitization, γ-phase crystallized at about 1000 °C. Mullitization started above 1100 °C. Judging from this mullitization reaction the formation of double alkoxide complexes is doubtful, or the complexes may have decomposed during the hydrolysis.

Komarneni et al. (1988) used TEOS and aluminum nitrate gels and applied microwave heating to prepare mullite powders and subsequently Ravi et al. (1998) also used microwave irradiation on a mixture of aluminum butoxide, TEOS and ethyl acetate solution to prepare mullite powders. The latter authors reported mullite formation at 976 °C. Nogami et al. (1990) reported the importance of the amount of water for the formation of Al–O–Si double alkoxide complexes. Since the hydrolysis rate of $Al(O^iPr)_3$ is very high, the excess amount of water (H_2O/ TEOS > 1) present in the prehydrolysis solution is consumed only to hydrolyze aluminum alkoxides but not to form double alkoxide complexes during cohydrolysis. However, if silicon alkoxide is prehydrolyzed in an adequate amount of water and is then mixed with aluminum alkoxide, double alkoxide complexes are formed, which do not decompose in the final hydrolysis step.

Wang et al. (1993) investigated the effect of steam on mullite formation from single phase gels and reported nucleation enhancement of pseudo-tetragonal mullite (see above) at about 980 °C using dynamic X-ray diffractometry. Wang and Thomson (1995) studied the mullite formation from non-stoichiometric single phase gels that were slowly hydrolyzed. They found that metastable pseudo-tetragonal mullite (see above) and γ-phase crystallized below 1000 °C in the gels with aluminum to silicon ratios Al/Si < 8/1, but at higher aluminum to silicon ratios γ-phase crystallized below 1000 °C and orthorhombic mullite formed directly at temperatures above 1250 °C. Chen and Vilminot (1995) used TEOS and aluminum butoxide to prepare two series of mullite gels, one involving a chemical modification of alkoxide with ethylacetoacetate and the other by reacting both alkoxides in isopropanol or 2-methoxyethanol. The first series led to the formation of pseudo-tetragonal mullite (see above) after the 1000 °C exotherm while the second series proceeds through the crystallization of a spinel-type phase in the same temperature range. Their differences were attributed to differences in levels of mixing, with the latter series of gels being inhomogeneous relative to the latter. Domrochev et al. (1999) used alkoxyaluminum silicates with the same and different alkoxyaluminium groups at the aluminum and silicon atoms, which were prepared by alcoholysis of di-isobutylaluminum silicate. The solid residues apparently crystallized to single phase mullite in the temperature range 500 to 1000 °C after vacuum thermolysis of the precursors. Ghosh and Pramanik (2001) prepared mullite gels using TEOS and aluminum formate as precursors with water as solvent, and claim to have produced low-cost nanometer-sized powders after calcination.

Sin et al. (2001) used a very complex solution route by first dissolving aluminum metal powder in nitric acid and granular silica in 40% hydrofluoric acid (HF) to which ethylenediaminetetraacetic acid (EDTA) and acrylamide monomers were added. These precursors were then heated in the temperature range 1100 to 1350 °C with and without starch. They crystallized mullite platelets with starch and acicular crystals without starch at 1315 °C after treatment for 6 h.

From the literature data it may be concluded that the following points are important for mullite synthesis by hydrolysis using aluminum and silicon alkoxides: (1) Hydrolysis of alkoxides should be carried out with an amount of water which is just sufficient for complete hydrolysis; (2) An aqueous solution diluted by organic

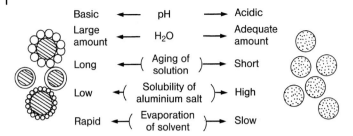

Fig. 4.1.10 Schematic microstructural development of mixed silicon alkoxide and aluminum salt starting materials depending on reaction-controlling factors (from Schneider et al. 1994).

solvents should be used in order to avoid local high water concentrations, which may cause excess hydrolysis locally; (3) An acidic catalyst is preferable to a basic one.

Mullite syntheses using different kinds of metal organic compounds instead of alkoxides have also been reported. Pouxviel et al. (1986) examined syntheses with alumino siloxane ($Al(O^iPr)_2OSiMe_3$) and silicon aluminum ester ($(EtO)_3Si-O-Al(OBu)_2$)) as starting materials (Me = CH_3, Bu = C_4H_9). Mullitization at around 1000 °C was found in the latter preparation method but not in the former one. Mizukami et al. (1990) used TEOS and aluminum dibutoxide ethylacetoacetate ($Al(OBu)_2(AcAcEt)$), and examined the effect of ligands and solvents on the mullitization. They suggested that mullitization among the specimens prepared from different ligands and solvents changed owing to their different levels of mixing. Low temperature mullitization is possible when using organic ligands because they have high reactivities to form Si–O–Al bonds. Meng and Huggins (1983) used silicon acetate and aluminum nitrate and dissolved them in water or methanol. A xerogel was obtained by evaporating the solvent. Mullite appears after heat treatment at 1324 °C for 18 h. Various experimental conditions influencing the chemical homogeneity of xerogels prepared from the combination of silicon alkoxides and aluminum salts are schematically shown in Fig. 4.1.10 (Okada et al. 1991).

4.1.3.2 Solution-plus-sol Process

There are many reports on the use of admixtures of sols and salts as starting materials for mullite synthesis. These preparation methods are considered to be advantageous compared with those of admixtures of sols, because composite particles and not simply mixed particles are expected to form under certain conditions. Even if this is not the case, precipitation of the salt component is expected on the surface of the sol particles, which act as heterogeneous nucleation sites. Composite particles can be prepared by adsorption, hydrolysis and precipitation reactions of a salt component on the surface of sol particles. Yoldas (1980) prepared boehmite first from aluminum butoxide and dispersed it in TEOS to make Al–O–Si bonds by polymerization. Mullite formation was reported at 1300 °C and above.

Hoffman et al. (1984) used a diphasic method where a boehmite sol was dispersed in an alcohol solution of TEOS and caused gelling by heating. They also mixed silica sol with aluminum nitrate solution to prepare gels, which were investigated by differential thermal analysis (DTA) and showed an exotherm at about 1350 °C for the boehmite plus TEOS gel, reflecting the formation of mullite. Sueyoshi and Contreras Soto (1998) prepared fine mullite powders using homogeneous precipitation of fumed silica and a mixture of aqueous solutions of aluminum sulfate and ammonium bisulfite. This precursor transformed principally to γ-phase and trace mullite at 950 °C but transformed to mullite completely above 1250 °C. Kubota and Takagi (1987, 1988) and Kazakova et al. (1999) used aluminum nitrate (or another salt) and the silica sol method of Hoffman et al. (1984) and studied crystallization of mullite. Kubota and Takagi (1988) reported crystallization of mullite at 1250 °C. Imose et al. (1998) used a boehmite solid-silica solution gel synthesized by a hydrazine method, which yields stoichiometric mullite of high surface area at high temperatures. The as-prepared powder and powders heated below 1200 °C showed very fine needle-like particles whereas the mullite particles are of prismatic morphology. The crystallized mullite powder after heating at 1300 °C showed a high surface area (87 m^2 g^{-1}).

Anilkumar et al. (1997) used boehmite sol and TEOS to prepare mullite following the method of Hoffman et al. (1984). Nishu et al. (1989) prepared a pseudo-boehmite sol by adding ammonium hydroxide to aluminum chloride solution, and adjusting the pH of the solution to 8. Simultaneously a monosilicic solution was prepared by dissolving silica gel in boiling water and adjusting the pH of the solution to 9. The two solutions were mixed and stirred for 5 h at 25 °C at a pH of 9 to yield adsorption of the silica component on pseudo-boehmite sol particles. The precipitates were filtered and dried at room temperature. Mullitization of this material was observed at 980 °C with no γ-phase formation. Nishu et al. (1989) investigated the coordination number of aluminum in these samples by ^{27}Al MAS NMR spectroscopy and found that fourfold and sixfold coordinated aluminum occurs, while aluminum atoms are only sixfold coordinated in pseudo-boehmite sols. Judging from the mullitization reaction it was considered that the silica component is not only adsorbed on the surface but also penetrates into the inner parts of the samples.

Sacks et al. (1991) proposed a new preparation method for composite particles with α-alumina coated by amorphous silica layers. The concept is schematically shown in Fig. 4.1.11. In a first step the authors prepared α-alumina particles about 0.2 μm in size by elutriation and suspended them in ethanol. This suspension was mixed with ethanol solution containing dissolved TEOS. The TEOS was then hydrolyzed by adding ammoniated water and the silica component precipitated on the surface of α-alumina particles to form composite particles. Compacts of the composite particles sinter to almost full density at temperatures as low as 1300 °C due to the rearrangement of particles by viscous deformation of the amorphous silica layers ("Transient viscous sintering, TVS", Sacks et al. 1991). However, firing at 1500 °C for 2 h is necessary to convert the Al$_2$O$_3$/SiO$_2$ composite particles extensively to mullite. Since the interdiffusion of aluminum, silicon and oxygen

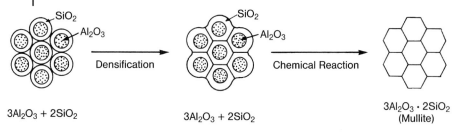

Fig. 4.1.11 Reaction principles of mullite synthesis using the transient viscous sintering (TVS) technique, starting from α-alumina grains coated by thin silica glass layers (according to Sacks et al. 1991, from Schneider et al. 1994).

atoms (considered as diffusion of silicon atoms into α-alumina in this case) at a 0.1-μm scale is necessary for mullite formation, the reaction is very sluggish. Tang et al. (2002) used a core-shell approach with aluminum hydroxide core and silica shell by a heterogeneous nucleation and growth process in an ethanol solution containing ammonia. Fine mullite powders were crystallized by heating at 1500 °C for 2 h using this approach. Dense mullite ceramics have successfully been produced even at temperatures below 1300 °C from amorphous silica coated γ-alumina particle nanocomposites making use of the transient viscous sintering technique (Bartsch et al. 1999).

Shiga et al. (1991) tried to prepare composite particles by the homogeneous precipitation method. Colloidal silica particles were dispersed in aluminum sulfate solution with concentrations of 0.005 to 1.0 mol l^{-1} (M). Urea (0.5 to 2.5 M) was added, and the admixture was refluxed by heating. Mullitization of the precipitates is observed at temperatures as low as 1000 °C but with a large amount of γ-phase formed, and extensive mullitization was accomplished below 1400 °C. Intimate contact of silica and alumina components, and/or a partial formation of Si–O–Al bonds accelerated the formation of mullite. However, a complete stoichiometric reaction could not be achieved by this preparation method. Lee et al. (2002) and Kim et al. (2003) investigated the effects of precursor pH and sintering temperature on the synthesis and morphology of mullite using a coprecipitation of aluminum nitrate and colloidal silica sols. They obtained mullite above 1200 °C and the morphology depended on pH. Needle-like mullite results under acidic conditions while rod-like or granular mullite are obtained at a pH above 8 as the temperature of sintering is increased. Ivankovic et al. (2003) prepared four different types of precursors, two of them using a sol and solution approach, and showed that this type of diphasic approach yields mullite at 1300 °C. Two major processing methods exist for the conversion of the above mentioned sols into gels. One method is to adjust the solution to pH values of 5 to 7, which causes flocculation. The other method is to evaporate the solvent and concentrate the sol in order to transform it into gel.

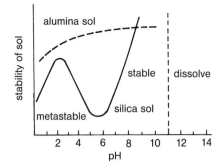

Fig. 4.1.12 Stability of silica and alumina sols plotted versus acidity (changing pH-values), (from Schneider et al. 1994).

4.1.3.3 Sol-plus-sol Process

This method applies for preparation techniques using starting materials of admixtures of sols. It is basically very similar to the conventional method using oxides or silicates as mentioned in Section 4.1.1, if the materials are mixed under wet conditions. The particle size, however, differs significantly for the two preparation methods. The particle size of sols is around ten to several tens of nanometers (nm) and therefore are much smaller than those used in the conventional method, which are in the range of several microns (μm).

Sols can be obtained by many preparation methods. Silica sols can be prepared by dispersion of ultrafine particles such as fumed silica and colloidal silica, and by the hydrolysis of silicon alkoxides. Alumina sol can be prepared by dispersion of ultrafine particles of γ-alumina, by dispersion of pseudo-boehmite (γ-AlO(OH)), which in turn is produced by precipitation of aluminum salt solution by adjusting to neutral conditions, and also by hydrolysis of aluminum alkoxides. The approximate particle size of these sols is around 10 nm. The stability of sols changes with the pH of the solution as shown in Fig. 4.1.12. Alumina sol is stable for a wide pH range except for strongly acidic and strongly basic conditions. The stability range of silica sol is complex. It shows a maximum at a pH of around 2 and decreases as the pH of the solution increases. At pH 5 to 6 a minimum is achieved and flocculation occurs easily. Stability of the silica sol increases again at higher pH values (above 7), and silica dissolves easily under strongly basic conditions above a pH of 12. Therefore, adequate pH ranges to mix silica sol and alumina sols are around 2 or 9 to 10. Schneider et al. (1993b) demonstrated the role of pH range of the silica sol for the development of mullite precursors. The diluted silica sols adjusted to 10 < pH < 13 results in formation of large silica-rich areas, and therefore retard mullitization.

Ghate et al. (1973) utilized the properties of silica and alumina sols for mullite synthesis. They prepared alumina sol by dispersing γ-alumina particles in hydrochloric acid solution. Silica sol was added slowly to this suspension and adjusted to pH 6 to 7. In this pH range the surfaces of alumina particles are positively charged while those of silica particles are negative. Therefore, heteroflocculation, which causes intimate mixing of two sol particles, can be expected. Aqueous suspensions of high-surface-area γ-alumina and silica sols were mixed and gelled by solvent evaporation rather than by pH change (Sacks and Pask, 1978, 1982a, 1982b). These authors reported complete mullitization at 1450 °C after treatment for 24 h and these results are similar to those of Ghate et al. (1973). Sacks et al. (1997) used a core-shell approach by precipitating hydrous silica on α-alumina cores and investigating their mullitization behavior, finding that the first stage of the reaction is controlled by the dissolution of alumina in silica. The two-sol technique was also used by Metcalfe and Sant (1975) who reported mullite formation at 1550 °C after treatment for 3 h. These authors simply mixed the sols and dried them without a gelling step and this process probably led to segregation of the two components. This is perhaps the reason why the mullitization required a higher temperature than that of Sacks and Pask (1978, 1982a, 1982b).

A different type of mullite precursor, the so-called "diphasic gel" or type II precursor (see Section 1.4.2) was synthesized by Hoffman et al. (1984) by mixing aqueous silica sol with boehmite sol dispersed in ethanol. These gels yielded no exotherm at 960 °C in differential thermal analysis (DTA), but a slight peak appeared at about 1250 °C. X-ray diffraction data of the samples showed the presence of boehmite in the as-prepared state which reacted to γ-alumina above about 1000 °C. Mullitization was first achieved above 1300 °C. The crystallization behavior of this precursor group was ascribed to the diphasic character of the gels. Komarneni et al. (1986b) prepared diphasic gels of different aluminum to silicon ratios by initially peptizing boehmite in nitric acid and mixing silica sol with the peptized boehmite. Gelling was accomplished at room temperature and the gels were later dried at 60 °C. These gels were characterized by ^{29}Si and ^{27}Al MAS NMR spectroscopy, which revealed the nanoscale mixing of components, as expected. Ismail et al. (1986, 1987) used γ-alumina and silica sols, mixing them under acidic conditions. They first hydrolyzed γ-alumina at 90 °C and peptized it using concentrated nitric acid at 95 °C. Then, they gelled this sol after mixing with acidified silica sol by evaporating the solvent, and dried the gel to powder. They reported extensive mullitization at 1300 °C after 1 h. Boehmite sol prepared from aluminum ethoxide and silica sol prepared from tetramethyloxysilane (TMOS) were used by Hamano et al. (1985) for mullite synthesis. They mixed the two sols and dried them by adding to a hot oil bath at 160 °C prior to crystallization studies. These mixed sols showed initial mullite formation at about 1250 °C. Debely et al. (1985) prepared a uniform aluminosilicate powder by heterocoagulation of fine γ-alumina powder and silica sol and studied the densification behavior of these diphasic powders. Mroz and Laughner (1989) prepared mullite from colloidal sols of pseudo-boehmite and silica sol and investigated their densification behavior by mullite seeding.

Somiya et al. (1985) first treated alumina and silica sols hydrothermally in the temperature range between 300 to 600 °C and used these powders for subsequent mullite crystallization by solid-state heating. Mullitization was observed at about 1300 °C with the hydrothermally treated sol precursors. Pask et al. (1987) reported similar behavior in the phase development of mullite precursors. Wei and Halloran (1988a, 1988b) prepared diphasic aluminosilicate gels by gelling a mixture of colloidal pseudo-boehmite and a silica sol prepared from acid hydrolyzed TEOS, and studied their phase transformation. They found that the two discrete phases transformed independently up to 1200 °C and then reacted to form mullite. Mizuno and Saito (1989) prepared highly pure mullite powders using aluminum salts or boehmite and fumed silica at 1350 °C. Yokoyama et al. (1997) prepared mullite composition powders by adsorption of monosilicic acid on amorphous aluminum hydroxide, and mullite crystallized from these powders at 1000 °C without the formation of a spinel phase. Kaya and Boccaccini (2002) used diphasic sols of boehmite and silica in their investigation of microstructural variations in mullite ceramics obtained by different forming techniques. Mullite fiber mats were prepared (Chatterjee et al. 2002) by a single step sol-gel method from spinnable sols with no organic binders, and crystallization studies revealed crystallization of γ-alumina at 900 °C and mullite at 1250 °C, confirming the results obtained earlier by others. Using diphasic sols, Li and Thomson (1991) determined that a metastable, pseudo-tetragonal mullite (see Sections 1.1.3.7 and 2.5.1) was not crystallized up to 1200 °C while it formed in single phase gels. Many others have used diphasic sols to determine various aspects of mullite formation and sintering (Li and Thomson, 1990, Schneider et al. 1993b, 1994, Hong and Messing, 1998, 1999, Boccaccini et al. 1996, 1999, Kara et al. 2000, Tang et al. 2002, Kim et al. 2003, Kaya et al. 2002, 2003, Chakraborty and Das, 2003).

Hong and Messing (1997) investigated the transformation kinetics of diphasic mullite gels doped with phosphorus oxide (P_2O_5), titania (TiO_2), and boria (B_2O_3) to mullite using quantitative X-ray diffraction (XRD) and differential thermal analysis (DTA). Nucleation and growth of mullite whiskers from diphasic gels doped with lanthanum oxide (La_2O_3) has been investigated by Regiani et al. (2002). Using DTA they showed that the mullite crystallization temperature decreased from 1350 °C to 1240 °C (see also Section 6.1). Naskar et al. (2002) obtained hollow mullite microspheres from emulsified diphasic sols by an ion-extraction method and investigated their crystallization behavior by DTA and XRD. The gel microspheres crystallized to an aluminum silicon spinel (γ-phase) at 900 to 970 °C and orthorhombic mullite at 1200 °C.

According to the kind of starting materials used, the precursors that transform to mullite from the amorphous state at about 980 °C were called "polymeric" and those that transform at higher temperatures were called "colloidal" (Pask et al. 1987). A schematic structure model for the precursors prepared by admixtures of sols is shown in Fig. 4.1.8. The mullitization routes of these samples are essentially the same as for those prepared by conventional mixing using alumina and silica particles. The mullitization temperature, however, changed to around 1200 °C and mullitization is accomplished up to 1350 to 1400 °C. These temperatures are apparently lower compared with those of the conventional methods, because of the small particle size and the intimate mixing of the two components of the sols.

4.1.4 Spray Pyrolysis Approach

This preparation method uses droplets formed from a solution which is sprayed into a furnace heated at high temperatures. Fig. 4.1.13 shows a schematic illustra-

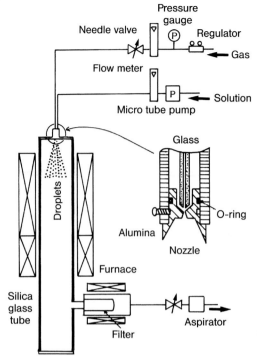

Fig. 4.1.13 Schematic outline of spray pyrolysis equipment (after Takigawa et al. 1990, from Schneider et al. 1994).

tion of the spray pyrolysis equipment (Takigawa et al. 1990). Reactions such as evaporation of solvents, precipitation of compounds and thermal decomposition occur instantly. Therefore, it is considered to be an appropriate preparation method for the synthesis of multicomponent ceramics. The powders obtained by this process have characteristic spherical shapes in submicrometer to micrometer sizes. Droplets have mainly been prepared by two different methods using atomizers and ultrasonicators, respectively.

Kanzaki et al. (1985) dissolved TEOS and aluminum nitrate in a water-methanol (1:1 in volume) solution and sprayed the droplets into a furnace heated at 350 to 650 °C. They used a borosilicate glass atomizer with compressed air to form droplets. The as-sprayed specimens are amorphous and the DTA curves of specimens show very sharp exothermic peaks at around 1000 °C. Only mullitization is observed but no γ-phase. Hamano et al. (1985) examined similar processes as Kanzaki et al. (1985) did, and distinguished two cases: (1) Powders prepared by spraying a solution dissolving TEOS and aluminum nitrate in ethanol at 250 °C. This material displays similar mullitization to that reported by Kanzaki et al. (1985); (2) Powders prepared by spraying a solution at 350 °C. The solution is derived by dissolving TEOS and aluminum sulfate in an ethanol/water solution. γ-Phase formation but no mullitization was obtained below 1200 °C. Precipitation of aluminum sulfate crystals is observed in the as-sprayed specimens in this case. These data and experimental results from the combination of TEOS and aluminum ni-

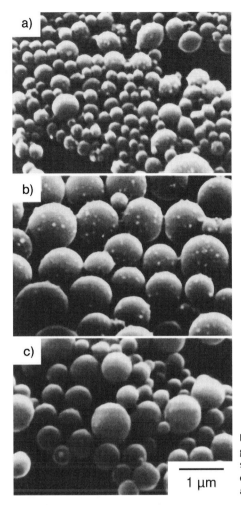

Fig. 4.1.14 Scanning electron micrographs of spray pyrolyzed mullite precursor powders prepared under different experimental conditions (after Sakurai et al. 1988, from Schneider et al. 1994).

1 μm

trate (Okada and Otsuka, 1986, Hoffman et al. 1984), and from TEOS and aluminum chloride (Al-Jarsha, 1985), imply that the type of aluminum salt used is an important factor in obtaining chemically homogeneous (molecularly mixed) specimens from the combination of silicon alkoxides and aluminum salts.

Sakurai et al. (1988) prepared mullite powders as shown in Fig. 4.1.14 by the spray pyrolysis method using an ultrasonicator. Starting materials were TEOS and Al(O^{i}Pr)$_3$. The solution was prepared as follows: TEOS was dissolved in a solution prepared by mixing dehydrated ethanol, water and a catalyst and prehydrolyzed by refluxing at 75 °C for 6 h in a nitrogen atmosphere. The prehydrolyzed TEOS was mixed with an ethanol-water solution, Al(O^{i}Pr)$_3$ was dissolved in this solution and the mixture was refluxed again under the same conditions to cohydrolyze both alkoxides. The experimental conditions and crystalline phases detected for corresponding materials are listed in Table 4.1.3. Comparing the results of C-series

Table 4.1.3 Experimental conditions of the spray pyrolysis method using both alkoxides and their crystalline phase obtained by firing. Reproduced by permission of the Ceramic Society of Japan.

Specimen	Prehydrolysis			Cohydrolysis		State	Crystalline phase	
	H_2O/TEOS ratio	Catalyst	Refluxing	H_2O/TEOS ratio	Refluxing		1000 °C	1200 °C
A-1	–	–	–	2.78	No	Insoluble	Spinel phase	Mullite + θ-Al_2O_3
B-1	–	–	–	<0.3	Yes	Precipitate	Spinel phase	Mullite + θ-Al_2O_3
C-1	0.8	HNO_3	Yes	2.78	No		Mullite	Mullite
C-2	0.8	NH_4OH	Yes	2.78	No		Spinel phase	Mullite + θ-Al_2O_3
D-1	0.8	HNO_3	Yes	0.06	Yes		Mullite + spinel phase	Mullite + 0-Al_2O_3
D-2	0.8	HNO_3	Yes	3.19	Yes		Mullite	Mullite + θ-Al_2O_3
D-3	0.8	HNO_3	Yes	4.44	Yes		Mullite	Mullite

Concentration of TEOS = 0.02 mol l^{-1}, Al(O^{i}Pr)$_3$; TEOS ratio = 3 : 1.
From Schneider et al. 1994.

specimens, the acidic catalyst was found to be important in order to obtain chemically homogeneous specimens. Obviously the reaction conditions under which hydrolysis is promoted and polymerization is suppressed are decisive in obtaining chemically homogeneous precursors. The results of the D-series show that the amount of water has a great influence on the chemical homogeneity of the precursors. Good results were found for specimens prepared with an H_2O/TEOS ratio of around 5. Incomplete hydrolysis due to an insufficient amount of water leaves some residual non-hydrolyzed silicon alkoxide which evaporates during the spray pyrolysis. The appropriate ratio of H_2O/TEOS for this process is apparently higher than those for other hydrolysis methods (Suzuki et al. 1988, Hirata et al. 1989, Nogami et al. 1990). It may be concluded that controlled hydrolysis is necessary in order to obtain chemically homogeneous specimens, and that the appropriate water amount is different for the individual preparation methods.

Moore et al. (1992) used a high temperature aerosol decomposition technique to produce high-purity mullite powders of 0.6 μm. Aluminum nitrate and fumed silica were used as precursors and depending on the reaction conditions either amorphous or crystalline mullite powders were obtained. Janackovic et al. (1996) produced submicrometer spherical particles of mullite powder by ultrasonic spray pyrolysis of emulsion and solutions, with TEOS or silicic acid and aluminum nitrate as precursors. The synthesis of mullite from TEOS emulsion occurred by crystallization of γ-alumina from the amorphous phase and its subsequent reaction with amorphous silica, and crystallization of pseudo-tetragonal mullite below 1000 °C (see above and Sections 1.1.3.7 and 2.5.1). Powders resulting from silicic acid solutions first yielded γ-alumina between 900 and 1000 °C, then reacted with amorphous silica between 1100 and 1200 °C to yield mullite. Thus mullite crystallization depends on the starting precursors as has been found by many others. Janackovic et al. (1998) synthesized nanostructured spherical particles of mullite powders by ultrasonic spray pyrolysis of TEOS, aluminum nitrate and nitric acid and reported γ-alumina formation at 1100 °C and mullite formation at 1200 °C. Baranwal et al. (2001) used flame pyrolysis of alcohol-soluble precursors to produce 60 to 100-nm mullite composition nanopowders. The soluble precursors were obtained from aluminum hydroxide and fumed silica by a novel dissolution process in triethanolamine and ethylene glycol. Crystallization of mullite occurred at 900 °C using these nanopowders.

4.1.5
Hydrothermally Produced Mullite

One of the earliest hydrothermal syntheses of mullite was done by Roy and co-workers (Roy and Osborn, 1952, 1954) using alumina silica gels sealed in platinum foil, placed in a high-pressure vessel and held at a constant temperature and water pressure. The alumina silica gels were prepared by the coprecipitation technique described by Ewell and Insley (1935). Roy and Osborn (1952) in their first study reported stable assemblages of mullite plus α-alumina and water and mullite plus quartz and water at temperatures above 575 °C in the Al_2O_3–SiO_2–H_2O system.

In a later study, Roy and Osborn (1954) reported the formation of mullite from high alumina mixtures at temperatures above about 425 °C in the system Al_2O_3–SiO_2–H_2O. Roy and Roy (1955) also reported the formation of mullite under hydrothermal conditions in the system MgO–Al_2O_3–SiO_2–H_2O. Aramaki and Roy (1962) reported the formation of mullite using gels under hydrothermal conditions at 945 °C and above in their investigation of the revised phase diagram for the system Al_2O_3–SiO_2. However, no attempts were made to make pure mullite for ceramic applications in any of the above studies.

Suzuki et al. (1984) treated admixtures of alkoxides or sols hydrothermally, and examined the thermal phase changes of the obtained powders. A pseudo-boehmite-like phase containing some silica besides alumina was obtained when the mixtures were treated under 50 MPa at 300 to 400 °C for 2 h. On the other hand, some aluminosilicates as well as a small amount of mullite were formed by the hydrothermal treatment of admixtures of sols at 500 to 600 °C and 50 MPa for 2 h. Komarneni et al. (1986c) treated single and diphasic gels of mullite composition under hydrothermal conditions in the temperature range 300 to 700 °C and a pressure range of 69 to 100 MPa for 4 to 12 h with the intention of preparing fine mullite powders. Fine mullite powder could not be produced under various hydrothermal conditions. Hydrasilite was found to be the stable phase under all of the conditions used. Only in the case of a diphasic mullite gel was some mullite detected by X-ray diffractometry. The crystallization of mullite in the diphasic gel, unlike in the single-phase gel, has been attributed to the slow reaction in the former to form hydralsite because of discrete silica and alumina phases (Komarneni et al. 1986c). Mullite needles of about 5 µm were detected by transmission electron microscopy. These authors suggested that rapid initial heating and heating to higher temperatures may be necessary to prevent the formation of hydralsite before the crystallization of mullite.

Mullite precursors were first prepared from aluminum isopropoxide and TEOS by hydrothermal treatment by Somiya et al. (1985). Aluminum isopropoxide and TEOS were mixed by dissolving in benzene and refluxing for 5 h at about 80 °C followed by their hydrolysis/condensation under hydrothermal conditions. The hydrothermally treated precursor powders were washed to remove organics such as alcohols and benzene and subsequently subjected to another hydrothermal treatment at 600 °C under a pressure of 20 MPa for different durations. The amount of crystallized mullite increased with increasing duration, and formed aggregates with 1- to 2-µm acicular crystals. These authors also varied the above procedure as follows: They first hydrolyzed TEOS and aluminum isopropoxide at a pH of 10, the precipitate was dried at 120 °C and calcined at 600 °C followed by hydrothermal treatment at 600 °C under a pressure of 20 MPa for 2 to 72 h. These calcined precursors also yielded the same type of mullite as above. Thus it appears that hydrothermal treatment is not a viable option for fine mullite powder preparation directly although precursor chemicals can be first treated by hydrothermal methods to achieve hydrolysis followed by solid-state calcinations to obtain mullite (Suzuki et al. 1984, Somiya et al. 1985, Kutty and Nayak, 2000, Kaya et al. 2002).

4.1.6
Vapor-state-derived Mullite

Almost all the processes described above are syntheses from a solution phase. On the other hand, the chemical vapor deposition (CVD) method utilizing the vapor-phase process has also been applied for mullite synthesis. Hori and Kurita (1990) used silicon and aluminum chlorides as starting materials. The chlorides were separately evaporated and transported by nitrogen gas to the mixing zone. In the mixing zone they were quickly heated by a hydrogen-oxygen combustion flame in which the temperature is estimated to be 1900 °C. The temperature gradually decreased through the reaction zone and was about 900 °C at the exit. The process time was estimated to be approximately 65 ms. The obtained powders were spherical and about 40 to 70 nm in size. Small amounts of mullite as well as γ-phase were detected in the as-prepared specimens. Extensive mullitization occurred after firing at 1000 °C for 10 min but complete mullitization required temperatures as high as 1500 °C. Double exothermic peaks are observed at about 1000 °C in the DTA curve. The reason for this inhomogeneity is attributed either to a metastable subliquidus relation between the liquid and the metastable alumina solid solution at high temperatures or to a metastable liquid immiscibility in the Al_2O_3–SiO_2 system during the cooling process. Chung et al. (1992) described a similar mullite powder production route. They used the counter-flow diffusion flame burner, in which fuel (hydrogen and nitrogen), chlorides and the oxidant flows (oxygen and nitrogen) are mixed to a combustion flame. The obtained powders are spherical and about 20 to 30 nm in size. The powders prepared by a low-temperature flame are amorphous in the as-prepared samples, while those prepared by a high temperature flame consist of similar phases to those reported by Hori and Kurita (1990). Itatani et al. (1995) produced the best mullite powders at 1200 °C using a chemical vapor deposition technique based upon reaction among aluminum and silicon chlorides in oxygen. The as-prepared powders contained mullite, a small amount of γ-alumina and amorphous material which reacted at 1300 °C to give only mullite. Monteiro et al. (1997) reported the deposition of mullite and mullite-like coatings on silicon carbide by dual-source metal plasma immersion ion implantation and deposition. The as-deposited films were amorphous but could be crystallized to mullite by annealing at 1100 °C for 2 h. Mullite coatings were made on silicon carbide (SiC) and silicon nitride (Si_3N_4) substrates (Mulpuri and Sarin, 1996, Hou et al. 1999, Armas et al. 2001, Sotirchos and Nitodas, 2002). The topic of coatings is dealt with in Chapter 5.

4.1.7
Mullite Produced by Miscellaneous Methods

The self-combustion method has been extensively investigated, mainly for the purpose of peparing various non-oxide ceramic powders. Ogawa and Abe (1990) prepared mullite powders by this method. They used silicon and aluminum metal powders as the starting materials. The particles obtained were very small spheres with sizes in the range between 0.01 and 10 μm. In spite of their very small particle sizes the powders display little agglomeration. No information on the mullitization rates of these powders is available so far.

4.1.8
General Remarks on the Different Chemical Synthesis Methods of Mullite

The various mullitization routes reported for each preparation process are summarized in Fig. 4.1.15. The type-A route corresponds to the cases of conventional mixing and the sol-gel methods. Mullitization temperatures change according to the particle sizes of the starting materials but are not lower than 1200 °C. The type-B route corresponds to precipitation methods. The mullitization temperatures are slightly lower than those of the sol-gel method because some degree of molecular level mixing of the two components is considered, and also the sizes of the precipitates are generally smaller than those of sols. Since the reaction controlling step is interdiffusion of aluminum, silicon and oxygen atoms, however, it is not possible

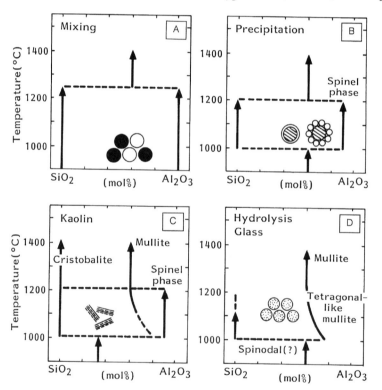

Fig. 4.1.15 Schematic illustration of different mullite synthesis routes using various starting materials. (a) Conventional mixing and sol-gel method, respectively; (b) Precipitation method; (c) Thermal decomposition of kaolinite; (d) Hydrolysis technique (after Okada et al. 1991, from Schneider et al. 1994).

to decrease the mullitization temperature substantially below 1200 °C. The type-C route corresponds to thermal decomposition of kaolin minerals and also to the hydrolysis methods prepared without catalyst with an appropriate amount of water. This route can therefore be considered as an intermediate state between the type-B and type-D routes. The type-D route corresponds to hydrolysis reactions under controlled hydrolysis conditions with an acidic catalyst and an appropriate amount of water. The spray pyrolysis method is further preferable if solutions are used that are prepared under precisely controlled hydrolysis conditions.

In a comprehensive study, Schneider et al. (1993b) brought a better understanding to the classification of mullitization routes. Using the thermal phase developments observed in the precursors, they established the existence of three principal types of precursors (designated as types I, II and III, see Section 1.4). Schneider et al. believe that these principles are valid for any mullite precursor development. The difference in the various precursor types do not originate from their starting materials, but lie in the scale of homogeneity and distribution of aluminum and silicon species within the precursors. Because the reactions that occur during synthesis are sensitive to small changes in the preparation conditions, most pre-mullites reported in the literature consist of a combination of different types of precursors, rather than just the intended end members. Therefore a diphasic gel should not be regarded as the pure representative of a certain precursor type. Similarly it is possible to synthesize a precursor that exhibits the same mullitization route as a so-called "colloidal" precursor, although it is obtained from alkoxides, which are regarded as polymeric sources.

In summary, a type I precursor is amorphous in the as-prepared state and yields mullite as the only crystalline phase at about 980 °C. Type II precursors contain pseudo-boehmite and non-crystalline silica in the as-prepared state. Pseudo-boehmite transforms to spinel phase (γ-phase: either pure γ-alumina or γ-alumina containing more or less high amounts of silicon, see Section 1.4) above about 400 °C and to mullite above about 1250 °C. A type III precursor is also amorphous in the as-prepared state; however, it partially converts to spinel phase above about 980 °C which yields mullite above about 1100 °C. The DTA data of these precursors also differ strikingly. Type I displays a sharp and single exotherm at about 980 °C, type II shows no exotherm at 980 °C, but a broad and weak exothermic peak at about 1250 °C, while type III exhibits a sharp but less intense exothermic peak at 980 °C as well as a broad exotherm at 1250 °C (see also Section 1.4).

The most important condition for obtaining chemically homogeneous mullite precursors is to suppress polymerization of the silica component. This condition seems to be compatible with the structural arrangement of SiO_4 tetrahedra in the crystal structure of mullite (Sadanaga et al. 1962), in which SiO_4 tetrahedra have discrete monomeric structure and shared corners with AlO_4 tetrahedra. In order to satisfy these conditions, the following procedures are considered to be effective: (1) preparation from solutions with low concentration; (2) choice of the proper combination of salts and solvents to achieve high solubility; (3) hydrolysis under acidic conditions; (4) hydrolysis with an appropriate amount of water; (5) formation of the Al–O–Si double alkoxide complexes by prehydrolysis of silicon alkoxide.

The mullite formed at around 1000 °C shows differences in some points compared with those synthesized at higher temperatures. This type of mullite is called transitional mullite, primary mullite, tetragonal-like mullite or pseudo-tetragonal mullite (see also Sections 1.1.3.7 and 2.5.1). Brindley and Nakahira (1959a, 1959b, 1959c) considered the chemical composition of this type of mullite as $Al_2O_3 \cdot SiO_2$. This approach has been contradicted by the determination of the chemical composition using the relationship between Al_2O_3 content and the lattice constants of mullite (Cameron, 1977, Okada et al. 1991, Okada and Otsuka, 1988), by chemical analysis with analytical electron microscopy (Okada and Otsuka, 1988), and also by structure refinements by means of the Rietveld method (Ban and Okada, 1992, Fischer et al. 1994, 1996). All these investigations confirmed the observation that these mullites display Al_2O_3 contents (much) higher than 60 mol% Al_2O_3. The Al_2O_3 composition of these mullites gradually shifts towards that of stoichiometric mullite (60 mol%) with the firing temperature, and almost coincides with the bulk composition at temperatures above 1200 °C (see also Sections 1.1.3.7 and 2.5.1).

4.2
Processing of Mullite Ceramics
S. Komarneni and H. Schneider

4.2.1
General Sintering Characteristics

Basic investigations on the sintering mechanisms of mullite powders were performed by Sacks and Pask (1982a, 1982b). According to these authors the sintering process of mullite can be divided into three different temperature stages, based on geometrical changes occurring in the powder compacts during densification. The intermediate stage is of greatest importance, since it is characterized by grain growth and the evolution of isolated closed pores from formerly open-pore channels. Factors influencing sintering are powder compact characteristics, heating conditions and the Al_2O_3/SiO_2-ratio of the starting material. Sacks and Pask describe three composition ranges of the starting materials:

- Compositions between about 60 and 65 wt. % Al_2O_3 with very high densities of the sintering products, due to a large amount of liquid phase present at sintering temperature ("Liquid phase sintering").
- Compositions between about 71.8 and 74 wt. % Al_2O_3 with high densities of the sintering products, but lower sintering velocities, due to a lower amount of liquid (Fig. 4.2.1).
- Compositions with more than 75 wt. % Al_2O_3 with low densities of the sintering products and low sintering velocities, due to the lack of a coexisting liquid phase ("Solid-state sintering").

Sacks and Pask (1982a) concluded that the primary mechanism of densification in mullite is grain boundary mass transport or diffusion, and that the sintering rates

Fig. 4.2.1 Density (% TD = percentage theoretical density) of mullite powder compacts plotted as a function of the annealing time at different sintering temperatures (the bulk Al_2O_3 content of samples is 73 wt.%, after Sacks and Pask, 1982a, from Schneider et al. 1994).

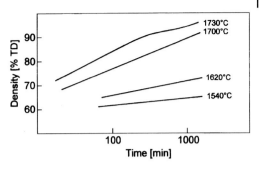

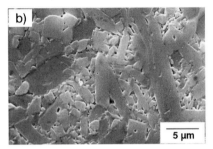

Fig. 4.2.2 Scanning electron micrographs of mullite ceramics fabricated from precalcined precursors. The samples have been sintered at 1700 °C (4 h), and later hot isostatically pressed at 1600 °C and 1200 bar (0.5 h) to near 100% density. Chemical bulk composition: (a) "Al_2O_3-rich": 78 wt.% Al_2O_3 and 22 wt.% SiO_2; (b) "SiO_2-rich": 72 wt.% Al_2O_3 and 28 wt.% SiO_2 (Kanka and Schneider, 1994, from Schneider et al. 1994).

are essentially controlled by the presence or absence of a liquid film. Metcalfe and Sant (1975) established a direct correlation between the mean grain sizes of the mullite powder compacts and the densities. The presence of a liquid silicate phase is associated with the development of microstructures having prismatic mullite grains. Equiaxed grain structures are observed when the glass is eliminated by the increase of the Al_2O_3/SiO_2-ratio of the material (see also Figs. 4.2.2 and 4.2.3).

Sacks and Pask (1982b) evaluated the effect of powder particle agglomerations on the sintering characteristics of mullite compacts. They found that the density achieved depends on the size and packing of the agglomerates, and that increase in the grinding time of the powders causes increased densification rates. The reason is certainly the breakdown of agglomerates producing additional surface areas and smaller effective particle sizes. A homogeneous and uniform distribution of fine particles in the green state of powder compacts is necessary to achieve good sintering values.

The sintering atmosphere can also affect the densification and microstructural development of mullite powder compacts. Mullite decomposes at high temperature in a reducing atmosphere (e.g. Davis et al. 1972, Provost and Le Doussal,

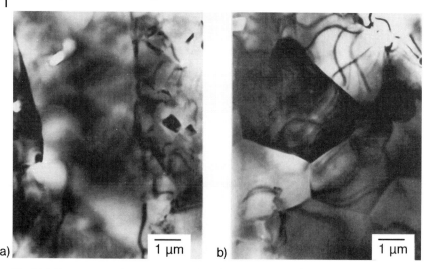

Fig. 4.2.3 Transmission electron micrographs of mullite ceramics fabricated as described in Fig. 4.2.2. a) "SiO$_2$-rich": 72 wt.% Al$_2$O$_3$ and 28 wt.% SiO$_2$, b) "Al$_2$O$_3$-rich": 78 wt.% Al$_2$O$_3$ and 22 wt.% SiO$_2$ (from Schneider et al. 1994)

1974a, 1974b, Sacks, 1979, Sacks et al. 1990; see Section 3.2.1). Sacks (1979) suggested that sintering rates are higher under low oxygen partial pressures owing to enhanced diffusion along the grain boundaries having high defect densities.

A large variety of mullite ceramic fabrication techniques has been described in the literature. Important processing routes are:
- Sintering of mullite powder compacts
- Reaction sintering of Al$_2$O$_3$- and SiO$_2$-containing reactants
- Reaction bonding of mullite
- Reaction sintering of chemically produced mullite precursors
- Transient viscous sintering of composite powders

The details of the different processing routes described below provide generally valid features of mullite processing. However, the cited studies represent a selection only, because it was impossible to refer to the very large number of papers published in this research field.

4.2.2
Sintering of Powder Compacts

Crystalline mullite powders are not easily sintered (Nurishi and Pask, 1982, Rana et al. 1982, Rodrigo and Boch, 1985, Boch et al. 1986a, 1986b). This has been related to the slowness of diffusion in mullite systems. Activation enthalpy calculations on mullite densification and grain growth yielded very high values (about 700 kJ mol^{-1}, Ghate et al. 1975, Dokko et al. 1977, Sacks and Pask 1982a, 1982b) which

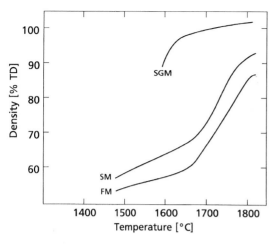

Fig. 4.2.4 Sintering curves of powder compacts fabricated from commercially produced sinter-mullit (SM) and fused-mullite (FM), and from laboratory-synthesized chemical-mullite (SGM), (% TD = percentage theoretical density; from Schneider et al. 1994).

is on the order of the activation energy of Si^{4+} lattice diffusion (702 kJ mol^{-1}, Aksay 1973). Owing to the slow rate of diffusion of aluminum and silicon species, high sintering temperatures (above 1700 °C) are required to achieve full density. One suitable way to improve the sintering behavior of mullite powders is to reduce the mean diffusion distances of aluminum and silicon species by closer packing of mullite particles. Another way is to increase the driving force for sintering densification by using fine powder with high surface areas and with suitable grain size distributions (see, for example, Sacks et al. 1990). In order to avoid residual porosity and exaggerated mullite grain growth in the sintered bodies, soft and hard agglomerates occurring in the starting powders must be broken down. Sacks et al. (1990) and Müller and Schneider (1993a) stated that mullite ceramics with high densities and homogeneous microstructures are obtained with suspension techniques (slip or tape casting) because corresponding green bodies have high densities.

Albers (1993) investigated the sintering behavior of commercial sinter-mullite and fused-mullite powder compacts and found that above about 1500 °C both materials first shrink slowly ($\leq$ 1650 °C) and later ($\geq$ 1650 °C) intensively (Fig. 4.2.4). Solid-state sintering mechanisms were taken into account in both cases. In spite of this assumption sinter-mullite powder compacts exhibit a higher sinter activity than fused-mullites. Albers (1993) explained this by proposing the occurrence of a large number of small "reactive" mullite crystals in individual sinter-mullite powder grains. Fused-mullite grains usually consist of large single crystal fragments (Fig. 4.2.4), and therefore are less "reactive".

A detailed study on the sintering behavior of slip-cast commercial fused-mullite powders has been performed by Müller and Schneider (1993a). By sintering between 1650 and 1800 °C, the pore volume of samples decreases as the pore size gradually increases (Fig. 4.2.5). Suitable microstructures with homogeneous pore size distributions develop at 1700 °C (Fig. 4.2.5b). Above this temperature limit exaggerated growth of large tabular mullite crystals is observed (Fig. 4.2.5c,d), which is likely due to secondary crystallization (see Kingery et al. 1976). The pres-

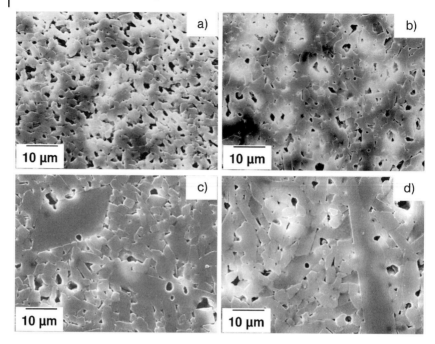

Fig. 4.2.5 Scanning electron micrographs of ceramics fabricated from commercially fused-mullite with the slip-casting technique. Samples were sintered at different temperatures: (a) 1650 °C; (b) 1700 °C; (c) 1750 °C; (d) 1800 °C (holding time: 4 h). Note the occurrence of exaggerated mullite grain growth at temperatures above 1700 °C (from Schneider et al. 1994).

ence of grain boundary glass phase, macropores and of large secondary grains in fused-mullite ceramics sintered at such high temperatures have negative effects on the mechanical properties. Grain boundary glass phases are responsible for viscous grain sliding deformation at high temperature, whereas macropores and large mullite grains may serve as defect nuclei (flaws).

Hot isostatic pressing (HIP) runs performed on pressureless sintered fused-mullite ceramics yielded samples with very homogeneous microstructures with no intergranular pores (Fig. 4.2.6). Transmission electron micrographs indicate the existence of some glass phase, especially at three-grain junctions. However, most mullite grain boundaries exhibit no intergranular glassy film. Schneider et al. (1993c) have shown that hot isostatically pressed materials can be optically translucent (see also Sections 2.3.1 and 4.6.3).

Numerous papers have been published on the sintering behavior of highly reactive mullite powders. Hirata et al. (1990) investigated mullite powders produced by chemical vapor deposition (CVD). Sintering of such powders takes advantage of the extremely fine particle sizes (about 40 to 65 nm). Hirata and coworkers showed that pore size distributions of the green bodies play a major role on densification,

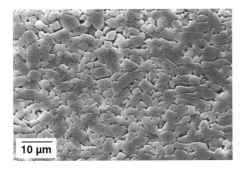

Fig. 4.2.6 Microstructure of a commercially produced fused-mullite ceramic, sintered at 1700 °C (4 h) and hot isostatically pressed at 1650 °C and 1200 bar (0.5 h) to near 100% density (from Schneider et al. 1994).

since smaller pores are more easily eliminated than larger ones. Pore size distributions are affected by the dispersion degree of the colloidal suspension, and by the particle cluster microstructures. Maximum densities are obtained for monosize pore distributions.

Kanzaki et al. (1990) investigated the sintering behavior of spray pyrolyzed mullite (bulk compositions: 60 to 78 wt. % Al_2O_3), and found that densification of compacts with low Al_2O_3 contents is enhanced, owing to the presence of a liquid phase at sintering temperature. Sintered specimens are characterized by prismatic mullite crystals which are embedded in a finer grained matrix containing glass as a secondary phase. The coexisting glass phase is almost absent in Al_2O_3-rich ceramics. This may be the reason for the retarded sintering velocity and the development of an equiaxed mullite crystal morphology (see also Fig. 4.2.2a,b).

Ismail et al. (1990) published data on sintering of sol-gel-derived mullites (bulk compositions: 70 to 75 wt. % Al_2O_3) between 1600 and 1700 °C. Above about 1650 °C the density of all powder compacts became high, although there existed a reciprocal dependence on the Al_2O_3 content. The higher post-sintering density of SiO_2-rich compositions was again explained by liquid phase sintering in a similar way as other authors did (e. g. Kanzaki et al. 1990). A typical sol-gel mullite sintering curve is drawn in Fig. 4.2.4.

Huang et al. (2000) used a high purity, fine-grained mullite powder for the fabrication of dense samples by hot pressing. Nearly stoichiometric mullite powder was hot pressed at 1550 °C to produce an almost fully dense microstructure of fine, nearly uniaxial grains. Samples were annealed in the temperature range 1550 to 1750 °C and mullite grain growth was determined. Grain growth was found to be relatively slow at 1550 °C and the mullite grains remained nearly equiaxed. Rapid anisotropic grain growth, however, occurred by annealing at temperatures above the eutectic temperature (about 1590 °C, see also Section 2.2.5).

4.2.3
Reaction Sintering of Alumina and Silica

A suitable way to avoid the problems involved with densification of mullite powder compacts is reaction sintering of alumina and silica reactants, because it allows a high degree of densification prior to mullitization. A further advantage of reaction sintering is the relatively low processing costs (see, for example, Aksay et al. 1991, Somiya and Hirata, 1991). Reaction sintering of alumina- and silica-containing reactants includes processing of alumina and silica powders, and of clays and refractory-grade bauxite, admixed with alumina and silica, respectively. Temperatures and rates of mullite formation depend on the starting materials. Of special importance are the alumina and silica reactants used, their chemical purity, their particle size and whether the particles are single crystals or poly-crystalline (Nurishi and Pask, 1982, Rana et al. 1982, Rodrigo and Boch, 1985, Rodrigo, 1986, Boch et al. 1990).

Wahl et al. (1961) showed that the mechanisms of mullitization in mixtures of various forms of alumina and silica are influenced by the type of the reactants, while small variations of the bulk chemical composition have less influence. Wahl et al. stated that diaspore combines more readily with silica to form mullite than gibbsite does. Pankratz et al. (1963) described incomplete mullitization of α-alumina and quartz at temperatures between 1500 and 1540 °C. Miller et al. (1966) synthesized a dense mullite material from vacuum hot pressed mixtures of silica glass and α-alumina at about 1650 °C. Another technique was used by Locsei (1963) in reacting aluminum fluoride (AlF_3) and silica in the presence of water. Mullite is thereby formed as the second step of a two-step reaction. The first step consists of the emission of silicon tetrafluoride (SiF_4) and the formation of topaz above about 600 °C. Above about 900 °C crystallization of mullite occurs.

Johnson and Pask (1982) used α-alumina and quartz mixtures, and found that mullite crystals form on α-alumina grains by interdiffusion of aluminum and silicon atoms. Initially the mullite crystals are pseudomorphic after α-alumina grains until the system reaches chemical equilibrium. With subsequent growth by a solution precipitation process, they become rectangular with rounded edges, and elongate further with the duration of heat treatment. Boch et al. (1990) mentioned that the processing parameters during reaction sintering have to be tailored exactly in order to initiate complete mullitization and high densification. Boch and coworkers found that the use of submicron powders sintered at high heating rates produce almost theoretical density and a very high mullite content at 1600 °C. The alumina to silica (Al_2O_3/SiO_2) ratio of the material is a critical parameter. Highest densification rates are achieved for "low" Al_2O_3 contents near to that of 3/2-mullite (about 72 wt. % Al_2O_3), whereas "Al_2O_3-rich" compositions (>75 wt. % Al_2O_3) exhibit poor densification and frequently exaggerated mullite grain growth.

Saruhan et al. (1994), Müller and Schneider (1993b) and Schmücker et al. (1993) reinvestigated temperature-dependent mullite formation processes using α-alumina plus quartz, α-alumina plus cristobalite and α-alumina plus silica glass reaction couples. They found that mullite formation is a multiple-step process with more or less extended regions of mullite nucleation (stage I), regions of high mullitization rates (stage II), a temperature field of low mullitization (stage III) and finally an area of high mullite formation rates (stage IV, Fig. 4.2.7). In spite of these similarities, significant differences of the mullite formation mechanisms can be determined for each of the alumina plus silica reaction systems.

Reaction of quartz plus α-alumina systems starts early (<1100 °C) with periph-

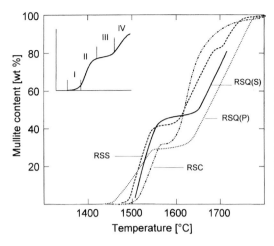

Fig. 4.2.7 Mullite formation by reaction sintering of alumina plus silica. RSQ: Quartz + α-alumina, RSC: Cristobalite + α-alumina, RSS: Silica glass + α-alumina. Reaction stages I, II, III and IV refer to those described in the text. RSQ (P), RSC, and RSS values are from pressed compacts, RSQ (S) values are from a slip-cast specimen (from Schneider et al. 1994).

Quartz + α-Al₂O₃

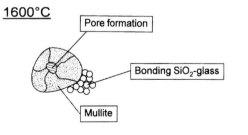

Fig. 4.2.8 Simplified mechanisms of reaction sintering steps of quartz + α-alumina (RSQ, see Fig. 4.2.7) at different temperatures. Details are given in the text (from Schneider et al. 1994).

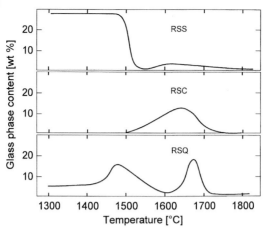

Fig. 4.2.9 Glass phase formation by reaction sintering of alumina plus silica. RSQ: Quartz + α-alumina, RSC: Cristobalite + α-alumina, RSS: SiO$_2$ glass + α-alumina. Values are calculated by subtracting the total of crystalline phase contents from 100% (from Schneider et al. 1994).

eral melting of quartz grains (Fig. 4.2.8). Partial melting of quartz (Fig. 4.2.9) is probably the result of diffusion of aluminum into quartz. Formation of a sub-solidus liquid at that low temperature can be explained with the existence of a metastable Al$_2$O$_3$–SiO$_2$ phase diagram without mullite, in which the liquidus curves originating from pure alumina and silica extend below 1100 °C (Davis and Pask, 1972, Risbud and Pask, 1978, Nurishi and Pask, 1982). The presence of impurities, especially alkali oxides shifts the pseudo-eutectic point even to lower temperatures. Impurities occurring in the quartz raw materials also have a controlling influence on the amount and viscosity of the liquid phase.

Mullite formation curves display an induction or nucleation period (between 1450 and 1470 °C, stage I), followed by a steep branch of the curve with high reaction rates (1470 to 1530 °C, stage II). Schneider et al. (1994) suggested that the latter is to be explained by a liquid-phase-assisted reaction, taking place by penetration of the SiO$_2$-rich melt into the α-alumina grain aggregates (Fig. 4.2.8, see also Rana et al. 1982). With increasing temperature (1530 to 1580 °C, stage III) mullite formation rates decrease rapidly. The reason for this is the crystallization of the silica glass to less reactive cristobalite. Furthermore, gradually growing mullite reaction interfaces at the points of contact of silica and α-alumina compounds may act as diffusion barriers. Finally at high temperature (>1650 °C, stage IV) mullitization rates increase again. Probably a rapid liquid phase-controlled mullite crystallization takes place (Figs. 4.2.7 and 4.2.8), which is enabled by the melting of residual cristobalite and dissolution of α-alumina (Fig. 4.2.9). The comparison of mullite formation curves demonstrates that mullitization significantly depends on the grain packing in the green body. Since green densities and grain size distributions are more favorable in slip-cast bodies, their mullite formation curve is shifted towards lower temperatures with respect to that of pressed specimens (Fig. 4.2.7).

The temperature-dependent mullitization of α-alumina plus cristobalite systems differs significantly from that of α-alumina plus quartz. Since the material contains small amounts of impurities, transient glasses form, which enable early sin-

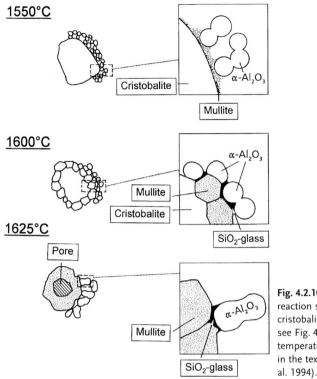

Cristobalite + α-Al$_2$O$_3$

1550°C

Cristobalite

α-Al$_2$O$_3$

Mullite

1600°C

α-Al$_2$O$_3$

Mullite

Cristobalite

1625°C

SiO$_2$-glass

Pore

Mullite

α-Al$_2$O$_3$

SiO$_2$-glass

Fig. 4.2.10 Mechanisms of reaction sintering steps of cristobalite + α-alumina (RSC, see Fig. 4.2.7) at different temperatures. Details are given in the text (from Schneider et al. 1994).

tering. Initial mullitization is observed at about 1470 °C. Mullite formation rates are low at first, but become faster (between 1470 and 1570 °C, stages I and II, Fig. 4.2.7). The rate-controlling step of mullitization is certainly a solid-state reaction at the point of contact of α-alumina particles and silica grains (Schneider et al. 1994, Fig. 4.2.10). With increasing duration of the process mullite areas between alumina and silica gradually become thicker, causing longer diffusion paths for aluminum and silicon species and an associated slow down of mullite formation (1580 to 1600 °C, stage III). Finally, at higher temperatures (>1600 °C, stage IV) melting of the reactants enables rapid liquid-phase-assisted mullitization (Figs. 4.2.7, 4.2.9 and 4.2.10).

Mullite formation in α-alumina plus silica glass in many terms resembles that of α-alumina plus cristobalite. Schneider et al. (1994) suggested that prior to mullitization an almost complete crystallization of the silica glass to cristobalite takes place. This would mean that mullite mainly develops by solid-state reaction between α-alumina and cristobalite in the low-temperature branch (stages I and II) and not between α-alumina and silica glass (see also Boch et al. 1990). Mullite formation rates of α-alumina plus silica glass reaction couples are higher in region II than those of α-alumina plus cristobalite reaction couples, probably because of

SiO$_2$ Glass + α-Al$_2$O$_3$

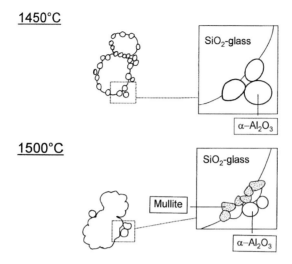

__1450°C__

SiO$_2$-glass

α–Al$_2$O$_3$

__1500°C__

SiO$_2$-glass

Mullite

α–Al$_2$O$_3$

__1600°C__

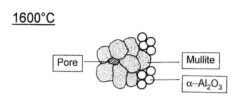

Pore

Mullite

α–Al$_2$O$_3$

Fig. 4.2.11 Mechanisms of reaction sintering steps of SiO$_2$ glass + α-alumina (RSS, see Fig. 4.2.7) at different temperatures. Details are given in the text (from Schneider et al. 1994).

an intense pre-mullite viscous flow densification of powder compacts. The dense particle packing shortens diffusion paths and, hence, accelerates mullitization (Figs. 4.2.7, 4.2.9 and 4.2.11).

Sintering curves of α-alumina plus quartz, α-alumina plus cristobalite and α-alumina plus silica glass reaction couples are shown in Fig. 4.2.12 (Schneider et al. 1994, Müller and Schneider, 1993b). α-alumina plus quartz compacts display intense shrinkage between about 1000 and 1400 °C, which was explained by transient liquid-phase densification along the liquid boundary layers around quartz. Above about 1520 °C, shrinkage slows down to nearly zero, probably owing to a mullitization-related volume increase working against shrinkage. Above about 1650 °C densification increases again by liquid-phase sintering. It is interesting to note that almost total final densification is achieved prior to mullite formation. Slip-cast α-alumina plus quartz systems (RSQ(S)) display much stronger sintering densifications than pressed green bodies (RSQ(P)). The reason may be a denser packing of particles and favorable particle and pore size distributions of the former.

Weak shrinkage is observed in α-alumina plus cristobalite systems between about 1100 and 1450 °C enabled by liquid-phase-assisted sintering due to the presence of a small amount of silica glass. Above about 1450 and up to 1520 °C shrink-

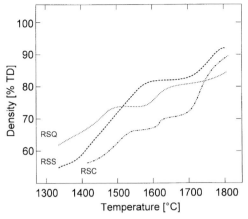

Fig. 4.2.12 Reaction sintering curves of alumina plus silica. RSQ: Quartz + α-alumina, RSC: Cristobalite + α-alumina, RSS: SiO₂ glass + α-alumina. (% TD = percentage theoretical density), (from Schneider et al. 1994).

age rates become higher, probably owing to solid-state diffusion (Fig. 4.2.12). Above about 1520 °C, shrinkage rates decrease for similar reasons to those quoted for the α-alumina plus quartz reaction couples. Above about 1650 °C samples again shrink rapidly owing to liquid-phase sintering enabled by fusion of cristobalite (see Fig. 4.2.9). Contrary to relationships in α-alumina plus quartz reaction couples a much lower densification is achieved prior to mullitization.

The densification of α-alumina plus silica glass is already high at low temperatures (Fig. 4.2.12). A densification mechanism involving viscous flow of soft silica glass grains may be the densifying step. Since the silica glass phase content drops between about 1450 and 1520 °C by transformation into cristobalite (see Fig. 4.2.9), the further shrinkage behavior of samples resembles that of the α-alumina plus other silica systems.

An important result of the reaction sintering studies of Schneider et al. (1994) and Müller and Schneider (1993b) is that the largest proportion of the final density is achieved prior to mullitization, if a liquid-phase or liquid-phase-assisted sintering process is involved at relatively low temperature (≤ 1500 °C, α-alumina plus quartz, α-alumina plus silica glass). If solid-state diffusion is the major sintering mechanism (α-alumina plus cristobalite) only part of the densification occurs prior to mullitization, while a further part takes place together with mullite formation. Therefore, reaction sintering of α-alumina plus cristobalite often produces a lower densification than that of α-alumina plus quartz or α-alumina plus silica glass systems (Fig. 4.2.12). The results of reaction sintering of alumina and silica compounds lead to the conclusion that sintering behavior, reaction processes, microstructural development and associated properties of mullite ceramics strongly depend on the alumina and silica sources. Moreover, powder processing routes and sintering procedures must be considered.

4.2.4
Reaction Bonding of Different Starting Materials

A characteristic feature of reaction-bonded ceramics such as silicon nitride (Si_3N_4), silicon carbide (SiC) and alumina is low shrinkage during fabrication (e.g. Haggerty and Chiang, 1990, Wu et al. 1993), which has been designated as "near-net-shape" processing. The low shrinkage characteristics refer to the volume increase during nitridation, carbonization or oxidation which works against sintering-induced shrinkage. Mullite ceramics fabricated by reaction bonding can start from mechanically alloyed aluminum, silicon carbide and alumina powder mixtures (Wu and Claussen, 1991). Wu and Claussen pointed out that fabrication of these reaction-bonded mullites involve a two-step process: During the first step at 1200 °C, aluminum and silicon carbide particles are oxidized to form alumina and silica. In a second step at 1550 °C sintering occurs. Reaction-bonded mullite ceramics consisting of mullite plus α-alumina have outstanding mechanical bending strength (>500 MPa; Claussen, personal communication) at high sintering densities (>97% theoretical density). This, however, is not an inherent mullite property, but is due to zirconia (ZrO_2) impurities introduced in the material by milling.

Mechnich et al. (1998) processed mullite ceramics and mullite-matrix composites with the reaction-bonding technique. They used mullite-seeded silicon metal and α-alumina precursors without and with low amounts of yttria (Y_2O_3) and ceria (CeO_2). They showed that improved microstructural homogeneity and better mechanical properties can be achieved with these powder mixtures than with aluminum metal or aluminum-silicon alloys plus silica or silicon carbide. Small amounts of yttria and ceria accelerated silicon oxidation and mullite formation through transient and metastable melting, and yielded homogeneous, glass-free and relatively dense ceramics after long-term heat treatment at 1350 °C instead of >1550° in normal reaction-bonded mullite systems. Mechnich et al. (1999) found that the amount of ceria added to the starting powders must be tailored carefully. Exaggerated ceria contents produce large amounts of low viscosity CeO_2–Al_2O_3–SiO_2 liquids which may have the disadvantage of sealing the open porosity. This slows the oxygen diffusion into the specimen considerably, with the consequence that non-oxidized silicon and a residual CeO_2–Al_2O_3–SiO_2 glass coexist in the ceramics after processing. A solution to this problem is to simultaneously enhance mullite crystal growth through seeding, which works against excessive liquid-phase-induced shrinkage of the samples. This in turn enables complete oxidation and recrystallization of all liquid phases. Ceria-doped reaction-bonded mullite ceramics have microstructures in which the micrometer-scale mullite crystals are decorated by nanometer-scale ceria precipitates (Fig. 4.2.13). It has been suggested that such microstructures significantly reduce grain growth by hindering the mobility of grain boundaries.

Suttor et al. (1997) synthesized monolithic mullite with low sintering shrinkage from siloxane plus alumina and siloxane plus aluminum mixtures. The synthesis was based on the reaction-bonding process of amorphous silica, which forms when siloxane is oxidized, with alumina filler at temperatures >1250 °C.

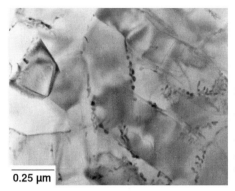

Fig. 4.2.13 Transmission electron micrograph of ceria- (CeO_2)-doped and seeded reaction-bonded mullite (RBM) after heat treatment of an silicon metal-α-alumina mixture at 1350 °C (5 h). Note that the transiently formed glass has completely crystallized and that nanosized ceria crystallites decorate the mullite grains. (Courtesy P. Mechnich).

0.25 μm

4.2.5
Reaction Sintering of Chemically Produced Mullite Precursors

The starting materials are mullite precursor powders prepared from organic and/ or inorganic sources, or gel pieces or layers. It has been shown in many studies that the temperature for preparing mullite ceramics by reaction sintering depends on the particle size and on the scale of mixing of the aluminum- and silicon-bearing reactants. Conventional powder metallurgically processed mullite ceramics require very high temperatures for reaction sintering (>1700 °C). However, lower temperatures (<1400 °C) are necessary if chemical routes with colloidal or sol-gel mullite fabrication techniques are applied. Another major advantage of the chemical route is the production of highly pure and homogeneous materials (e. g. Sacks et al. 1990). The basic principles of reaction sintering of mullite precursors synthesized by coprecipitation of inorganic starting compound will be discussed below.

Mazdiyasni and Brown (1972) prepared mullite powders by the hydrolytic decomposition of mixed metal alkoxides, i.e. aluminum isopropoxide and silicon isopropoxide. They obtained highly dense translucent polycrystalline bodies of stoichiometric mullite by vacuum hot pressing of high purity submicron mixed oxide powders at 1500 °C. Prochazka and Klug (1983) used a sol-gel method in which aluminum isopropoxide and ethylmetasilicate were dissolved in a six-fold volume of cyclohexane and hydrolyzed in a blender with the theoretical amount of water to obtain silica and aluminum hydroxide ($Al(OH)_3$). Mullite powders have been prepared between 1200 and 1400 °C, wet-milled and then hot pressed at 1630 °C in order to achieve infrared transparent mullite ceramics. Yoldas (1980) produced monolithic materials of varying aluminum to silicon ratios from silicon tetraethoxide, ($Si(OC_2H_5)_4$) and aluminum *sec*-butoxide, ($Al(OC_4H_9)_4$) by first hydrolyzing the aluminum butoxide to aluminum oxide hydroxide (AlOOH) and then condensing it with unhydrolyzed or partially hydrolyzed silicon ethoxide. Dense microstructures were obtained with 20 % SiO_2 to 80 % Al_2O_3 and 35 % SiO_2 to 65 % Al_2O_3 compositions with these precursors. Although the main objective of this investigation was to study the effect of heat treatment on optical transparency,

Yoldas apparently obtained high densities in some compositions, which are close to stoichiometric mullite, by unwittingly using a diphasic route.

Hoffman et al. (1984) introduced diphasic xerogels as a new class of materials with mullite as an example. These diphasic xerogels are compositionally different nanocomposites according to our classification (Komarneni, 1992). The use of diphasic gels of boehmite and silica sol was a breakthrough in achieving high densities of stoichiometric mullite (Komarneni et al. 1986b, Roy and Komarneni, 1989, Komarneni and Roy, 1990, Komarneni and Rani, 1992) at temperatures as low as 1200 °C because the two components are mixed on a nanoscale in the gel (Komarneni et al. 1986b). The high densification of these gels was attributed to excess free energy from the heat of reaction as well as to processes of densification and crystallization occurring almost simultaneously (Roy and Komarneni, 1989, Komarneni and Roy, 1990). Subsequent to the work of Hoffman et al. (1984) and Komarneni et al. (1986b) for which a U.S. patent was issued in 1989 (Roy and Komarneni, 1989), there has been an intensive effort to exploit and explain the diphasic gel route in the processing of highly dense mullite (Ismail et al. 1986, 1987, Wei and Halloran, 1988a, 1988b, Sundaresan and Aksay, 1991, Li and Thompson, 1990, 1991). Ismail et al. (1986, 1987) used a boehmite sol formed from peptization of γ-alumina and a silica sol to obtain a diphasic gel which was calcined at 1400 °C to obtain mullite. This mullite was milled to obtain an average particle size of 1.3 μm, which was then sintered at 1650 °C to achieve 98% theoretical density. A somewhat similar sol-gel procedure was used by Ghate et al. (1973) to prepare high purity, fine grained mullite by mixing a fumed alumina powder consisting predominantly of γ-alumina with an ammonia-stabilized aqueous dispersion of colloidal silica. However, no densification studies of this mullite powder were reported by these authors. Debely et al. (1985) used a heterocoagulation technique with a fine-grained alumina powder and colloidal silica sol and obtained relative densities of 83 and 98% for samples sintered at 1200 °C and at 1400 °C, respectively. Full density was only achieved at 1550 °C.

High purity stoichiometric mullite powders were prepared by calcination at 1300 °C of the products precipitated from an aqueous solution of aluminum nitrate and colloidal silica and sintered at 1650 °C but did not achieve full density (Kubota and Takagi, 1987). Hirata et al. (1985, 1989) prepared mullite powders from aluminum propoxide and silicon ethoxide but were unable to achieve good densification below 1600 °C. Mizuno and Saito (1989) prepared mullite powders from four different alumina sources and with fumed silica and found that aluminum sulfate and fumed silica give the best powders. They could achieve about 97.5% densification after sintering at 1650 °C. Sonuparlak (1988) prepared an infrared transparent, dense mullite at 1250 °C by controlling key sol-gel processing parameters such as molecular structure of the gel and packing of particles.

Huling and Messing (1989, 1990) proposed hybrid gels consisting of diphasic and single phase (or polymeric) gels for microstructure control as well as a small reduction in mullite crystallization temperature. They also used mullite crystalline seeds in diphasic gels and showed a 30 °C reduction in crystallization temperature by differential thermal analysis (DTA), (Huling and Messing, 1991). Komarneni

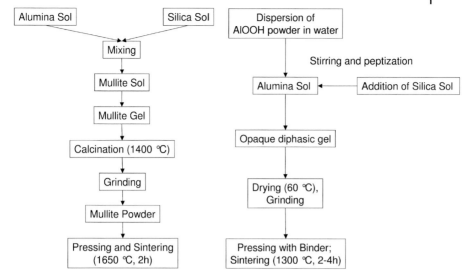

Fig. 4.2.14 Flowchart for mullite fabrication developed by Chichibu Cement Company.

Fig. 4.2.15 Flowchart for mullite synthesis as presented by Hoffman et al. (1984).

and Roy (1990) used mullite, anatase and kaolinite seeds in two types of diphasic and one single phase gel but did not find increased densification. The diphasic gels, when seeded with mullite or other crystals, can be construed as both compositionally and structurally different nanocomposites as per the earlier classification (Komarneni, 1992). Mroz and Laughner (1989) used the both compositionally and structurally different nanocomposites of mullite and showed different microstructures with different seed concentrations.

Although there are numerous other studies (Sacks et al. 1991, Sporn and Schmidt, 1989, Ismail et al. 1990, Sacks et al. 1990, Schneider et al. 1994), the conceptual innovation of diphasic mullite gels (Hoffman et al. 1984, Komarneni et al. 1986a, Roy and Komarneni, 1989) is the key for the commercialization of mullite powders by the Chichibu Cement Company at the rate of five tons per month (Somiya and Hirata, 1991). The flow chart for mullite powder preparation by the Chichibu Cement Company is given in Fig. 4.2.14, while the flow chart for diphasic mullite gel powder by Roy, Komarneni and others (Hoffman et al. 1984, Komarneni et al. 1986b, Roy and Komarneni, 1989) is given in Fig. 4.2.15. The Chichibu Cement Company produces crystalline mullite powder via the diphasic route at 1400 °C. These mullite powders have been used for fabricating various structural ceramics including ceramic belts for high-temperature sintering applications. Komarneni et al. (1986b), however, showed that the calcined diphasic gel can be directly used in the fabrication of structural mullite ceramics because they intensively densify upon sintering between 1300 and 1400 °C (Fig. 4.2.16) because of simultaneous densification and crystallization (Pach et al. 1995, 1996).

Kanka and Schneider (1994) processed non-crystalline mullite precursors produced by coprecipitation of sodium aluminate and silica sols with two bulk com-

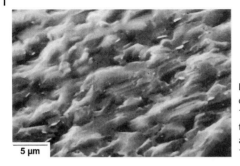

5 µm

Fig. 4.2.16 Mullite ceramic directly produced from diphasic gels after sintering at 1300°C. Note the high degree of densification because of simultaneous sintering and crystallization (see Komarneni et al. 1986a and Pach et al. 1995, 1996).

10 nm

Fig. 4.2.17 High-resolution transmission electron micrograph of a typical grain boundary between mullite crystals in an Al_2O_3-rich chemical-mullite produced from coprecipitated precursors (bulk composition: 78 wt.% Al_2O_3 and 22 wt.% SiO_2, see Figs. 4.2.2 and 4.2.3). Note that no grain boundary glass phase is observed (from Schneider et al. 1994).

positions ("Al_2O_3-rich": 78 wt. % Al_2O_3, 22 wt. % SiO_2 and "SiO_2-rich": 72 wt. % Al_2O_3, 28 wt. % SiO_2) by pressureless sintering and hot isostatic pressing (HIP) techniques. Microstructural development and sintering mechanisms of samples were found to be dependent on the Al_2O_3/SiO_2 ratios of the starting powders, the calcination temperatures, sintering temperatures and hot isostatic pressing conditions. Kanka and Schneider (1994) pointed out that at early stages of reaction sintering (between about 900 and 1200 °C) the presence or absence of a liquid phase with high SiO_2 contents has a decisive influence on sintering. It has been suggested that, especially at later stages of reaction sintering (above about 1200 °C), liquid-phase sintering of high SiO_2 compositions produces large prismatic mullite crystals embedded in a fine-grained matrix. High Al_2O_3 compositions develop an equiaxed mullite microstructure with some additionally occurring α-alumina grains as a result of solid-state sintering (Figs. 4.2.2 and 4.2.3) but without any glass grain boundary phase (Fig. 4.2.17).

The different mullite grain growth modes in mullite ceramics with high Al_2O_3 or SiO_2 in turn directly influence the densification process in heat treatments with longer durations. Reaction sintering of high SiO_2 compositions produces a stiff skeleton of interlinked elongated mullite crystals. By increasing the sintering temperatures, or after holding longer at high temperature, the fine grained matrix between the prismatic mullite network may further sinter, producing dense areas with a small number of relatively large pores. However, the stiff mullite skeleton does not allow any further bulk volume shrinkage of sample specimens, thus

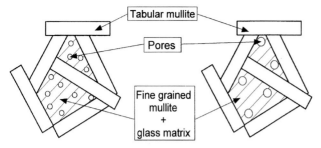

Fig. 4.2.18 Progress in sintering of SiO₂-rich mullite ceramics, fabricated as described in Fig. 4.2.2 (72 wt. % Al₂O₃ and 28 wt. % SiO₂). The microstructure consists of tabular mullite crystals embedded in a fine-grained matrix. In an early sintering step the porosity is relatively large and a great number of pores occurs (left). In a later sintering step the number of pores is lower although individual pore volumes increase (right), (from Schneider et al. 1994)

limiting the overall densification (Fig. 4.2.18). Reaction sintering of high Al₂O₃ compositions produces microstructures consisting of fine grained equiaxed mullite plus α-alumina. Other investigations yielded similar high final densities of high Al₂O₃ and SiO₂ compositions.

The composition-controlled microstructural development of mullite ceramics has a direct influence on their mechanical properties. The interlinked network of prismatic mullite crystals occurring in high SiO₂ compositions produces favorable mechanical strength by a kind of mullite-mullite (self-)reinforcement up to temperatures of about 1200 °C. The equiaxed microstructure of high Al₂O₃ compositions, on the other hand, may explain why these materials have slightly lower mechanical strengths at room temperature than high SiO₂ materials.

The study of Kanka and Schneider (1994) supports the reciprocal relationship between sintering densification and Al₂O₃ content discussed by Kanzaki et al. (1990), Ismail et al. (1990) and Sacks et al. (1990) if precursor powders calcined at high temperature (above about 1500 °C) prior to sintering are used. Sintering densities of such samples are significantly higher for high SiO₂ (about 96% of the theoretical density) than for high Al₂O₃ compositions (about 85% of the theoretical density) at a sintering temperature of 1700 °C. Low pre-sintering calcination temperatures (below 1100 °C) cause reversed relationships, with high Al₂O₃ materials producing higher sintering densification (about 98% of the theoretical density) than high SiO₂ compositions (about 94% of the theoretical density; sintering temperature in both cases: 1700 °C). This discrepancy is explained by the prefixing of the microstructural development during calcination. High SiO₂ compositions calcined at relatively low temperatures produce elongated mullite crystals with high aspect ratios, while high Al₂O₃ materials form equiaxed crystals. During sintering the mullite crystals grow further but certainly without a basic change of their contours. The densification of high SiO₂ compositions is thus limited by the formation of stiff mullite skeletons, whereas high Al_2O_3 materials show increasing temperature-dependent sintering densification.

Hot isostatic pressure densification experiments on the above described pressureless pre-sintered mullite ceramics have been performed by Kanka and Schneider (1994). The microstructures of the hot isostatically pressed samples are very similar to the pressureless pre-sintered materials but show nearly no intragranular pores (Figs. 4.2.2 and 4.2.3). Densification curves of hot isostatically pressed high Al_2O_3 and SiO_2 mullite compositions are plotted in Fig. 4.2.19. The relative densification of the SiO_2-rich material is stronger and approaches more rapidly theoretical density (100%) than high Al_2O_3 compositions. Furthermore, its shrinkage curve is shifted towards lower temperature with respect to the high Al_2O_3 curve. Two main reasons may be responsible for the different densification curves of high Al_2O_3 and SiO_2 materials. High SiO_2 materials have lower densities prior to hot isostatic pressing than high Al_2O_3 materials. Consequently, high SiO_2 sample specimens can be densified more easily than high Al_2O_3 compositions. Furthermore, SiO_2-rich mullite materials contain significantly larger amounts of glass phase during sintering than Al_2O_3-rich compositions. Therefore, pressure-aided viscous flow is a main driving force for rapid densification in the case of high SiO_2 compositions during hot isostatic pressing, whereas slower solid-state diffusion-controlled sintering is a main factor for the latter.

A correlation between the microstructural development of ceramics and bulk Al_2O_3 content (72 to 78 wt. %) of heat-treated mullite precursor compacts has been established by Kanka and Schneider (1994). They produced mullite ceramics via a

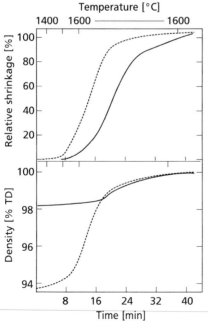

Fig. 4.2.19 Density (% TD = percentage theoretical density, below) and relative shrinkage curves (above) of mullite ceramics fabricated from coprecipitated precursors calcined at 1100 °C during hot isostatic pressing (HIP, 1600 °C, 1200 bar, 0.5 h). Powder compacts were sintered (1700 °C, 4 h) prior to hot isostatic pressing. Full lines: High Al_2O_3 material: 78 wt. % Al_2O_3 and 22 wt. % SiO_2, broken lines: High SiO_2 material: 72 wt. % Al_2O_3 and 28 wt. % SiO_2 (from Schneider et al. 1994, after Kanka and Schneider, 1994).

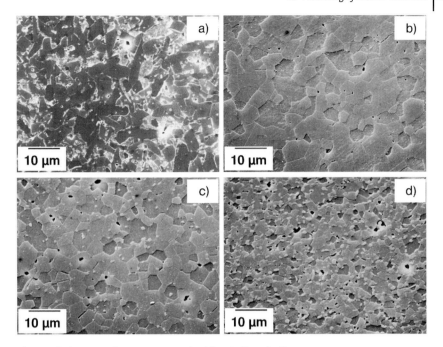

Fig. 4.2.20 Scanning electron micrographs (chemically etched) of chemical-mullite ceramics sintered at 1750 °C (2 h). Bulk compositions of sample are (a) 72 wt.% Al_2O_3 and 28 wt.% SiO_2; (b) 74 wt.% Al_2O_3 and 26 wt.% SiO_2; (c) 76 wt.% Al_2O_3 and 24 wt.% SiO_2; (d) 78 wt.% Al_2O_3 and 22 wt.% SiO_2. Samples (c) and (d) contain some additional α-alumina (light color) (from Schneider et al. 1994).

sol-gel process using tetraethyloxysilane (TEOS) and aluminum *sec*-butylate as starting compounds. Kanka and Schneider again pointed out that the grain morphology of the sintered samples strongly depends on the bulk composition of the materials. Compositions with 72 wt.% Al_2O_3 exhibit elongated mullite grain morphology. As the bulk Al_2O_3 content of the sample specimens increases the mullite grains gradually become smaller and equiaxed (Fig. 4.2.20). Schneider et al. (1994) believe that the change of the microstructure with the Al_2O_3 content is controlled by the amount of glass phase formed during sintering. The glass phase content during sintering is supposed to be relatively high in the 72 wt.% Al_2O_3 composition, but gradually becomes lower with increasing Al_2O_3 content. Therefore the amount of glass phase present in the ceramics during but not necessarily after sintering may determine whether liquid-phase or solid-state sintering is the effective densification mechanism or a mixture of both. If liquid-state sintering is predominant, large prismatic mullite grains are produced (Fig. 4.2.20a), whereas solid-state sintering develops smaller equiaxed mullite grains (Fig. 4.2.20d; see also Fig. 4.2.2a,b).

Osendi and Miranzo (1997) studied the thermal evolution of a mullite gel with a bulk composition $2Al_2O_3 \cdot SiO_2$, prepared from a commercial precursor. The densification of gel powders was investigated up to temperatures of 1780 °C. The microstructure of dense materials always showed the presence of residual Al_2O_3 grains. Hong and Messing (1998) investigated densification and anisotropic grain growth in diphasic gel-derived, titania-doped mullite. The densification in diphasic mullite gels increases with the titania content by reducing the viscosity of the coexisting glass. The onset temperature for anisotropic mullite grain growth decreases with increasing titania concentration because the sintering temperature for final-stage densification decreases. With a 5 wt. % titania-doped mullite, the lowest onset temperature for anisotropic grain growth is about 1500 °C. Hong and Messing (1999) prepared highly textured mullite ceramics by enhancing anisotropic grain growth by titania doping and by templating grain growth on oriented acicular mullite seed particles in a diphasic mullite gel precursor. The mullite precursor crystallizes and densifies to an equiaxed microstructure of 1 to 2 μm mullite grains initially, but the mullite seeds grow rapidly in the length direction upon further heating to produce a highly textured microstructure. A range of oriented microstructures and anisotropic grains were produced by using different concentrations of mullite seeds.

Kamiya et al. (1998) prepared ultrafine mullite precursor powders by co-polymerization of alkoxides, calcined in the range from 800 to 1200 °C and consolidated by ultra-high cold isostatic pressing up to 1 GPa. The maximum density of the green compacts reaches 70 % of theoretical and the green bodies can be sintered to more than 95 % of theoretical at 1500 °C. This sintering below the liquid formation temperature leads to a fine microstructure of the mullite ceramic with a grain size below 300 nm. Kara et al. (2000) prepared mullite ceramics by reaction sintering of pseudo, boehmite, colloidal silica and aluminum sulfate plus colloidal silica powder mixtures. They found by transmission electron microscopy that these mullite ceramics are free from glassy phases at triple junctions and grain boundaries. The wetting of grain boundaries by glass in the mullite ceramics is either due to incomplete reaction between alumina and silica components or to release of silica from the mullite structure with increasing temperature, and thus depends on the prior thermal history of the ceramics.

4.2.6
Transient Viscous Sintering of Composite Powders

An innovative mullite fabrication technique, transient viscous sintering (TVS), has been introduced by Sacks et al. (1991). Sacks and coworkers obtained highly densified mullite ceramics with controlled microstructures by sintering of powder compacts consisting of α-alumina particles coated with amorphous SiO_2 layers in a temperature range between about 1100 and 1300 °C. Between about 1500 and 1700 °C the material is converted to mullite by reaction of alumina and silica. The samples remain almost fully dense after mullitization and exhibit an equiaxed fine-grained microstructure. The excellent densification behavior of transient viscous sintered mullite is due to viscous flow of amorphous silica prior to cristobalite formation and mullitization. Sacks et al. (1991) stated that the main advantages of transiently viscous sintered mullite in comparison to sol-gel-processed mullite ceramics are lower weight loss and shrinkage during drying and sintering of powder compacts. Transient viscous sintered mullite compacts are also suitable for shape forming at sintering temperatures by pressure-assisted viscous deformation (Xue and Chen, 1991). Dense mullite ceramics have also been produced via transient viscous sintering from silica-coated γ-alumina particle nanocomposites (Figs. 1.4.11 and 4.1.11, Bartsch et al. 1999). This method reduces the processing temperatures as much as 300 °C (from about 1600 to 1300 °C) with respect to silica-coated α-alumina particle microcomposites and to other alumina-silica reaction couples. The excellent densification behavior and the low mullite formation temperature make these nanocomposites a superior raw material for mullite ceramics and composites. The process has also been used to produce zirconia- ($ZrO_{2,P}$/mullite), α-alumina- ($Al_2O_{3,P}$/mullite) and silicon carbide-reinforced mullite matrix composites (SiC_P/mullite, see Sections 7.1 and 7.3).

The effect of the green density of powder compacts on crystallization and mullitization of transiently viscous sintered mullite using uniaxial die pressing and rotary forging at room temperature was investigated by Wang et al. (1992). Wang and coworkers showed that the highly compacted green specimens displayed lower silica glass-cristobalite transformation temperatures. The highly densified α-alumina particle/silica glass layer compacts also show an increased degree of mullitization. This was related to an increased concentration of mullite nuclei and to improved contact areas between cristobalite and α-alumina crystals.

4.3
Mechanical Properties of Mullite Ceramics
K. Okada and H. Schneider

Mullite displays moderate mechanical strength and rather low fracture toughness at room temperature (Fig. 4.3.1). Its mechanical properties are thus inferior to those of non-oxide ceramics such as silicon nitride (Si_3N_4) and silicon carbide (SiC), and of oxide ceramics such as α-alumina (Al_2O_3) and zirconia (ZrO_2). However, in comparison to these materials, mullite shows only little degradation of mechanical properties at high temperatures. In this section, the mechanical properties of mullite ceramics are evaluated at room temperature and at high temperature with the focus on (1) mechanical strength and fracture toughness, (2) elastic modulus, (3) hardness, (4) thermal shock resistance, (5) wear resistance, (6) fatigue behavior, and (7) creep resistance. A discussion of the crystal structure control of mechanical properties of mullite is given in Section 2.1.

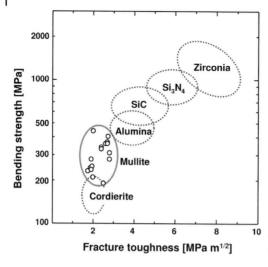

Fig. 4.3.1 Correlation between bending strength and fracture toughness of mullite in comparison to other advanced ceramics.

4.3.1
Mechanical Strength and Fracture Toughness

Mullite is the only stable compound in the system Al_2O_3–SiO_2 from medium up to high temperature at ambient pressure. It has therefore become a main constituent of many ceramics. Conventional mullite-bearing ceramics, such as porcelain, pottery and many refractories, contain a more or less high amount of glass phase, which causes low thermomechanical stability of the ceramics at elevated temperature. On the other hand, high purity mullite ceramics as first processed by Dokko et al. (1977) display very good creep resistance. Starting from the pioneering work of Dokko et al. mullite has emerged as an attractive engineering material for use at high temperatures in air. Kanzaki et al. (1985) measured bending strengths of mullite ceramics as high as 360 MPa and fracture toughnesses of 2.8 MPa $m^{0.5}$ at room temperature. The bending strength of these mullite ceramics decreases only slightly up to 1400 °C, while other advanced ceramics such as silicon carbide, silicon nitride, alumina and zirconia display a rather strong reduction of mechanical properties at elevated temperature, especially in air atmosphere.

4.3.1.1   Mechanical Strength and Fracture Toughness at Room Temperature

The high mechanical strength (360 MPa) of mullite ceramics published by Kanzaki et al. (1985) can be explained by spray pyrolysis-derived alkoxides used for the processing. Ismail et al. (1987), Sivakumar et al. (2001) and Ohira et al. (1996) reported even higher bending strengths (400 to 440 MPa) starting from commercial high purity sol-gel mullite powders. The same is true for Itoh et al. (1990) who published bending strengths of 415 MPa for mullite prepared from a mixture of kaolinite and aluminum hydroxide. Keys to the achievement of good mechanical properties are the particle size and the homogeneity of precursors, which can be

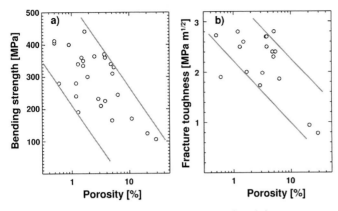

Fig. 4.3.2 Dependence between (a) Bending strength and (b) Fracture toughness of mullite ceramics plotted as a function of porosity.

achieved by long term wet ball milling. There is less dependence on whether synthetic or natural raw materials are used, although differences may occur in the mechanical properties at high temperatures owing to the presence of impurities in the latter. Bending strength and fracture toughness of mullite ceramics are plotted as a function of porosity in Fig. 4.3.2. Although the data are fairly scattered, a tendency towards increasing strength and toughness with the decrease of porosity is observed.

Much work has been performed in order to understand the relationships between mechanical properties and microstructural parameters of ceramics, such as grain size and porosity. Yamade et al. (1990) examined the relationships between bending strength and fracture toughness on the one hand and grain size and porosity of mullite ceramics on the other. These dependences can be formulated by the Knudsen equations:

$$\sigma = A\, d^{-a} \exp\left(-bP\right) \qquad (8)$$

$$K_{IC} = B\, d^{c} \exp\left(-fP\right) \qquad (9)$$

where σ and K_{IC} are the bending strength (MPa) and fracture toughness (MPa m$^{0.5}$), A, B, a, b, c and f are constants, d is the grain size (μm), and P is the porosity (%). The bending strength data shown in Fig. 4.3.2 have been fitted using Eq. (8). A good correlation exists between bending strength (σ) and porosity represented by $\sigma = 500 \exp\left(-0.225P\right)$, whereas the correlation between bending strength and grain size (d) is poor. The data center into two groups, which can be described by $\sigma = 600\, d^{-0.5}$ and $\sigma = 228\, d^{-0.16}$. The grain size data show a better correlation with the Hall-Petch equation,

$$\sigma = \sigma_0 + K d^{-0.5} \qquad (10)$$

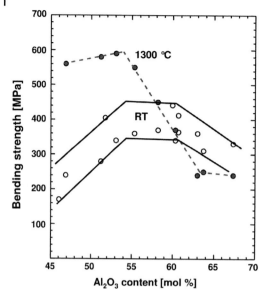

Fig. 4.3.3 Bending strength of mullite ceramics at room temperature (RT) and at 1300°C in air plotted versus chemical composition (Al_2O_3 content).

where σ_0 and K are constants. Using the Hall-Petch equation, the data shown in Fig. 4.3.2 are fitted to two groups, i.e. high strength: $\sigma = 600\, d^{-0.5}$ and low strength: $\sigma = 150 + 71.4\, d^{-0.5}$. Thus, the former group depends more strongly on grain size than the latter.

The influence of the Al_2O_3 content of mullite ceramics on the mechanical properties has systematically been examined by Kumazawa et al. (1988). Changes in bending strength and fracture toughness of mullite ceramics at room temperature as a function of the Al_2O_3 content are shown in Figs. 4.3.3 and 4.3.4, respectively (Ismail et al. 1987, Sivakumar et al. 2001, Ohira et al. 1996, Kumazawa et al. 1988, Mizuno, 1991). Bending strengths increase from 150 to 250 MPa at 46 mol% Al_2O_3 to 350 to 450 MPa between 54 and 61 mol% Al_2O_3, and slightly decrease to 250 to 350 MPa at 67 mol% Al_2O_3. Multiple factors such as porosity, mullite, α-alumina, and especially glass phase contents may be responsible for this development. Compared with the changes in bending strength versus Al_2O_3 content of mullite ceramics only little variation is observed in the fracture toughness (2.5–3 MPa m$^{0.5}$, Fig. 4.3.4).

4.3.1.2 Mechanical Strength and Fracture Toughness at High Temperatures
Many studies have been carried out on the bending strength of mullite ceramics at high temperatures after the pioneering work of Kanzaki et al. (1985). The investigations involved variations in the bulk Al_2O_3 contents and different raw materials and sintering conditions (Fig. 4.3.5). The results can be divided into two groups. One group displays a strength maximum at high temperatures (samples with 55 mol% and 60 mol% (commercial) Al_2O_3, see Kumazawa et al. 1988, Ohnishi et al. 1990), while the other group shows no distinct strength maximum and a slight

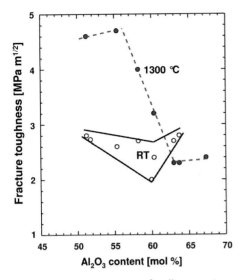

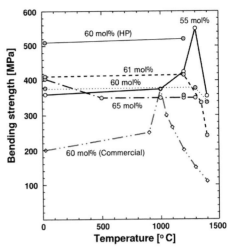

Fig. 4.3.4 Fracture toughness of mullite ceramics at room temperature (RT) and at 1300 °C in air plotted versus chemical composition.

Fig. 4.3.5 Bending strength of mullite ceramics with different chemical composition (Al$_2$O$_3$ contents) plotted as functions of temperature.

decrease in strength above 1200 to 1300 °C. Mullite ceramics of the first group are fabricated by liquid-phase sintering owing to their high SiO$_2$ contents and because they contain high amounts of impurities. Both factors enhance glassy phase formation. The microstructure of these ceramics typically consists of elongated mullite grains and small amounts of glassy phase at the grain boundaries. The strength maximum of the 60-mol% Al$_2$O$_3$ sample is near 1000 °C. The high temperature strength increase has been explained by stress relaxation and/or crack healing caused by softening of the glassy phase at the grain boundaries. The strength maximum of the 55 mol% Al$_2$O$_3$ sample is near 1300 °C. The different behavior of the two materials is mainly attributed to different impurity contents in the coexisting glassy phase. Sodia, potassia and magnesia impurities amount to about 1.35 wt. % in the 60-mol% sample, while it is only 0.03 wt. % in the 55 mol% Al$_2$O$_3$ sample.

The hot-pressed (HP) mullite ceramics belonging to the second group (60, 61 and 65 mol%, and 60 mol% Al$_2$O$_3$, Ismail et al. 1987, Mizuno, 1991, Kanzaki and Tabata, 1985) display no strength maxima. This has been explained by the specific microstructures of the three materials. They consist of equiaxed mullite grains with a very small amount of glassy phase located at the triple points of mullite grains but not at the grain boundaries. The bending strengths of these mullite ceramics show significant strength decrease above 1200 to 1300 °C only. The high-temperature properties are thus superior to those of other advanced oxide ceramics such as α-alumina and zirconia, although they are lower than those of non-oxide ceramics such as silicon nitride and silicon carbide. However, since the use of non-oxide ceramics is limited at high temperatures in air, mullite is thought to be a suitable candidate material for many applications.

The relationship between Al_2O_3 content and mechanical properties of mullite ceramics at high temperature has been investigated in detail by Kumazawa et al. (1988). The mechanical properties at 1300 °C are different from those at room temperature: Bending strength and fracture toughness clearly increase with decreasing Al_2O_3 content of ceramics (Figs. 4.3.3 and 4.3.4). This enhancement has been related to the presence of a coexisting glass phase. The fraction of glassy phase in the 55 mol% Al_2O_3 sample is estimated to be about 8 vol.% leading to the formation of relatively thick grain boundary glassy phases. This grain boundary glass phase, being of high viscosity at 1300 °C, strongly glues individual mullite grains, and thus enhances strength and toughness. Since the glass phase content decreases with increasing Al_2O_3 content, this effect then becomes less intense. Ohnishi et al. (1990) describe different fracture modes below and above the temperature where maximum strength is achieved. At low temperatures, fracture is essentially intergranular, while it is intragranular at higher temperatures, where slow crack growth does occur.

4.3.2
Elastic Modulus

The elastic modulus describes the elastic change of the shape of components by applying stress. The deformation refers to the stress to strain ratio if processes are linearly elastic. Four equations derived from the different applied stress modes can be used to describe deformation (see Fig. 4.3.6):

$$E = \sigma_x / \varepsilon_x \tag{11}$$

$$\gamma = -\varepsilon_y / \varepsilon_x = -\varepsilon_z / \varepsilon_x \tag{12}$$

$$G = \tau / \gamma \tag{13}$$

$$B = p / (\Delta V / V) \tag{14}$$

where E, ν, G and B are Young's modulus, Poisson's ratio, shear modulus and bulk modulus, while σ, ε, τ, γ, p, ΔV and V are tensile stress, elongation per unit length, shear stress, shear strain, pressure, volume change and volume, respectively. Young's modulus, Poisson's ratio and shear and bulk moduli are related to each other in an isotropic elastic body as follows:

$$E = 9BG/(3B + G) = 2G(1 + \nu) = 3B(1 - 2\nu) \tag{15}$$

The elastic constants of mullite single crystals and mullite ceramics have been measured by Hildmann et al. (2001) and Ledbetter et al. (1998) by means of the acoustic technique and by Schreuer et al. (2005) by resonance ultrasonic spectros-

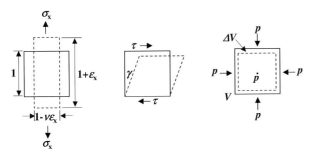

Fig. 4.3.6 Schematic drawing of the different deformation modes of ceramics caused by various types of stress.

Table 4.3.1 Elastic constants of single crystal mullite and of mullite ceramics in comparison to some related materials.

Material	Density [g cm^{-3}]	E [GPa]	Poisson's ratio	G [GPa]	B [GPa]
Mullite ceramics	3.156	227.5	0.28	88.9	172.4
Mullite, void free	3.168	229.1	0.28	89.5	173.9
Mullite single crystal	3.096	224.7	0.2787	87.86	169.2
Mullite whisker	3.1	226.3	0.2803	88.37	171.7
Sillimanite single crystal	3.241	236	0.2705	92.86	171.4
α-alumina single crystal	3.986	403	0.234	163	252
Quartz single crystal	2.649	95.6	0.077	44.4	37.7
Fused silica	2.202	73	0.168	31	37

E = Young's modulus, G = shear modulus, B = Bulk modulus
Data of sillimanite, α-alumina, quartz and fused silica are given for comparison (see Ledbetter et al. 1998, Ceramic Society of Japan, 2002).

copy (see Section 2.1.2). Data are listed in Table 4.3.1 together with related materials. Since single crystal and polycrystalline mullite ceramics yield similar results, although the former are richer in Al$_2$O$_3$, the effect of composition of ceramics can not play a major role.

The porosity of samples is known to have a definite influence on the Young's modulus, according to the equation:

$$E = E_0(1-bP) \tag{16}$$

where E_0 is the Young's modulus of dense pore-free ceramics, b a material constant, and P the porosity. From E and P data of Ismail et al. (1987), Sivakumar et al. (2001), Mizuno, (1991), Ohnishi et al. (1990) and Ledbetter et al. (1998), the constant b is calculated to be 3.8 for mullite. This value is similar to those reported for α-alumina (Al$_2$O$_3$, 3.95), zirconia (ZrO$_2$, 3.54) and silicon carbide (SiC, 3.2), (Ceramic Society of Japan, 2002) and is related to an elongation of pores (ideal b value of 2.0). The temperature dependence of the Young's modulus E up to high temperatures has been reported by Ohnishi et al. for mullite ceramics without glass phase (bulk composition: 60 mol% Al_2O_3) and with glass (bulk composition: 46 mol% Al_2O_3). The temperature coefficients $(dE/dT)/E$ are -0.94×10^{-4} (298–1473 K) and -0.8×10^{-4} (298–1193 K). These values are apparently larger than those reported by Ledbetter and coworkers (-0.49×10^{-4} K^{-1}, 260–298 K).

4.3.3
Hardness

Microhardness data of mullite single crystals and ceramics have been measured on the basis of Vickers indentation experiments. The microhardness of single-crystal mullite was determined by Kollenberg and Schneider (1989), Pitchford et al. (2001) and by Kriven et al. (1999, 2004). Single crystal studies yield a microhardness of mullite near 15 GPa (see Section 2.1.3, Fig. 2.1.5). The microhardness of polycrystalline mullite ceramics, however, can vary considerably from about 10 to 14.5 GPa. The data scatter is attributed to differences in the porosity and amount of co-existing glassy phase in the samples. The microhardness of dense mullite ceramics without a coexisting glass phase is similar to that of single crystal mullite. At room temperature the microhardness of mullite is lower than that of other engineering ceramic materials such as silicon carbide (SiC, 25.5 GPa), α-alumina (Al_2O_3, 18 GPa), (see, for example, Ceramic Society of Japan, 2002, and Section 2.1.3).

The temperature dependence of the microhardness of mullite up to 1000 °C is expressed by the following formula:

$$H = H_0 \exp(-kT) \tag{17}$$

where H is the hardness at a given temperature T (K), H_0 is the hardness at 0 K and k is a material constant. Change of hardness of mullite (Pitchford et al. 2001) as a function of temperature is shown in Fig. 4.3.7 (see also Section 2.1.3, Fig. 2.1.5) together with data for silicon carbide (SiC), silicon nitride (Si_3N_4), alumina (Al_2O_3) and zirconia (ZrO_2), (Ceramic Society of Japan, 2002). The data for single crystal mullite are not shown in this figure because its temperature dependence is similar to that of mullite ceramics. Since the data are plotted on a logarithmic scale the slopes of the straight lines correspond to the constant k, which in turn is a measure of the degree of hardness degradation with temperature. The temperature-dependent evolution of the microhardness of mullite ceramics can be divided into three straight lines in the temperature ranges 25 to 1000, 1000 to 1300 and 1300 to 1400 °C. Corresponding k values are 3.8×10^{-4}, 7.9×10^{-4} and 49×10^{-4}, respectively. The changes in the k values at the highest temperature interval are believed to correspond to changes of the softening mechanisms of ceramics with softening of the coexisting glass at grain boundaries, and enhanced grain boundary diffusion (see Section 2.1.3), while the reason for the small k value change at 1000 °C could be attributed to a possible structural transformation of

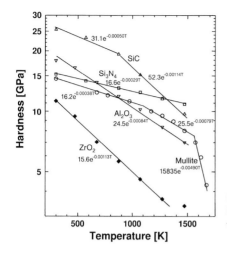

Fig. 4.3.7 Temperature-induced changes of hardness of mullite ceramics in comparison to silicon carbide (SiC), silicon nitride (Si₃N₄), alumina (Al₂O₃) and zirconia (ZrO₂).

mullite occurring in this temperature range (see Section 2.5.2). The k value of mullite up to 900 °C is lower than those of silicon carbide (5.0×10^{-4}), alumina (8.4×10^{-4}) and zirconia (11.3×10^{-4}), and slightly higher than that of silicon nitride (2.9×10^{-4}). The favorable temperature-dependent evolution of the microhardness of mullite ceramics up to 1000 °C has been interpreted as the result of the low mobility of dislocations in mullite (Kollenberg and Schneider, 1989, see Section 2.1.3).

4.3.4
Thermal Shock Resistance

Thermal shock fracture is caused by high temperature-induced stress. From both theoretical and experimental points of view, it is difficult to predict the changes in thermal stresses generated in a sample. Thus the prediction of thermal shock failure of mullite ceramics is difficult. Most data are derived from water quenching experiments. This method yields the critical thermal shock temperature differences (ΔT) but no direct information on the thermal shock resistance. Thermal shock temperature differences ΔT of mullite ceramics consisting of 46 mol% Al₂O₃ with a coexisting glass phase and those without glass phase consisting of 60 mol% Al₂O₃ are 300 and 350 °C, respectively (Ohnishi et al. 1990). A thermal shock temperature difference of 300 °C was reported for 50 vol.% Al₂TiO₅/mullite composites by Huang et al. (2000). The critical thermal shock temperature differences are related to bending strength (σ_{max}), Poisson' ratio (n), Young's modulus (E) and thermal expansion coefficient (α) by:

$$\Delta T = \sigma_{max}(1-n)/(E\alpha) \tag{18}$$

It has to be taken into account, however, that the ΔT evaluated from this equation often differs from the experimental results. Generally it can be stated that the thermal shock temperature differences of mullite ceramics are similar to those of zirconia and silicon carbide ceramics. They are slightly higher than α-alumina ceramics, but are significantly lower than those of silicon nitride, cordierite ($Mg_2Al_4Si_5O_{18}$) and tieillite (Al_2TiO_5) ceramics (Ceramic Society of Japan, 2002, Huang et al. 2000). This is due to the different thermal expansions of the materials.

4.3.5
Wear Resistance

The wear resistance of mullite ceramics has been investigated between room temperature and 1000 °C by Senda et al. (1994, 1995) using the material fabricated by Ohnishi et al. (1990). The coefficient of friction at room temperature is about 0.75 in a mullite ceramic without glass and a bulk composition of 60 mol% Al_2O_3, and about 0.9 in one with glass phase and with 46 mol% Al_2O_3. The friction coefficient gradually increases to 1.0 with increasing temperature in the glass-free sample, while there is only a small change in samples with glass phase. On the other hand, specific wear losses of mullite ceramics with and without glass phase are about 7×10^{-4} mm^3 N^{-1} m^{-1} and are apparently higher than that of α-alumina (3.5×10^{-5} mm^3 N^{-1} m^{-1}). Wear losses increase with temperature up to 600 °C (1.2×10^{-3} mm^3 N^{-1} m^{-1}). Above 600 °C and up to 1000 °C the specific wear loss of the glass-free mullite ceramic becomes distinctly smaller to reach a final value of about 3×10^{-5} mm^3 N^{-1} m^{-1}, while the sample with glass phase shows little change in this temperature range. The strong lowering of the wear loss in glass-free mullite ceramics corresponds to the behavior of alumina ceramics. The fine-grained surfaces together with dynamic re-crystallization at high temperatures is thought to be the reason for decreasing wear loss in these samples.

4.3.6
Fatigue Behavior

Static fatigue has been investigated for mullite ceramics consisting of 71, 72 and 74 wt. % Al_2O_3 (59, 60 and 63 mol% Al_2O_3, Kubota et al. 1994). Processes can be described by:

$$t_f = B\sigma^{-N} \tag{19}$$

where t_f is the time-to-failure at a constant applied stress (σ), B is a constant and N is the crack growth parameter (fatigue parameter). The fatigue parameter N directly relates to the fatigue resistance of a material. It normally decreases with the amount of coexisting glass phase in a ceramic. Fatigue parameters and fracture toughnesses of mullite ceramics are listed in Table 4.3.2. Obviously the fatigue resistance increases with the bulk Al_2O_3 content of the mullite ceramic, which is compatible with the amount of glassy phase, which decreases simultaneously. In contrast to the increase of the fatigue resistance, the fracture toughness decreases

Table 4.3.2 Dependence of the fatigue parameter and fracture
toughness on the composition of mullite ceramics at 1200 °C.

Composition [wt.%]	71	72	74
Fatigue parameter N	17	23–24	26
Fracture toughness K_{IC} [MPa m$^{0.5}$]	4.3	2.4	1.7

Reference: Kubota et al. (1994)

with increasing Al_2O_3 content. This is considered to be brought about by blunting
of the crack tip in the glass of the low Al_2O_3 composition. The fatigue resistance of
mullite ceramics is high, even if compared with silicon nitride ($N = 12.8$ at
1200 °C, Quinn, 1990).

The crack healing behavior of mullite ceramics was examined by Hvizdos and
Kasiarova (2002). Crack healing occurs mainly by glass phase wetting of the crack.
The healing proceeds effectively at 1300 °C and is related to the softening tem-
perature of the glass.

4.3.7
Creep Resistance

The excellent creep resistance of mullite and mullite ceramics was first reported by
Dokko et al. (1977), (see Section 2.1.1). The steady-state creep rate (e) of mullite
ceramics is described by the following equation (Kumazawa et al. 1987):

$$e = (ADGb)/(kT)(b/d)^m(s/G)^n \tag{20}$$

where A is a dimensionless constant, D is the diffusion coefficient, G is the shear
modulus, b is Burger's vector, k is the Boltzmann constant, T is the absolute temperature, d is the grain size, m is the inverse grain size exponent, s is the stress and
n is the stress exponent. Three different creep mechanisms have been described
for polycrystals: (a) diffusion creep, (b) dislocation creep, and (c) grain boundary
sliding creep.

If creep is controlled by diffusion n becomes 1 and the steady-state creep rate is
represented by:

$$e = (ADbs)/(kT)(b/d)^m \tag{21}$$

The main diffusion reaction routes can be separated into intergranular (lattice)
and grain boundary diffusion, and the diffusion coefficient can be represented
as

$$D = D^l + (\pi/d)dD^b \tag{22}$$

where D^l is the lattice diffusion coefficient, d is the effective grain boundary width for diffusion and D^b is the grain boundary diffusion coefficient. The value of m becomes 2 if $D^l \gg D^b$ and the creep rate e given by

$$e = (14 \; \Omega D^l s)/(d^2 kT) \tag{23}$$

which is called the Nabarro-Herring equation, where Ω is the atomic volume.

By contrast, m becomes 3 if $D^b \gg D^l$ and the creep rate is given by the Coble equation:

$$e = (14 \; \Omega D^b s)/(d^3 kT) \tag{24}$$

When creep is controlled by the movement of dislocations (dislocation mechanism), m and n become 0 and $3 \leq n \leq 5$, respectively. In this mechanism, creep rate is independent of grain size. When creep is controlled by grain boundary sliding, m becomes 1 but n changes with the microstructure of the sample. For example, n is 1 in the presence of glass at the grain boundaries, while it is 2 if there is no glass phase at the grain boundaries.

The creep characteristics of mullite ceramics are summarized in Table 4.3.3. The ranges of m, n and Q (activatim energy) values that control the kinetics and mechanisms of creep are $m = 1.2$–2.5, $n = 1$–2 and $Q = 357$–$1051 \; kJ \; mol^{-1}$. There is considerable scatter of values, which may be due to variations of raw materials in chemical composition (Al$_2$O$_3$ content, amount of impurities), the preparation method and the different firing conditions, but can be also caused by the different creep testing methods.

The dependence of creep on the Al$_2$O$_3$ content of mullite ceramics has been investigated by Tomatsu et al. (1987, Fig. 4.3.8) at 1500 °C with samples previously fired at 1600 and 1700 °C, respectively. The strain rates are high in Al$_2$O$_3$-poor samples, decrease between about 60 and 62 mol% Al$_2$O$_3$, and then slightly increase up to 65 mol% Al$_2$O$_3$. Changes in creep are explained by the different microstructures of materials. From about 56 to 60 mol% Al$_2$O$_3$ bulk composition, elongated mullite grains coexist with a glassy phase at the mullite's grain boundaries. When the Al$_2$O$_3$ content changes from about 60 to 62 mol%, the amount of glassy phase decreases, and the phase is present at triple-point junctions only, with no glassy phase being present at the grain boundaries. Nearly no glass phase occurs in the 62 mol% Al$_2$O$_3$ sample with a microstructure consisting of equiaxed mullite grains. With further increase of Al$_2$O$_3$ content, α-alumina coexists with equiaxed mullite. Obviously the strain rate has a minimum in samples containing neither α-alumina nor glassy phase at the grain boundaries (62 mol% Al$_2$O$_3$). On the other hand, the occurrence of α-alumina increases strain rate slightly as it has a higher deformation rate than mullite (Okamoto et al. 1990).

In Fig. 4.3.8, the strain rates of the samples pre-fired at 1700 °C are about one order of magnitude lower than those at 1600 °C. Although no grain size data are given by Tomatsu et al., the lowering of strain rates can be attributed to a difference in the grain sizes because grain growth should be enhanced at higher firing tem-

Table 4.3.3 Creep parameters of mullite ceramics.

Al$_2$O$_3$ content [mol%]	n	m	activation energy [kJ/mol]	Creep temperature [°C]	Raw material	Sintering temperature [°C]	Creep testing method	Creep mechanism	Reference
55.6	1.9	-	837–963	1400	Spray	1650	Tension	-	Kumazawa et al. (1987)
55.6	1.6, 1.5	-	-	1400–1500	Sol-gel	1600, 1700	Bending	-	Tomatsu et al. (1987)
57.9	1.8, 1.4	-	-	1400–1500	Sol-gel	1600, 1700	Bending	-	Tomatsu et al. (1987)
59	1.1–1.2	2,4	578	1350–1500	Precipitation	1.650	Bending	Grain boundary sliding	Ashizuka et al. (1989)
60	1.1–1.2	-	410	1100–1300	-	1750	Bending	Diffusional flow	Rhamin et al. (1997)
60	1.1–1.2	-	357	1177–1287	Sol-mixing	HP 1600	Compression	Grain boundary sliding	Nixon et al. (1990)
60	1.1	-	360	1250–1290	-	-	-	-	Koester et al. (1988)
60	-	-	1051	1287–1357	Sol-mixing	HP 1600	Compression	Cavity formation	Nixon et al. (1990)
60	1.1	-	1050	1290–1360	-	-	-	-	Koester et al. (1988)
60	1.1–1.2	-	731–930	1300–1450	-	1750	Bending	Grain boundary sliding	Rhamin et al. (1997)
60	-	-	381	1300–1350	Sol-mixing	1650	Compression	Diffusional flow	Arellano-Lopez et al. (2002)
60	1.1, 1.2	-	748	1307–1427	Sol-gel	1600	Compression	Grain boundary sliding	Lessing et al. (1975)
60	-	-	390	1350–1400	Sol-mixing	1650	Compression	Diffusional flow	Arellano-Lopez et al. (2002)
60	0,95	-	687	1350–1450	Alkoxide	HP 1500	Bending	Diffusion	Lessing et al. (1995)
60	-	-	754	1357–1427	Sol-mixing	HP 1600	Compression	Grain boundary sliding	Nixon et al. (1990)
60	1.1	-	750	1360–1430	-	-	-	-	Koester et al. (1988)
60	1	2,5	810	1365–1480	Sol-gel	1650	Bending	Diffusion	Okamoto et al. (1990)
60	1,6	-	837–963	1400	Spray	1650	Tension	-	Kumazawa et al. (1987)
60	1	2	710	1400–1500	Alkoxide	HP 1550	Compression	-	Dokko et al. (1977)
60	1,3	-	-	1400–1500	Sol-gel	1700	Bending	-	Tomatsu et al. (1987)
60	1,3	-	1030	1400–1550	Sol-gel	1650	Compression	Diffusion	Ohira et al. (1991)
60.2	1.2–1.3	2	716	1350–1500	Precipitation	1650	Bending	Grain boundary sliding	Ashizuka et al. (1991)
61.4	1.3, 1.0	-	-	1400–1500	Sol-gel	1600, 1700	Bending	-	Tomatsu et al. (1987)
62.6	1.3–1.5	1,2	703, 707	1350–1500	Precipitation	1650	Bending	Grain boundary sliding	Ashizuka et al. (1991)
62.6	1.3, 1.0	-	-	1400–1500	Sol-gel	1600, 1700	Bending	-	Tomatsu et al. (1987)
62.8	1.1–1.4	-	700	1400–1500	-	-	Bending	Grain boundary sliding	Party and Hasselman (1972)
63.8	1.3, 1.0	-	-	1400–1500	Sol-gel	1600, 1700	Bending	-	Tomatsu et al. (1987)
65.1	1.4, 1.0	-	-	1400–1500	Sol-gel	1600, 1700	Bending	-	Tomatsu et al. (1987)
67.6	1.1, 1.2	-	498	1350–1500	Precipitation	1650	Bending	Grain boundary sliding	Ashizuka et al. (1989)
67.6	1,8	-	837–963	1400	Spray	1650	Tension	-	Kumazawa et al. (1987)

n = stress exponent; m = grain size exponent; HP = hot pressing

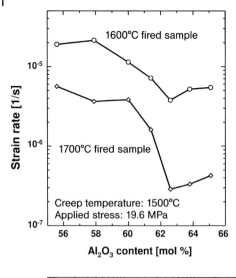

Fig. 4.3.8 Strain rates of mullite ceramics at 1500°C under applied stress of 19.6 MPa plotted as a function of composition (Al_2O_3 content). Prior to the experiments the samples were fired at 1600 and 1700 °C, respectively.

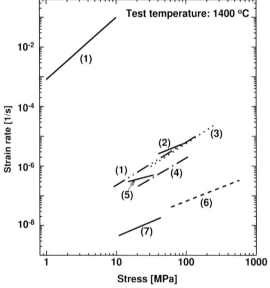

Fig. 4.3.9 Stress-strain rate relationships of mullite ceramics from different starting materials at 1400°C. (1) Xue and Chen (1992), (2) Tkalcec et al. (1998), (3) Yoon (1990), (4) Lessing et al. (1975), (5) Ashizuka et al. (1991), (6) Dokko et al. (1977), (7) De Arellano-Lopez et al. (2002).

perature, which in tun reduces creeps. Fig. 4.3.9 shows strain rate versus stress relations at 1400 °C for various mullite samples with different grain sizes. As expected, strain rates increase with a decrease of grain size. Consequently, pre-mullite displays very high strain rates due to its nanometer-range particle size.

The activation energies of creep, Q, range widely. Data are summarized in Fig. 4.3.10 as a function of creep temperature for samples consisting of 60 mol% Al_2O_3. The Q values obtained at lower temperatures lie near 400 kJ mol^{-1}, while those obtained at higher temperatures range from about 700 to 1000 kJ mol^{-1}. Low temperature creep has been attributed to diffusional flow and grain boundary slid-

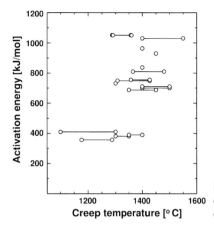

Fig. 4.3.10 Relationship between activation energy of creep and temperature of mullite ceramics, (see Table 4.3.3 for references).

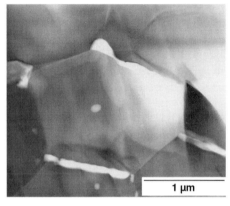

Fig. 4.3.11 Grain boundary cavity formation in a mullite ceramic subjected to strong creep deformation.

ing mechanisms of the glass phase, and the small values of Q are explained by viscous glass flow. High temperature creep is also attributed to diffusional flow and grain boundary sliding mechanisms. However, these processes occur in the grain boundaries but without a significant contribution from a glassy phase. In that case formation of cavities in the grain boundaries has been considered as a further creep mechanism (Fig. 4.3.11).

4.4
Thermal Properties of Mullite Ceramics
K. Okada and H. Schneider

This section focuses on thermal conductivity and thermal expansion data. Thermo-chemical data, such as heat capacity (C_p), enthalpy (H), entropy (S), Gibbs energy (G), enthalpy of formation (ΔH_f) and Gibbs energy of formation (ΔG_f) are discussed in Section 2.2.1.

Table 4.4.1 Thermal diffusivity and conductivity of mullite.

Thermal conductivity $[W\,m^{-1}\,K^{-1}]$	Heat capacity $[J\,g^{-1}\,deg^{-1}]$	Thermal diffusitivity $[m^2\,s^{-1}]$	Bulk density $[g/cm^{-3}]$	Reference
6.9	0.93	2.4×10^{-6}	3.17	Sivakumar et al. (2001)
5.9	0.77	2.4×10^{-6}	3.08	Hayashi et al. (1989)
5.2	0.767	2.15×10^{-6}	3.158	Russell et al. (1996)

4.4.1
Thermal Conductivity

Generally, the thermal conductivity (λ) of ceramics is small, which has made the materials widely used for thermal insulating purposes. Thermal conductivity is obtained from measurements of heat capacity (C_p) and thermal diffusitivity (D_{th}) with the laser flash or hot wire methods and is calculated from the equation $k = C_p D_{th} \rho$ (ρ = bulk density). The thermal conductivity of mullite ceramics ranges from about 5 to 7 $W\,m^{-1}\,K^{-1}$, as reported by Sivakumar et al. (2001), Hayashi et al. (1998) and Russell et al. (1996), (see Table 4.4.1). The thermal conductivity of mullite is significantly smaller than that of α-alumina (about 40 $W\,m^{-1}\,K^{-1}$) but is similar to that of silica (about 7 $W\,m^{-1}\,K^{-1}$). New accurate thermal diffusivity values have been obtained by Hildmann and Schneider (2004) from mullite single-crystal measurements. These data are discussed in Section 2.2.3.

4.4.2
Thermal Expansion

Thermal expansion data are crucial, because they give information on the high-temperature shape stability and on temperature-produced strains in ceramic bodies. The thermal expansion of mullite is relatively low with respect to other advanced oxide materials such as α-alumina, zirconia, spinel ($MgAl_2O_4$), or magnesia, but is slightly higher than that of tieillite (Al_2TiO_5), cordierite ($Mg_2Al_4Si_5O_{18}$) and silica. Structure-related thermal expansions (intrinsic expansions) of mullite have been measured by high-temperature X-ray and neutron diffractometry (e. g. Schneider and Eberhard 1990, Brunauer et al. 2001, Schneider and Schneider 2004). Effects are described in detail in Section 2.2.2. For the characteristics of ceramics bulk thermal expansions (extrinsic expansions), including microstructural influences such as phase and grain-size distributions and porosity, have to be considered as well.

The linear thermal expansion of mullite ceramics up to 1000 °C is in the range of 4.5 to 6 ($\times 10^{-6}$ °C^{-1}), (Camerucci et al. 2001, Ishitsuka et al. 1987, Huang et al. 2000). The thermal expansion curve of a mullite ceramic is shown in Fig. 4.4.1 (Ishitsuka et al. 1987, see also Section 7.3.2.3). The thermal expansion coefficient

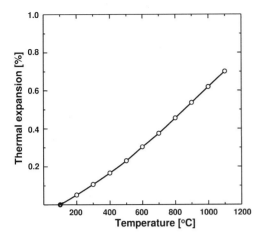

Fig. 4.4.1 Linear thermal expansion of mullite ceramics (after Ishitsuka et al. 1987).

of this mullite ceramic is about 6×10^{-6}, which is rather high with respect to the lattice thermal expansion (about 5×10^{-6} °C^{-1}) in the same temperature range (see Section 2.2.2). The difference between both data sets is attributed to the influence of the microstructure (grain boundary phase, pores, thermal stress) on the bulk thermal expansion.

4.5
Miscellaneous Properties of Mullite Ceramics
K. Okada and H. Schneider

4.5.1
Electrical Properties

The dielectric constants (ε) of mullite ceramics have been measured by several workers (e.g. Somiya and Hirata, 1991), since mullite is a potential substrate material for integrated circuit (IC) packages (see also Section 4.6.2). Dielectric constants between 6.7 and 7.5 (1 MHz) have been reported. The large spread of ε data is related to differences in chemical compositions and microstructures of the mullite samples, with the amount of glassy phase and the porosity playing important roles. The dielectric constants of mullite ceramics are lower than those of α-alumina ($\varepsilon \approx 9.5$), aluminum nitride ($\varepsilon \approx 8.9$) and silicon nitride ($\varepsilon \approx 8.1$), but are higher than those of magnesium silicate ($\varepsilon \approx 6.0$), beryllium oxide ($\varepsilon \approx 6.5$) and cordierite ($Mg_2Al_4Si_5O_{18}$, $\varepsilon \approx 5.0$). Ruh and Chizever (1998) published the results of dielectric measurements up to 14 GHz and showed that the values are more or less constant over a wide frequency range. The electrical resistivity of mullite at room temperature is below 10^{14} Ω cm, reflecting its insulating character.

For any material to be used as an intercircuit (IC) package substrate, it must have a very low dielectric constant because the root of ε is proportional to the signal transmission-delay time, which must be small to increase the speed of computer. Many studies have been performed with a view to lowering the ε value of mullite ceramics. Good solutions are composites with silica glass (Somiya and Hirata, 1991) or cordierite (Camerucci et al. 2001).

The dielectric loss tangent (tan Δ) is another important property of electrical insulators, especially at high frequencies. Since insulating materials have low dielectric constants, the dielectric loss tangent is low too. The dielectric loss tangent of mullite is that of a good insulating material and is similar to that of magnesium silicate, a common high frequency insulator.

4.5.2
Optical Properties

The optical properties of mullite single crystals are given in Section 2.3.1. Data on the optical behavior of mullite ceramics are rare. Schneider et al. (1993) studied the optical translucency of hot isostatically pressed mullite ceramics and found transmittance in the infrared (3 to 5 μm wavelength, see Fig. 2.3.1) and in the visible region (see also Section 4.6.3).

4.5.3
Chemical Corrosion Behavior

The chemical resistance of mullite has been investigated in various aggressive solutions, melts and gases. Mullite has good corrosion resistance against acids except for hydrofluoric acid (HF), while fair corrosion resistance is observed with alkali solutions. Takei et al. (1998) investigated porous mullite ceramics with various aggressive leachants and found similar dissolution rates of mullite grains and glassy phase.

In molten salts, mullite exhibits good corrosion resistance to alkali carbonates, alkali sulfates and alkali nitrates, while its stability is low in the presence of sodium and potassium hydroxide. The corrosion behavior of mullite is better than that of silica and of other silicates in molten sodium sulfate and sodium carbonate (Jacobson et al. 1996), and has been explained by its relatively low SiO_2 content. Mullite also displays a higher resistance in hot sodium chloride and sodium sulfate than cordierite (Takahashi et al. 2002), but a lower one than α-alumina. In metal melts, mullite is rather unstable, except for a few metals such as aluminum and nickel, (Table 4.5.1, see also Section 7.4.1).

The corrosion resistance of mullite in aggressive gases is generally good up to high temperatures. Since the gas permeability of mullite is low, mullite is a good refractory and insulating material under various severe gas atmospheres up to high temperatures. Strength degradation of mullite, however, is reported after long term hot hydrogen gas exposure at 1250 °C (Herbell et al. 1998). The degradation is related to the highly reducing environment with decomposition of the glassy grain boundaries but also of mullite itself (see Section 3.2.1).

Table 4.5.1 Corrosion resistance of mullite in comparison to other important ceramics.

Reactant		Mullite	Alumina	Magnesia	Zirconia	Silicon nitride	Silicon carbide	Boron nitride
Acid	HCl	Good	Good	Bad	Fair	Good	Good	Fair
	HNO_3	Good	Good	Bad	Fair	Good	Good	Fair
	H_2SO_4	Good	Good	Bad	Bad	Good	Good	Bad
	HF	Bad	Bad	Bad	Bad	Bad	Good	Fair
Alkali	NaOH		Fair			Good	Fair	Fair
	KOH	Fair	Good	Good	Good	Good	Fair	Bad
Molten salt	NaOH	Bad (500)	Good (500)	Bad (500)	Good (500)	Bad	Bad	Bad (500)
	KOH	Bad (500)	Fair (500)	Bad (500)	Fair (500)	Bad	Bad	Bad (900)
	Na_2CO_3		Good (1000)					
	K_2CO_3	Good (1000)	Good (1000)		Good (1000)	Good (1000)		Bad (1000)
	Na_2SO_4		Good (1140)	Good (1140)	Good (1140)		Bad (1000)	
	K_2SO_4		Good (400)			Good (350)		Good (350)
	$NaNO_3$		Good (400)			Good (350)		Good (350)
	KNO_3	Good (400)		Good (1000)	Fair (1000)	Good (400)	Good (400)	Good (400)
	PbO	Bad (1000)	Fair (1000)		Bad (800)			
	V_2O_5	Bad (800)	Bad (800)					
Slag	Acidic	Good	Good	Bad	Good			
	Neutral	Good	Good		Good			
	Basic	Fair	Fair	Good	Good			

Table 4.5.1 (continued).

Reactant		Mullite	Alumina	Magnesia	Zirconia	Silicon nitride	Silicon carbide	Boron nitride
Molten metal	Al	Good	Fair	Fair	Good	Good	Good	Good
	Fe	Bad	Fair	Good	Good	Good		Good
	Ti	Bad	Bad	Bad	Good			
	V	Bad	Good	Good	Good			
	Zn	Bad	Fair	Fair	Bad		Good	
	Ni	Good	Fair	Good	Good	Good	Good	
	Zr	Bad	Bad	Bad	Good			
	Si	Bad	Bad	Bad	Good			Good
Gas	Vacuum	Good (1500)	Good (1700)	Good (1700)		Good (900)		
	H₂		Good (1700)	Good (1700)	Fair (2200)			
	N₂	Good (1500)	Good (1700)	Good (1700)	Fair (2200)	Good (1800)		
	O₂	Good (1550)	Good (1900)	Fair (2300)	Fair (2300)	Fair (1000)		
	Ar	Good	Good (1700)	Good (1700)				
	H₂O	Good	Good (1700)	Bad	Bad (1800)	Good (800)		
	NH₃	Good		Good (1700)	Bad (2200)			
	H₂S					Good (1000)		
	HCl	Good	Fair	Bad	Good	Good		
	S	Good (930)	Good	Fair	Fair			
	CO	Good	Good (1700)	Good (1700)	Good (1400)			
	CO₂	Good	Good (1200)	Good (1200)	Good (1200)			
	SO₂	Good	Good	Fair	Good			

Numbers in parentheses: temperature [°C]

4.6
Application of Mullite Ceramics
K. Okada and H. Schneider

Mullite is a major constituent of traditional ceramics such as whitewares, porcelains, structural clay minerals and refractories and also of advanced materials with various new applications. This does not apply to mullite ceramics only, but to mullite matrix composites (see Section 7.2.6).

4.6.1
Engineering Materials

4.6.1.1 Refractory Materials
As a refractory, mullite has many advantageous properties, which make it an important material for many industrial fields: (1) high melting point (about 1890 °C); (2) excellent mullite-structure-inherent creep resistance; (3) very good thermal shock and spalling resistance because of its low thermal expansion coefficient; (4) high shear modulus; (5) strong heat insulating character, due to its low thermal conductivity; (6) very good electrical insulation; (7) high gas impermeability; (8) excellent resistance against corrosive media.

Mullite is a main constituent of fireclay and high alumina refractories. Fireclay materials have Al_2O_3 contents below 45 wt. % and have usually been designated as *schamotte* products. Since fireclay materials have lower Al_2O_3 contents than mullite, they consist of mullite and silica, often with glassy phases at the grain boundaries, which restricts the high-temperature application of corresponding materials. The benefits of fireclay refractories are light weight, good heat insulation, low thermal expansion and high corrosion resistance against slags. High alumina refractories consist of well developed mullite and/or α-alumina crystals interlocked with glassy phase. In refractories with more than 72 wt. % Al_2O_3, the α-alumina content gradually increases with the Al_2O_3 content. Mullite enhances the spalling resistance and the stability under load of refractories at high temperature, while α-alumina has a positive influence on the corrosion resistance and mechanical properties of materials. Commercial mullite is made from a variety of raw materials: Bayer process-derived alumina plus silica, diaspore and gibbsite plus kaolin, aluminosilicate minerals kyanite, sillimanite and andalusite plus alumina and refractory-grade bauxite (Schneider et al. 1982, 1986, 1987). Different types of commercial mullite have been distinguished, sinter-mullite and fused-mullite. Chemically produced mullite because of its high price has only limited importance for traditional refractory products.

Sinter-mullite. Often mixtures of Bayer alumina and clay and refractory-grade bauxite have been used as starting materials. The raw materials are crushed, blended, pressed to pellets or briquets and are calcined in tunnel or rotary kilns often at temperatures above 1750 °C (Table 4.1.2). After calcination the material is crushed again to the required grain size distribution. Typical compositions and properties of refractory-grade sinter-mullites are given in Table 4.6.1.

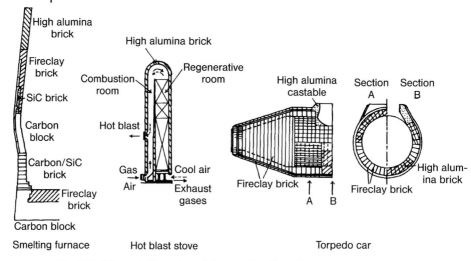

High alumina
brick

High alumina brick

Fireclay
brick

Combustion
room

Regenerative
room

SiC brick

High alumina
castable

Section
A

Section
B

Carbon
block

Hot blast

Carbon/SiC
brick

Gas

Cool air

Air

Exhaust
gases

High alum-
ina brick

Fireclay
brick

Fireclay brick

A   B

Fireclay brick

Carbon block

Smelting furnace

Hot blast stove

Torpedo car

Fig. 4.6.1 Schematic illustration of the use of mullite refractories in the steel making industry, with special reference to high alumina materials (Ceramic Society of Japan, 1990).

Fused-mullite. Alumina (e. g. Bayer alumina) and silica (e. g. silica sand) mixtures have been used as raw materials. The fusion of the mixed starting materials is performed in electric arc furnaces. The molten starting batch is cast in ingot molds at about 1900 °C, cooled to room temperature and then processed according to the later use.

The steel making industry is the largest consumer of mullite-based refractories. Mullite-based bricks have frequently been used for the lining of the upper parts of melting furnaces, hot blast stoves, coke making furnaces, hot iron runners, continuous casting furnaces, and torpedo ladles (Fig. 4.6.1, Ceramic Society of Japan, 2002). Mullite refractories are also common linings in the calcination zone of cement rotary kilns. Although fused-mullite brick linings of glass tanks have often been replaced by fused alumina-zirconia-silica (AZS) materials nowadays, mullite-based bricks are still a common refractory material in the forehearth of furnaces (Fig. 4.6.2). Mullite ceramics have also been used for shelves, saggers and props in the ceramic industry together with silicon carbide and cordierite ceramics. This is because of the low creep deformation, good thermal shock resistance and good corrosion resistance during firing of mullite ceramics. For information on technical applications of mullite fibers and mullite matrix composites the reader is referred to Sections 6.4, 7.2.6 and 7.3.

4.6.1.2 High Temperature Engineering Materials

Mullite ceramics display excellent chemical corrosion resistance and very low gas permeability together with favorable refractoriness. Mullite ceramics, therefore, have been widely used for crucibles, thermal protection systems and for thermo-

Table 4.6.1 Composition and properties of commercial mullites.

	Sinter-mullite[a]					Fused-mullite[b]	
	1	2	3	4	5	6	7
Chemical composition [wt %]							
Al_2O_3	67.3	74.4	73.0	72.7	73.0	76.0	76.6
SiO_2	31.0	24.0	25.3	23.1	25.0	23.4	22.8
Fe_2O_3	0.23	0.34	0.54	1.80			0.07
TiO_2	0.79	0.45	0.15	1.61			0.01
CaO	0.11	0.04	0.10	0.09	2.0	0.6	0.10
MgO	0.36	0.21	0.13	0.08			0.03
$Na_2O + K_2O$	0.29	0.41	0.45	0.42			0.41
Physical properties							
Open porosity [%]	9	14	11–13	8–11	9	2	1.5
Phase composition [wt %]							
Mullite	87	94	93	96	85	85	n.d.
α-alumina	3	4	4	2	2	5	8.5
Glass phase	9	2	3	2	13	10	n.d.

a) Sinter-mullite
   1–3: Mixes of Bayer alumina with clay
   4: Bauxite
   5: Mixes of kaolin with alumina
b) Fused-mullite
   6: Mixes of kaolin with alumina
   7: unknown
   Data an Samples 1–4 are from Hawkes (1962), data from samples 5
   and 6 are from König (1978), data from sample 7 are from Majdic et
   al. (1980).

couple tubes. Such ceramics generally are relatively pure, with mullite contents above 95 %. Data on the chemical corrosion resistance of mullite ceramics against solutions, melts, slags and gases is summarized in Table 4.5.1 in comparison to other important ceramics (Lay, 1984, Ceramic Society of Japan, 1990). The stability of mullite ceramics in the presence of molten salts and slags is similar to that of α-alumina ceramics, while the resistance of mullite against molten metals is lower than that of zirconia. By contrast, the corrosion resistance of mullite ceramics to gases, and their gas impermeability, are excellent.

4.6.1.3 Materials for Heat Exchangers

Heat exchangers are important tools to save exhaust energy in technical systems. Since the efficiency of heat exchangers increases with the operation temperature, high temperature stability of materials is required. Other critical properties are

Fig. 4.6.2 Fused-mullite refractory. (above) Section through the refractory brick; note the contraction cavities in the middle of the brick. (below, a to d) Microprobe element distribution patterns, showing the occurrence of mullite (M), zirconia (B = baddeleyite = ZrO_2) and silicate glass (G).

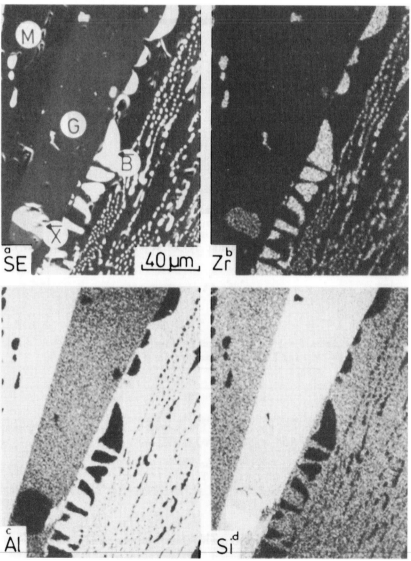

thermal shock resistance, corrosion resistance, oxidation resistance and suitable thermomechanical properties. Since high thermal shock resistance is the crucial property, materials with low thermal expansion coefficients, such as cordierite ($Mg_2Al_4Si_5O_{18}$), tieillite (Al_2TiO_5), lithium aluminosilicate ($Li_2Al_2Si_nO_{2n+4}$), and especially mullite have been considered.

Mullite has also become a key material in heat regenerating systems exposed to high gas temperatures. Spherical particles (about 0.5 to 2 mm in diameter) of mullite have been employed for heat regenerating processes at reaction temperatures between 1300 and 1400 °C, well beyond the working temperature of cordierite (about 1200 °C). Thus mullite extends the working range of heat regenerating systems to temperatures that have not hitherto been available.

4.6.1.4 Structural Materials

Mullite is a good high temperature engineering material, though its application in, for example, engines, testing machines and gas turbines often requires improvement with respect to its mechanical properties. The advantage of mullite ceramics for these applications is derived from the very good creep behavior and the high compressive strength of mullite up to elevated temperatures. Mullite crucibles, plates and tubes for various applications (Fig. 4.6.3) have been fabricated by different techniques (see Section 4.2). A special production line is water-stabilized plasma spraying (Fig. 4.6.3). Structures produced by water-stabilized plasma spraying display excellent shock resistance, due to their favorable microstructures with defined porosities.

4.6.2
Electronic Packaging Materials

Owing to the rapid progress in digital processing, electronic devices have to be miniaturized, lightened, densified and consolidated. High performance properties are necessary for electronic packaging materials, which are important parts of electronic devices. Critical requirements are: (1) enlargement of silicon wafers, (2) multilayering, (3) increase in heat radiation efficiency, (4) package condensation by

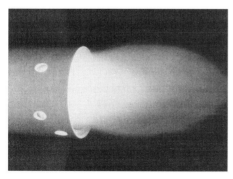

Fig. 4.6.3 Burner tube made of mullite by means of the water-stabilized plasma spraying technique (WAPLOC, from Schneider et al. 1994).

decreasing large scale integration (LSI) device errors, and, last but not least, (5) lowering of production costs. To satisfy these demanding requirements, any package or substrate should have a thermal expansion similar to that of silicon, good insulation resistivity, low dielectric constant, good chemical stability, high heat resistivity, high thermal conductivity, high mechanical strength, high accuracy of dimensions of the sintered bodies, good adhesion to the electrodes, thick film and thin film processibility and low production costs. No material is available so far that satisfies all these criteria. Therefore the material choice refers to the specific substrate, i.e. whether high thermal conductivity, large size or other properties are envisaged. For substrates with high thermal conductivity, aluminum nitride is widely used because of its excellent thermal conductivity. On the other hand, mullite is a candidate for large dimension substrates with high density packaging. This is because its thermal expansion is close to that of silicon and because of its relatively low dielectric constant. Both properties are crucial in order to minimize stress generating mismatches in silicon chips and packaging materials and to reduce the signal delay time of the electronic circuits. For further enlarging the silicon wafer size, it is, however, necessary to adjust the thermal expansions even more closely, which can be realized by admixture of silica glass and mullite. Thermal expansions are in the range of 3.5×10^{-6} °C^{-1}, which is very close to that of silicon (3.5×10^{-6} °C^{-1}) in large scale integration (LSI) chips (Horiuchi et al. 1988, Ceramic Society of Japan 2002, Ishihara, 1999).

Mullite-glass materials are also effective in lowering the dielectric constant (ε). The dielectric constant is directly correlated with the signal transmission delay time (T_{pd}) as represented by the equation $T_{pd} = \sqrt{\varepsilon}/c$ (c = light velocity). Since packages are becoming larger and more and more dense, lowering the delay time is an important issue. The relationship between dielectric constants and signal transmittance delay time of mullite and mullite-glass composites is shown in Fig. 4.6.4 in comparison to values of other substrate materials. Single phase mullite ceramics show about 14% lowering of the delay time compared with alumina

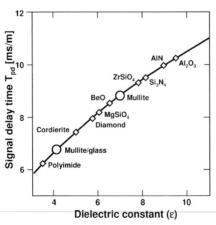

Fig. 4.6.4 Relationship between dielectric constant and signal transmittance delay time of mullite and mullite-glass composites used as substrates in electronic devices. Data are given in comparison to other relevant substrate materials (according to Schneider et al. 1994).

Table 4.6.2 Relevant properties for electric package substrate materials.

Material	Dielectric constant [at 1 MHz]	Dielectric loss tangent [at 1 MHz]	Thermal expansion [$\times 10^{-6} K^{-1}$]	Thermal conductivity [W m^{-1} K^{-1}]	Bending strength [MPa]
Mullite	7.0	0.0003	4.4	0.06	300
Al$_2$O$_3$	9.5	0.0003	8	0.3	500
AlN	8.9	0.001	4.5	2.6	400
MgSiO$_3$	6.0	0.0002	7.8	0.03	130
BeO	6.5	0.0005	7.8	2.4	200
ZrSiO$_4$	7.8	0.001	4.8	0.05	320
Cordierite	5.0	0.003	1.8	0.01	140
SiC	45.0	–	4	1.5	600
Si$_3$N$_4$	8.1	0.0007	3.8	0.3	1000
Diamond	5.7	–	2.3	20	–
Polyimide	3.5	–	66	0.002	–
Glass ceramics	5.8	0.005	11.5	0.02	100

From Schneider et al. (1994).

ceramics but improvements up to about 35% are achieved by mullite-glass materials (Horiuchi et al. 1988, Ceramic Society of Japan, 2002, Ishihara, 1999).

By means of mullite-glass composites the dimensional accuracy of sintered bodies can be improved, which is a pre-condition for enlarging the chip size. An accuracy of about 0.5% can be achieved with mullite magnesium aluminosilicate (MgO–Al$_2$O$_3$–SiO$_2$) glass composites (Table 4.6.2, Ishihara, 1999). A further advantage of this composite is the availability of tungsten metal as wiring material. Tungsten has a similar thermal expansion to the substrate, and thus allows high-density wiring. Corresponding composites have been used for substrates at Hitachi Supercomputers (Kobayashi et al. 1990, see Fig. 4.6.5). On the other hand, mullite yttrium silica (Y$_2$O$_3$–SiO$_2$) glass composites (Horiuchi et al. 1988) are suitable substrates for a pin grid array package as documented by Sinko Electric Industries. Obviously, mullite-glass composites have many advantages as package substrate materials. It has to be kept in mind, however, that high sintering temperatures can have a negative effect.

In electronic devices, high temperature is becoming an important problem. For this reason, the amounts of aluminum nitride and mullite ceramics for substrates have increased. Mullite is a suitable cap material for aluminum nitride substrates because the two materials display similar thermal expansion coefficients.

4.6.3
Optical Materials

Different mullite and mullite-glass ceramics for optical applications have been examined. The center of interest has been the applicability as a window material

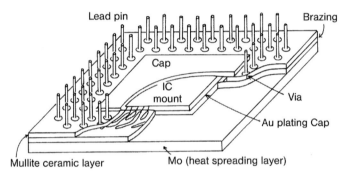

Fig. 4.6.5 Photograph and drawing illustrating the use of mullite in a ceramics-tungsten package in the Hitachi computer M-880 (Courtesy Hitachi Co., Japan).

for the mid-infrared range (3 to 5 μm wavelength) and in visible light, under chemically harsh, high temperature, and mechanically stressful environments (see Schneider et al. 1993a). Optically translucent mullite has been considered suitable for optical windows in the visible spectral range (Fig. 4.6.6) up to elevated temperatures. The optical application of mullite for luminescent window materials was studied by Wojtowicz Lempicki (1988). They prepared Cr^{3+}-doped mullite-glass ceramics, which are considered to be a potential host material for solid-state laser activators. More information on the optical properties of mullite are given in Sections 2.3.1 and 4.5.2.

4.6.4
Tribological Materials

Although the hardness of mullite is lower than that of alumina and silicon carbide, unlike these materials it shows little hardness degradation up to about 1000 °C (see

a)    b)

c)

Fig. 4.6.6 Optical translucent mullite ceramics. (a) Fused-mullite sintered at 1700°C and subsequently hot isostatically pressed at 1650°C and 1200 bar (see Fig. 4.2.6); (b) fused-mullite sintered as under (a) but not hot isostatically pressed; (c) reference photograph without any mullite ceramic (from Schneider et al. 1994).

Section 2.1.3). Therefore, mullite components have been successfully applied for brake linings and ceramic guides. Cermets containing mullite have been used for aircraft and rapid transit railway brake linings, generating high friction heat (Hanazawa, 1987). These brake linings consist of copper-tin or copper-iron alloys as matrices with mullite as dispersion phase, for enhancement of the friction resistance (friction agent). They also contain graphite or molybdenum disilicide ($MoSi_2$) as lubricating agents. Mullite-containing cermets display good long term heat resistance and thus are suitable for applications such as clutch facings for automobiles, where conventional brake lining materials made from composite-reinforced polymers cannot be applied.

Ceramic guides are used to suppress wear by friction between guides and running strings. Corresponding materials are requested to have high wear resistance and durability. Although alumina ceramics are mainly used for this application, mullite ceramics and porcelains have also been considered.

4.6.5
Porous Materials for Filters and Catalyst Supports

Porous mullite ceramics are important for applications such as filters and catalyst supports. Corresponding components and structures must be stable against high temperature cycling and should not react with polluting oxides in the exhaust gas. At the present time, most ceramics used for these purposes are honeycombs made of cordierite. However, because of the low corrosion resistance of cordierite, mullite and mullite-cordierite ceramics come into play for a new generation of ceramic honeycombs with improved properties. Mullite honeycombs for diesel particulate traps have been fabricated with the corrugation technique (Matsushita Electric Co., 1991) and as semi-closed-cell mullite foams from slurries by using the replication process, e. g. with polyurethane sponges as templates (Tulliani et al. 1999). These systems exhibit collection efficiencies of more than 90% at working temperatures up to 1200 °C. Further extension towards higher working temperatures is envisaged for these honeycombs. Dario and Bachiorrini (1999) described the interaction of polluting oxides of the exhaust gases with mullite in diesel vehicle filters. They found that mullite is stable in the presence of the oxides of calcium, zinc and vanadium but is attacked by sodium and lead oxides.

Other applications of porous mullite ceramics with high technical potential are high temperature gas filters for combustion furnaces such as incinerators and electric power generation stations. For these applications, high chemical corrosion resistance, high mechanical strength, high separation efficiency and low pressure drop during the dust separation must be guaranteed (see also Section 7.2.6).

Porous ceramics have also been foreseen as carriers for bioreactors (Horitsu, 1991). These porous ceramic components must be chemically stable in a wide range of solution acidity (pH), and must have appropriate pore size distributions and the capability to immobilize enzymes. Toriyama et al. (1992) investigated the suitability of porous zirconia, titania, alumina, zircon ($ZrSiO_4$) barium titanate ($BaTiO_3$), tieillite (Al_2TiO_5), forsterite (Mg_2SiO_4) and cordierite ($Mg_2Al_4Si_5O_{18}$) as carriers of enzymes. Actually porous mullite ceramics are found to be most suitable materials for the immobilization of and stabilization of enzymes. According to Horitsu (1991) porous mullite ceramics have also been applied as filters in breweries and for wastewater disposal plants. The pore size suitable for these applications is about 10 μm (Toriyama et al. 1992).

4.6.6
Materials for Miscellaneous Applications

Beads prepared from mullite have been used to replace cast iron sands, placing sands, sands for fluidizing bed furnaces and filters (Nagai Ceramics Co., 1992). The reason for this is the good thermal stability and low thermal conductivity of mullite.

To expand the application of solar cells, various low cost procedures have been examined: Polycrystalline silicon thin films on low cost substrates are considered as an alternative to bulk silicon-based cells. Slaoui et al. (2002) produced polycrystalline silicon thin films on mullite substrates by rapid thermal chemical vapor deposition at temperatures above 1000 °C. The performance of these systems was enhanced by zone melting recrystallization making use of the light trapping effect by the mullite substrate. Their highest efficiency was 8.2% on a 1 cm^2 cell size with a 20 μm thick silicon film.

Information on the oxygen ion conductivity of mullite was given by Fischer and Janke (1969) and Rommerskirchen et al. (1994). They mentioned that mullite is an oxygen ion conductor at temperatures above 1600 °C, satisfying the Nernst equation. This information has been supported by Suito et al. (1992). The authors suggested a use of mullite as an oxygen sensor, which works at 1600 °C even in very low oxygen concentration (< 10 ppm). As mentioned in Section 2.2.4.1 measurements of the oxygen diffusion in the indicated temperature range do not confirm this idea.

The application of ceramic coatings to prevent oxidation of metals at high temperatures has been extensively studied, especially with zirconia coatings. However, owing to the high oxygen ion diffusion in zirconia, oxidation of the metals and spallation of the ceramics can become a serious problem. Ramaswamy et al. (1998) investigated the effectiveness of mullite coatings on stainless steel. Mullite coatings being 425 μm thick actually show excellent stability and withstand more than 1000 cycles of rapid quenching from 1000 °C. This makes it a good candidate to protect metals from oxidation. More information on mullite protective coatings is given in Chapter 5.

References

Aksay, I. A. (1973). Diffusion and phase relationship studies in the alumina silica system. Ph. D. Thesis, University of California, Berkeley.

Aksay, I. A. and Pask, J. A. (1975). Stable and metastable equilibria in the system SiO_2–Al_2O_3. J. Am. Ceram. Soc. **58**, 507–512.

Aksay, I. A., Dabbs, D. M. and Sarikaya, M. (1991). Mullite for structural, electronic, and optical applications. J. Am. Ceram. Soc. **74**, 2343–2358.

Albers, W. (1993). Reaction sintering of quartz, cristobalite, and silica glass with α-Al_2O_3. Ph. D. Thesis, Univ. Hannover.

Al-Jarsha, Y.M.M., Biddle, K. D. and Das, A. K. (1985). Mullite formation from ethyl silicate and aluminium chlorides. J. Mater. Sci. **20**, 1773–1781.

Anilkumar, G. M., Mukundan, P., Damodaran, A. D. and Warrier, K.G.K. (1997). Effect of precursor pH on the formation characteristics of sol-gel mullite. Mat. Lett. **33**, 117–122.

Aoki, C., Ban, T., Hayashi, S. and Okada, K. (1992). Analysis of mullitization process by NMR-analysis for sol-gel solution. In: Abstracts of the Annual Meeting of the Ceramic Society of Japan, Paper No. 2E16, Ceramic Society of Japan, Tokyo.

Aramaki, S. and Roy, R. (1962). Revised phase diagram for the system Al_2O_3–SiO_2. J. Am. Ceram. Soc. **45**, 229–242.

Aramaki, S. and Roy, R. (1963). A new polymorph of Al_2SiO_5 and further studies in the system Al_2O_3–SiO_2–H_2O. Amer. Min. **48**, 1322–1347.

Arellano-Lopez, A. R. de, Melendez-Martinez, J.J., Cruse, T. A., Koritala, R. E., Routbort, J. L. and Goretta, K. C. (2002). Compressive creep of mullite containing Y_2O_3. Acta Mater. **50**, 4325–4338.

Armas, B., Sibieude, F., Mazel, A., Fourmeaux, R. and de Icaza Herrera, M. (2001). Low-pressure chemical vapor deposition of mullite layers using a cold-wall reactor. J. Surf. Coat. Techn. **141**, 88–95.

Ashizuka, M., Okuno, T. and Kubota, Y. (1989). Creep of mullite ceramics. J. Ceram. Soc. Jpn. **97**, 662–668.

Ashizuka, M., Honda, T. and Kubota, Y. (1991). Effect of grain size on creep in mullite ceramics. J. Ceram. Soc. Jpn. **99**, 292–295.

Ban, T. and Okada, K. (1992). Structure refinement of mullite by the Rietveld method and a new method for estimation of chemical composition. J. Am. Ceram. Soc. **75**, 227–230.

Baranwal, R., Villar, M. P., Garcia, R. and Laine, R. M. (2001). Flame spray pyrolysis of precursors as a route to nano-mullite powder: powder characterization and sintering behavior. J. Am. Ceram. Soc. **84**, 951–961.

Bartsch, M., Saruhan, B., Schmücker, M. and Schneider, H. (1999). Novel low-temperature processing route of dense mullite ceramics by reaction sintering of amorphous SiO_2-coated Al_2O_3 particle nanocomposites. J. Am. Ceram. Soc. **82**, 1388–1392.

Bauer, W. H. and Gordon, I. (1951). Flame fusion synthesis of several types of silicate structures. J. Am. Ceram. Soc. **34**, 250.

Bauer, W. H., Gordon, I. and Moore, C. H. (1950). Flame fusion synthesis of mullite single crystals. J. Am. Ceram. Soc. **33**, 140–143.

Boccaccini, A. R., Trusty, P. R. and Telle, R. (1996). Mullite fabrication from fumed silica and boehmite sol precursors. Mat. Lett. **29**, 171–176.

Boccaccini, A. R., Bucker, M., Kahlil, T. K. and Ponton, C. B. (1999). Fabrication of mullite ceramics by rotary forging and pressureless sintering. J. Europ. Ceram. Soc. **19**, 2613–2618.

Boch, P., Giry, J. P. and Rodrigo, P.D.D. (1986a). Reaction sintering of mullite-based ceramics. Workshop on Advanced Ceramics, Tokyo Institute of Technology, Tokyo.

Boch, P., Giry, J. P. and Rodrigo, P.D.D. (1986b). Reaction sintering of mullite and zirconia-mullite ceramics. In: Bunk, W.

and Hausner, H. (eds.) Ceramic materials and components for engines. Deutsche Keramische Gesellschaft, Köln, pp. 307–314.

Boch, P., Chartier, T. and Rodrigo, P.D.D. (1990). High-purity mullite ceramics by reaction-sintering. Ceramic Trans **6**, 353–374.

Bowen, N. L. and Greig, J. W. (1924). The system: Al_2O_3–SiO_2. J. Am. Ceram. Soc. **7**, 238–254.

Brindley, G. W. and McKinstry, H. A. (1961). The kaolinite-mullite reaction series: Part IV, The coordination number of aluminium. J. Am. Ceram. Soc. **44**, 506–507.

Brindley, G. W. and Nakahira, M. (1959a). The kaolinite-mullite reaction series: Part I, A survey of outstanding problems. J. Am. Ceram. Soc. **42**, 311–314.

Brindley, G. W. and Nakahira, M. (1959b). The kaolinite-mullite reaction series: Part II, Metakaolin. J. Am. Ceram. Soc. **42**, 314–318.

Brindley, G. W. and Nakahira, M. (1959c). The kaolinite-mullite reaction series: III. The high-temperature phases. J. Am. Ceram. Soc. **42**, 319–324.

Brinker, J. C. and Scherer, G. W. (1990). Sol-gel science, the physics and chemistry of sol-gel processing. Academic Press, New York, p. 908.

Brunauer, G. Frey, F., Boysen, H. and Schneider, H. (2001). High temperature thermal expansion of mullite: An in situ neutron diffraction study up to 1600 °C. J. Europ. Ceram. Soc., **21**, 2563–2567.

Bulens, M. and Delmon, B. (1977). The exothermic reaction of metakaolinite in the presence of mineralizers. Influence of crystallinity. Clay Clay Minerals **25**, 271–277.

Bulens, M., Leonard, A. and Delmon, B. (1978). Spectroscopic investigations of the kaolinite-mullite reaction sequence. J. Am. Ceram. Soc. **61**, 81–84.

Cameron, W. E. (1977). Composition and cell dimensions of mullite. Amer. Ceram. Soc. Bull. **56**, 1003–1011.

Camerucci, M. A., Urretavizcaya, G., Castro, M. S. and Cavalieri, A. L. (2001). Electrical properties and thermal expansion of cordierite and cordierite-mullite materials. J. Europ. Ceram. Soc., **21**, 2917–2923.

Ceramic Society of Japan (ed.) (1990). Hand-

book of Ceramics (2nd edition). Gihodo Publishing, Tokyo, pp. 2081–2084.

Ceramic Society of Japan (ed.) (2002) Handbook of Ceramics (2nd edition). Gihodo Publishing, Tokyo, Vol. 1, pp. 313–314.

Ceramic Society of Japan (ed.) (2002). Handbook of Ceramics (2nd edition). Gihodo Publishing, Tokyo, Vol. 2, pp. 1249–1250.

Chakraborty, A. K. and Das, S. (2003). Al–Si spinel phase formation in diphasic mullite gels. Ceram. Intern. **29**, 27–33.

Chatterjee, N. B. and Panti, B. N. (1965). Mullite refractories from clay-bauxite or quartz-bauxite mixes. Trans. Indian Ceram. Soc. **24**, 116–121.

Chatterjee, N. D. and Schreyer, W. (1972). The reaction enstatite$_{ss}$ + sillimanite = sappharine$_{ss}$ + quartz in the system MgO–Al_2O_3–SiO_2. Contrib. Mineral. Petrol. **36**, 49–62.

Chatterjee, M. , Naskar, M. K, Chakrabarty, P. K and Ganguli, D (2002). Mullite fibre mats by a sol-gel spinning technique. J. Sol-Gel Sci. Tech. **25**, 169–174.

Chaudhuri, S. P. (1969). X-ray study of induced mullitization of clay. Trans. Indian Ceram. Soc. **28**, 24.

Chaudhuri, S. P. (1982). Crystallization of glass in the system $K_2O(Na_2O)$–Al_2O_3–SiO_2. Ceramics Intern. **8**, 27–33.

Chen, Y. F. and Vilminot, S. (1995). Characterization of sol-gel mullite powders. Mat. Res. Bull. **30**, 291–298.

Chung, S.-L., Sheu, Y.-C. and Tsai, M.-S. (1992). Formation of SiO_2, Al_2O_3 and $3Al_2O_3 \cdot 2SiO_2$ particles in a counterflow diffusion flame. J. Am. Ceram. Soc. **75**, 117–123.

Dario, M. T. and Bachiorrini, A. (1999). Interaction of mullite with some polluting oxides in diesel vehicle filters. Ceramics Intern. **25**, 511–516.

Davis, R. F. and Pask, J. A. (1972). Diffusion and reaction studies in the system Al_2O_3–SiO_2. J. Am. Ceram. Soc. **55**, 524–531.

Davis, R. F., Aksay, I. A. and Pask, J. A. (1972). Decomposition of mullite. J. Am. Ceram. Soc. **55**, 98–101.

Debely, P. E., Barringer, E. A. and Bowen, H. K. (1985). Preparation and sintering behavior of fine grained Al_2O_3–SiO_2 composite. J. Am. Ceram. Soc. **68**, C76–C78.

Dokko, P. C., Pask, J. A. and Mazdiyasni, K. S.

(1977). High temperature mechanical properties of mullite under compression. J. Am. Ceram. Soc. **60**, 150–155.

Domrochev, G. A., Shcherbakov, V. I., Basova, G. V., Malysheva, I. P., Matveev, A. P. and Vasilevskaya, I. L. (1999). Mixed aluminum silicon alkoxides $(RO)_2AlOSi(OET)_3$. Russ. J. Gen. Chem. **69**, 1410–1413.

Ewell, R. H. and Insley, H. (1935). Thermal behavior of the kaolin minerals. J. Res. Nat. Bur. Standards **14**, 615–627.

Fenstermacher, J. E. and Hummel, F. A. (1961). High temperature mechanical properties of ceramic materials: Part IV, Sintered mullite bodies. J. Am. Ceram. Soc. **44**, 284–289.

Fischer, R. X., Schneider, H., Voll, D. (1996). Formation of aluminium rich 9:1 mullite and its transformation to low alumina mullite upon heating. J. Europ. Ceram. Soc. **16**, 109–113.

Fischer, R. X., Schneider H., and Voll, D. (1994). Crystal structure of Al-rich mullite. Amer. Min. **79**, 983–990.

Fischer, W. A. von and Janke, D. (1969). Aluminiumoxyd und Mullit als feste Elektrolyte in Sauerstoffmesszellen. Archiv Eisenhüttenwesen **40**, 707–716.

Ganguly, J. (1972). Staurolite stability and related paragenesis: Theory, experiments, and applications. J. Petrol. **13**, 335–365.

Ghate, B. B., Hasselman, D.P.H. and Spriggs, R. M. (1973). Synthesis and characterisation of high purity, fine grained mullite. Amer. Ceram. Soc. Bull. **52**, 670–672.

Ghate, B. B., Hasselman, D.P.H. and Spriggs, R. M. (1975). Kinetics of pressure-sintering and grain growth of ultra fine mullite powder. Ceramics Intern. **1**, 105–110.

Ghosh, N. N. and Pramanik, P. (2001). Aqueous sol-gel synthesis of nano-sized ceramic composite powders with metal-formate precursors. Mat. Sci. Eng. **C 16**, 113–117.

Grofczik, J. and Tamas, F. (1961). Mullite, its structure, formation and significance. Publ. House. Hungarian Acad. Sci., Budapest.

Guse, W. and Mateika, D. (1974). Growth of mullite single crystals $(2Al_2O_3SiO_2)$ by the Czochralski method. J. Crystal Growth **22**, 237–240.

Guse, W. and Saalfeld, H. (1990). X-ray characterization and structure refinement of a new cubic alumina phase $(\Omega\text{-}Al_2O_3)$ with

spinel-type structure. N. Jb. Min. Mh. **5**, 217–226.

Haggerty, J. S. and Chiang, Y. M. (1990). Reaction-based processing methods of ceramics and composites. Ceram. Eng. Sci. Proc. **11**, 757–794.

Hamano, K., Nakagawa, Z., Cun, G. and Sato, T. (1985). Formation process of mullite from kaolin minerals and various mixtures. In: Somiya, S. (ed.) Mullite. Uchida Rokakuho Publishing, Tokyo, pp. 37–49.

Hamano, K., Sato, T. and Nakagawa, Z. (1986). Properties of mullite powder prepared by co-precipitation and microstructure of fired bodies. Yogyo-Kyokaishi **94**, 818–822.

Hamano, K., Okada, S., Nakajima, H. and Okuda, F. (1988). Preparation of mullite ceramics from kaolin and aluminium hydroxide. In: Abstracts of the Annual Meeting of the Ceramic Society of Japan, Paper No. 2F05, Ceramic Society of Japan, Tokyo.

Hanazawa, T. (1987). Cermets for friction materials. Tetsu Hagane **73**, 786–795.

Hawkes, W. H. (1962). The production of synthetic mullite. Trans. Brit. Ceram. Soc. **61**, 689–703.

Hayashi, K., Kyaw, T. M. and Okamoto, Y. (1998). Thermal properties of mullite/partially stabilized zirconia composites. High Temp. High Press. **30**, 283–290.

Herbell, T. P., Hull, D. R. and Garg, A. (1998). Hot hydrogen exposure degradation of the strength of mullite. J. Am. Ceram. Soc. **81**, 910–916.

Hildmann, B. and Schneider, H. (2004). Structure control of the temperature conductivity in mullite. J. Am. Ceram. Soc., in press.

Hildmann, B., Ledbetter, H., Kim, S. and Schneider, H. (2001). Structural control of elastic constants of mullite in comparison to sillimanite. J. Am. Ceram. Soc. **84**, 2409–2414.

Hirata, Y., Minamizono, H. and Shimada, K. (1985). Property of SiO_2–Al_2O_3 powders prepared from metal alkoxide. Yogyo-Kyokaishi **93**, 46–54.

Hirata, Y., Sakeda, K., Matsushita, Y., Shimada, K. and Ishihara, Y. (1989). Characterization and sintering behaviour of alkoxide-derived aluminosilicate powders. J. Am. Ceram. Soc. **72**, 995–1002.

Hirata, Y., Aksay, I. A., Kurita, R., Hori, S. and

Koji, H. (1990). Processing of mullite with powders processed by chemical vapour deposition. Ceramic Trans. **6**, 323–338.

Hoffman, D. W., Roy, R. and Komarneni, S. (1984). Diphasic xerogels, a new class of materials: Phases in the system Al_2O_3–SiO_2. J. Am. Ceram. Soc. **67**, 468–471.

Hong, S.-H. and Messing, G. L. (1997). Mullite transformation kinetics in P_2O_5-, TiO_2-, and B_2O_3-doped aluminosilicate gels. J. Am. Ceram. Soc. **80**, 1551–1559.

Hong, S.-H. and Messing, G. L. (1998). Anisotropic grain growth in diphasic-gel-derived titania-doped mullite. J. Am. Ceram. Soc. **81**, 1269–1277.

Hong, S.-H. and Messing, G. L. (1999). Development of textured mullite by templated grain growth. J. Am. Ceram. Soc. **82**, 867–872.

Hori, S. and Kurita, R. (1990). Characterization and sintering of Al_2O_3–SiO_2 powders formed by chemical vapour deposition. Ceramic Trans. **6**, 311–322.

Horitsu, H. (1991). Ceramics bioreactor. Kinozairyou, **11**, 14–28.

Horiuchi, M., Mizushima, K., Takeuchi, Y. and Wakabayashi, S. (1988). New mullite ceramic packages and substrates. IEEE Trans. Comp. Hybrids Manuf. Technol. **11**, 439–446.

Horte, V.C.H. and Wiegmann, J. (1956). Study on the reaction between amorphous SiO_2 and Al_2O_3. Naturwiss. **43**, 9–10.

Hou, P., Basu, S. N. and Sarin, V. K. (1999). Nucleation mechanisms in chemically vapor-deposited mullite coatings on SiC. J. Mat. Res. 14, 2952–2958.

Huang, T., Rahaman, M. N., Mah, T.-I. and Parthasarathay, T. A. (2000). Anisotropic grain growth and microstructural evolution of dense mullite above 1550 °C. J. Am. Ceram. Soc. **83**, 204–210.

Huling, J. C. and Messing, G. L. (1989). Hybrid gels for homoepitactic nucleation of mullite. J. Am. Ceram. Soc. **72**, 1725–1729.

Huling, J. C. and Messing, G. L. (1990). Hybrid gels designed for mullite nucleation and crystallization control. In: Better Ceramics Through Chemistry IV, Materials Research Society Symposium Proceedings, Vol. 180 Zelinski, B.J.J., Brinker, C.J., Clark, D. E. and Ulrich, D. R. (eds.). Materials Research Society, Pittsburgh, PA, pp. 515–526.

Huling, J. C. and Messing, G. L. (1991). Epitactic nucleation of spinel in aluminum silicate gels and effect on mullite crystallization. J. Am. Ceram. Soc. **74**, 2374–2381.

Hvizdos, P. and Kasiarova, M. (2002). Indentation crack healing in low glass-content mullite. Key Eng. Mater. **223**, 257–260.

Iler, R. K. (1979). The Chemistry of Silica. Wiley, London and New York, pp. 172–177.

Imose, M., Takano, Y., Yoshinaka, M., Hirota, K. and Yamaguchi, O. (1998). Novel synthesis of mullite powder with high surface area. J. Am. Ceram. Soc. **81**, 1537–1540.

Ishihara, S. (1999). Doctoral thesis, Tokyo Institute of Technology, Tokyo, Japan.

Ishitsuka, M., Sato, T., Endo, T. and Shimada, M. (1987). Sintering and mechanical properties of Ytbria-doped tetragonal ZrO_2 polycryal/mullite composites. J. Am. Ceram. Soc. **70**, C-342–C-346.

Ismail, M.G.M.U., Nakai, Z., Minegishi, K. and Somiya, S. (1986). Synthesis of mullite powder and its characteristics. Intern. J. High. Technol. Ceram. **3**, 123–134.

Ismail, M.G.M.U., Nakai, Z. and Somiya, S. (1987). Microstructure and mechanical properties of mullite prepared by the sol-gel method. J. Am. Ceram. Soc. **70**, C7–C8.

Ismail, M.G.M.U., Nakai, Z. and Somiya, S. (1990). Sintering of mullite prepared by sol-gel method. Ceramic Trans. **6**, 231–241.

Itatani, K., Kubozono, T., Howell, F. S., Kishioka, A. and Kinoshita, M. (1995). Some properties of mullite powders prepared by chemical vapor deposition. J. Mat. Sci. **30**, 1158–1165.

Itoh, M., Hamano, K. and Okada, S. (1990). Preparation of mullite ceramics from kaolin and aluminum hydroxide. In: Abstracts of the 3rd Autumn Symposium on Ceramics, Paper No.6-2A13. Ceramics Society of Japan, Tokyo.

Ivankovic, H., Tkalcec, E., Nass, R. and Schmidt, H. (2003). Correlation of the precursor type with densification behavior and microstructure of sintered mullites. J. Europ. Ceram. Soc. **23**, 283–292.

Jacobson, N. S., Lee, K. N. and Yoshio, T. (1996). Corrosion of mullite by molten salts. J. Am. Ceram. Soc. **79**, 2161–2167.

Janackovic, D., Jokanovic, V., Kostic-Gvozde-novic, L., Zivkovic, L. and Uskokovic, D. (1996). Synthesis, morphology, and formation mechanism of mullite particles produced by ultrasonic spray pyrolysis. J. Mat. Res. **11**, 1706–1716.

Janackovic, D., Jokanovic, V., Kostic-Gvozde-novic, L. and Uskokovic, D. (1998). Synthesis of mullite nanostructured spherical powder by ultrasonic spray pyrolysis. Nanostructured Mater. **10**, 341–348.

Jaymes, I. and Douy, A. (1995). Homogeneous precipitation of mullite precursors. J. Sol-Gel Sci. Tech. **4**, 7–13.

Johnson, S. and Pask, J.A. (1982). Role of impurities on formation of mullite from kaolinite and $Al_2O_3–SiO_2$ mixtures. Amer. Ceram. Soc. Bull. **61**, 838–842.

Kamiya, H., Suzuki, H., Ichikawa, T., Cho, Y.I. and Horio, M. (1998). Densification of sol-gel derived mullite ceramics after cold isostatic pressing up to 1 GPa. J. Am. Ceram. Soc. **81**, 173–179.

Kanka, B. and Schneider, H. (1994). Sintering mechanisms and microstructural development of coprecitated mullite. J. Mater. Sci. **29**, 1239–1249.

Kansal, P., Laine, R.M. and Babonneau, F. (1997). A processable mullite precursor prepared by reacting silica and aluminum hydroxide with triethanolamine in ethyleneglycol: structural evolution on pyrolysis. J. Am. Ceram. Soc. **80**, 2597–2606.

Kanzaki, S. and Tabata, H. (1985). Sintering and mechanical property of spray pyrolized mullite powder. In: Somiya, S. (ed.). New Materials. Series-Mullite. Uchida-Roka-kuho, Tokyo, pp. 51–61.

Kanzaki, S., Tabata, H., Kumazawa, T. and Ohta, S. (1985). Sintering and mechanical properties of stoichiometric mullite. J. Am. Ceram. Soc. **68**, C6–C7.

Kanzaki, S., Tabata, H. and Kumazawa, T. (1990). Sintering and mechanical properties of mullite derived via spray pyrolysis. Ceramic Trans. **6**, 339–351.

Kara, F., Turan, S., Little, J.A. and Knowles, K.M. (2000). Microstructural characterization of reaction sintered mullites. J. Am. Ceram. Soc. **83**, 369–376.

Kawai, S., Yoshida, M. and Hashizume, G. (1990). Preparation of mullite from kaolin by dry-grinding. J. Ceram. Soc. Jpn. **98**, 669–674.

Kaya, C. and Boccaccini, A.R. (2002). Microstructural variations in mullite ceramics derived from diphasic sols using different forming techniques. J. Mat. Res. **17**, 3000–3003.

Kaya, C., Kaya, F. and Boccaccini, A.R. (2002). Colloidal processing of glassy-phase-free mullite from heterocoagulated boehmite/silica nanocomposite sol particles. Adv. Eng. Mater. **4**, 21–28.

Kaya, C., Butler, E.G. and Lewis, M.H. (2003). Microstructurally controlled mullite ceramics produced from monophasic and diphasic sol-derived pastes using extrusion. J. Mat. Sci. **38**, 767–777.

Kazakova, I.L., Vol'khin, V.V., Pongratz, P. and Halfaks, E. (1999). Formation of mullite phases in oxide mixtures prepared from SiO_2 sols and aluminum salts. Russ. J. Appl. Chem. **72**, 1073–1078.

Keefer, K.D. (1984). The effect of hydrolysis conditions on the structure and growth of silicate polymers, In: Brinker, C.J., Clark, D.E. and Ulrich, D.R. (eds.) Better ceramics through chemistry. North-Holland, New York, pp. 15–24.

Kim, J.W., Lee, J.E., Jung, Y.G., Jo, C.Y., Lee, J.H. and Paik, U. (2003). Synthesis behavior and grain morphology in mullite ceramics with precursor pH and sintering temperature. J. Mat. Res. **18**, 81–87.

Kingery, W.D., Bowen, H.K. and Uhlmann, D.R. (1976). Introduction to Ceramics. 2nd edn. Wiley, New York.

Kobayashi, F., Murata, S., Watanabe, Y., Zushi, Y., Yamamoto, M., Kobayashi, T. and Kodaka, M. (1990). Packaging systems and hardware technology on large-sized computer M-880. Nikkei Electronics, **12**, 209–224.

König, G. (1978). Synthetische keramische Rohstoffe in der Feuerfestindustrie. Ber. Dt. Keram. Ges. **55**, 229–232.

Kollenberg, W. and Schneider, H. (1989). Microhardness of mullite at temperatures up to 1000 °C. J. Am. Ceram. Soc. **72**, 1739–1740.

Komarneni, S. (1992). Nanocomposites. J. Mat. Chem. **2**, 1219–1230.

Komarneni, S. and Rani, L. (1992). Nanocomposite sol-gel route to low K ceramic substrates. In: Emerging Optoelectronic Technologies: Proceedings of the Conference on Emerging Optoelectronic Technologies Bangalore, India, Dec. 16–20, 1991 (eds. A. Selvarajan et al.) pp. 147–150.

Komarneni, S. and Roy , R. (1990). Mullite derived from diphasic nanocomposite gels. In Mullite and Mullite Matrix Composites, edited by S. Somiya, R. F. Davis, and J. A. Pask, Ceramic Trans., **6**, pp. 209–220.

Komarneni, S., Roy, R., Fyfe, C. A., Kennedy, G. J. and Strobl, H. (1986a). Solid-state ^{27}Al and ^{29}Si magic-angle spinning NMR of aluminosiliate gels. J. Am. Ceram. Soc. **69**, C42–C44.

Komarneni, S., Suwa, Y. and Roy, R. (1986b). Application of compositionally diphasic xerogels for enhanced densification: The system Al_2O_3–SiO_2. J. Am. Ceram. Soc. **69**, C-155–C-156.

Komarneni, S., Roy, R., Breval, E., Ollinen, M. and Suwa, Y. (1986c). Hydrothermal route to ultrafine powders utilizing single and di-phasic gels. Ad. Ceramic Mater. **1**, 87–92.

Komarneni, S., Breval, E. and Roy, R. (1988). Microwave preparation of mullite powders processing of materials. Mat. Res. Soc. Proc., Pittsburgh, PA, pp. 235–238.

Kong, L. B., Gan, Y. B., Ma, J., Zhang, T. S., Boey, F. and Zhang, R. F. (2003). Mullite phase formation and reaction sequences with the presence of pentoxides. J. Alloys Comp. **351**, 264–272.

Kriven, W. M. and Pask, J. A. (1983). Solid solution range and microstructure of melt-grown mullite. J. Am. Ceram. Soc. **66**, 649–654.

Kriven, W. M., Palko, J. W., Sinogeikin, S., Bass, J. D., Sayir, A., Brunauer, G., Boysen, H., Frey, F. and Schneider, J. (1999). High temperature single crystal properties of mullite. J. Europ. Ceram Soc. **19**, 2529–2541.

Kriven, W. M., Siah, L. F., Schmücker, M. and Schneider, H. (2004). High temperature microhardness of single crystal mullite. J. Am. Ceram. Soc. **87**, 970–972.

Kubota, Y. and Takagi, H. (1987). Preparation and mechanical properties of mullites and mullite-zirconia composites. Sci. Report of Toyo Soda **31**, 11–21.

Kubota, Y. and Takagi, H. (1988). Preparation and mechanical properties of mullite-zirconia composites. In: Somiya, S., Yamamoto, N. and Yanagida, H. (eds.). Advances in Ceramics, Vol. **24**, Science and Technology of Zirconia III, pp. 999–1005.

Kubota, Y., Ashizuka, M. and Ishida, E. (1994). Static fatigue and fracture toughness of mullite ceramics at 1200 °C. J. Ceram. Soc. Jpn., **102**, 805–809.

Kumazawa, T., Kanzaki, S., Wakai, T. and Tabata, H. (1987). Creep of mullite ceramics. In: Abstracts of the 25th Symposium on the Basic Science of Ceramics, Paper No. 1E08. Ceramic Society of Japan, Tokyo.

Kumazawa, T., Kanzaki, S., Ohta, S. and Tabata, H. (1988). Influence of chemical composition on the mechanical properties of SiO_2–Al_2O_3 ceramics. J. Ceram. Soc. Jpn., **96**, 85–91.

Kutty, T.R.N. and Nayak, M. (2000). Photoluminescence of Eu^{3+}-doped mullite ($xAl_2O_3.ySiO_2$; z/y=3/2 and 2/1) prepared by a hydrothermal method. J. Mater. Chem. Phys. **65**, 158–165.

Lay, L. A. (1984). Corrosion Resistance of Technical Ceramics. National Physical Laboratory, Teddington, Middlesex, UK.

Ledbetter, L., Kim, S., Balzar, D., Crudele, S. and Kriven, W. (1998). Elastic properties of mullite. J. Am. Ceram. Soc. **81**, 1025–1028.

Lee, J. E., Kim, J. W., Jung, Y. G., Jo, C. Y. and Palk, U. (2002). Effects of precursor p_H and sintering temperature on synthesizing and morphology of sol-gel processed mullite. Ceramics Intern. **28**, 935–940.

Lessing, P. A., Gordon, R. S. and Mazdiyasni, K. S. (1975). Creep of polycrystalline mullite. J. Am. Ceram. Soc. **58**, 149.

Li, D. X. and Thomson, W. J. (1990). Mullite formation kinetics of a single-phase gel. J. Am. Ceram. Soc. **73**, 964–969.

Li, D. X. and Thomson, W. J. (1991). Tetragonal to orthorhombic transformation during mullite formation. J. Mat. Res. **6**, 819–824.

Locsei, B. P. (1968) Possible technical developments in the manufacture of fireclay refractory materials: III, Mechanisms of the synthesis of mullite in the system kaolinite-aluminum fluoride. Keram. Z. **20**, 362–367.

Locsei, B. (1963). Solid phase kinetics of the reaction between AlF_3 and SiO_2. Proceed. 6th Conf. Silicate Ind., 291, Budapest.

Loughnan, F. C. (1969). Chemical Weathering of the Silicate Minerals. Elsevier, pp. 32–34.

Lysak, S. V. and Drizheruk, M. E. (1977).

Effect of gaseous atmosphere on mullitization of alumo-silicates. Neorgan. Mater. (Moskva) **13**, 1686–1690.

Majdic, A., Routschka, G., Köhler, E. K., Schulte, K., Wecht, P. (1980). Studien an Mullit- und Sinter-Korundsteinen: Teil I: Technologische Eigenschaften. Keram. Z. **31**, 494–497, 547–550.

Matsushita Electric Co. (1991). Catalogue. Matsushita Electric Co., Kadoma, Japan.

Mazdiyasni, K. S. and Brown, L. M. (1972). Synthesis and mechanical properties of stoichiometric aluminum silicate (Mullite). J. Am. Ceram. Soc. **55**, 548–552.

Mechnich, P., Schneider, H., Schmücker, M. and Saruhan, B. (1998). Accelerated reaction bonding of mullite. J. Am. Ceram. Soc. **81**, 1931–1937.

Mechnich, P., Schmücker, M. and Schneider, H. (1999). Reaction sequence and microstructrual development of CeO_2-doped reaction-bonded mullite. J. Am. Ceram. Soc. **82**, 2517–2522.

McGee, T. D. and Wirkus, C. D. (1972). Mullitization of aluminium-silicate gels. Amer. Ceram. Soc. Bull. **51**, 577–581.

Meng, G.-Y. and Huggins, R. A. (1983). A new chemical method for preparation of both pure and doped mullite. Mat. Res. Bull. **18**, 581–588.

Metcalfe, B. L. and Sant, J. H. (1975). The synthesis, microstructure, and physical properties of high purity mullite. Trans. Brit. Ceram. Soc. **74**, 193–201.

Miller, D. G., Singleton, R. H. and Wallace, A. V. (1966). Metal fiber reinforced ceramic composites. Amer. Ceram. Soc. Bull. **45**, 513–517.

Mitachi, S., Matsuzawa, M., Kaneko, K., Kanzaki, S. and Tabata, H. (1990). Characterization of SiO_2–Al_2O_3 powders prepared from metal alkoxides. In: Somiya, S., Davis, R. F. and Pask, J. A. Mullite and Mullite Matrix Composites. Ceramic Trans. **6**, 275–286.

Mizukami, F., Maeda, K. and Toba, M. (1990). Effect of organic ligands on the formation of mullite. In: Abstracts of the 3rd Autumn Symposium of Ceramics, Paper No. 6-1A01, Ceramic Society of Japan, Tokyo.

Mizuno, M. (1991). Microstructure, microchemistry and flexural strength of mullite ceramics. J. Am. Ceram. Soc. **74**, 3017–3022.

Mizuno, M. and Saito, H. (1989). Preparation of highly pure fine mullite powder. J. Am. Ceram. Soc. **72**, 377–382.

Monteiro, O. R., Wang, Z. and Brown, I. G. (1997). Deposition of mullite and mullite-like coatings on silicon carbide by dual-source metal plasma immersion. J. Mat. Res. **12**, 2401–2410.

Moore, K. A., Cesarano, J., Smith, D. M. and Kodas, T. T. (1992). Synthesis of submicrometer mullite powder via high temperature aerosol decomposition. J. Am. Ceram. Soc. **75**, 213–215.

Moya, J. S., Valle, J. and Aza, de S. (1982). The sintering behavior of an active premullite powder obtained from kandites (Pejovnik, S. and Ristic, M. M. eds.). In: Sintering-Theory and Practice, Proc. Of the 5th Int. Round Table Conf. On Sintering Materials Science Monographs, Vol. 14, Elsevier, Amsterdam, pp. 409–415.

Mroz, T. J. Jr and Laughner , J. W. (1989). Microstructures of mullite sintered from seeded sol-gels. J. Am. Ceram. Soc. **72**, 508–509.

Müller, B. and Schneider, H. (1993a). Density and microstructure of slip-casted fused-mullite ceramics obtained from pressureless sintering and hot isostatic pressing (HIP) techniques. Unpublished results.

Müller, B. and Schneider, H. (1993b). Reaction sintering of slip casted quartz plus α-Al_2O_3. Unpublished results.

Mulpuri, R. P. and Sarin, V. K. (1996). Synthesis of mullite coatings by chemical vapor deposition. J. Mat. Res. **11**, 1315–1324.

Murthy, M. K. and Hummel, F. A. (1960). X-ray study of the solid solution of TiO_2, Fe_2O_3 and Cr_2O_3 in mullite ($3Al_2O_3 \cdot 2SiO_2$). J. Am. Ceram. Soc. **43**, 267–272.

Naigai Ceramics Co. (1992). Catalogue. Naigai Ceramics Co., Seto, Japan.

Naskar, M. K., Chatterjee, M. and Lakshmi, N. S. (2002). Sol-emulsion-gel synthesis of hollow mullite microspheres. J. Mat. Sci. **37**, 343–348.

Nishio, T., Kijima, K., Kajiwara, K. and Fujiki, Y. (1994). The influence of preparation procedure in the mullite preparation by solution method to the mixing of Al and Si and the crystallization behavior. J. Ceram Soc. Jpn. **102**, 462–470.

Nishu, K., Yokoyama, T., Watanabe, T. and Tarutani, T. (1989). Characterization of amorphous aluminosilicate formed by adsorption of silicic acid on aluminium hydroxide. In: Abstracts of the 27th Symposium of the Basic Science of Ceramics, Paper No. 1B08. Ceramic Society of Japan, Tokyo.

Nixon, R. D., Chevacharoenkul, S., Davis, R. F. and Tiegs, T. N. (1990). Creep of hot-pressed SiC-whisker-reinforced mullite. Ceramic Trans. **6**, 579–603.

Nogami, M., Shan, C., Moriya, S. and Nagasaka, K. (1990). Effect of hydrolysis conditions on the structure of gels prepared from metal alkoxides in the system Al_2O_3–SiO_2. J. Ceram. Soc. Jpn. **98**, 93–97.

Nurishi, Y. and Pask, J. A. (1982). Sintering of α-Al_2O_3 amorphous silica compacts. Ceramics Intern. **8**, 57–59.

Ogawa, M. and Abe, S. (1990). Synthesis of oxide powders by metal powders combustion method. In: Abstracts of the Annual Meeting of the Ceramic Society of Japan, Paper No. 2G06, Ceramic Society of Japan, Tokyo.

Ohira, H., Shiga, H., Ismail, M. G. M. U., Nakai, Z., Akiba, T. and Yasuda, E. (1991). Compressive creep of mullite ceramics. J. Mater. Sci. Lett. **10**, 847–849.

Ohira, H., Ismail, M. G. M. U., Yamamoto, Y., Akiba, T. and Somiya, S. (1996). Mechanical properties of high purity mullite at elevated temperatures. J. Europ. Ceram. Soc. **16**, 225–229.

Ohnishi, H., Kawanami, T., Nakahira, A. and Niihara, K. (1990). Microstructure and mechanical properties of mullite ceramics. J. Ceram. Soc. Jpn. **98**, 541–547.

Okada, K. (1991). Synthetic ceramic raw materials. In: Handbook of Advanced Ceramics. Ceramic Society of Japan, Tokyo, pp. 132.

Okada, K. and Otsuka, N. (1986). Characterization of the spinel phase from SiO_2–Al_2O_3 xerogels and the formation process of mullite. J. Am. Ceram. Soc. **69**, 652–656.

Okada, K. and Otsuka, N. (1987). Change in chemical composition of mullite formed from $2SiO_2$ $3Al_2O_3$ xerogel during the formation process. J. Am. Ceram. Soc. **70**, C245–C247.

Okada, K. and Otsuka N. (1988). Chemical composition change of mullite during formation process. Sci. Ceramics **14**, 497–502.

Okada, K. and Otsuka, N. (1990a). Formation process of mullite. Ceramic Trans. **6**, 375–387.

Okada, K. and Otsuka, N. (1990b). Preparation of transparent mullite films by dip-coating method. Ceramic Trans. **6**, 425–434.

Okada, K., Otsuka, N. and Ossaka, J. (1986a). Characterization of the spinel phase formed in the kaolin-mullite thermal sequence. J. Am. Ceram. Soc. **69**, C251–C253.

Okada, K., Hoshi, Y. and Otsuka, N. (1986b). Formation reaction of mullite from SiO_2–Al_2O_3 xerogels. J. Mater. Sci. Lett. **5**, 1315–1318.

Okada, K., Otsuka, N. and Somiya, S. (1991). Review of mullite synthesis routes in Japan. Amer. Ceram. Soc. Bull. **70**, 1633–1640.

Okamoto, Y., Fukudome, H., Hayashi, K. and Nishikawa, T. (1990). Creep deformation of polycrystalline mullite. J. Europ. Ceram. Soc. **6**, 161–168.

Osendi, M. I. and Miranzo, P. (1997). Thermal evolution and sintering behavior of a 2:1 mullite gel. J. Am. Ceram. Soc. **80**, 1573–1578.

Ossaka, J. (1961). Tetragonal mullite-like phase from coprecipitated gels. Nature **19**, 1000–1001.

Pach, L., Komarneni, S. and Liu, C. (1995). Porous mullite ceramics through diphasic gels of large boehmite and small silica particles. J. Porous Mater. **1**, 155–163.

Pach, L., Iratni, A., Kovar, V., Mankos, P., and Komarneni, S. (1996). Sintering of diphasic mullite gel. J. Europ. Ceram. Soc. **16**, 561–566.

Pankratz, L. B., Weller, W. W. and Kelley, K. K. (1963). Low-temperature heat content of mullite. U. S. Bur. Mines, Rept. Invest. No. 6287, 7.

Pascual, J., Zapatero, J., de Haro, M.C.J., Varona, I., Justo, A., Perez-Rodriguez, J. L. and Sanchez-Soto, P. J. (2000). Porous Mullite and mullite-based composites by chemical processing of kaolinite and aluminum metal wastes. J. Mat. Chem. **10**, 1409–1414.

Pask, J. A., Zhang, X. W. and Tomsia, A. P.

(1987). Effect of sol-gel mixing on mullite microstructure and phase equilibria in the α-Al$_2$O$_3$–SiO$_2$ system. J. Am. Ceram. Soc. **70**, 704–707.

Paulick, L.A., Yu, Y-F. and Mah, T-I. (1987). Ceramic powders from metal alkoxide precursors in ceramic powder science, Advances in Ceramics, Vol. 21. (Messing, G.L., McCauley, J.W. and Haber, R.A. Eds.), Amer. Ceram. Soc., pp. 121–129.

Penty, R.A. and Hasselman, D.P.H. (1972). Creep kinetics of high purity, ultrafine grain polycrystalline mullite. Mater. Res. Bull. **7**, 1117–1123.

Pitchford, J.E., Stearn, R.J., Kelly, A. and Clegg, W.J. (2001). Effect of oxygen vacancies on the hot hardness of mullite. J. Am. Ceram. Soc. **84**, 1167–1168.

Pouxviel, J.C., Boilot, J.P., Dauger, A. and Huber, L. (1986). Chemical route to alumino-silicate gels, glasses and ceramics. In: Brinker, C.J., Clark, D.E. and Ulrich, D.R. (eds.) Better Ceramics through Chemistry II. Materials Research Society, Pittsburgh, pp. 259–274.

Prochazka, S. and Klug, F.J. (1983). Infrared-transparent mullite ceramic. J. Am. Ceram. Soc. **66**, 874–880.

Provost, G. and Le Doussal, H. (1974a). Untersuchung des Fließverhaltens feuerfester Werkstoffe bei unterschiedlichen Ofenatmosphären (Teil 1). Tonind. Ztg. **98**, 103–109.

Provost, G. and Le Doussal, H. (1974b). Untersuchung des Fließverhaltens feuerfester Werkstoffe bei unterschiedlichen Ofenatmosphären (Teil 2). Tonind. Ztg. **98**, 168–172.

Quinn, G.D. (1990). Fracture mechanism maps for advanced structural ceramics. J. Mater. Sci. **25**, 4361–4376.

Ramaswamy, P., Seetharamu, S., Varma, K.B.R. and Rao, K.J. (1998). Thermal shock characteristics of plasma sprayed mullite coatings. J. Therm. Spray Techn. **7**, 497–504.

Rana, A.P.S, Aiko, O. and Pask, J.A. (1982). Sintering of α-Al$_2$O$_3$/quartz, and α-Al$_2$O$_3$/cristobalite related to mullite formation. Ceramics Intern. **8**, 151–153.

Ravi, B.G., Praveen, V., Panneer Selvam, M. and Rao, K.J. (1998). Microwave-assisted preparation and sintering of mullite and

mullite-zirconia composites from metal organics. Mat. Res. Bull. **33**, 1527–1536.

Regiani, I., Magalhaes, W.L.E., Ferreira de Souza, D.P., Paiva-Santo, C.O. and Ferreira de Rhanim, H., Olagnon, C., Fantozzi, G. and Torrecillas, R. (1997). Experimental characterization of high temperature creep resistance of mullite. Ceramics Intern. **23**, 497–507.

Risbud, S.H. and Pask, J.A. (1978). Mullite crystallization from SiO$_2$–Al$_2$O$_3$ melts. J. Am. Ceram. Soc. **61**, 63–67.

Rodrigo, P.D.D. (1986). Reaction sintering of mullite-based ceramics. Ph.D. thesis, University of Limoges, Limoges.

Rodrigo, P.D.D. and Boch, P. (1985). High purity mullite ceramics by reaction sintering. Int. J. High Techn. Ceramics **1**, 3–30.

Rommerskirchen, I., Chavez, F. and Janke, D. (1994). Ionic conduction behaviour of mullite (3Al$_2$O$_3$ 2SiO$_2$) at 1400 to 1600 °C. Solid State Ionics **74**, 179–187.

Rossini, A.R., Arazi, S.C. and Krenkel, T.G. (1970). Mullitization of mixtures of kaolinitic clay and aluminium hydroxides. Bol. Soc. Espan. Ceram. **9**, 579–591.

Roy, D.M. and Roy, R. (1955). Synthesis and stability of minerals in the system MgO–Al$_2$O$_3$–SiO$_2$–H$_2$O. Amer. Min. **40**, 147–178.

Roy, R. (1956). Aids in hydrothermal experimentation; II, Methods of making mixtures for both "dry" and "wet" phase equilibrium studies. J. Am. Ceram. Soc. **39**, 145–146.

Roy, R. and Komarneni, S. (1989). U.S. Patent No. 4, 828,031

Roy, R. and Osborn, E.F. (1952). Studies in the system alumina–silica–water. In: Proc. Problems of Clay and Laterite Genesis symposium, Amer. Inst. Mining and Metallurgical Engineers, New York, pp. 76–80.

Roy, R. and Osborn, E.F. (1954). The system Al$_2$O$_3$–SiO$_2$–H$_2$O. Amer. Min. **39**, 853–885.

Ruh, R. and Chizever, H.M. (1998). Permittivity and permeability of mullite-SiC whisker and spinel-SiC-whisker composites. J. Am. Ceram. Soc., **81**, 1069–1070.

Ruh, R., Mazdiyasni, K.S. and Mendiratta, M.G. (1988). Mechanical and microstructural characterisation of mullite and mullite-SiC-whisker and ZrO$_2$-toughened-mullite-SiC-whisker composites. J. Am. Ceram. Soc. **71**, 503–512.

Russell, L. M., Donaldson, K. Y., Hasselman, D. P. H., Ruh, R. and Adams, J. W. (1996). Thermal diffusicity/conductivity and specific heat of mullite-zirconia-silicon carbide whisker composites. J. Am. Ceram. Soc. **79**, 2767–2770.

Sacks, M. D. (1979). Sintering behaviour of mullite-containing materials. Ph. D. Thesis, University of California, Berkeley.

Sacks, M. D. and Pask, J. A. (1978). Sintering of mullite. In: Palmour III, H., Davis, R. F. and Hare, T. M. (eds.). Processing of Crystalline Ceramics, Materials Science Research, Vol. II, Plenum, New York, pp. 193–203.

Sacks, M. D. and Pask, J. A. (1982a). Sintering of mullite-containing materials: I, Effect of composition. J. Am. Ceram. Soc. **65**, 65–70.

Sacks, M. D. and Pask, J. A. (1982b). Sintering of mullite-containing materials: II, Effect of agglomeration. J. Am. Ceram. Soc. **65**, 70–77.

Sacks, M. D., Lee, H.-W. and Pask, J. A. (1990). A review of powder preparation methods and densification procedures for fabricating high density mullite. Ceramic Trans. **6**, 167–207.

Sacks, M. D., Bozkurt, N. and Scheiffele, G. W. (1991). Fabrication of mullite and mullite-matrix composites by transient viscous sintering of composite powders. J. Am. Ceram. Soc. **74**, 2428–2437.

Sacks, M. D., Wang, K., Scheiffele, G. W. and Bozkurt, N. (1997). Effect of composition on mullitization behavior of α-alumina/silica microcomposite powders. J. Am. Ceram. Soc. **80**, 663–672.

Sadanaga, R., Tokonami, M. and Takeuchi, Y. (1962). The structure of mullite, $2Al_2O_3 \cdot SiO_2$, and relationship with the structures of sillimanite and andalusite. Acta Cryst. **15**, 65–68.

Sakurai, O., Mizutani, N. and Kato, M. (1988). Preparation of mullite powders from metal alkoxides by ultrasonic spray pyrolysis. J. Ceram. Soc. Jpn. **96**, 639–645.

Sales, M. and Alarcon, J. (1996). Synthesis and phase transformations of mullites obtained from SiO_2–Al_2O_3 gels, J. Europ. Ceram. Soc. **16**, 781–789.

Saruhan, B., Voss, U. and Schneider, H. (1994). Solid solution range of mullite up to 1800 °C and microstructural development of ceramics. J. Mater. Sci. **29**, 3261–3268.

Schmücker, M., Albers, W. and Schneider, H. (1993). Mullite formation by reaction sintering of quartz and α-Al_2O_3 – A TEM study. J. Europ. Ceram. Soc. **15**, 511–515.

Schneider, H. and Eberhard, E. (1990). Thermal expansion of mullite. J. Am. Ceram. Soc., **73**, 2073–2076.

Schneider, H., Seifert-Kraus, U. and Majdic, A. (1982). Microchemistry of refractory-grade bauxites. Amer. Ceram. Soc. Bull. **61**, 741–745.

Schneider, H., Wang, J. and Majdic, A. (1986). Thermal expansion of refractory-grade bauxites at high temperature. Ber. Dtsch. Keram. Ges. **63**, 461–470.

Schneider, H., Wang, J. and Majdic, A. (1987). Firing of refractory-grade Chinese bauxites under oxidizing and reducing atmospheres. Ber. Dtsch. Keram. Ges. **64**, 28–31.

Schneider, H., Schmücker, M., Ikeda, K. and Kaysser, W. A. (1993a). Optically translucent mullite ceramics. J. Am. Ceram. Soc. **76**, 2912–2914.

Schneider, H., Saruhan, B., Voll, D., Merwin, L. and Sebald, A. (1993b). Mullite precursor phases. J. Europ. Ceram. Soc. **11**, 87–94.

Schneider, H., Fischer, R. X. and Voll, D. (1993c). A mullite with lattice constants $a > b$. J. Am. Ceram. Soc. **76**, 1879–1881.

Schneider, H., Saruhan, B., Voll, D., Merwin, L. and Sebald, A. (1993d). Mullite precursor phases. J. Europ. Ceram. Soc. **11**, 87–94.

Schneider, H., Okada, K. and Pask, J. A. (1994). Mullite and Mullite Ceramics. Wiley, Chichester, pp. 1–251.

Schneider, J. and Schneider , H. (2004). Temperature-induced strucural expansion of mullite. Unpublished results.

Schreuer, J., Hildmann, B. and Schneider, H. (2005). New data on the elastic constants of mullite up to 1400 °C. Unpublished results.

Senda, T., Saruta, M. and Ochi, Y. (1994). Tribology of mullite ceramics at elevated temperatures. J. Ceram. Soc. Jpn. **102**, 556–561.

Senda, T., Drennan, J. and McPherson, R. (1995). Sliding wear of oxide ceramics at elevated temperatures. J. Am. Ceram. Soc. **78**, 3018–3024.

Shiga, H., Ismail, M.G.M.U. and Katayama, K. (1991). Sintering of ZrO_2 toughened mullite ceramics and its microstructure. J. Ceram. Soc. Jpn. **99**, 798–802.

Shindo, I. (1990). The melting nature and growth of mullite single crystals by the modified floating zone method. Ceramic Trans. **6**, 103–113.

Simendic, B. and Radonjic, L. (1998). Low temperature transformation of alumino-silicate gels. Ceramics Intern. **24**, 553–557.

Sin, A., Picciolo, J.J., Lee, R.H., Gutierrez-Mora, F. and Goretta, K.C. (2001). Synthesis of mullite powders by acrylamide polymerisation. J. Mat. Sci. Lett. **20**, 1639–1641.

Sivakumar, R., Doni Jayaseelan, D., Nishikawa, T., Honda, S. and Awaji, H. (2001). Influence of MgO on microstructure and properties of mullite–Mo composites fabricated by pulse electric current sintering. Ceramics Intern., **27**, 537–541.

Slaoui, A., Bourdais, S., Beaucarne, G., Poortmans, J. and Reber, S. (2002). Polycrystalline silicon solar cells on mullite substrates. Solar Energy Mater. Solar Cells **71**, 245–252.

Somiya, S. and Hirata, Y. (1991). Mullite powder technology and applications in Japan. Amer. Ceram. Soc. Bull. **70**, 1624–1632.

Somiya, S., Yoshimura, M., Suzuki, M. and Yamaguchi, T. (1985). Hydrothermal processing of mullite powders from alkoxides in "Mullite" S. Somiya (ed.), Uchida Rokakuho Publishing Co., Tokyo, Japan, pp. 63–87.

Sonuparlak, B. (1988) Sol-gel processing of infrared-transparent mullite. Adv. Ceram. Mater. **3**, 263–267.

Soro, N., Aldon, L., Olivier-Fourcade, J., Jumas, J.C., Laval, J.P. and Blanchart, P. (2003). Role of iron in mullite formation from kaolins by Mossbauer spectroscopy and Rietveld refinement. J. Am. Ceram. Soc. **86**, 129–134.

Sotirchos, S.V. and Nitodas, S.F. (2002). Factors influencing the preparation of mullite coatings from metal chloride mixtures in CO_2 and H_2. J. Crystal Growth **234**, 569–583.

Sporn, D. and Schmidt, H. (1989). Synthesis of mullite and cordierite powders by sol-gel process and their sintering behavior. Euro-Ceramics **1**, 120–124.

Sueyoshi, S.S. and Contreras Soto, C.A. (1998). Fine pure mullite powder by homogeneous precipitation. J. Europ. Ceram. Soc. **18**, 1145–1152.

Sundaresan, S. and Aksay, I.A. (1991). Mullitization of diphsic aluminosilicate gels. J. Am. Ceram. Soc. **74**, 2388–2392.

Suttor, D., Kleebe, H-J. and Ziegler, G. (1997). Formation of mullite from filled siloxanes. J. Am. Ceram. Soc. **80**, 2541–2548.

Suzuki, M., Hiraishi, S., Yoshimura, M. and Somiya, S. (1984). Preparation of mullite powder by calcination of the products hydrothermally treated from mixed alkoxide or mixed sol. Yogyo-Kyokaishi **92**, 320–327.

Suzuki, H., Tomokiyo, Y., Suyama, Y. and Saito, H. (1988). Preparation of ultra-fine mullite powder from metal alkoxides. J. Ceram. Soc. Jpn. **96**, 67–73.

Suzuki, H., Saito, H., Tomokiyo, Y. and Suyama, Y. (1990). Processing of ultrafine mullite powder through alkoxide route. In: Somiya, S., Davis, R.F. and Pask, J.A. (eds.). Mullite and Mullite Matrix Composites, Ceramic Trans., **6**, 263–274.

Takahashi, J., Kawai, Y. and Shimada, S. (2002). Hot corrosion of cordierite/mullite composites by Na-salts. J. Europ. Ceram. Soc. **22**, 1959–1969.

Takei, T., Hayashi, S., Yasumari, A. and Okada, K. (1998). Preparation of mesoporous mullite fibers by selective leaching method. Proc. Fabr. Adv. Mater. VI **1**, 901–906.

Takigawa, K., Nonaka, K., Okada, K. and Otsuka, N. (1990). Preparation of PLZT fine powders by the spray pyrolysis method. Trans. Brit. Ceram. Soc. **89**, 82–86.

Tang, Y., Lu, Y., Li, A., Li, X., Shi, S. and Ling, Z. (2002). Fabrication of fine mullite powders by α-Al(OH)$_3$–SiO_2 core-shell structure precursors. J. Appl. Surf. Sci. **202**, 211–217.

Taylor, A. and Holland, D. (1993). The chemical synthesis and crystallization sequence of mullite. J. Non-Cryst. Solids **152**, 1–17.

Temuujin, J., Okada, K. and MacKenzie, K.J.D. (1998). Characterization of alumino-silicate (mullite) precursors prepared by a mechanochemical process. J. Mater. Res. **13**, 2184–2189.

Thim, G.P., Bertran, C.A., Barlette, V.E., Macedo, M.I.F. and Oliveira, M.A.S. (2001). Experimental Monte Carlo simulation: the role of urea in mullite synthesis. J. Europ. Ceram. Soc. **21**, 759–763.

Tkalcec, E., Nass, R., Krajewski, T., Rein, R. and Schmidt, H. (1998). Microstructure and mechanical properties of slip cast sol-gel derived mullite ceramics. J. Europ. Ceram. Soc. **18**, 1089–1099.

Tomatsu, H., Maeda, K., Ohnishi, H. and Kawanami, T. (1987). High temperature creep of mullite ceramics. In: Abstracts of the Annual Meeting of the Ceramics Society of Japan, Paper No. 1C14. Ceramic Society of Japan, Tokyo.

Toriyama, M., Kawamoto, Y., Suzuki, T., Yokogawa, Y., Nishizawa, K., Nagae, H., Kato, M. and Fukushima, T. (1992). Preparation and evaluation of mullite ceramics carrier for immobilization of enzyme. J. Ceram. Soc. Jpn., **100**, 1376–1380.

Tulliani, J.-M., Montanaro, L., Bell. T.J. and Swain, M.V. (1999). Semiclosed-cell mullite foams: Preparation and macro- and micro-mechanical characterization. J. Am. Ceram. Soc. **82**, 961–968.

Wahl, F.M., Grim, R.E. and Graf, R.B. (1961). Phase transformations in silica-alumina mixtures as examined by continuous X-ray diffraction. Amer. Min. **46**, 1064–1076.

Wang, J.G., Ponton, C.B. and Marquis, P.M. (1992). Effects of green density on crystallization and mullitization in the transiently sintered mullite. J. Am. Ceram. Soc. **75**, 3457–3461.

Wang, Y. and Thomson, W.J. (1995). Mullite formation from nonstoichiometric slow hydrolyzed single phase gels. J. Mat. Res. **10**, 912–917.

Wang, Y., Li D.X., and Thomson, W.J. (1993). The influence of steam on mullite formation from sol-gel precursors. J. Mat. Res. **9**, 195–205.

Wei, W.-Ch. and Halloran, J.W. (1988a). Phase transformation of diphasic aluminosilicate gels. J. Am. Ceram. Soc. **71**, 166–172.

Wei, W.-Ch. and Halloran, J.W. (1988b). Transformation kinetics of diphasic aluminum silicate gels. J. Am. Ceram. Soc. **71**, 581–587.

West, R.R. and Gray, T.J. (1958). Reactions in silica-alumina mixtures, J. Am. Ceram. Soc. **41**, 132–136.

Wilson, H.H. (1969). Mullite formation from the sillimanite group minerals. Amer. Ceram. Soc. Bull. **48**, 796–797.

Wojtowicz, A.J. and Lempicki, A. (1988). Luminescence of Cr^{3+} in mullite transparent glass ceramics (II). J. Lumin. **39**, 189–203.

Wu, S. and Claussen, N. (1991). Fabrication and properties of low-shrinkage reaction-bonded mullite. J. Am. Ceram. Soc. **74**, 2460–2463.

Wu, S., Holz, D. and Claussen, N. (1993). Mechanisms and kinetics of reaction-bonded aluminium oxide ceramics. J. Am. Ceram. Soc. **76**, 970–980.

Xue, L.A. and Chen, I.-W. (1992). Fabrication of mullite body using superplastic transition phase. J. Am. Ceram. Soc. **75**, 1085–1091.

Yamada, H. and Kimura, S. (1962). Studies on the co-precipitates of alumina and silica gels and its transformations at higher temperatures. Yogyo-Kyokaishi **70**, 87–93.

Yamade, Y., Kawaguchi, Y., Takeda, N. and Kishi, T. (1990). Effect of porosity and grain size on bending strength and fracture toughness of mullite ceramics. In: Abstracts of the 3rd Autumn Symposium on Ceramics, Paper No. 2-1D14. Ceramics Society of Japan, Tokyo.

Yamuna, A., Devanarayanan, S. and Lalithambika M. (2002). Phase-pure mullite from kaolinite, J. Am. Ceram. Soc. **85**, 1409–1413.

Yokoyama, T., Nishu, K., Torii, S. and Ikeda, Y. (1997). Mullite precursor Part I. Characterization of mullite precursor formed by a reaction of monosilicic acid on aluminum hydroxide. J. Mater. Res. **12**, 2111–2116.

Yoldas, B.E. (1980). Microstructure of monolithic materials formed by heat treatment of chemically polymerized precursors in the Al_2O_3–SiO_2 binary. Amer. Ceram. Soc. Bull. **59**, 479–483.

Yoon, C.K. (1990). Superplastic flow of mullite/2Y TZP composites. Ph. D. Thesis. University of Michigan, Ann Arbor.

5
Mullite Coatings

Many metals and ceramics are susceptible to degradation when exposed to harsh environments at elevated temperatures, and require protective environmental barrier coatings (EBCs). As reported by Lee (2000b), these coatings must resist reacting with the aggressive environments to avoid degrading themselves, and must act as excellent diffusion barriers to effectively separate the substrate from the environment. The coatings must have a coefficient of thermal expansion (CTE) compatible with the substrate in order to avoid excessive residual stresses on thermal cycling. The coatings must maintain phase stability under prolonged high temperature, particularly avoiding the formation of new phases that are accompanied by large volumetric changes. And, finally, the coatings must be chemically compatible with the substrates to avoid detrimental chemical reactions at the coating/substrate interface during high-temperature exposure.

Mullite has received considerable attention since the late 1980s as a candidate material for EBCs, especially for protecting silicon-based ceramics and composites used in hot section structural components in gas turbines (Butt et al. 1990, Price et al. 1992). These silicon-based ceramics, such as silicon carbide (SiC) and silicon nitride (Si_3N_4), form dense silica scales when exposed to clean, dry oxygen environments. However, sinter additives such as yttrium, magnesium and aluminum can migrate from the grain boundary phase of the substrate to the silica self oxidation layers, causing pore evolution and crystallization. These lead to microcracking of the protective oxide layer, compromising its effectiveness as a diffusion barrier (Göring, 1991). These silicon-based ceramics are also susceptible to hot-corrosion and recession in more aggressive combustion environments. The presence of elements such as sodium, vanadium and sulfur, lead to the formation of corrosive oxides such as Na_2O, V_2O_5, SO_2 and SO_3. These oxides react with the protective silica scales formed on the silicon-based ceramics forming non-protective low-melting silicates, leading to severe pit formation, material loss and increased porosity (Jacobson, 1993). Also, in the presence of high-pressure water vapor, the protective silica scale volatilizes to gaseous Si–O–H species, exposing the ceramic surface (Robinson and Smialek, 1999). This leads to an accelerated oxidation of the ceramic surface to silica, which in turn volatilizes. This repeated cycling of oxidation and volatilization leads to a rapid recession of the surface of the silicon-based ceramic.

Mullite has superior corrosion resistance, creep resistance and high-temperature strength and toughness, compared to the other silicon-based ceramics. It can exist over a wide compositional range, enabling the deposition of functionally graded coatings. Additionally, mullite has a good CTE match with silicon-based ceramics, especially with silicon carbide. Several techniques have been explored to deposit protective coatings of mullite. These include:
– Chemical vapor deposition (CVD)
– Plasma and flame spraying
– Physical vapor deposition (PVD)
– Deposition from an aqueous slurry
– Self oxidation by oxygen implantation of metal alloys

These techniques are discussed in detail below. Of the processes listed above, the CVD and plasma/flame-spraying processes have shown the most promise.

5.1
Chemical Vapor-deposited Coatings (CVD Coatings)
S. Basu and V. Sarin

The chemical vapor deposition (CVD) process involves the formation of a coating on a heated surface by means of a chemical reaction between reagents in the vapor phase at a specified temperature and pressure. Chemical vapor deposition is a process that is well suited for tailoring interfaces, controlling interfacial reactions and controlling microstructure and composition of the film during growth. The CVD technique lends itself readily to control of coating uniformity and thickness, especially for complex parts with edges, corners and curvatures. Additionally, it allows the deposition of coatings at temperatures well below the melting or sintering temperature of mullite.

5.1.1
Thermodynamics of Chemical Vapor-deposited Coatings

Detailed thermodynamic analysis of the $AlCl_3$–$SiCl_4$–CO_2–H_2 system has been performed to identify process parameters for the growth of CVD mullite coatings (Mulpuri and Sarin, 1995, 1996). The overall reaction for mullite formation has been reported to be:

$$6\ AlCl_3 + 2\ SiCl_4 + 13\ CO_2 + 13H_2 \rightarrow 3Al_2O_3 \cdot 2SiO_2 + 13\ CO + 26\ HCl \qquad (1)$$

Based on this reaction, ternary phase diagrams can be constructed between $AlCl_3$, $SiCl_4$, and CO_2 for various partial pressures of H_2. In addition, the mullite deposition efficiency (fraction of aluminum and silicon in the input chlorides ending up as mullite) was calculated under different conditions to help establish initial process parameters. Since chemical equilibrium is rarely achieved under most CVD flow conditions, these phase diagrams are basically used as guidelines to determine trends.

A typical ternary phase diagram of $AlCl_3$–$SiCl_4$–CO_2–H_2 system at 1000 °C and 75 torr (Fig. 5.1.1) shows that oxides form at high CO_2 concentrations, oxides and carbon form at medium CO_2 concentrations, and oxides and carbides form at low CO_2 concentrations (Mulpuri and Sarin, 1996). Thus, in order to obtain carbon-free mullite coatings, it is necessary to operate close to the CO_2 corner of the ternary phase diagram. A study of the effect of the input metal chlorides ($AlCl_3$/$SiCl_4$ ratio) along line AA in the CO_2 corner in Fig. 5.1.1 shows that both the amount of mullite formation and the deposition efficiency are optimal when the input molar gas-phase $AlCl_3$/$SiCl_4$ ratio is 3. This ratio of gas-phase reactants corresponds to the molar Al/Si ratio of 3 in stoichiometric mullite. The calculations also predicted sillimanite formation instead of mullite at the CO_2 end of the ternary phase diagram at temperatures below 800 °C. Increasing the temperature to 1200 °C gave no clear benefits from a deposition efficiency standpoint. The calculations also showed that increasing the total reactor pressure widens the region in the phase diagram over which mullite formation is predicted. However, increasing the pressure too much led to the inclusion of carbon in the solid phases being deposited. Thus, the optimal choice of pressure is a compromise between mullite deposition efficiency and carbon-free coating deposition. Finally, the thermodynamic analysis shows that there is no underlying benefit by using bromides and fluorides as alternate halide sources for aluminum and silicon. Zemskova and

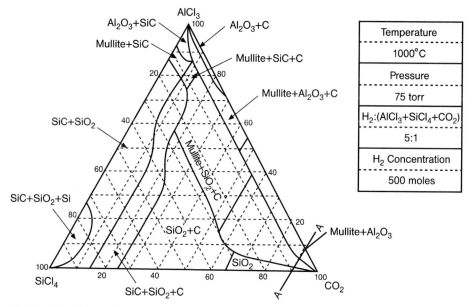

Fig. 5.1.1 The $AlCl_3$, $SiCl_4$, CO_2 CVD ternary phase diagram at 1000 °C and 75 torr chamber pressure. The figure suggests deposition in the CO_2-rich regime to avoid formation of C and carbides (from Mulpuri, 1996).

co-workers (2000) have explored the use of alumino siloxanes as a single-source precursor for metalorganic chemical vapor deposition (MOCVD) of alumino silicate coatings.

It should be noted that input thermodynamic data were based on mullite being a line compound and therefore its projected formation in the phase diagram is along the 3:1 Al/Si line. However, it has been conclusively established that non-stoichiometric mullite can be formed both in the bulk (Cameron, 1977) and via the CVD coating process (Doppalapudi and Basu, 1997, Hou et al. 2001), as will be discussed.

5.1.2
Growth Kinetics of Chemical Vapor-deposited Coatings

Auger and Sarin (2001) studied the kinetics of growth of chemical vapor-deposited mullite. They reported that under these typical deposition conditions the rate-limiting step is the intermediate water-gas shift reaction, given as (Tingey, 1966):

$$CO_{2\ (g)} + H_{2\ (g)} \rightarrow H_2O_{\ (g)} + CO_{\ (g)} \tag{2}$$

This water-gas shift reaction has also been reported to be the rate-limiting step for the formation of CVD Al_2O_3 coatings (Lindstorm and Schacher, 1980). The study showed that increasing the substrate temperature and/or the total pressure of the CVD reactor (without a change in gas-phase composition) leads to an increase in the coating growth rate and the coating surface becomes increasingly faceted. Increasing the total reactor pressure was also accompanied by a clear increase in the grain size of the coating, as seen in Figs. 5.1.2a,b. At high temperatures (1000 °C and above) and total reactor pressures of 100 torr and above, unstable secondary growth occurs leading to the deposition of non-uniform coatings (see Fig. 5.1.2c). Below a critical metal chloride concentration (sum of partial pressures of $AlCl_3$ and $SiCl_4$), non-crystalline aluminosilicate coatings formed with high

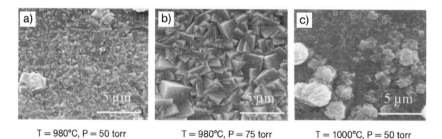

| T = 980°C, P = 50 torr | T = 980°C, P = 75 torr | T = 1000°C, P = 50 torr |

Fig. 5.1.2 SEM micrographs showing the surface morphology of CVD mullite coatings grown under different conditions of temperature and pressure with the gas phase $AlCl_3/SiCl_4$ ratio constant at 3. (a) T = 980 °C, P = 50 torr, (b) T = 980 °C, P = 75 torr, and (c) T = 1000 °C, P = 50 torr (from Auger and Sarin, 2001).

growth rates and with virtually no accompanying homogeneous powder formation. This has been also observed by Nitodas and Sotirchos (2000). Increasing the metal chloride concentration above this value led to the formation of crystalline mullite coatings at a growth rate significantly lower than that of the non-crystalline case. Increasing the metal chloride partial pressure led to a further decrease in the coating growth rate due to significant homogeneous gas-phase powder nucleation. Also, within the regime of crystalline mullite coating deposition, a significant increase in growth rates was observed when the aluminum to silicon molar ratio in the gas-phase reactants was close to 3, which is the molar ratio of Al/Si in stoichiometric mullite.

5.1.3
Microstructure of Chemical Vapor-deposited Coatings

Mullite coatings have been successfully deposited by CVD on a variety of ceramic substrates, including silicon carbide (SiC), silicon nitride (Si_3N_4), alumina (Al_2O_3) and mullite (Fig. 5.1.3), although detailed studies have been only carried out for the silicon carbide and silicon nitride substrates (Mulpuri and Sarin, 1996, Auger and Sarin, 1997, Sarin and Mulpuri, 1998, Basu et al. 1999a, Hou et al. 2001). The fracture cross-section of a typical CVD mullite coating on silicon carbide is shown in Fig. 5.1.3a, while Fig. 5.1.4 shows the cross-sectional transmission electron micrograph of such a coating (Hou, 2000). These coatings have been found to be uniform, adherent and free of pores and cracks. Fig. 5.1.5 shows an X-ray diffrac-

Fig. 5.1.3 Fracture cross-sections of CVD mullite coatings deposited on (a) SiC, (b) Al_2O_3 and (c) mullite substrates (from Auger, 1999).

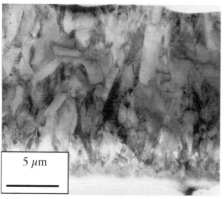

5 μm

Fig. 5.1.4 TEM cross-sectional bright-field micrograph of dense, pore and crack-free CVD mullite coating (from Hou et al. 2001).

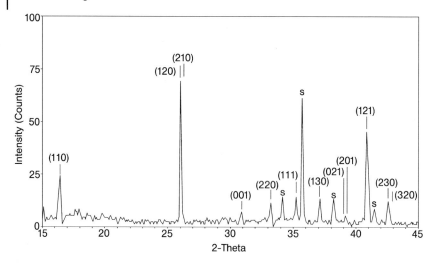

Fig. 5.1.5 X-ray diffraction (XRD) scan of a CVD mullite coating on SiC. The SiC substrate peaks are marked S, and all non-substrate peaks are indexed as mullite. The tetragonal structure of the as-deposited mullite coating is evidenced by the lack of splitting of the 120/210 peaks (from Hou et al. 2001).

Fig. 5.1.6 SEM micrographs of (a) the surface and (b) a fracture cross-section of a uniform 2-μm thick CVD mullite coating on a 15-μm diameter Nicalon SiC fiber (from Varadarajan 1999). The partial separation at the fiber/coating interface is an artifact due to mechanical stresses in creating the fracture cross-section.

tion (XRD) scan of a typical coating, in which all non-substrate peaks were found to match with mullite (Hou et al. 2001). As mentioned, the CVD process has the ability to deposit uniform coatings on parts with complex shapes having large curvatures. Figs. 5.1.6a and b show the surface and the cross-section of a 15-μm diameter Nicalon SiC fiber uniformly coated with an aapproximately 2-μm thick CVD mullite layer (Varadarajan, 1999).

Changing the input gas-phase AlCl$_3$/SiCl$_4$ ratio into the CVD reactor led to a

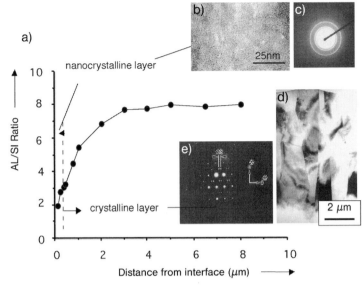

Fig. 5.1.7 (a) Composition gradation (expressed as the Al/Si ratio) across a functionally graded CVD mullite coating. The surface of this coating has an Al/Si ratio of 8; (b) cross-sectional TEM micrograph of the nanocrystalline layer, formed when the Al/Si ratio is below 3.2 ± 0.3; (c) SAED pattern of γ-Al$_2$O$_3$ in the nanocrystalline layer; (d) cross-sectional TEM micrograph showing the columnar mullite grains with increasing Al/Si ratio in the crystalline layer of the as-deposited coating; (e) [010] mullite SAED pattern from a mullite grain, showing superlattice spots with spacing 2S.

change in the composition of the growing coating. Thus, the composition of CVD mullite coatings can be functionally graded. Fig. 5.1.7a shows such a composition gradation in a coating on silicon carbide (Hou et al. 2001), with the composition expressed as the Al/Si molar ratio in mullite (the Al/Si ratio of stoichiometric mullite is 3).

Interestingly, for a given AlCl$_3$/SiCl$_4$ ratio, a coating on a silicon carbide substrate starts off as relatively silica-rich (low Al/Si ratio), while that on an alumina substrate starts out as relatively aluminum-rich (high Al/Si ratio), as illustrated in Fig. 5.1.8 (Basu et al. 1999a, Basu and Sarin 2000, Hou et al. 1999). Thus, the substrate plays an important initial role in determining the relative adsorption/desorption rates of the aluminum- and silicon-containing species from the gas phase. In all cases, if the composition of the coating was not close to stoichiometric mullite, the initial growth occurred as an intimate mixture of vitreous SiO$_2$ and nanocrystalline (about 5-nm grain size) γ-alumina phases, and not as mullite (Hou et al. 1999). Fig. 5.1.7 b shows the transmission electron micrograph of such a "nanocrystalline" layer on a silicon carbide substrate, along with a selected-area electron diffraction (SAED) pattern from the region (Fig. 5.1.7 c), consistent with γ-alumina. The thickness of this nanocrystalline layer for mullite coatings on silicon carbide has been reported to be typically less than 1 μm, and to decrease with increasing $AlCl_3/SiCl_4$ ratio in the gas phase (Basu et al. 1999).

Nucleation of crystalline mullite has been observed to occur only when the composition gradation (Al/Si ratio increasing for silicon carbide and decreasing for alumina substrates) allowed the surface of the growing coating to be within a narrow composition range of 3.2 ± 0.3, which is close to that of stoichiometric mullite (Hou et al. 1999). This is shown schematically in Fig. 5.1.8. This observation was further supported by the lack of a nanocrystalline zone in coatings grown on mullite substrates, where nucleation of crystalline mullite occurs directly on the substrate having an Al/Si ratio of 3. It should be noted that nucleation of mullite during CVD coating growth, at compositions exclusively close to stoichiometry, is in direct contrast to reports of mullite nucleation on annealing atomically mixed precursors, where Al_2O_3-rich mullite has been formed at about 900 °C independent of the composition of precursors. The lack of nucleation of mullite in the Al_2O_3-rich intimately mixed nanocrystalline layer adjacent to the interface of the chemically vapor-deposited mullite coatings on alumina suggests that the mechanism of surface nucleation on a growing CVD coating may not be identical to that of bulk nucleation in the atomically mixed precursors. As will be discussed in more detail later, annealing these coatings at 1100 °C or higher for 100 h leads to complete mullitization of the nanocrystalline layer, if there is sufficient γ-alumina present (Hou et al. 2001).

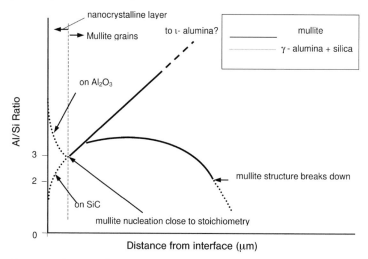

Fig. 5.1.8 Schematic of composition variation in CVD mullite coatings on SiC and Al_2O_3 substrates. Coatings on SiC start off as Si-rich while those on Al_2O_3 start off as Al-rich. In both cases, nucleation of crystalline mullite occurs when the composition comes within a narrow range close to stoichiometry (Al/Si ≈ 3). Once nucleated, the Al/Si ratio in the mullite grains can be increased to highly Al-rich compositions (approaching ι-alumina), but not to highly Si-rich compositions.

Once the mullite grains nucleate, they can then be made to grow to highly Al_2O_3-rich non-stoichiometric compositions. Rapid, low-activation-energy surface diffusion during the CVD process allows the ad-atoms to use the coating grain surface as a template for continued growth, even when the ratio of the arrival rates of the aluminum- and silicon-containing species are far from stoichiometry. The transmission electron micrograph (Fig. 5.1.7 d) shows some of the columnar mullite grains, whose Al/Si ratio was found to increase towards the coating surface. An SAED pattern from one of the grains is shown in Fig. 5.1.7 e (Hou et al. 2001).

Fig. 5.1.8 schematically shows that the increase in the Al/Si ratio in mullite would lead towards the formation of alumina with mullite structure – termed as iτ-alumina by Cameron (1977). With suitable manipulation of the input $AlCl_3/SiCl_4$ ratio, Al/Si ratios as high as 15 have been achieved at the coating surface. These compositions are among the most Al_2O_3-rich mullite structures that have been reported to date (Fischer et al. 1996). Because of anticipated hot-corrosion and recession problems (Jacobson and Lee, 1996, Haynes et al. 2000), there is a strong practical motivation to substantially reduce or eliminate the silica component from the coating surface in direct contact with corrosive atmospheres, especially those containing steam. Functionally graded coatings, with stoichiometric mullite at the contact with the silicon carbide substrate and Al_2O_3-rich mullite at the top would have an excellent match of thermal expansion coefficients at the coating/substrate interface, and excellent hot-corrosion and recession resistance, owing to the lack of silica at the outer surface of the coating. Thus, these functionally graded chemically vapor-deposited mullite coatings on silicon-based ceramics are expected to have the hot-corrosion and recession resistance of alumina, while maintaining the thermal shock resistance of mullite (Basu and Sarin, 2002). However, this requires a high-temperature stabilization of these Al_2O_3-rich mullites. It has been shown that Al_2O_3-rich mullites transform to stoichiometric mullite above about 1200 °C (see Section 2.5.2). These premises are currently being researched at Boston University.

It is interesting to study the SiO_2-rich end of chemical vapor-deposited mullites by decreasing the input $AlCl_3/SiCl_4$ ratio once mullite grains have nucleated. Such experiments have been conducted by growing a coating on silicon carbide, in which the Al/Si ratio was increased until mullite grains nucleated at an Al/Si ratio of about 3. The growth of these mullite grains was then continued with decreasing input $AlCl_3/SiCl_4$ ratio. The mullite structure could not be sustained to an Al/Si ratio of about 2, below which continued growth occurred as a nanocrystalline structure of silica and γ-alumina (Basu and Sarin, 2002). This is shown schematically in Fig. 5.1.8, which shows that it is possible to grow non-stoichiometric chemical vapor-deposited mullite to highly Al_2O_3-rich compositions, but not to SiO_2-rich compositions. The results confirm the observation that mullite mixed-crystal formation is limited to $x = 0.20$ (referring to the general formula of mullite; $Al_{4+2x}Si_{2-2x}O_{10-x}$), which corresponds to an Al_2O_3 content of about 70 wt.% (see Section 1.1.3).

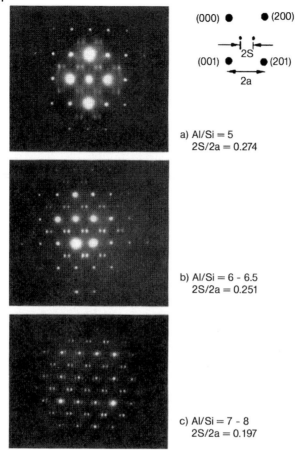

(000) ●   ● (200)

2S

(001) ●   ● (201)

2a

a) Al/Si = 5
2S/2a = 0.274

b) Al/Si = 6 - 6.5
2S/2a = 0.251

c) Al/Si = 7 - 8
2S/2a = 0.197

Fig. 5.1.9 Mullite [010] SAED patterns showing the change in superlattice spacing (expressed as S/a) with coating composition (from Doppalapudi and Basu, 1997).

5.1.4
The Structure of Chemical Vapor-deposited Al_2O_3-rich Mullite

Current knowledge of the crystal structure of Al_2O_3-rich mullite is given in Section 1.1.3. Things are especially interesting for mullites with $x > 0.67$, where all possible oxygen vacancies have been formed and complex structural arrangements are necessary to compensate the electrical charge disequilibrium caused by further aluminum-for-silicon substitution. It is suggested that the real structures of sol-gel-derived and chemically vapor-deposited mullites are the same, although the sizes of lattice domains of the latter appear to be larger. This follows from the distance $2S$ between the two superlattice spots located symmetrically around the 1 0 1/2 position along the a* axis in a [010] selected area diffraction pattern, which is directly related to the domain sizes (Fig. 5.1.9). Cameron (1977) plotted the pa-

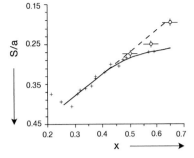

Fig. 5.1.10 Variation of S/a with the oxygen vacancy concentration, x. The solid curve is plotted from Cameron's (1977) data, marked as '+'. Data from the functionally graded CVD mullite are marked as 'o', and fit well with the extrapolation of the linear portion of Cameron's plot (from Doppalapudi and Basu, 1997).

rameter S/a (a = lattice constant). Fig. 5.1.10 shows this plot with data of Cameron (1977) and Doppalapudi and Basu (1997). The plot shows that the chemically vapor-deposited mullite display x values up to about 0.65, corresponding to near pseudo-tetragonal composition (about 85 wt.% Al_2O_3), and that the domain sizes in the chemically vapor-deposited Al_2O_3-rich mullites are larger (smaller S/a ratios) than those reported by Cameron in samples of similar compositions.

5.1.5
High Temperature Phase Transformations in Chemical Vapor-deposited Mullite

The phase stability of chemically vapor-deposited mullite coatings has been investigated using high-temperature annealing studies. It has been reported that the composition profile across a graded mullite coating showed little variation after a 100-h anneal at 1250 °C (Hou et al. 2001). This suggests that long-range-diffusion processes are not occurring in the coating, indicating that mullite is a good diffusion barrier.

Since the a and b lattice parameters of orthorhombic mullite are slightly different, there is typically a splitting of (hkl) and (khl) peak pairs (e.g. 120/210, 230/320, etc) in X-ray diffraction scans of mullite. However, X-ray diffraction scans of as-deposited mullite coatings (Fig. 5.1.5) almost never exhibit a splitting of such pairs, indicating that the as-deposited mullite coatings are (near) pseudo-tetragonal. The mullite structure can move towards a pseudo-tetragonal configuration owing to compositional effects alone (see Section 1.1.3). Annealing studies of as-deposited coatings for 100 h at temperatures of 1200 °C and above led to a pseudo-tetragonal to orthorhombic transformation of mullite, with the splitting of the peak pairs becoming more evident at higher temperatures (see Section 2.5.1).

Annealing mullite coatings for 100 h at 1150 °C and above resulted in crystallization of the nanocrystalline precursor layer. For coatings on silicon carbide substrates, precursor layers deposited using an $AlCl_3/SiCl_4$ input ratio of 2 and above transformed completely to equiaxed mullite grains after 100-h anneals at temperatures between 1150 °C and 1300 °C (Hou et al. 2001). The transmission electron micrograph in Fig. 5.1.11a shows equiaxed mullite grains in the precursor layer of a coating grown using an input $AlCl_3/SiCl_4$ ratio of 2, with an accompanying [214] mullite SAED pattern (Fig. 5.1.11b) from one such grain formed after a

[214]

Fig. 5.1.11 (a) TEM micrograph showing complete mullitization of a nanocrystalline layer deposited with an input AlCl₃/SiCl₄ ratio of 2, as a result of a 100-h anneal at 1300 °C; (b) the SAED pattern from one of the equiaxed grains is consistent with [214] mullite (from Hou et al. 2001).

100-h anneal at 1300 °C (Hou et al. 2001). It is clearly evident from the figure that the crystallization occurs with no accompanying microcracking or porosity formation.

In contrast, a precursor layer deposited using an input $AlCl_3/SiCl_4$ ratio of 1 did not convert completely to mullite on annealing. The cross-sectional transmission electron micrograph in Fig. 5.1.12a shows two distinct sublayers formed within the precursor layer after a 100-h anneal at 1250 °C (Hou et al. 2001). The SAED pattern (Fig. 5.1.12b) from the bottom layer identifies it as cristobalite. The high-resolution electron micrograph (HREM) in Fig. 5.1.12c shows the upper layer to consist of nano-sized crystalline mullite grains embedded in a SiO_2-rich vitreous matrix. The small-angle electron diffraction pattern from the upper layer did not show the ring pattern of γ-alumina, indicating that the nanocrystallites of γ-alumina had dissolved into the vitreous silica-rich matrix on annealing. This observation is consistent with the studies of Wei and Halloran (1988a, 1988b) on mullitization mechanisms in diphasic gels, which transformed into an intimate mixture of γ-alumina and vitreous silica. They reported that the nucleation mechanism consisted of three serial steps, dissolution of γ-alumina in the vitreous silica matrix, nucleation of mullite after an incubation period and growth of mullite grains by diffusion in the matrix phase.

Since the average Al/Si ratio in the precursor layer increases with increasing input $AlCl_3/SiCl_4$ ratio, it is concluded that the presence of a sufficient Al_2O_3 content in the nanocrystalline precursor layer is necessary to ensure complete mullitization. A complete mullitization of the nanocrystalline precursor layer is highly desirable in order to avoid the devitrification of silica to cristobalite. There is a large mismatch between the coefficient of thermal expansion of the cristobalite phase (10.3×10^{-6} °C^{-1}; Lynch 1981) and that of the silicon carbide substrate (4.7×10^{-6} °C^{-1}). Devitrification of silica has been known to cause spallation of

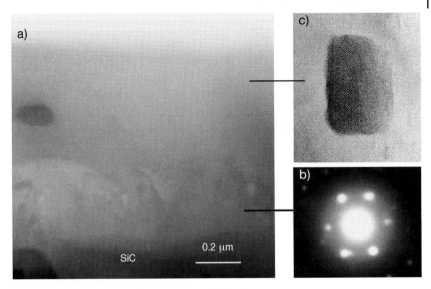

Fig. 5.1.12 (a) Cross-sectional TEM micrograph showing two discernable layers formed in a nanocrystalline layer deposited with an input $AlCl_3/SiCl_4$ ratio of 1 after a 100-h anneal at 1250 °C; (b) [111] SAED pattern from the bottom layer, showing cristobalite formation by devitrification of silica; (c) high-resolution TEM micrograph of the upper layer, showing the presence of a mullite crystallite in a silica-rich vitreous matrix (from Hou et al. 2001).

coatings due to the large stresses that accompany the 3.3 % volume reduction that occurs when vitreous silica converts to β-cristobalite, and a further 2.2 % volume reduction when β-cristobalite transforms to α-cristobalite (Lynch, 1981). In fact, the cracking and spallation of mullite coatings (grown with $AlCl_3/SiCl_4$ ratios greater than 2) annealed for 100 h at 1400 °C is directly attributed to the formation of cristobalite in the nanocrystalline layer. Fig. 5.1.13b shows the presence of cristobalite peaks in the X-ray diffraction scans of the coating (Hou et al. 2001).

However, if the same as-deposited coatings were first pre-annealed for 100 h at 1250 °C, they remained completely adherent and crack free after a subsequent 100-h anneal at 1400 °C. Fig. 5.1.14a shows a low magnification scanning electron micrograph of the surface of such a coating (Hou et al. 2001). A higher magnification micrograph (see Fig. 5.1.14b) of the coating surface shows an absence of microcracks as well as a lack of any substantial grain growth, even after the 1400 °C exposure. It is conjectured that the complete mullitization of the nanocrystalline layer during the 1250 °C pre-anneal, prevented the devitrification of silica during the 1400 °C anneal. This was supported by the lack of a cristobalite peak in the X-ray diffraction scan. Thus, pre-treatment of chemically vapor-deposited mullite coatings at 1250 °C to induce complete mullitization of the nanocrystalline precursor layer is necessary before exposure to temperatures as high as 1400 °C for long-term applications.

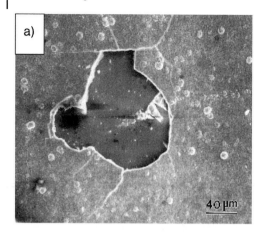

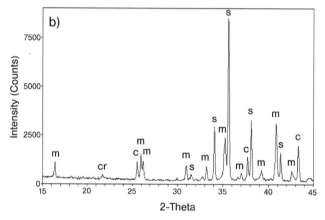

Fig. 5.1.13 (a) Cracking and spallation in a CVD mullite coating annealed at 1400 °C for 100 hours; (b) θ–2θ XRD scan from coating, showing the presence of cristobalite (cr) and α-Al₂O₃ (corundum) (C) phases in the annealed mullite (m) coating (from Hou et al. 2001).

The adhesion of chemically vapor-deposited mullite coatings has been evaluated by cycling them between 1250 °C and room temperature. The samples were also subjected to substantial thermal shock by rapid insertion and removal of the samples from the hot zone of the furnace after holding the sample for 1 h at temperature. Fig. 5.1.15a shows a fracture cross-section of a coating after 500 cycles (Hou et al. 2001). The coating exhibited no signs of cracking and/or spallation. The excellent adhesion of the coating can be partially attributed to two reasons. The first is the formation of equiaxed mullite grains in the nanocrystalline layer surface leading to a close match of thermal expansion coefficients at the coating/substrate interface. The second is the gradation of the thermal expansion coefficients across the thickness of the mullite coating, which avoids any abrupt changes across the coating thickness, while allowing the coating surface to be highly Al₂O₃-rich. Fur-

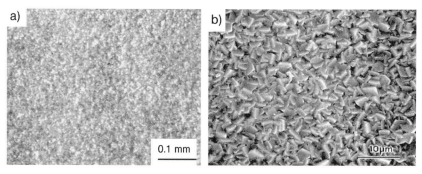

Fig. 5.1.14 (a) Low-magnification SEM micrograph showing no cracking or spallation in a CVD mullite coating pre-annealed at 1250° for 100 h, followed by a 100 h anneal at 1400 °C; (b) higher magnification SEM micrograph of coating surface, showing a lack of microcracking and grain growth, even after the 100-h exposure at 1400 °C (from Hou et al. 2001)

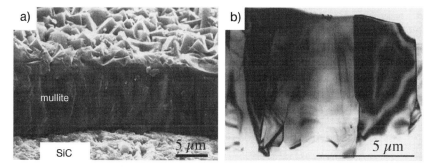

Fig. 5.1.15 (a) Fracture cross-section of CVD mullite coating showing excellent adhesion after 500 cycles of cyclic oxidation at 1250 °C; (b) cross-sectional TEM micrograph showing no phase separation after the 500-h exposure at 1250 °C (from Hou et al. 2001).

thermore, as can be seen in the cross-sectional transmission electron micrograph (Fig. 5.1.15b), the high-Al_2O_3 coating surface (Al/Si ≈ 8) showed no signs of phase separation after a total of 500 h of exposure at 1250 °C during this cyclic oxidation test (Hou et al. 2001).

To better understand the effect of coating composition on the crystallization at 1400 °C, a compositionally graded coating (surface Al/Si ratio ≈ 8) that had been subjected to the two-step anneal (100 h at 1250 °C followed by 100 h at 1400 °C) was carefully examined by transmission electron microscopy. In general, other than the previously described pseudo-tetragonal to orthorhombic transformation and mullitization of the nanocrystalline precursor layer, no other morphological changes were observed in regions where the Al/Si ratio was below 5. However, in regions where the Al/Si ratio was between 5 and 6, 100- to about 300-nm sized

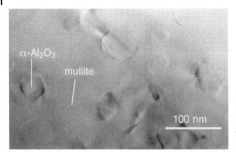

Fig. 5.1.16 Crack-free precipitation of nano-sized α-Al$_2$O$_3$ in the high Al-mullite (5 < Al/Si < 6) after the two-step anneal of 100 h at 1250 °C, followed by 100 h at 1400 °C (from Hou et al. 2001).

precipitates are observed, as shown in Fig. 5.1.16 (Hou et al. 2001). It is significant that this precipitation phenomenon was not accompanied by the formation of any microcracks. Microanalysis in the scanning transmission electron microscope along with X-ray diffraction scans of the coatings identified these precipitates to be α-alumina. The precipitation of α-alumina in Al$_2$O$_3$-rich mullite is consistent with reports by other researchers (Fischer et al. 1996), suggesting that annealing Al$_2$O$_3$-rich chemically vapor-deposited mullite may be a viable method of producing uniform finely dispersed alumina/mullite nanocomposite coatings. At even higher aluminum contents (6 < Al/Si < 8), precipitation of α-alumina was accompanied by the formation of twinned mullite. This is in agreement with the observations of Nakajima and Ribbe (1981), who reported twinning in high Al$_2$O$_3$-mullite.

5.1.6
Oxidation, Hot Corrosion and Recession Protection of Chemical Vapor-deposited Mullite

Mullite coatings have been found to serve as excellent oxidation barriers for silicon carbide substrates. Fig. 5.1.17 shows a plot of weight gain for mullite-coated (Fig. 5.1.6) and uncoated Nicalon silicon carbide fibers oxidized in flowing oxygen at 1300 °C (Varadarajan et al. 2001). The uncoated fibers were found to gain weight owing to the formation of silica at the surface, part of which spalled when the sample was cooled. In contrast, the mullite-coated silicon carbide fibers showed almost no weight gain, and exhibited no signs of cracking or spallation after oxidation.

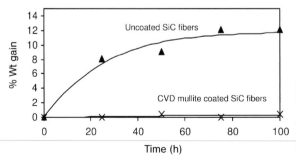

Fig. 5.1.17 Weight gain versus time plots for uncoated (Δ) and mullite coated (X) SiC fibers oxidized in flowing oxygen at 1300 °C (from Varadarajan et al. 2001).

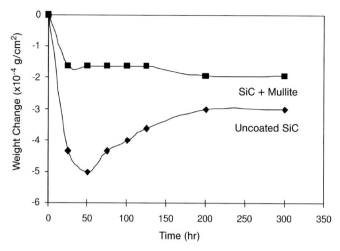

Fig. 5.1.18 Weight change versus time plots for CVD mullite coated and uncoated SiC subjected to hot-corrosion (Na$_2$SO$_4$ + O$_2$) at 1100 °C (from Pattanaik et al. 1998).

Mullite-coated and uncoated silicon carbide substrates have also been subjected to hot-corrosion tests by loading the surface with about 5 mg cm^{-2} of sodium sulfate (Na$_2$SO$_4$) and subjecting the samples to flowing oxygen (200 standard cm^3 min^{-1}) at 1100 °C for 300 h (Pattanaik and Sarin, 1998). Fig. 5.1.18 compares the change in weight of a coated and an uncoated sample as a function of time. Substantial weight gain occurred in the uncoated sample after an initial weight loss. Weight loss occurs owing to the formation of gaseous products while weight gain can be attributed to the formation of silica by the oxidation of silicon carbide. The silica scale formed by oxidation of silicon carbide reacts with sodium sulfate (Na$_2$SO$_4$) to form a liquid phase, through which transport is rapid enough to expose the surface of the non-reacted silicon carbide to further oxidation (Jacobson, 1993). The depth of hot-corrosion attack was found to be in excess of 20 µm for the uncoated silicon carbide substrate (Fig. 5.1.19a, Pattanaik et al. 1998). In direct contrast, the mullite-coated silicon carbide sample exhibited no weight gain. Examination of the sample after oxidation showed no formation of silica. The Al$_2$O$_3$-rich composition of mullite allowed the coating to remain unreacted in the presence of molten sodium sulfate (Fig. 5.1.19b), indicating that the chemically vapor-deposited mullite coating acted as an effective hot-corrosion barrier (Basu et al. 1999b).

Lee and co-workers (1996) have reported that chemically vapor-deposited mullite coatings were also effective in protecting silicon nitride substrates after a 100-h exposure to a corrosive environment containing sodium sulfate and oxygen at 1000 °C. The coating surface was found to be unaffected by hot-corrosion, although a small degree of sodium penetration through the mullite grain boundaries was observed to a depth of 1 µm from the coating surface.

The effectiveness of chemically vapor-deposited mullite coatings against corro-

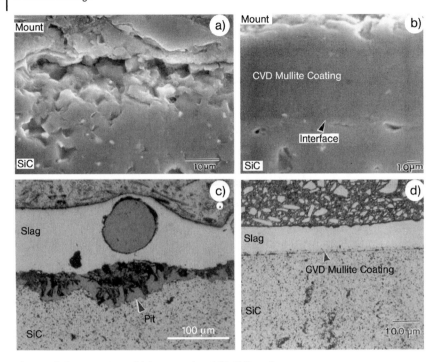

Fig. 5.1.19 Cross-sections of (a) uncoated and (b) CVD mullite coated SiC substrates after hot-corrosion (Na₂SO₄ + O₂) at 1100 °C for 300 h (Pattanaik et al. 1998). The uncoated sample had a 20-μm depth of hot-corrosion attack, while the mullite coated sample was practically unaffected; (c) optical cross-sectional micrograph of pitting in uncoated SiC substrate in contact with an acidic Fe-based coal slag after 300 h at 1260 °C; (d) the presence of a CVD mullite coating effectively protected the SiC substrate (from Auger et al. 2000).

sion attack by an acidic iron-based coal slag was investigated by Auger et al (2000). After a 300-h exposure at 1260 °C, the uncoated silicon carbide substrates suffered severe material loss and pitting due to coal slag corrosion (Fig. 5.1.19c). Uniform chemically vapor-deposited mullite coatings were found to be very effective in protection against the coal slag. The coating did not degrade in the presence of the liquid slag and did not allow liquid slag seepage to the silicon carbide substrate, thereby protecting the substrate from pitting (Fig. 5.1.19d). Although some diffusion of iron to the coating/silicon carbide interface was seen, no cracking or spallation was observed in the uniform mullite coatings.

Haynes et al. (2000) investigated the resistance of silicon carbide fiber-reinforced silicon carbide matrix composites (SiC_F/SiC) overlaid with chemically vapor-deposited silicon carbide seal coats, in high-temperature, high-pressure tests in air/water-vapor atmospheres. They reported that the chemically vapor-deposited silicon carbide rapidly formed silica scales with a dense amorphous inner layer and a thick porous outer layer of cristobalite. Spallation of the oxide occurred during thermal cycling owing to the high/low temperature cristobalite transformation. Thin, dense chemically vapor-deposited mullite coatings on the silicon carbide seal coats effectively suppressed the silicon carbide oxidation. There was also no microstructural evidence of the volatilization of the silica phase from the surface of the mullite coatings. Haynes et al. concluded that dense crystalline high-purity chemically vapor-deposited mullite is stable in high-temperature and -pressure, moisture-containing environments, at least for low gas velocities found in the steam rig in which the volatile $SiO_y(OH)_x$ decomposition products are not carried away rapidly. The potential for mullite volatility is much greater at the higher gas velocities found in burner rigs and turbine engines (see Section 3.2). It is conjectured that functionally graded mullite coatings with virtually no silica at the outer coating surface would resist volatilization, even at higher gas velocities.

5.2
Plasma- and Flame-sprayed Coatings
S. Basu and V. Sarin

Plasma spraying and flame spraying are widely used techniques for deposition of mullite coatings. These processes involve injecting fine mullite particles into a torch, where they are melted and accelerated to high speeds by a plasma or flame towards the substrate. On contacting the substrate these particles flatten and rapidly solidify as thin splats that pile up upon each other to form the coating. Both processes are relatively inexpensive and lead to high deposition rates.

5.2.1
Microstructural Characteristics and Stability of Coatings

Plasma-sprayed mullite coatings have been used successfully to improve the oxidation resistance of silicon-based ceramics (silicon nitride, silicon carbide). Additionally, such coatings have also been found to be suitable for protection of silicon carbide heat exchangers against corrosion from sodium sulfate vapors at temperatures above 900 °C. Butt et al. (1990) have reported that plasma-sprayed single-phase mullite or two-phase mullite-zirconia coatings on silicon carbide heat exchanger tubes extended the lifetimes of these components considerably. Good adhesion strength of such plasma-sprayed coatings was attributed to the mechanical bonding between the mullite coating and the silicon carbide substrate.

Braue et al. (1996) studied the microstructures of plasma-sprayed mullite layers by scanning and transmission electron microscopy. They found that the mullite layers could be characterized by distinct states of disequilibria in a nanometer scale, with extremely heterogeneous phase developments and pore distributions. The microstructure of the plasma-sprayed material consisted of large spherical mullite grains embedded in a mixture of fine-grained recrystallized mullite and a glass matrix. The matrix contained numerous small, rounded pores. Since the

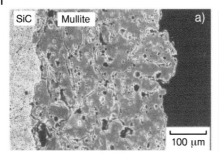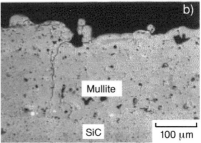

Fig. 5.2.1 SEM micrographs of cross-sections of plasma-sprayed mullite coatings on (a) unheated and (b) heated SiC substrates, after two 24-h thermal cycles between room temperature and 1000 °C (from Lee et al. 1995). Heating of the substrate reduces the presence of amorphous phases in the coating, thereby reducing the propensity for crack formation on thermal cycling.

large spherical mullite grains have a grain size of the order of the starting powders, it was conjectured that they represent only partially molten powders. The heat flux perpendicular to the surface of the large grains favored the development of columnar mullite crystallites in the liquid film surrounding the large grains. The microstructure of the matrix changed significantly from one region to the other. Areas with droplet-like grains containing isolated pockets of amorphous phase alternated with regions of dendritic crystals with interdendritic glass phase or with mosaic-like ideomorphic (often equiaxed) crystals intergrown with glass. The presence of a small frequency of spherical pores indicates that they were formed at a temperature where most of the matrix material consists of a low viscosity silicate melt.

Lee and co-workers (Lee et al. 1995, Lee and Miller, 1996, Lee, 1998, 2000a, 2000b, Lee et al. 2003) have developed several generations of plasma-sprayed mullite coatings on silicon carbide. They have reported that coatings deposited by conventional plasma spraying tended to crack and debond during thermal cycling. Fig. 5.2.1a shows such a coating after two 24-h thermal cycles between room temperature and 1000 °C (Lee et al. 1995). This crack formation and debonding was attributed to the formation of non-crystalline aluminosilicate phases due to the rapid quenching of molten splats on the cold silicon carbide substrates. Cracks form due to crystallization of the glass to mullite at about 1000 °C, which leads to a volumetric shrinkage.

In order to eliminate cracking of the mullite coatings, Lee and coworkers etched the silicon carbide surface with molten sodium carbonate (Na_2CO_3) to create a rough surface for good mechanical bonding, and then heated the substrate during deposition. They found that heating the substrate to temperatures above 1050 °C produces fully crystallized coatings. Fig. 5.2.1b shows such a coating after two 24-h thermal cycles between room temperature and 1000 °C (Lee et al. 1995). The reduction in the crack density is obvious from the figure, although some through-thickness vertical cracks still occur.

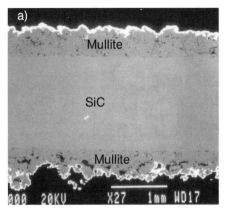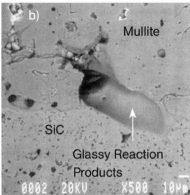

Fig. 5.2.2 (a) Cross-section of plasma-sprayed mullite coating on SiC after 50 h in a hot-corrosion rig at 1000 °C showing good protection of the substrate; (b) in areas where through-thickness vertical cracks intersected the interface, glassy sodium silicate phases formed (from Lee, 2000a).

Lee and Miller (1996) studied the durability of fully crystalline mullite plasma-coated silicon carbide and silicon carbide fiber-reinforced silicon carbide matrix composites (SiC$_F$/SiC) under thermal cycling conditions. These coatings exhibit good adherence on the substrates and resistance to cracking. The crack resistance, however, appeared to decrease with increasing temperature under higher frequency cycling with an accompanying enhancement in the oxidation rates.

Mullite coatings also performed well in hot-corrosion rig experiments. The coatings were first subjected to 600 1-h thermal cycles at 1200 °C in air, which led to the formation of some vertical cracks. The samples were then exposed in a hot-corrosion burner rig for 50 h at 1000 °C. Although uncoated silicon carbide substrates showed severe hot-corrosion attack, the plasma-sprayed mullite coatings protected the substrates as seen in Fig. 5.2.2a (Lee 2000b). There was limited attack of the silicon carbide at locations where the vertical cracks intersected the coating/substrate interface, leading to the formation of sodium silicate glasses (see Fig. 5.2.2b).

The performance of these plasma-sprayed mullite coatings was however not satisfactory in water-vapor-containing atmospheres. Plasma-sprayed mullite-coated silicon carbide samples exhibit considerable weight loss in a high-pressure burner rig operating a rich-burn condition at 6 atm and 1230 °C (Fig. 5.2.3, Lee 2000a). A similar weight loss was also observed after a 2-h cycle exposure in 50% water-vapor/oxygen atmosphere at 1300 °C. Fig. 5.2.4 shows the cross-section of a mullite-coated silicon carbide sample after the high-pressure burner rig test (Lee, 2000a). As is evident in the figure, the selective volatilization of silica results in the formation of a porous α-alumina skeleton on the surface.

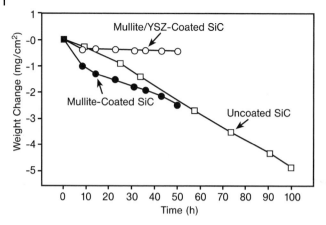

Fig. 5.2.3 Weight gain versus time plots for uncoated, mullite coated and mullite/YSZ coated SiC after exposure at 1230 °C to a high-pressure burner rig at 6 atm for 50 h (from Lee, 2000a).

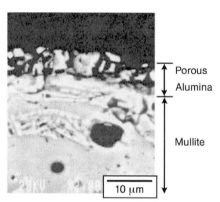

Fig. 5.2.4 Formation of a porous alumina layer on the surface of a plasma-sprayed mullite coating after a 50-h exposure at 1230 °C in a high pressure burner rig at 6 atm, due to selective volatilization of silica from mullite (from Lee, 2000b).

5.2.2
Plasma-sprayed Environmental Barrier Coatings (EBCs)

In order to prevent the selective volatilization of silica from plasma-sprayed mullite, Lee and co-workers developed multi-layered coatings. They found that deposition of a plasma-sprayed yttria-stabilized zirconia (YSZ) overlay coating over the mullite reduced silica volatilization considerably in a high-pressure (6 atm) burner rig at 1230 °C, (Fig. 5.2.3). The coating also performed reasonably well when subjected for a total of 100 h to a 2-h cycle exposure in a 90 % water-vapor/10 % oxygen atmosphere at 1300 °C (Lee 2000a). Although most interfacial areas exhibited excellent adherence with limited oxidation, some interfacial areas, especially at locations where vertical cracks intersected the interface, exhibited accelerated oxidation leading to the formation of a thick porous silica scale.

Lee (1998) reported that the purity of starting mullite powders used in the

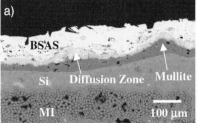

Fig. 5.2.5 Mullite-based EBC systems consisting of (a) Si/mullite/BSAS and (b) Si/mullite+BSAS/BSAS layers on melt infiltrated (MI) SiC/SiC fiber-reinforced composites, after a 100-h exposure to 1-h cycles at 1300 °C in a 90% H_2O–O_2 atmosphere (from Lee et al. 2003).

plasma-spray process plays a key role in determining the lifetimes of the coatings. Specifically, the presence of alkali or alkaline earth metal oxides in the powders leads to contamination of the silica scales formed between the coating and the silicon carbide substrate, accompanyied by the formation of pores in the silica. The ability of alkali and alkaline earth metal oxides to "open-up" the silica network and enhance oxygen permeability is well known (Lamkin, 1992). Pores were believed to have formed when gases (CO or CO_2) generated by the oxidation of silicon carbide by oxygen were bubbled through the low viscosity melt. Lee found that the porosity at the mullite/silicon carbide interface increased with increasing temperature and contamination, suggesting that a key to reduced pore formation and oxidation is to use chemically pure mullite powders for plasma spraying.

In order to improve the crack resistance of the coating, a more crack-resistant overlay coating was developed (Lee, 2000a, Lee et al. 2003). This barium strontium aluminum silicate [BSAS: $(1-x)$BaO–xSrO–Al_2O_3–SiO_2, $0 \leq x \leq 1$] topcoat has a lower level of silica activity to minimize the selective volatilization of silica. The BSAS layer was also more resistant to cracking, presumably due to its low modulus, which enhances the compliance of the coating. Replacing the YSZ topcoat with a BSAS layer more than doubled the time of onset of accelerated oxidation in water-vapor-containing atmospheres (Lee, 2000a). Even further improvement was achieved by adding a silicon bond layer between the mullite and the silicon carbide (Fig. 5.2.5a). According to Lee and coworkers (2003) the silicon layer improves the bond strength between silicon carbide and mullite. These environmental barrier coatings showed excellent protection of the substrate after a 100-h exposure to 1-h cycles at 1300 °C in a 90% water-vapor/10% oxygen atmosphere. Evidently, the formation of an extensive diffusion zone between the mullite and the BSAS (Fig. 5.2.5a) did not appear to affect the durability of the coating system adversely.

Another variant of these multi-layered environmental barrier coatings included a silicon bond coat, a mullite plus BSAS intermediate coat and a BSAS top coat, as seen in Fig. 5.2.5b (Lee et al. 2003). Again, the coating system exhibited excellent protection of the substrate in water-vapor-containing atmospheres at 1300 °C.

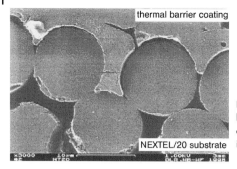

Fig. 5.2.6 Flame-sprayed mullite thermal barrier coatings on ceramic/ceramic composites showing excellent mechanical interlocking at the interface with the fibers (from Steinhauser et al. 2000).

However, some areas in the mullite/BSAS and silicon interface showed the formation of glassy reaction products. This degradation was much more severe at temperatures above 1400 °C. Lee et al. (2003) concluded that the silicon/(mullite+BSAS)/BSAS environmental barrier coating (Fig. 5.2.5b) was more robust than the silicon/mullite/BSAS system (Fig. 5.2.5a) owing to its superior crack resistance. However, the former is much more susceptible to formation of glassy phases, especially at higher temperatures.

5.2.3
Plasma-sprayed Thermal Barrier Coatings (TBCs)

Plasma- or flame-sprayed mullite coatings have also been used as thermal protection systems (TPS) for oxide fiber/mullite matrix composites (Steinhauser et al. 2000). These coatings reduce hot points in the composite structures thereby enhancing the high-temperature mechanical stability of the components.

Flame-sprayed mullite coatings, deposited using an acetylene/oxygen torch, exhibited substantial open porosity since the particle velocities are substantially lower than in the plasma-sprayed case. However, as shown in Fig. 5.2.6, the velocities are sufficiently high to ensure sufficient penetration of the molten particles into the top layer of the fibers. The adhesion of the mullite coating can be attributed to the excellent mechanical interlocking at the interface with the fibers. Temperature measurements in burner tests have shown that the peak temperature at the hot site of the ceramic matrix composite can be reduced by more than $100 \, C° \, mm^{-1}$ mullite layer thickness, assuming free connection cooling on the cold side of the hybrid system. Thermal and elastic mismatches between thermal protection systems and ceramic composite substrates in that case are low, owing to the similar phase compositions.

5.3
Deposition of Mullite Coatings by Miscellaneous Techniques
S. Basu and V. Sarin

Ultrathin aluminosilicate multi-layers (2, 5, 9, 30 nm thick) have been produced by Schmücker et al. (2001) by physical vapor deposition using a double-source jumping beam equipment. The as-deposited non-crystalline layers, when heat-treated at 1000 °C, behave differently. The 30-nm thick layer forms γ-alumina only. γ-Alumina plus a minor amount of mullite have been identified in the 5- and 9-nm thick layers, while only mullite formation occurs in the 2-nm thick layer. The crystallization of physically vapor-deposited aluminosilicate films with thicknesses above about 5 nm shows similar behavior to diphasic (type II) mullite precursors, while double layers with thicknesses below about 5 nm astonishingly behave like single-phase mullite precursors (type I, see Sections 1.4.1 and 1.4.2).

Pulsed-laser deposition (PLD) using high-energy CO_2 lasers has been used to produce protective mullite coatings. Fritze et al. (1998) studied the oxidation resistance of such coatings on carbon fiber-reinforced carbon/silicon carbide composites (C_F/C–SiC) composites. They found that thin (900 nm) pulsed-laser-deposited layers do not markedly improve the oxidation resistance of the material. However, significant improvement was observed in hybrid systems with 25-μm thick mullite coatings. From ^{18}O diffusivity measurements, they concluded that inward oxygen diffusion across the outer mullite layer controls the kinetics of oxidation. Thermal cycling produced cracks within the silicon carbide intermediate layer and spallation from the C_F/C–SiC substrate was found to be responsible for the failure of the oxidation protection of the hybrid system.

Federer (1990) developed an aqueous slurry process to apply refractory oxide coatings with varying concentrations of Al_2O_3 (between 50 to 90 wt.%) on silicon carbide. The study showed that the coatings containing mullite as the primary phase performed the best, although all coatings cracked and spalled during high-temperature exposure.

Kisly and Kodash (1989) produced protective mullite layers on the surface of aluminum-alloyed molybdenum disilicide ($MoSi_2$). They found that the oxidation of aluminum and molybdenum disilicide produces silica and aluminum molybdate, which in time decompose to molybdenum oxide (MoO_3) and alumina. Finally, above about 1300 °C, dense mullite coatings formed, which adhered quite firmly to the substrate. The system has the advantage of being self-healing when annealed in air at high temperatures.

Welsch and Schneider (unpublished) have discussed the possibility of producing mullite protection layers on silicon aluminum alloys by oxygen ion implantation. This technique may open a wide field of application, since silicon aluminum alloys can be easily sputter deposited on many metal substrates.

References

Auger, M. L. (1999). Development of CVD mullite coatings for si-based ceramics. Ph. D. Dissertation, Boston University, Boston.

Auger M. L. and Sarin, V. K. (1997). The development of CVD mullite coatings for high temperature corrosive applications. *Surface and Coatings Tech.* 94–95, 46–52.

Auger, M. L. and Sarin, V. K. (2001). A kinetic investigation of CVD mullite coatings on Si-based ceramics. *Int. J. Refr. Metals & Hard Mater.* 19, 479–494.

Auger, M. L., Sengupta, A. and Sarin, V. K. (2000). Coal slag protection of SiC with chemically vapor deposited mullite coatings. *J. Am. Ceram. Soc.* 83 [10] 2429–2435.

Basu, S. N. and Sarin, V. K. (2000). Development of CVD mullite coatings. *Proceedings of the Fifteenth International Conference on Chemical Vapor Deposition: CVD XV.* Allendorf M. D., Besmann, T. M., Hitchman, M. L. and Shimogaki, Y. eds., The Electrochemical Society Series 2000–13, 308–316.

Basu S. N. and Sarin, V. K. (2002). Functionally graded CVD mullite coatings. *Mater. Res. Soc. Symp. Proc.* 697, 7.2.1–7.2.6.

Basu, S. N., Hou, P. and Sarin, V. K. (1999a). Formation of mullite coatings on silicon based ceramics by chemical vapor deposition. *Journal of Refractory Metals and Hard Materials* 16 [4–6], 343–352.

Basu, S. N., Pattanaik, A. K. and Sarin, V. K. (1999b). Oxidation and corrosion resistant CVD mullite coatings", *Proceedings of Emerging Trends in Corrosion Control*, Vol. 2, Khanna, A. S., Sharma, K. S. and Sinha, A. K. eds., NACE International, Academia Books International, New Delhi, India. 1016–1025.

Braue, W., Paul, G., Pleger, R., Schneider, H. and Decker, J. (1996). In-plane microstructure of plasma sprayed Mg–Al spinel and 2/1-mullite based protective coatings: an electron microscopy study. *J. Eur. Ceram. Soc.* 16, 85–97.

Burnham, C. W. (1963). Crystal structure of mullite, Carnegie Institution Washington Year Book. 158–162.

Butt, D. P., Mecholsky, J., Roode, M. and Proce, J. R. (1990). Effects of plasma-sprayed ceramic coatings on the strength distribution of silicon carbide materials. *J. Am. Ceram. Soc.* 73, 2690–2696.

Cameron, W. E. (1977). Mullite: a substituted alumina. *American Minerologist*, 62, 747–755.

Doppalapudi, D. and Basu, S. N. (1997). Structure of mullite coatings grown by chemical vapor deposition. *Materials Science and Engineering*, A231, 48–54.

Doppalapudi, D., Basu S. N. and Sarin, V. K. (1996). Structural evolution of mullite coatings on silicon based ceramics. *Proceedings of the Thirteenth International Conference on Chemical Vapor Deposition*, Besmann, T. M., Allendorf, M. D., Robinson McD. and Ulrich R. K. eds., The Electrochemical Society, 664–669.

Federer, J. I. (1990), Alumina based coatings for protection of SiC ceramics, *J. Mater. Eng.* 12, 141–149.

Fischer, R. X., Schneider, H. and Voll, D. (1996). Formation of alumina rich 9 : 1 mullite and its transformation to low aluminum mullite upon heating. *J. Eur. Ceram. Soc.* 16, 109–113.

Fritze, H., Jojic, J., Witke, T., Rüscher, C., Weber, S., Scherrer, S., Weiss, R., Schultrich, B. and Borchardt, G. (1998). Mullite based oxidation protection for SiC–C/C composites in air at temperatures up to 1900 K. *J. Eur. Ceram. Soc.* 18 , 2351–2364.

Göring, J. (1991). Zusammenhang zwischen mechanischen Eigenschaften und gefüge-spezifischen Versagensmechanismen unterschiedlicher Silicumnitrid- und Silicumcarbid-materilien. *Dissertation*, Universität Karlsruhe, Karlsruhe.

Haynes, J. A., Lance, M. J., Cooley, K. M., Ferber, M. K., Lowden, R. A. and Stinton, D. P. (2000). CVD mullite coatings in high-temperature, high-pressure air H_2O. *J. Am. Ceram. Soc.* 83 [3], 657–659.

Hou, P. (2000). Microstructural evolution in chemically vapor deposited mullite coatings. *Ph. D. Dissertation*, Boston University, Boston, MA.

Hou, P., Basu, S. N. and Sarin, V. K. (1999). Nucleation mechanisms in chemically vapor-deposited mullite coatings on SiC", *J. Materials Research* 14 [7], 2952–2958.

Hou, P., Basu S. N. and Sarin, V. K. (2001).

Structure and high temperature stability of compositionally graded CVD mullite coatings. *International Journal of Refractory Metals and Hard Materials* 19 [4–6], 467–477.

Jacobson, N. S. (1993). Corrosion of silicon-based ceramics in combustion environment. *J. Am. Ceram. Soc.* 76 [1] 3–28.

Jacobson N. S., and Lee, K. N. (1996). Corrosion of mullite by molten salts, *J. Am. Ceram. Soc.* 79 [8], 2161–2167.

Kisly P. S. and Kodash V. Y. (1989). The mullite coatings on heaters made of molybdenum disilicide. *Ceram. Int.* 15, 189–191.

Lee, K. N. (1998). Contamination effects on interfacial porosity during cyclic oxidation of mullite-coated silicon carbide. *J. Am. Ceram. Soc.* 81 [12], 3329–3332.

Lee, K. N. (2000a). Key durability issues with mullite-based environmental barrier coatings for Si-based ceramics. *Transactions of the ASME.* 122, 632–636.

Lee, K. N. (2000b). Current status of environmental barrier coatings for Si-based ceramics", *Surface and Coatings Tech.* 133–134, 1–7.

Lee, K. N. and Miller, R. A. (1996). Oxidation behavior of mullite-coated SiC and SiC/SiC composites under thermal cycling between room temperature and 1200°–1400 °C", *J. Am. Ceram. Soc.* 79 [3], 620–626.

Lee, K. N., Miller, R. A. and Jacobson, N. S. (1995). New generation of plasma-sprayed mullite coatings on silicon carbide. *J. Am. Ceram. Soc.* 78 [3] 705–710.

Lee, K. N., Fox, D. S., Eldridge, J. I., Zhu, D., Robinson, R. C., Bansal, N. P. and Miller, R. A. (2003). Upper temperature limit of environmental barrier coatings based on mullite and BSAS. Accepted for publication in *J. Am. Ceram. Soc.* 86 [8].

Lee, W. Y., More, K. L., Stinton, D. B. and Bae, Y. W. (1996). Characterization of Si_3N_4 coated with chemically-vapor-deposited mullite after Na_2SO_4-induced corrosion. *J. Am. Ceram. Soc.* 79 [9], 2489–2492.

Lindstorm J. N., and Schacher, H. (1980). Non-Equilibrium Conditions for CVD of Alumina. Proceedings of the 3rd European Conference on CVD. Hintermann, H. E. ed., Neuchatal, Switzerland, 208–217.

Lynch J. F. (1981). Editor. *Engineering Property Data on Selected Ceramics Volume III, Single Oxides*. Metals and Ceramics Information

Center Report MCIC-HB-07. Battelle Columbus Laboratories, Columbus, Ohio.

Mulpuri, R. P. (1996). Chemical vapor deposition of mullite coatings on silicon based ceramics for high temperature applications. *Ph. D. Dissertation.* Boston University, Boston.

Mulpuri, R., and Sarin, V. K. (1995). Thermodynamic analysis of cvd mullite coatings. Presented at the 19th Annual Cocoa Beach Conference and Exposition on Engineering Ceramics, Cocoa Beach, FL, The American Ceramic Society, Westerville, OH.

Mulpuri, R. P. and Sarin, V. K. (1996). Synthesis of mullite coatings by chemical vapor deposition. *J. Mater. Res.* 11 [6], 1315–1324.

Nakajima Y. and Ribbe, PH. (1981). Twinning and superstructure of Al-rich mullite. *Am. Mineral.* 66, 142–147.

Nitodas, S. F., and Sotirchos, S. V. (2000). Chemical vapor deposition of aluminosilicates from mixtures of $SiCl_3$, $AlCl_3$, CO_2 and H_2", *J. Elechem. Soc.* 147 [3], 1050–1058.

Pattanaik, A. K. and Sarin, V. K. (1998). High temperature oxidation and corrosion of CVD mullite coated SiC. *Surface Modification Technologies XII*, Sudarshan, T. S., Khor, A. K. and Jeandin, M. eds., ASM International, Materials Park, Ohio, 91–101.

Price, J. R., van Roode, M. and Stala, C. (1992). Ceramic oxide-coated silicon carbide for high temperature corrosive environments, *Key Eng. Maters.* 72–74, 71–84.

Robinson, R. C. and Smialek, J. L. (1999). SiC recession caused by SiO_2 scale volatility under combustion conditions: I, experimental results and empirical model. *J. Am. Ceram. Soc.* 82 [7], 1817–27.

Sarin, V. K. and Mulpuri, R. P. (1998). Chemical vapor deposition of mullite coatings. U. S. Patent No. 576008.

Schmücker, M., Hoffbauer, W. and Schneider, H. (2001). Constitution and crystallization behavior of ultrathin physical vapor deposited (PVD) Al_2O_3/SiO_2 laminates. *J. Eur. Ceram. Soc.* 21, 2503–2507.

Schneider, H. and Rymon-Lipinski T. (1988). Occurrence of pseudotetragonal mullite. *J. Am. Ceram. Soc.* 71 [3], C162–164.

Steinhauser, U., Brau, W., Göring, J., Kanka,

B. and Schneider, H. (2000). A new concept for thermal protection of all-mullite composites in combustion chambers. *J. Eur. Ceram. Soc.* 20, 651–658.

Tingey, G. L. (1966). Kinetics of the water-gas equilibrium reaction. I; the reaction of carbon dioxide with hydrogen", *J. Physical Chemistry* 70 [6], 1406–1412.

Varadarajan, S. (1999). Development of CVD mullite interfacial barrier coatings. *MS Thesis*, Boston University, Boston.

Varadarajan, S., Pattanaik, A. K. and Sarin, V. K. (2001). Mullite interfacial coatings for SiC fibers. *Surf & Coatings Technol.* 139, 153–160.

Wei, W. C. and Halloran, J. W. (1988a) Transformation kinetics of diphasic aluminosilicate gels. *J. Am. Ceram. Soc.* 71 [7], 581–587.

Wei, W. C. and Halloran, J. W. (1988b). Phase transformation of diphasic aluminosilicate gels. *J. Am. Ceram. Soc.* 71 [3], 166–172.

Zemskova, S. M., Haynes, J. A., Besmann, T. D., Hunt, R. D., Beach, D. B. and Golovlev, V. N. (2000). Preparation and characterization of some aluminosiloxanes as single-source MOCVD precursors for aluminosilicate coatings, *J. De Physique IV*, 10 [P2], 35–42.

6
Mullite Fibers

Mullite fibers (Mu_F) have gained increasing attention in recent years because of their high creep resistance and stability under harsh environments at high temperatures. This is in spite of the fact that above 1200 °C, unlike carbon (C_F), silicon carbide (SiC_F), and silicon boron nitrogen carbon ($SiBNC_F$) fibers, all current commercial oxide fibers show a lack of microstructural and mechanical stability. The limited stability of oxide fibers is explained by the mixed ionic and covalent atomic bonds in oxides and silicates. These are weaker than the pure covalent bonds in advanced non-oxide materials. As a consequence the oxides exhibit high reactivity and excessive grain growth at temperatures at which the non-crystalline fibers still remain unchanged. Polycrystalline and single phase oxide fibers are particularly susceptible to grain growth and strength reduction at temperatures as low as 1200 °C. The problem is often less severe in two-phase (e. g. α-alumina-mullite) systems. Possibly the decoration of the mullite grains with smaller-sized α-alumina crystals, slows down grain growth of mullite (see, for example, Kingery et al. 1976). The big advantage of oxide and aluminosilicate fibers is their inherent oxidation resistance, which makes them superior to any non-oxide ceramic fiber, especially if long-term high-temperature stability under harsh cycling conditions as in combustors of gas turbine engines is required (see Sections 6.4 and 7.2.6).

Mullite whiskers and continuous mullite fibers have been described.

6.1
Mullite Whiskers
H. Schneider

6.1.1
Whisker Formation from Melts

It has long been known that mullite in porcelain bodies has a needle-like, acicular shape. Perera and Allot (1985) grew mullite whiskers by firing kaolinite between 1400 and 1600 °C, the whisker dimension being directly correlated to the impurity content of the system. These whiskers are 0.5 to 5 μm long and display aspect ratios between 3 and 8. Katsuki et al. (1988) applied a similar processing route at

Fig. 6.1.1 Scanning electron micrograph of melt-grown mullite whiskers produced from aluminum sulfate ($Al_2(SO_4)_3$) and silica (SiO_2) with potassium phosphate (K_3PO_4) as a flux (heat treatment: 1100 °C, 3 h, with subsequent 1N HCl leaching. Courtesy A. Yamaguchi).

5 µm

temperatures up to 1700 °C and identified a dependence between whisker length and firing temperature. At 1700 °C the mullite whiskers are up to about 20 µm long, with aspect ratios of about 15. Li et al. (2001) synthesized mullite whiskers from kaolinite with addition of ammonium aluminum sulfate nonahydrate ($NH_4Al(SO_4)_2 \cdot 9H_2O$) and sodium phosphate dihydrate ($Na_3PO_4 \cdot 2H_2O$) as foaming and flux agents. While sulfate also serves as an aluminum source, phosphate promotes the formation of a secondary glass phase, from which the whiskers grow. The phosphate also enhances the leaching of the coexisting glass in hydrofluoric acid, which is necessary to isolate the whiskers. Good results are obtained after heat treatment at 1500 °C, where the acicular mullite crystals are up to 50 µm long. Hashimoto and Yamaguchi (2000) obtained mullite needles in a flux-assisted process using aluminum sulfate ($Al_2(SO_4)_3$), silica (SiO_2), and potassium phosphate (K_3PO_4) as starting materials. After firing at 1100 °C, mullite whiskers 2 to 5 µm long and 0.2 to 0.5 µm thick have been collected (Fig. 6.1.1). More data on melt-grown mullite whiskers are published by De Souza et al. (2000a,b) and by Regiani et al. (2002). They provide information on nucleaction and growth of aluminosilicate starting materials doped with 3 mol% rare earth oxides (Nd_2O_3, Yb_2O_3, Y_2O_3, La_2O_3). Owing to the presence of the rare earth oxides, glass layers of low viscosity develop at the surface of sample compacts, in which the whiskers grow (Fig. 6.1.2a). The newly formed mullite whiskers have their long axes (i.e. the **c** axes) parallel to the sample surfaces. Whisker thicknesses are independent of the annealing time of the experiments, while whisker lengths strongly depend on it. Regiani et al. showed that lanthania-doped systems give rise to rosette-like microstructures (Fig. 6.1.2b). The authors believe that the rosette-like arrangements are correlated to the whisker layers at the very surface of samples in such a way that the rosette centers correspond to the whiskers' nucleation sites.

6.1.2
Whisker Formation via Gas-transport Reactions

The major mullite whisker formation process is based on gas transport reactions, e.g. by means of volatile fluorine compounds. Okada and Otsuka (1989) synthesized mullite whiskers by firing mixtures of silica (SiO_2) gel and aluminum fluoride (AlF_3) with stoichiometric mullite bulk composition in an airtight container

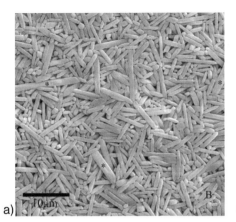 

Fig. 6.1.2 Scanning electron micrograph of a lanthania
(La$_2$O$_3$)-doped mullite pellet after firing at 1600 °C for 3 h.
(a) Untreated sample surface, showing the development of
mullite whiskers; (b) Microstructure of rosette-shaped mullite
agglomerates in a zone just below the whisker layer, which
was removed by acid treatment (Courtesy D. P. F. De Souza).

between 900 and 1600 °C. The length of these whiskers ranges from 6 to 9 μm as
the reaction temperature rises from 1100 to 1600 °C, while the aspect ratio simultaneously decreases from about 25 to 10. The whiskers produced below 1200 °C are
Al$_2$O$_3$-rich, whereas above about 1200 °C they become stoichiometric (about 72
wt. % Al$_2$O$_3$). This is in agreement with the observation that mullites that crystallize below about 1200 °C are enriched in Al$_2$O$_3$ (see Section 2.5.1). Different microstructures with either random whisker distribution or forming chestnut, burr-like
spherulites are observed if the method of fluorine addition is changed: Ismail et al.
(1990) prepared their mullite whiskers by gelling boehmite (Al(OH)$_3$) and silica sol
in a solution containing hydrofluoric acid (HF). Zaykoski et al. (1991) obtained
mullite whiskers from an anhydrous mixture of aluminum fluoride, alumina and
silica working in a closed system (Fig. 6.1.3). Mullite whiskers several hundred
microns long with small diameters been synthesized by Choi and Lee (2002)
from a mixture of silica and silicon in an alumina tube reactor under hydrogen and
carbon tetrafluoride (CF$_4$) gas flow. The gas transport reaction can be represented
as:

$$Al_2O_3(s) + 3\ CF_4(g) + 3\ H_2(g) \rightarrow 2\ AlF_3\ (g) + 3\ CO(g) + 6\ HF(g) \tag{1}$$

$$3\ SiO_2(s) + Si(s) + 3\ CF_4(g) + H_2(g) \rightarrow 3\ SiF_4(g) + 3\ CO(g) + H_2O(g) \tag{2}$$

Net reaction:
$$6\ AlF_3(g) + 2\ SiF_4(g) + 13\ H_2O(g) \rightarrow 3Al_2O_3 \cdot 2SiO_2(s) + 26\ HF(g) \tag{3}$$
$$\text{mullite}$$

s = solid, g = gaseous

40 µm

Fig. 6.1.3 Scanning electron micrograph of mullite whiskers grown from a mixture of aluminum fluoride (AlF₃) and silica (SiO₂) in a silicon fluoride (SiF₄) atmosphere at 1200 °C (Courtesy J. Zaykoski).

An easy and low cost way to produce mullite whiskers via gas-transport reactions is thermal decomposition of topaz ($Al_2SiO_4(F,OH)_2$) in closed systems. Moyer and Hughes (1994) and Moyer and Rudolf (1994) studied the decomposition of fluorotopaz to mullite at 1100 °C:

$$3Al_2SiO_4F_2(s) + 1/2\ O_2 \rightarrow 3Al_2O_3 \cdot 2SiO_2(s) + SiF_4(g) + F_2(g) \qquad (4)$$
topaz $\qquad\qquad\qquad\qquad$ mullite

s = solid, g = gaseous

They believe that the whiskers grow in a liquid phase. Although the presence of an intermediate liquid can enhance reactions the central step is certainly gas transport of aluminum and silicon fluoride species. Peng et al. (2003) mentioned mullite whisker formation by pyrolysis of hydroxyl fluorotopaz from a natural Australian concentration sand, according to the equation:

$$6\ Al_2SiO_4(F_{0.75}\ OH_{0.25})_2(s) \rightarrow 2\ (3Al_2O_3 \cdot 2SiO_2)(s) + 2\ SiF_4(g) + HF(g) + H_2O(g) \quad (5)$$
topaz $\qquad\qquad\qquad\qquad$ mullite

s = solid, g = gaseous

This mullite whisker formation route has been described as being simple and inexpensive, but the disadvantage is that the mullite whiskers are contaminated by impurities. Thus they are not suitable for the reinforcement of advanced composites unless they are purified, e.g. by chemical leaching.

Merk and Thomas (1991) reported details of the crystallography and composition of mullite whiskers derived from natural topaz. Their high resolution electron microscopy patterns show that the mullite crystals in these whiskers annealed at 1400 °C contain fine scaled, irregularly spaced twin lamellae perpendicular to [001]. The optical diffractograms from larger areas, containing several twin planes, show twinned superstructure spots and streaking parallel to c^*, due to a changing width of twin lamellae.

6.2
Sol-gel-derived Continuous Mullite Fibers
H. Schneider

Although the mechanical strength of continuous aluminosilicate fibers is inferior to, for example, that of α-alumina and silicon carbide fibers, they have attracted growing interest in recent years. This is due to the good creep resistance of mullite fibers up to high temperatures, along with an inherent oxidation stability. In order to improve the mechanical strength, fibers consisting of nano-sized polycrystalline mullite have successfully been developed. The mechanical strength of the polycrystalline mullite fibers can be derived from a modified Hall-Petch correlation ($\sigma = f(1/\sqrt{d})$, σ = strength, d = grain diameter, see Kingery et al. 1976, p. 794). This states that the strength of a ceramic correlates reciprocally with the grain size of constituents. Thus, although other factors like sub-grains and low-angle grain boundaries also have an influence on the stability of fibers, the nano-scale grain microstructure remains a key factor in tailoring the fibers' mechanical stability. On the other hand it has to be kept in mind that the creep resistance of fibers is directly correlated with the grain size of the constituents. Thus, depending on the application, specific grain-size distributions have to be adjusted during fiber fabrication and subsequent temperature treatments.

6.2.1
Laboratory Produced Fibers

Aluminosilicate melts have very low viscosities and thus exhibit a high tendency to crystallize, making it difficult to obtain continuous mullite fibers of good and reproducible quality. For that reason mullite fibers are normally produced by sol-gel techniques, starting from solutions or dispersions. The term sol-gel is thus understood in a wider sense: It refers also to fabrication routes in which the sol step is not realized. This is true, for example, for aluminum and silicon alkoxides with the general composition $M(OR)_n$ (M = Al, Si; R = C_xH_{2x+1}; n = 3 for Al and 4 for Si; compounds: Aluminum isopropoxide ($Al(^iOC_3H_7)_3$), aluminum *sec*-butoxide ($Al(OC_4H_9)_3$), tetraethyloxysilane (TEOS:$Si(OC_4H_9)_4$). Metal salts such as aluminum nitrate ($Al(NO_3)_3$) or aluminum acetate ($Al(O_2C_2H_3)_3$) are alternative starting compounds. Basic processes of the sol-gel route are hydrolysis and polymerization, represented by the equations:

$$Al(OR)_3 + H_2O \rightarrow Al(OR)_2OH + ROH \tag{6}$$

$$2Al(OR)_2OH \rightarrow (OR)_2Al\text{-}O\text{-}Al(OR)_2 + H_2O \tag{7}$$

and

$$Si(OR)_4 + H_2O \rightarrow Si(OR)_3OH + ROH \tag{8}$$

$$2Si(OR)_3OH \rightarrow (OR)_3Si\text{-}O\text{-}Si(OR)_3 + H_2O \tag{9}$$

The gelation of solutions or sols yields viscous products, suitable for the production of continuous fibers, e. g. by extrusion. After drying and calcination, the green fibers are sintered in a way that homogeneous and nano-sized microstructures are achieved. Details of the basic steps in sol-gel-derived ceramic fiber production have been given by several authors (for references see Naslain, 2000).

Continuous mullite fibers have been produced by dry spinning of single-phase (type I) or diphasic (type II, see Sections 1.4.1 and 1.4.2) mullite gels. Single-phase gels have the advantage that nearly ideal mixing of aluminum and silicon compounds exists in the precursor, and, as a consequence, mullitization occurs at very low temperatures (below 1000 °C). However, what at first glance appears as an advantage leads to serious problems during fiber production. A major disadvantage is the very slow sintering of mullite at the fiber production temperature (about 1300 °C). Other disadvantages of single phase mullite gels are low oxide yields and high shrinkage during calcinations. Further problems arise from the hydrolysis of aluminum alkoxide, which is much faster than that of silicon alkoxide, which makes single phase precursors difficult to prepare. Thus, it is better to work with diphasic mullite gels, consisting of transitional alumina ("γ-alumina") and non-crystalline silica. Mullitization of these materials takes place at higher temperatures (1200 to 1300 °C, see Section 1.4.2). This has the advantage that densification takes place prior to mullitization in a liquid-phase-assisted sintering process, comparable to transient viscous sintering (TVS, see Section 4.2.6).

In order to lower the mullitization temperature and to reduce the temperature-induced grain growth of fibers, boria (B_2O_3) has often been added to the alumino-silicate starting materials (e. g. Richards et al. 1991). However, at high temperature boria may evaporate, which causes fiber shrinkage. Moreover, addition of boria enhances the formation of glass phases, which has a negative influence on the creep behavior of fibers. A better way than boria addition is to strengthen the properties of boria-free mullite fibers by improving the microstructural homogeneity at a sub-micron level. Such approaches use aluminum and silicon alkoxides and organic solvents (Venkatachari et al. 1990, Al-Assafi et al. 1994, Yogo and Aksay, 1994), or aluminum nitrate plus alkoxides (Nishio and Fujiki, 1991). Venkatachari et al. were among the first to spin mullite fibers from boria-free systems

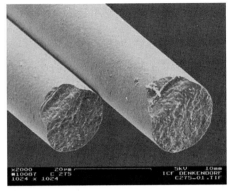

Fig. 6.2.1 Laboratory-produced continuous mullite fibers. Starting materials are basic aluminum chloride hydrate ($Al_2(OH)_5Cl \cdot 2$–$3H_2O$) and aqueous silica sol (SiO_2) with pyrolysis and sintering of the cross-wound fibers at 1100 and 1300 °C, respectively (Courtesy B. Clauss).

using the sol-gel technique. Their fibers sinter between 1000 and 1200 °C, and mullitization occurs at about 1300 °C. These fibers are reported to be very stable, retaining their nano-sized microstructure up to 1500 °C.

Venkatachari and coworkers also describe fibers consisting of mullite plus yttrium-stabilized zirconia. However, these fibers are less stable, and display enhanced grain growth at elevated temperatures, probably due to the presence of partial melts, which enhance diffusion. The fibers of Al-Assafi et al. (1994) form mullite at about 1000 °C. These fibers have grain sizes below 1 μm up to 1500 °C, and contain little or no grain boundary glass phase. Okada et al. (1998) combined aluminum and silicon alkoxides with aluminum nitrate, aluminum metal and silicon alkoxide in water solvent. They obtained dense and crack-free green fibers after heat treatment at 1100 °C. Their fiber preparation on the basis of aluminum chloride plus silicon alkoxide was less successful: Owing to the evaporation of chlorine the fibers are highly porous. Song (1998) used aluminum and silicon alkoxides plus aluminum nitrate to produce mullite fibers. Fibers of a suitable spinnability are obtained taking aluminum alkoxide to aluminum nitrate in the ratio from 3 to 5 in molecularly mixed single-phase gels. Chen et al. (1996) prepared transparent mullite fibers with smooth surfaces and uniform diameters of 15 μm from aluminum carboxylates and silicon alkoxide. The bulk Al_2O_3 content of these fibers after pyrolysis at 1250 °C is higher than that of the starting solution, possibly because of partial evaporation of unhydrolyzed silicon alkoxide. The Institute of Textile Chemistry and Chemical Fibers (Denkendorf, Germany) produces continuous mullite fibers from basic aluminum chloride hydrate ($Al_2(OH)_5Cl \cdot 2$–$3H_2O$) and nano-sized (20 nm) silica sols. A specific technique is used to cross-wind and store the green fibers. Pyrolysis and sintering of fibers are achieved at 1100 and 1300 °C. Homogeneous fibers of good quality, several hundred meters long and about 15 μm thick, have been obtained (Fig. 6.2.1).

Summarizing the actual knowledge of laboratory-produced polycrystalline mullite fibers, it shows that none of these fibers have the required thermo-mechanical properties, nor can they be consistently produced in the amounts necessary for fiber-reinforced ceramic matrix composites.

6.2.2
Commercially Produced Fibers

Commercial continuous aluminosilicate fibers have been widely used for thermal and electrical insulation, and for the reinforcement of ceramic and metal matrices (see Sections 6.4 and 7.2.6). These fibers are fabricated via chemical routes using the sol-gel technique (see above). The fibers must have small and constant diameters (<15 μm) and should be chemically and microstructurally homogeneous. In order to achieve satisfactorily high mechanical strength, grain diameters must be on the nano-scale (see above). There are three companies producing commercial aluminosilicate ceramic fibers: Sumitomo Chemicals (Japan), Denka-Nivity (Japan) and 3M (U.S.A.). Today by far the greatest proportion of aluminosilicate fibers used comes from 3M (see Table 6.2.1 and Fig. 6.2.2).

6.2.2.1 Altex Fibers (Sumitomo Chemicals, Japan)

Altex aluminosilicate fibers (F1 in Table 6.2.1 and Fig. 6.2.2) are obtained from green fibers by heat treatment in several steps up to 1400 °C. Their chemical composition is 85 wt. % Al_2O_3 and 15 wt. % SiO_2, corresponding to a phase content of about 30 wt. % mullite and 70 wt. % α-alumina. The fiber diameters range between 10 and 15 μm. Altex fibers are dense and display relatively high Young's

Table 6.2.1 Properties of commercial aluminosilicate fibers.

Fiber Key	Manufacturer (Brand name)	Chemical composition [wt.%]	Diameter [μm]	Density [g cm^{-3}]	Tensile strength [MPa]/ Young's modulus [GPa]	Price per kg [US$]
F1	Sumitomo (Altex)	Al_2O_3 : 85 SiO_2 : 15	10–15	3.3	1800/330	500
F2	Denka-Nivity (Nivity)	Al_2O_3 : 80 SiO_2 : 20	3–10	3.3	2000/170	600
F3	Denka-Nivity (Nivity)	Al_2O_3 : 70 SiO_2 : 30	11	2.9	2000/170	–
F4	3M (Nextel 720)	Al_2O_3 : 85 SiO_2 : 15	10–12	3.4	2100/260	820
F5	3M (Nextel 550)	Al_2O_3 : 73 SiO_2 : 27	10–12	3.03	2000/193	620
F6	3M (Nextel 480)	Al_2O_3 : 78 SiO_2 : 22	10–12	–	–/220	–
F7	3M (Nextel 440)	Al_2O_3 : 70 SiO_2 : 28 B_2O_3 : 2	10–12	3.05	2000/193	450
F8	3M (Nextel 312)	Al_2O_3 : 62 SiO_2 : 24 B_2O_3 : 14	10–12	2.7	1700/150	240

All quoted properties are from manufacturers' data. Prices refer to 2001 (after Newman and Schäfer, 2001).

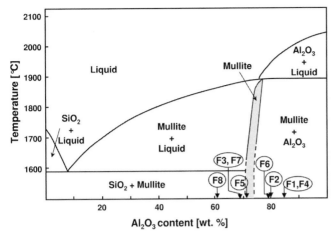

Fig. 6.2.2 Compositions of commercial aluminosilicate fibers plotted in an Al_2O_3–SiO_2 diagram (see Table 6.2.1).

moduli, but they suffer from low strength. Altex aluminosilicate fibers have mainly been used to reinforce aluminum alloys (see Abe et al. 1982).

6.2.2.2 Nivity Fibers (Denka-Nivity, Japan)

Nivity fibers (F2 and F3 in Table 6.2.1 and Fig. 6.2.2) consist of 70 to 80 wt. % Al_2O_3 and 20 to 30 wt. % SiO_2. The fibers have very small diameters, ranging between 3 and 11 μm, and densities varying between 2.9 and 3.3 $g\,cm^{-3}$. Nivity fibers have a strength in the same range as Altex fibers; however, they suffer from low stiffness. The price of Nivity fibers is relatively low and they have thus been used for low temperature applications at moderate loads.

6.2.2.3 Nextel Fibers (3M Company, U.S.A.)

Nextel aluminosilicate fibers (F4 to F8 in Table 6.2.1 and Fig. 6.2.2) cover a wide range of chemical compositions (from Nextel 312: 62 wt. % Al_2O_3 to Nextel 720: 85 wt. % Al_2O_3) with fairly constant diameters of about 12 μm. Although the main phase of all these fibers is mullite, temperature stabilities and mechanical properties can vary widely. Nextel 312 and Nextel 440 fibers contain boria (11 and 2 wt. % B_2O_3, respectively) to enhance mullitization and to reduce grain growth. However, the addition of boria is responsible for the occurrence of grain boundary glass phases, which can drastically reduce the stiffness of fibers by grain boundary gliding. Also boria gradually volatizes above about 1000 °C, which causes fiber shrinkage. Consequently, boria-containing Nextel fibers are suitable for low temperature applications only. Nextel 550 fibers consist of transition alumina ("γ-alumina") plus vitreous silica in the as-received state, and have mullitic overall composition. The phase assemblage transforms to single phase mullite with a mean grain size

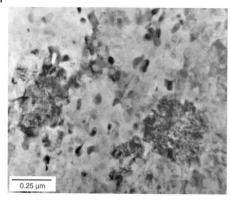

Fig. 6.2.3 Scanning electron micrograph of the microstructure of an as-received Nextel 720 aluminosilicate fiber. Note the mosaic-type shape of mullite grains, consisting of individual crystals of slightly different orientations. Small α-alumina crystallites are embedded in the mullite matrix (Courtesy M. Schmücker).

of about 0.4 μm at 1300 °C (see Schmücker et al. 2005). The drawback of Nextel 550 fibers again is the low stiffness (see Table 6.2.1).

The Nextel aluminosilicate fiber with the highest thermal stability is Nextel 720. Because of their technical importance for oxide fiber/oxide matrix composites and beyond, the temperature-controlled behavior of Nextel 720 fibers has been investigated by many research groups. Some specific results of these studies are referenced in what follows. As-received Nextel 720 aluminosilicate fibers are composed of 85 wt. % Al_2O_3 and 15 wt. % SiO_2, corresponding to about 70 wt. % mullite and 30 wt % α-alumina. Mosaic-type mullite "macrograins" occur with wavy contours (diameter: about 300 nm), each "macrograin" containing several sub-grains of slightly different orientation (Fig. 6.2.3). This irregular sub-grain structure is typical for sol-gel mullite produced by rapid pyrolysis. Small and elongated α-alumina crystals (length: About 70 nm) are embedded in the mullite matrix (see, for example, Schneider et al. 1998, Deleglise et al. 2001, Schmücker et al. 2005). Detailed descriptions of the temperature-induced microstructural development and of the grain growth of Nextel 720 aluminosilicate fibers has been given by Schneider et al. (1998) and by Schmücker et al. (2005). Their transmission electron microscopy studies indicate that the microstructure of the Nextel 720 fibers is stable up to about 1300 °C. Above 1300 °C the mosaic-type mullite grains disappear by recrystallization (Fig. 6.2.4). The driving force for this process is minimization of the interfacial energy by reduction of internal grain surfaces, elimination of small angle grain boundaries and polygonization of crystals. Heat treatment above 1500 °C yields an homogeneized microstructure, with the mullite and α-alumina crystals being about 0.25 μm in size (Fig. 6.2.5).

Schneider et al. (1998) found that the mullites in the as-received Nextel 720 fibers are enriched in Al_2O_3 (78 wt. %), typical for sol-gel derived mullites heat-treated below 1200 °C (see Section 2.5.1). At higher temperatures the Al_2O_3 content of mullite decreases and gradually approaches the value for stoichiometric mullite (72 wt. %). Owing to the Al_2O_3 exsolution from mullite the α-alumina fraction of the fiber increases from about 30 wt % in the as-received state to about 45 wt. % in the sample annealed at 1500 °C. This reaction causes shrinkage of

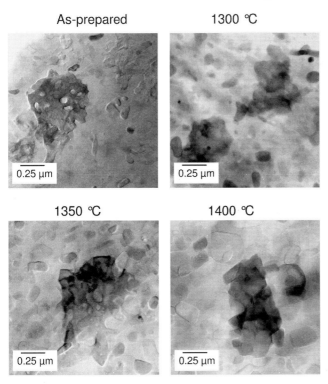

Fig. 6.2.4 Transmission electron micrographs of the temperature-induced microstructural development of Nextel 720 aluminosilicate fibers in the range up to 1400 °C. Note the recrystallization of mullite between 1300 and 1400 °C with disappearance of the mosaic-type mullite macrograin structure (Courtesy M. Schmücker).

about 1% of the fiber. Detailed studies on the temperature-produced microstructural development in Nextel 720 fibers have also been carried out by Bunsell and Berger (2000), Schmücker et al. (2001), and Deleglise et al. (2001, 2002). Their results basically confirm the earlier observations of Schneider and coworkers.

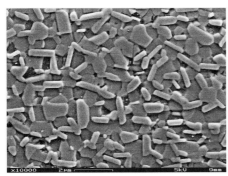

Fig. 6.2.5 Scanning electron micrograph of the microstructure of a Nextel 720 aluminosilicate fiber, annealed at 1600 °C for 1 h. Note the more or less isometric mullite grains (dark grey) and the columnar α-alumina crystals (light grey. Courtesy M. Schmücker).

a)

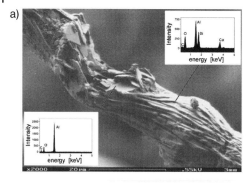

b)

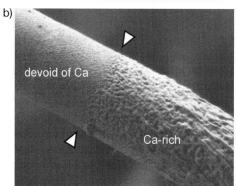

devoid of Ca

Ca-rich

Fig. 6.2.6 Scanning electron micrographs of the damage zone in a Nextel 720 aluminosilicate fiber caused by local impurity enrichments after annealing at 1400 °C for 2 h in air. (a) Central part of the damage zone, consisting of newly formed facetted α-alumina (0001) plates (spectrum bottom left). The newly formed peripheral crystals (spectrum top right) are phases of the system $CaO–Al_2O_3–SiO_2$. (b) Transitional range of the damage zone, showing high surface roughness due to exaggerated grain growth and the smooth surface of impurity-free fiber sections (Courtesy W. Braue).

Content and distribution of impurities play an important role in the understanding of the failure of Nextel 720 aluminosilicate fibers. Braue et al. (2001) determined traces of calcium, iron, zinc, copper, and occasionally sodium, homogeneously distributed in the bulk of fibers. Although the concentration of these impurities is too low to initiate fiber failure, it may negatively influence the creep behavior of fibers. Bunsell and Berger (2000) and Braue et al. (2001) also describe local enrichments of sodia (Na_2O) and calcia (CaO), with an amount several orders of magnitude higher than that of the bulk. Such locally enriched impurities may cause low melting eutectics in the ternaries $Na_2O/CaO–Al_2O_3–SiO_2$, giving rise to exaggerated grain growth of albite ($NaAlSi_3O_8$), anorthite ($CaAl_2Si_2O_8$) and α-alumina (Fig. 6.2.6). These fiber areas can act as flaws at which fibers fail. Schneider et al. (1998) estimated that the mean distance between such macroflaws is in the order of 200 mm.

Deleglise et al. (2002) provided a detailed study of the microstructural evolution of Nextel 720 aluminosilicate fibers under load and in the presence of sodium and calcium contaminants. At temperatures as low as 1100 °C, creep and associated dissolution of mullite is initiated. Strong α-alumina grain growth is observed at flaws on the fibers' surfaces, with the formation of large (0001) plates in the direction of tension (Fig. 6.2.7). Simultaneously the newly formed silicate liquid migrates towards the fiber periphery, thus producing thin fiber surface coatings.

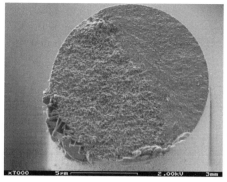

Fig. 6.2.7 Scanning electron micrograph of the fracture surface of a Nextel 720 aluminosilicate fiber after tensile load (15 MPa s^{-1} at 1200 °C). Note the occurrence of (0001) α-alumina plates at the big flaw (bottom left. Courtesy M.H. Berger).

Subsequent to creep at 1500 °C the fiber cores consist almost entirely of elongated α-alumina crystals, with their long axes being oriented in the load direction, and with thin glass layers developing between α-alumina grains. Deleglise et al. believe that these microstructural features are one reason for the relatively low creep resistance of Nextel 720 aluminosilicate fibers under these specific conditions. Fiber bundles annealed under the same conditions as single filaments behave differently: Although fibers from the periphery of bundles display similar effects to single filaments, those from the middle of bundles show no exaggerated growth of α-alumina. Deleglise et al. take this observation as evidence for the strong influence of "external" alkali or alkaline earth contaminants on the fiber stability.

The tensile strength of Nextel 720 aluminosilicate fibers has recently been measured by Schneider et al. (1998), Petry and Mah (1999), and by Wilson and Visser (2001). Wilson and Visser determined the statistical fracture distribution and the gauge length dependence of strength. They obtained a Weibull modulus of about 7 with a good fit of measured data to a linear Weibull relationship. Tensile strengths of Nextel 720 aluminosilicate fibers, annealed and tested in-situ at different temperatures are given in Fig. 6.2.8. The tensile strengths of annealed fibers are constant up to about 1100 °C. Above this temperature limit, strength gradually decreases to about 50% of the initial value at 1400 °C. In-situ measured high temperature strengths are lower than those collected from annealed fibers. This is reasonable, considering the influence of creep failure, which is effective in in-situ tests but not in annealed samples. Besides creep, fiber failure can be attributed to flaws produced by local enrichments of impurities on the fiber surfaces. According to Schneider et al. (1998) and Milz et al. (1999) the distance between flaws caused by local impurity enrichments in Nextel 720 fibers ranges between 130 and 200 mm. Since the fiber length relevant for failure in ceramic matrix composites is below 15 mm (Curtin, 1991) "impurity-produced" flaws in Nextel 720 aluminosilicate fibers should not have a controlling influence on the mechanical behavior of oxide/oxide composites.

The sub-critical crack growth of Nextel 720 aluminosilicate fibers has been studied by Milz et al. (1999) between 900 and 1200 °C. At 1000 °C intense sub-critical crack growth occurs. At 1100 °C this influence becomes weaker, and at 1200 °C,

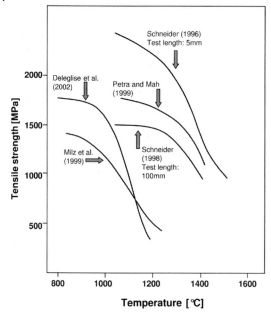

Fig. 6.2.8 Tensile strength of Nextel 720 aluminosilicate fibers measured at high temperatures (Milz et al. 1999, Deleglise et al. 2002) and after annealing (Schneider et al. 1998, Petry and Mah, 1999). Different fiber test lengths were used.

finally, the deformation is controlled by creep. Deleglise et al. (2002) re-studied the creep of Nextel 720 aluminosilicate fibers. Up to 1300 °C and under low stresses they describe "negative" deformation, i.e. shrinkage (Fig. 6.2.9). This fits well with the suggestion of Schneider et al. (1998) that the temperature-induced exsolution of Al_2O_3 from mullite and the associated formation of additional α-alumina causes densification of the fibers. With temperature and the level of applied stress the influence of "positive" creep-induced deformation increases. At 1500 °C, finally, creep becomes so intense that the fibers fail after a short time and under low stress load.

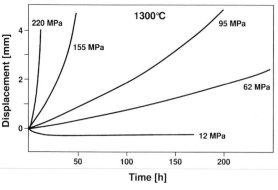

Fig. 6.2.9 Creep curves obtained from Nextel 720 aluminosilicate fiber bundles at 1300 °C and different stresses (according to Deleglise et al. 2002).

6.3

Continuous Melt-grown Mullite Fibers

H. Schneider

Single crystal and directionally crystallized mullite fibers, grown from aluminosilicate melts, should display very good creep resistance with a thermal stability far beyond that of polycrystalline mullite fibers. This can be derived from the absence of grain boundary sliding and grain growth. Composites based on such mullite fibers may open a wide field of applications, if homogeneous, flaw-free fibers with small diameters (<20 µm) can be made available in sufficient quantities at a reasonable price.

Sayir and Farmer (1995) used the laser heat flow zone (LHFZ) technique to produce mullite fibers from aluminosilicate melts. Source rods consisting of alumina and silica with organic additives have been produced with a CO_2 laser. This technique allows the preparation of fibers up to 1 m in length and about 200 µm in diameter. The fibers still contain gas bubbles and suffer from liquid-liquid demixing of the starting aluminosilicate melts. Single crystal mullite fibers have been obtained by using specific modulation frequencies of laser heating. Mullite crystallization is preceded by immiscibility of the aluminosilicate melt. The faceting of these fibers is very strong, producing strong changes of fiber diameters ("bamboo-type" fibers, Fig. 6.3.1).

Sayir and Farmer (1995) also grew polycrystalline mullite fibers from aluminosilicate melts of varying Al_2O_3/SiO_2 compositions by means of the laser heat flow zone technique under steady-state conditions. These fibers consisted of highly elongated and faceted mullite crystals, with their crystallographic **c** axes lying more-or-less parallel to the fiber axis. The mullite crystals are frequently enveloped by silicate glass films. The single crystalline and polycrystalline mullite fibers display tensile strengths between 550 and 1450 MPa. The low strength of fibers demonstrate the need of further improvement of the fiber production process, by which growth-produced flaws are minimized. Sayir and Farmer furthermore provided information on melt-grown polycrystalline mullite fibers using the edge-defined film-fed (EDF) technique, These fibers consist of block-shaped, highly elongated mullite crystals with excess glassy phase at the grain boundaries. The

Fig. 6.3.1 Scanning electron micrograph of a "bamboo-type" mullite single crystal fiber grown by the laser heat flow zone technique (LHFZ; Courtesy A. Sayir).

tensile strength of the edge-defined-film-fed-produced fibers is similarly low (800 to 900 MPa) to that of the laser heat flow zone-grown fibers.

Kriven and coworkers studied synthesis and properties of aluminosilicate fibers in the glassy state and after crystallization to mullite (Zhu et al. 1997, Kriven et al. 1998, Weber et al. 1999). The fibers are pulled from levitated drops of undercooled melts by inserting a tungsten stinger into the melt, which is subsequently withdrawn. The fiber diameter, which ranges between 1 and 50 μm, is controlled by the viscosity of the melt, which in turn depends on the degree of undercooling (70 to 100 °C below the melting point of mullite) and the fiber pulling rate (5 to 120 cm s^{-1}). The glass fibers have smooth surfaces with few flaws. The technique of Weber and coworkers also allows the preparation of cation-doped, colored aluminosilicate glass fibers with interesting optical properties. Crystallization of fibers takes place above about 950 °C. Dense and homogeneously crystallized mullite fibers with grain sizes between 0.5 and 1 μm have been produced under rapid heating conditions. The glass fibers exhibit strengths up to 6 GPa, while the crystallized fibers only display values below 1 GPa. According to Kriven and coworkers the main reason for the low strength of the crystallized fibers is the presence of crystallization-produced flaws. However, there is potential for improvement of the mechanical properties by modifying the fiber processing. Further improvement can be expected for directionally crystallized fibers, if a glassy grain boundary is avoided.

Xiao and Mitchell (1998, 1999) described mullite fibers grown from aluminosilicate melts with 72 wt. % Al_2O_3 by means of the inviscid melt spinning technique. These fibers are polycrystalline and consist of Al_2O_3-rich mullite (about 78 wt. % Al_2O_3), α-alumina and SiO_2-rich glass phase. Typically, the fibers are up to 450 μm thick and have crude surfaces. They require further intense development work for a potential application.

Mileiko and coworkers used the so-called internal crystallization method to produce mullite fibers (e.g. Kiiko and Mileiko, 1999, Mileiko et al. 2001). A molybdenum matrix with a number of continuous channels is infiltrated with a melt of mullite composition by means of capillary forces. The next steps are crystallization of fibers and dissolution of the matrix in an appropriate acid. The fibers, up to 80 mm in length, are clear, chemically homogeneous (about 78 wt. % Al_2O_3) and inclusion free, but suffer from their biconcave cross-sections. The fibers consist of a patchwork of single crystal areas, being up to 5 mm in size. The c axes of these single crystal areas display a misorientation of about 3° with respect to the fiber axis (Fig. 6.3.2, Rüscher et al. 2003).

Single crystal and directionally crystallized mullite fibers, although providing a potential for oxide/oxide composites with high thermal stability and favorable thermo-mechanical properties, need further intense research and development work in order to become an alternative to conventional polycrystalline fibers. In particular, the fiber fabrication techniques are still too complex to provide fibers in suitable quantities with an acceptable price.

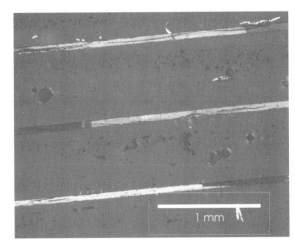

Fig. 6.3.2 Optical thin section micrograph of mullite single crystal fibers produced by the internal crystallization method. Note the extinction of the slightly different orientated areas having lengths of several millimeters (Courtesy C. Rüscher).

6.4
Application of Mullite Fibers
H. Schneider

Mullite fibers have been widely used to reinforce metal and ceramic matrices (see Sections 7.1 and 7.2). Another major application of fiber wool and fabrics is for

Fig. 6.4.1 Tube seals for a steam methane hydrogen reformer made with Nextel 312 aluminosilicate woven fiber fabrics. The seals prevent the ingress of cool air. Tube seals are mounted while the furnace is cold. When the furnace comes to temperature, the tubes will pull the seals tight (Courtesy 3M Deutschland, GmbH).

Fig. 6.4.2 Rotary kiln seal for a clinker furnace made with Nextel 312 aluminosilicate woven fiber fabrics. The seal covers the gap around the burner tip thus reducing the amount of cool air penetrating into the kiln (Courtesy 3M Deutschland GmbH).

temperature insulation purposes, e.g. as heat-resisting packings for motor car engines and sealing materials for heat resisting uses. Seals fabricated from aluminosilicate fiber insulations can help to save energy and to improve the safety of processes. As an example, tube seals for steam methane hydrogen reformers may be quoted. Steam methane hydrogen reformer facilities operate continuously at high temperatures for very long times. Fig. 6.4.1 shows a seal made from Nextel 312 aluminosilicate woven fabric. Another example are seals for rotary kilns. Fig. 6.4.2 shows the insulation of a clinker rotary furnace in a cement plant again using Nextel 312 aluminosilicate woven fabrics. Other applications are as fillers and for electrical insulation e.g for electrodes in electrical metal-smelting furnaces or for transportation bands operating at high temperatures.

References

Abe, Y., Horikiri, S., Fujimura, K. and Ichiki, E. (1982). High performance alumina fibers and alumina/aluminum composites. In: Hayashi, T., Kawata, K. and Umekawa, S. (eds.) Progress in Sci. Eng. Comp. Jap. Soc. Comp. Mater., pp. 1427–1434.

Al-Assafi, S., Cruse, T., Simmons, J.H., Brennan, A.B. and Sacks, M.D. (1994). Sol-gel processing of continuous mullite fibers. Ceram. Eng. Sci. Proc. **15**, 1060–1067.

Braue, W., Borath, R., Flucht, F., Göring, J. and Schneider, H. (2001). Failure analysis of Nextel™ 720 fibers subjected to high-temperature testing: The role of intrinsic fiber impurities. In: Krenkel, W., Naslain, R. and Schneider, H. (eds.) High Temperature Ceramic Matrix Composites. Wiley-VCH, Weinheim, pp. 90–95.

Bunsell, A.R. and Berger, M.-H. (2000). Fine diameter ceramic fibers. J. Europ. Ceram. Soc. **20**, 2249–2260.

Chen, L., Wang, B., Lui, S. and Yan, Y. (1996). Preparation of mullite fibers. J. Am. Ceram. Soc. **79**, 1494–1498.

Choi, H.-J. and Lee, J.-G. (2002). Synthesis of mullite whiskers. J. Am. Ceram. Soc. **85**, 481–483.

Curtin, W.A. (1991). Theory of mechanical properties of ceramic-matrix composites. J. Am. Ceram. Soc. **74**, 2837–2845.

Deleglise, F., Berger, M.H., Jeulin, D. and Bunsell, A.R. (2001). Microstructural stability and room temperature mechanical properties of the Nextel 720 fibre. J. Europ. Ceram. Soc. **21**, 569–580.

Deleglise, F., Berger, M.H. and Bunsell, A.R. (2002). Microstructural evolution under load and high temperature deformation mechanisms of a mullite/alumina fibre. J. Europ. Ceram. Soc. **22**, 1501–1512.

De Souza, M.F., Regiani, I. and De Souza, D.P.F. (2000a). Mullite whiskers from rare earth oxide doped aluminosilicate glasses. J. Mater. Sci. Lett. **19**, 421–423.

De Souza, M.F., Yamamoto, J., Regiani, I., Paiva-Santos, C.O. and De Souza, D.P.F. (2000b). Mullite whiskers grown from erbia-doped aluminum hydroxide-silica gel. J. Am. Ceram. Soc. **83**, 60–64.

Hashimoto, S. and Yamaguchi, A. (2000). Synthesis of needlelike mullite particles using potassium sulfate flux. J. Europ. Ceram. Soc. **20**, 397–402.

Ismail, M.G.M.U., Arai, H., Nakai, Z. and Akiba, T. (1990). Mullite whiskers from precursor gel powders. J. Am. Ceram. Soc. **73**, 2736–2739.

Katsuki, H., Furuta, S., Ichinose, H. and Nakao, H. (1988). Preparation and some properties of porous ceramic sheets composed of needle-like mullite. J. Ceram. Soc. Jpn. **96**, 1081–1086.

Kiiko, V.M. and Mileiko, S.T. (1999). Evaluation of the room-temperature strength of oxide fibres produced by the internal-crystallization method. J. Comp. Sci. Techn. **59**, 1977–1981.

Kingery, W.D. Bowen, H.K. and Uhlmann,

D. R. (1976). Introduction to Ceramics. 2nd edition. Wiley, New York.

Kriven, W. M., Jilavi, M. H., Weber, J.K.R., Cho, B., Felten, J. and Nordine, P. C. (1998). Synthesis and microstructure of mullite fibers grown from deeply undercooled melts. In: Tomsia, A. P. and Glaeser, A. (eds.) Ceramic Microstructure: Control at the Atomic Level. Plenum Press, New York, pp. 169–176.

Li, K., Shimizu, T. and Igarashi, K. (2001). Preparation of short mullite fibers from Kaolin via the addition of foaming agents. J. Am. Ceram. Soc. **84**, 497–503.

Merk, N. and Thomas, G. (1991). Structure and composition characterization of submicronic mullite whiskers. J. Mater. Res. **6**, 825–834.

Mileiko, S. T., Kiiko, V. M., Starostin, M. Y., Kolchin, A. A. and Kozhevnikoo, L. S. (2001). Fabrication and some properties of single crystalline mullite fibers. Scripta Mater. **44**, 249–255.

Milz, C., Göring, J. and Schneider, H. (1999). Mechnical and microstructural properties of Nextel^{TM} 720 fibers relating to its suitability for high temperature application in CMCs. Ceram. Eng. Sci. Proc. **20**, 191–198.

Moyer, J. R. (1995). Phase diagram for mullite-SiF$_4$. J. Am. Ceram. Soc. **78**, 3253–3258.

Moyer, J. R. and Hughes, N. N. (1994). A catalytic process for mullite whiskers. J. Am. Ceram. Soc. **77**, 1083–1086.

Moyer, J. R. and Rudolf, P. R. (1994). Stoichiometry of fluorotopaz and mullite made from fluorotopaz. J. Am. Ceram. Soc. **77**, 1087–1089.

Naslain, R. (2000). Ceramic oxide fibers from sol-gels and slurries. In: Wallenberger, F. T. (ed.) Advanced Inorganic Fibers. Processes, Structures, Properties, Application. Kluwer, Boston, pp. 205–232.

Newman, B. and Schäfer, W. (2001). Processing and properties of oxide, oxide composites for industrial applications. In: Krenkel, W., Naslain, R. and Schneider, H. (eds.). High Temperature Ceramic Matrix Composites. Wiley-VCH, Weinheim, pp. 600–609.

Nishio, T. and Fujiki, Y. (1991). Preparation of mullite fiber by sol-gel method. J. Ceram. Soc. Jpn. **99**, 654–659.

Okada, K., Otsuka, N., Brook, R. J. and

Moulson, A. J. (1989). Microstructure and fracture toughness of Y-TZP/mullite composites prepared by an in situ method. J. Am. Ceram. Soc. **72**, 2369–2372.

Okada, K. and Otsuka, N. (1989). Synthesis of mullite whiskers by vapour-phase reaction. J. Mater. Sci. Lett. **8**, 1052–1054.

Okada, K., Yasohama, S., Hayashi, S. and Yasumori, A. (1998). Sol-gel synthesis of mullite long fibers from water solvent systems. J. Europ. Ceram. Soc. **18**, 1879–1884.

Peng, P., Sorrel, C., Sharp, J. and Legge, P. (2003). Preparation of mullite whiskers from topaz decomposition. Mater. Lett. **58**, 1288–1291.

Perera, D. S. and Allot, G. (1985). Mullite morphology in fired kaolinite/halloysite clays. J. Mater. Sci. Lett. **4**, 1270–1272.

Petry, M. D. and Mah, T.-I. (1999). Effect of thermal exposures on the strengths of Nextel^{TM} 550 and 720 filaments. J. Am. Ceram. Soc. **82**, 2801–2807.

Regiani, I., Magalhaes, W.L.E., De Souza, D.P.F., Paiva-Santos, C. O. and De Souza, M. F. (2002). Nucleation and growth of mullite whiskers from lanthanum-doped aluminosilicate melts. J. Am. Ceram. Soc. **85**, 232–238.

Richards, E. A., Goodbrake, C. J. and Sowman, H. G. (1991). Reactions and microstructure development in mullite fibres. J. Am. Ceram. Soc. **74**, 2404–2409.

Rüscher, C. H., Mileiko, S. T. and Schneider, H. (2003). Mullite single crystal fibers produced by the internal crystallization method (ICM). J. Europ. Ceram. Soc. **23**, 3113–3117.

Sayir, A. and Farmer, S. C. (1995). Directionally solified mullite fibers. In: Lowden, R. A., Ferber, M. K., Hellmann, J. R., Chawla, K. K. and Di Pietro, S. G. (eds.) Ceramic Matrix Composites – Advanced High-Temperature Structural Materials. Mat. Res. Soc. Proc. **365**, 11–21.

Schmücker, M., Flucht, F. and Schneider, H. (2001). Temperature stability of 3 M Nextel^{TM} 610, 650 and 720 fibers – A microstructural study. In: Krenkel, W., Naslain, R. and Schneider, H. (eds.) High Temperature Ceramic Matrix Composites. Wiley-VCH, Weinheim, pp. 73–78.

Schmücker, M., Schneider, H., Mauer, T. and

Clauss, B. (2005). Kinetics of mullite grain growth in alumino silicate fibers. J. Am. Ceram. Soc., **88**, 488–490.

Schneider, H., Göring, J., Schmücker, M. and Flucht, F. (1998). Thermal stability of Nextel 720 alumino silicate fibers. In: Tomsia, A. P. and Glaeser, A. (eds.) Ceramic Microstructures: Control at the Atomic Level. Plenum Press, New York, pp. 721–729.

Song, K. C. (1998). Preparation of mullite fibers from aluminum isopropoxide-aluminum nitrate-tetraethylorthosilicate solutions by sol-gel method. Mater. Lett. **35**, 290–296.

Venkatachari, K. R., Moeti, L. T., Sacks, M. D. and Simmons, J. H. (1990). Preparation of mullite-based fibers by sol-gel processing. Ceram. Eng. Sci. Proc. **11**, 1512–1525.

Weber, J.K.R., Hampton, D.S., Merkley, D.R., Rey, C.A., Zatarski, M.M. and Nordine, P.C. (1994). Aero-acoustic levitation – A method for containerless liquid phase processing at high temperatures. Rev. Sci. Instrum. **65**, 456–465.

Weber, J.K.R., Cho, B., Hixon A.D., Abadie, J.G., Nordine, P.C., Kriven, W.M., Johnson, B.R. and Zhu, D. (1999). Growth and crystallization of YAG- and mullite-composition glass fibers. J. Europ. Ceram. Soc. **19**, 2543–2550.

Wilson, D. M. and Visser, L. R. (2001). High performance oxide fibers for metal and ceramic composites. Composites **A 32**, 1143–1153.

Xiao, Z. and Mitchell, B. S. (1998). The production of mullite fibers via inviscid melt spinning (IMS). Mater. Lett. **37**, 359–365.

Xiao, Z. and Mitchell, B. S. (1999). Optimization of process parameters in the production of mullite fibers via inviscid melt-spinning (IMS). Chem. Eng. Comm. **173**, 123–133.

Yogo, T. and Aksay, I. A. (1994). Synthesis of mullite fibre from an aluminosiloxane precursor. J. Mat. Chem. **4**, 353–359.

Zaykoski, J., Talmy, I., Norr, M. and Wuttig, M. (1991). Desiliconisation of mullite felt. J. Am. Ceram. Soc. **74**, 2419–2427.

Zhu, D. Jilavi, M. H. and Kriven, W. M. (1997). Synthesis and characterization of mullite and YAG fibers grown from deeply undercooled melts. Ceram. Eng. Sci. Proc. **18**, 31–38.

7
Mullite Matrix Composites

Mullite-based composites have been extensively investigated in recent years. The main aim of these studies was improvement of the mechanical properties by reducing the inherent brittleness of the material. In the 1980s, most attention was paid to mullite matrix composites toughened by addition of silicon carbide and zirconia particles. However, although there was progress in improving the material's characteristics no breakthrough has been achieved (see Section 7.3).

Nowadays most activities to increase the damage tolerance of mullite ceramics focus on fiber reinforcement. Fiber-reinforced mullite matrix composites are attractive for many advanced technical applications (Fig. 7.0.1). Examples are thermal protection systems for combustors of aero engines and stationary gas turbines and for re-entry space vehicles, heavy-duty burner tubes, heat exchangers, hot gas filters and catalytic convertor supports (see also, for example, Section 7.2.6). Con-

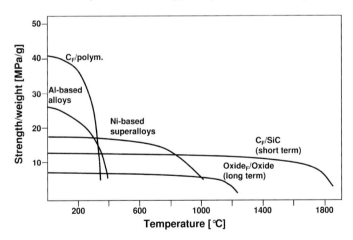

Fig. 7.0.1 Dependence of specific strength (strength/weight) of structural materials on temperature. C_F/polym: Carbon fiber-reinforced polymer matrix composite; oxide$_F$/ oxide: Oxide fiber-reinforced oxide matrix composite; C_F/SiC: Carbon fiber-reinforced silicon carbide composite; long term: Long term application (> 10000 h); short term: Short term application (<100 h), (after W. Schäfer, pers. comm.).

Table 7.0.1 Comparison of oxide/oxide and non-oxide continuous fiber-reinforced ceramic matrix composites (CMCs).

	Oxide/oxide CMCs	Non-oxide CMCs
Processing technique	Relatively easy	Complex
Processing costs	Relatively low	High
High temperature stability	Moderate	Very good (non-oxidizing environments) Very poor (oxidizing environments)
Mechanical behavior	Moderate	Good
Damage tolerance	Very good	Very good
Corrosion resistance	Good	Low
Thermal and electric conductivity	Low	High
Oxidation resistance	Excellent	Low

tinuous carbon fiber- or silicon carbide fiber-reinforced silicon carbide matrix composites (designated as C_F/SiC and SiC_F/SiC composites) have been proposed for these applications in past years. However, these non-oxide materials are susceptible to severe degradation in chemically aggressive, oxygen- and water vapor-rich environments. At that point the oxide fiber-reinforced oxide matrix composites (sometimes designated as oxide/oxide or $oxide_F$/oxide composites) come into play. Oxide fiber-reinforced oxide matrix composites are inherently stable against oxidation, but have less favorable thermo-mechanical properties than C_F/SiC or SiC_F/SiC systems. The advantages and disadvantages of non-oxide and oxide/oxide composites are summarized in Table 7.0.1. Among the oxide/oxide systems, composites with mullite matrices are characterized by stability in harsh environments, and high creep resistance. As a consequence, considerable research work is being expended to further improve fiber-reinforced mullite matrix composites, with particular emphasis on reliable and cost-effective fabrication. Although most research has been dedicated to continuous fiber-reinforced mullite matrix composites, work on whisker-reinforced mullite matrix materials has also been reported.

A new and innovative group of materials is metal-reinforced mullite matrix composites. Because of their specific ductility and electrical conductivity these composites have potential for use as catalyst supports, radiant burners and sensors (Section 7.4).

7.1
Whisker-reinforced Mullite Matrix Composites
H. Schneider

Much research has concentrated on silicon carbide whisker-reinforced mullite matrix composites (designated as SiC_W/mullite). This is due to the favorable mechanical properties of silicon carbide whiskers (Cooke, 1991, Wei and Becher, 1985, Claussen and Petzow, 1986, Ruh et al. 1988, Kumazawa et al. 1989). Other mullite composites combine silicon carbide whisker reinforcement with zirconia particle toughening (designated as $SiC_W/ZrO_{2,P}$/mullite, Claussen and Petzow, 1986, Ruh et al. 1988, Kamiaka et al. 1990). Finally, reinforcement of mullite matrices with mullite whiskers has also been described (designated as Mu_W/mullite, Meng et al. 1998).

7.1.1
Fabrication Routes

Wei and Becher (1985) produced 20 vol.% SiC_W/ mullite composites of nearly full density by hot pressing at 1600 °C and 70 MPa. Claussen and Petzow (1986) mixed ultrasonically treated silicon carbide whiskers and attrition-milled mullite powders, which were conventionally shaped and hot-pressed at 1600 °C. Kumazawa et al. (1989) produced SiC_W/mullite composites of high purity with the same technique (1500–1550 °C, about 40 MPa), while Sacks et al. (1991) applied the transient viscous sintering (TVS) technique (see Section 4.2.6). SiC_W/mullite composites have also been fabricated by Wu and Messing (1991) using lamination and hot pressing at temperatures between 1550 and 1850 °C and a pressure of 35 MPa.

Dense 50 vol.% silicon carbide whisker-reinforced mullite composites have been obtained by Wu and Messing (1994) by tape casting and hot pressing. Hot pressing yields a high degree of whisker orientation, which increases with the shear stress, whereas lamination controls the whisker placement. Lewinsohn et al. (1994) demonstrated the feasibility of fabricating laminated fully dense SiC_W/mullite ceramic tubes with improved strength via tape casting and hot isostatic pressing on a mandrel. The phase evolution of $SiC_W/ZrO_{2,P}$/mullite composites after long-term oxidation up to 1300 °C was described by Lin et al. (2000). Mullite whisker-reinforced mullite matrix composites (Mu_W/mullite) have been produced by Meng et al. (1998) using the in situ whisker growth method that was described by Okada and Otsuka (1991).

7.1.2
Mechanical Properties

Several studies have been dedicated to evaluation of the mechanical properties of whisker-reinforced mullite matrices, while studies on composites with chopped fibers are rare, because of the unfavorably low mechanical properties of these materials. The main reason for fabricating whisker-toughened mullite composites is to enhance the mechanical behavior by crack deflection and/or pull-out mechanisms. For that purpose high strength whiskers of silicon carbide are most suitable. The bending strength and fracture toughness of SiC_W/mullite composites are shown as a function of the whisker content and in comparison to zirconia particle ($ZrO_{2,P}$) toughened materials in Figs. 7.1.1 and 7.1.2. The composites show enhanced bending strength and fracture toughness with increasing silicon carbide

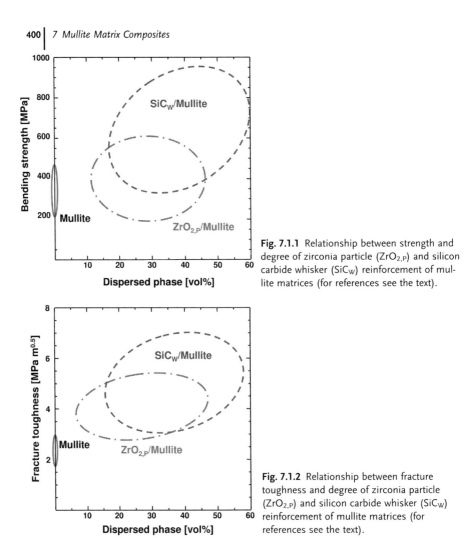

Fig. 7.1.1 Relationship between strength and degree of zirconia particle (ZrO$_{2,P}$) and silicon carbide whisker (SiC$_W$) reinforcement of mullite matrices (for references see the text).

Fig. 7.1.2 Relationship between fracture toughness and degree of zirconia particle (ZrO$_{2,P}$) and silicon carbide whisker (SiC$_W$) reinforcement of mullite matrices (for references see the text).

whisker content. Strengths between about 400 to 900 MPa and fracture toughnesses from 3.5 to 7 MPa m$^{0.5}$ have been observed, which are apparently higher than those of monolithic mullite (strength: 200 to 450 MPa, fracture toughness: 2 to 3 MPa m$^{0.5}$).

Among the various silicon carbide whisker-reinforced mullite ceramics, the laminated composites produced by Wu et al. (1991) exhibit the biggest improvement of properties. Bending strengths and fracture toughnesses as high as 940 MPa and 6.9 MPa m$^{0.5}$ have been achieved by addition of 40 vol.% silicon carbide whiskers (monolithic mullite: 320 MPa and 2.7 MPa m$^{0.5}$). The enhancement is more distinct for strength than for fracture toughness and this tendency is also observed in other silicon carbide whisker-reinforced mullite ceramics (Figs. 7.1.1 and 7.1.2). This is in contrast to the statement of Faber and Evans (1983), who say that, owing

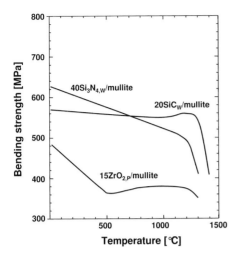

Fig. 7.1.3 Temperature-induced evolution of bending strength of particle ($ZrO_{2,p}$) – and whisker ($Si_3N_{4,w}$, SiC_w) – reinforced mullite matrix composites. The number in front of the composites' designation indicates the volume percentage of the reinforcing compound (for references see the text).

to crack deflection and/or fiber pull-out, whisker reinforcement is more effective in enhancing fracture toughness than in enhancing strength. Probably the interfacial bonding between whiskers and mullite matrix in the former composites is not adequate for the expected toughening mechanisms. However, an appropriate interface is crucial for the enhancement of mechanical properties of whisker-reinforced mullite matrix composites. Zhien et al. (1995) provided data on the mechanical properties of SiC_w/mullite composites containing LiO_2–Al_2O_3–SiO_2 (LAS) glass. By increasing the amount of the glass up to 10 wt. % the room temperature strength of the material rises from about 240 MPa to 310 MPa. This can be understood by the higher whisker-to-matrix bonding through the glass phase.

The Mu_w/mullite composites prepared by Meng et al. (1998) on the basis of the in situ mullite whisker growth method exhibit improved fracture toughness from about 1.9 to 3.0 MPa $m^{0.5}$, while the bending strength increases only slightly from 250 to 280 MPa. Obviously the interfacial bonding between matrix and whiskers is difficult to control, since whiskers and matrix both consist of mullite.

High temperature bending strengths of whisker-reinforced mullite composites are shown in Fig. 7.1.3 together with the properties of zirconia particle toughened (designated as $ZrO_{2,p}$/mullite) and silicon nitride whisker-reinforced composites (designated as $Si_3N_{4,w}$/mullite, see also Section 7.3.2). The figure demonstrates that a high level of bending strength is retained up to 1300 °C in 20 vol.% SiC_w /mullite materials (Kumazawa et al. 1989). At higher temperatures the strength drops rapidly, probably because of reactions between whiskers and the mullite matrix. Kumazawa et al. (1990) found increased strength at room temperature, if the grain boundary glass phase is crystallized to cristobalite. This material retains its strength up to 1300 °C.

The creep behavior of a 20 vol.% SiC_w/mullite composite is shown Fig. 7.1.4 at temperatures between 1450 and 1700 K (1177 and 1427 °C) under compressive stress (Nixon et al. 1990). Branches of the creep curve with different slopes are observed in the ranges 1450 to 1560 K (1177 to 1287 °C), 1560 to 1630 K (1287 to

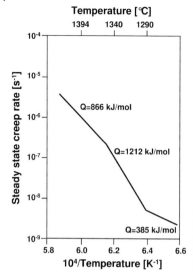

Fig. 7.1.4 Steady-state creep curve of a 20 vol. % silicon carbide whisker-reinforced mullite matrix composite (SiC_W/mullite) in the temperature range between 1230 and 1430 °C at a constant stress of 100 MPa (after Nixon et al. 1990).

1357 °C) and 1630 to 1700 K (1357 to 1427 °C). Related stress exponents (2.0 to 2.2, 2.3 to 2.9 and 3.0 to 3.2, respectively) are slightly higher than those reported for monolithic mullite. The same is true for the activation energies of creep, which are 385, 1212 and 866 kJ mol^{-1}, respectively. Nixon et al. explain creep with grain boundary sliding due to the viscous flow of inter-granular glassy phase, formation of cavities in the inter-granular glassy phase, and subsequent grain boundary sliding again, in going towards higher temperatures. The creep mechanism is similar to that of monolithic mullite containing a glassy phase. Nixon et al. mentioned that the creep rate at high temperature is one order of magnitude smaller than that of monolithic mullite containing glass, but is one order of magnitude higher than in glass-free polycrystalline mullite.

7.1.3
Thermal Properties

There is little information on the thermal properties of whisker-reinforced mullite matrix composites. The heat capacity c_p of SiC_W/mullite composites is lower than that of monolithic mullite, owing to the low c_p value of silicon carbide (Russell et al. 1996). The thermal conductivity of SiC_W/mullite composites has not yet been determined, but data exist for SiC_W/ZrO_{2,P}/mullite composites. The thermal conductivity of these materials is higher than that of monolithic mullite, because of the high thermal conductivity of silicon carbide (Russel et al. 1996). No experimental data on the thermal expansion of whisker-reinforced mullite matrix composites are available so far.

7.1.4
Miscellaneous Properties

Gebhardt and Ruh (2001) evaluated the electrical properties of SiC_W/mullite composites, with the idea that these materials could be promising temperature sensors. The presence of silicon carbide whiskers strongly influences the dielectric constant, resistivity and impedance of the mullite materials. The correlation between silicon carbide whisker fraction and electrical conductivity is explained by the increasing number of contacts between silicon carbide whiskers in the composite.

7.2
Continuous Fiber-reinforced Mullite Matrix Composites
H. Schneider

A major aim of research and development in the field of continuous fiber-reinforced mullite matrix composites is to reduce the inherent brittleness of the matrix by improvement of the toughness. Initially, at relatively low load, the composites show linear elastic deformation. At higher loads matrix cracks occur, but without significant fiber failure. At that stage the cracked mullite matrix is held together by the fibers, which carry the entire load. If the external load exceeds the fiber strength, the fibers successively break, leading finally to the failure of the composite. The resistance of the matrix against failure due to sudden changes of the mechanical and thermal loads has been designated "damage tolerance". Suitably high damage tolerance of fiber-reinforced mullite composites can be achieved by different strategies, which all have the aim of minimizing fiber failure. All approaches rely on the fact that the failure strain of fibers is higher than that of the mullite matrix, and, therefore, cracks preferably propagate in the matrix or in the fiber/matrix interface.

Different approaches have been employed for the fabrication of continuous fiber-reinforced mullite matrix composites: Matrices have been reinforced by endless single fibers (fiber diameters: 60–120 µm), fiber bundles containing up to 800 single filaments (fiber diameters: 10–12 µm), and two- (2D) or three-dimensional (3D) fiber webs and fabrics (fiber diameters: 10–12 µm or 60–120 µm). The best processing route depends on the size, shape and potential application of components and structures.

7.2.1
Fabrication Routes

Activities in the field of continuous fiber-reinforced mullite matrix composites started with unidirectional fiber reinforcements and has been extended towards two (2D) or three dimensional (3D) fiber architectures. Different fiber bundle or fiber fabric infiltration routes have been considered with the aim of achieving homogeneous composites:

- Vapor infiltration
- Sol-gel infiltration and polymer impregnation pyrolysis
- Metal melt infiltration with subsequent oxidation
- Slurry infiltration
- Electrophoretic deposition

Only the major processing routes will be described in more detail below. While monolithic structural mullite ceramics ideally have homogeneous and dense microstructures without pores and cracks, suitable fiber-reinforced mullite matrices can be either dense or porous, depending on fabrication and potential application. A critical point in both dense and porous composites is the bonding of the fibers in the matrix: It must be weak enough to allow the fracture energy dissipation necessary for toughening via crack deflection, fiber pull-out and fiber/matrix friction. On the other hand, the fiber/matrix bonding must be strong enough to allow load transfer between fibers and matrices.

Advantages of mullite-based composites include stability in oxidizing environments up to high temperatures low thermal expansion and low thermal and electrical conductivities, but also peculiarities such as transmittance of infrared and radar electromagnetic waves in wide frequency ranges. For many applications, mullite matrix composites are therefore superior to non-oxide materials such as C_F/SiC or SiC_F/SiC composites. This chapter focuses on silicon carbide fiber, alumina fiber and aluminosilicate ("mullite") fiber-reinforced mullite matrices (designated as SiC_F/mullite, $Al_2O_{3,F}$/mullite, and Mu_F/mullite composites). Other composites, e.g. with yttrium aluminum garnet (YAG_F), zirconia ($ZrO_{2,F}$), silicon nitride ($Si_3N_{4,F}$) or silicon boron nitrogen carbon fibers ($SiBNC_F$) at present have less or no technical importance, because of the difficulty of manufacturing them in reproducible quality and to a suitable price. The same is true for single crystal fibers, for example on the basis of alumina or mullite. Papers dealing with such composites will not be referenced here.

Infiltration of the matrix into uncoated or interphase coated fiber bundles or fabrics is typically performed with liquid phases, in the form of aqueous slurries or precursor sols. Ideally, the infiltration medium contains the matrix material in its final and consolidated form. In practice, however, matrix preforms have often been used with subsequent consolidation steps such as reaction sintering, reaction bonding or hot pressing (Fig. 7.2.1). Different fabrication techniques have been applied:

- Infiltration of fiber bundles with subsequent winding (Fig. 7.2.2a). A cross-ply fiber composite involves slurry infiltration by the matrix preform, and winding of the infiltrated fiber bundles onto a mandrel. The process relies on the following steps: After removal of the polymeric sizing of fibers, the fiber tow is infiltrated with the slurry, which contains ceramic particles in a water-based solution, by dipping, soaking and withdrawal. Prior to the sintering, the infiltrated fiber tows are wound on a mandrel. In order to achieve higher matrix densities multiple infiltrations can be applied. The pre-impregnated materials (prepregs) can be arranged in a variety of stacking sequences with unidirectional (1D) or

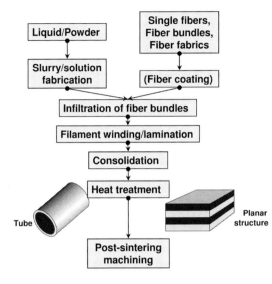

Fig. 7.2.1 Schematic flow sheet of the fabrication of continuous fiber-reinforced mullite matrix composites. Processes start from single fibers, fiber bundles or fiber fabrics, which are infiltrated with colloidal precursors or slurries and are formed by lamination or winding techniques, respectively.

two dimensional (2D) fiber arrangements. They can be cut, formed, laminated, joined and, after sintering, machined (Fig. 7.2.3, see, for example, Schneider et al. 2000, Kanka et al. 2001a,b).

- Fiber fabric processing (Fig. 7.2.2b). This is often based on slurry infiltration of the fiber prepregs under pressure or in a vacuum chamber. Following the pressure infiltration, the dried panels can be further vacuum impregnated with polymeric solutions or alkoxide precursors. This procedure is typically repeated several times in order to achieve the required high green density, or to introduce a second phase (see, for example, Saruhan, 2003, Carelli et al. 2002). The green and sintered composites can be processed in the same way as the wound materials (see Fig. 7.2.3)

- Electrophoretic impregnation of fiber fabrics (Fig. 7.2.4). Boccaccini and coworkers published a series of papers dealing with electrophoretic deposition infiltration of continuous fiber preforms with aluminosilicate ("mullite") matrices. Electrophoresis has been considered as a simple and inexpensive fabrication technique (e.g. Boccaccini and Panton, 1995, Boccaccini et al. 1996, Boccaccini et al. 1997). Electrophoretic impregnation primarily relies on the presence of nano-sized, charged particles in a colloidal suspension and deposition on electrically conducting fiber preforms serving as electrodes. The technique can also be applied to non-conducting oxides such as α-alumina and mullite in modified deposition cells. The acidity (pH) of the suspension is chosen in such a way that alumina particles are positively, and silica particles are negatively, charged. Owing to their electrical charges the silica and alumina particles coagulate and move in the electric field as single compounds (see Boccaccini et al. 2001). After electrophoretic impregnation of the preform the specimens are conventionally sintered and can be handled as shown in Fig. 7.2.3.

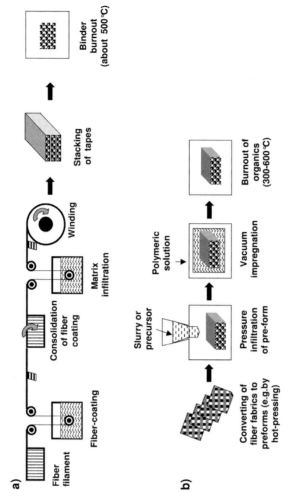

Fig. 7.2.2 Schematic flowcharts of green body fabrication of continuous fiber-reinforced mullite matrix composites. (a) Winding technique: It involves infiltration of fiber bundles, consolidation, winding, cutting of the as-wound tubes and stacking them. (b) Lamination technique: It involves the formation of fiber preforms e.g. by hot pressing of fiber webs, pressure and vacuum infiltration of fiber preforms. In both cases fibers are interphase-coated, if necessary (after Saruhan, 2003).

Wound green tubes
Laminated green fabrics

Forming
shaping, joining

Pressureless sintering
(<1350 °C)
Hot pressing
(1200-1400 °C)

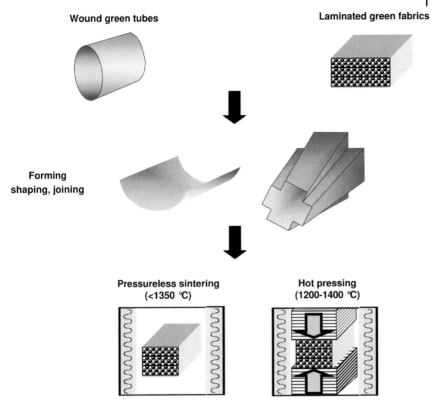

Fig. 7.2.3 Schematic flowchart showing forming, shaping, joining, consolidation and sintering of continuous fiber-reinforced mullite matrix composites produced by winding of bundles or lamination of fabrics.

7.2.1.1 Non-oxide Fiber-reinforced Composites

In a first approach one-dimensionally aligned continuous fiber-reinforced composites were produced to enhance the mechanical properties of the mullite matrix. Iwata et al. (1989) prepared carbon fiber-reinforced mullite composites (designated as C_F/mullite) by filament winding, while Asaumi et al. (1991) made silicon carbide fiber-reinforced mullite laminated composites (designated as SiC_F/mullite) by stacking of fiber webs and mullite matrix sheets.

Yamade et al. (1995) used carbon coated 140-μm thick silicon carbide fibers (Textron SCS-6). Unidirectionally stacked fiber layers in rectangular intervals have been infiltrated with mullite matrix and consolidated by hot pressing (up to 1640 °C and 20 MPa). Holmquist et al. (1997) cast a mullite slurry over silicon carbide fiber fabrics, which were then stacked, laminated and hot pressed isostatically. These SiC_F/mullite composites display high strength and non-brittle fracture with extensive fiber pull-out, caused by fiber-matrix gliding within the carbon (graphite) interphase (Fig. 7.2.5a). Higher process temperatures yield

Sol particles

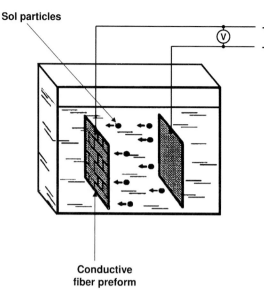

Fig. 7.2.4 Schematic drawing of an electrophoretic deposition cell used for the production of continuous fiber-reinforced mullite matrix composites. The fiber fabric is connected to a metal frame so that it acts as the positive electrode when particles in the suspension are negatively charged (after Boccaccini et al. 2001).

Conductive fiber preform

stronger fiber/matrix bonding and thus less favorable fracture behavior (Fig. 7.2.5b). Good results have also been achieved with the electrophoretic infiltration of densely woven silicon carbide fiber fabrics with mullite matrices (see Boccaccini and Panton, 1995, Boccaccini et al. 1996, 1997). Hirata et al. (2000) produced mullite matrix composites by repeated polymer precursor infiltration into a Si–Ti–C–O fiber-reinforced mullite matrix, and achieved final densities above 90%. The same group published results on the microstructure of a 10 vol.% Si–Ti–

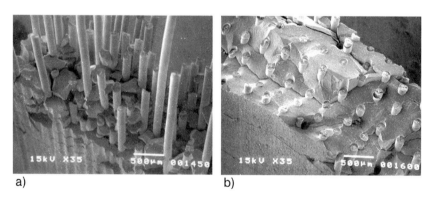

a)   b)

Fig. 7.2.5 Scanning electron micrographs of fracture surfaces of SiC$_F$/mullite composites (Textron SCS-6 fibers: Diameter ≈ 140 μm with ≈ 3-μm carbon coatings). The composites are hot isostatically pressed at (a) 1450 °C, 160 MPa; (b) 1600 °C, 20 MPa. Note the more intense fiber/matrix bonding (i.e. lower degree of fiber pull-out) in sample (b) with the higher process temperature (Courtesy M. Holmquist).

C–O fiber-reinforced mullite material obtained by hot pressing between 1500 and 1750 °C in nitrogen atmosphere (Hirata et al. 2001).

7.2.1.2 Oxide Fiber-reinforced Composites

The performance of oxide-based composites at high service temperatures relies on the stability of fibers and matrices. The development of high performance fibers, dense or porous mullite matrices and durable interphases opens a wide field of application for oxide/oxide components and structures (see Section 7.2.6). Research work in the field of oxide fiber-reinforced oxide matrix composites therefore has been dedicated to identifying adequate fibers, suitable matrices and efficient interphase systems. To date the most widely used commercial aluminosilicate fibers in the system Al_2O_3–SiO_2 come from the 3M Company under the trade name Nextel^{TM}. Preference is given to Nextel 610 alumina fibers if high strength at low and moderate temperatures is required, and to Nextel 720 aluminosilicate fibers if the composite must have good creep resistance and maintain its strength up to about 1200 °C (see also Section 6.2.2.3).

Considerable efforts have been spent on issues related to matrix development, interface tailoring and compound manufacturing of dense and porous mullite matrix composites. An overview of recent activities in the field of oxide fiber-reinforced mullite matrix composites is given in Table 7.2.1. At high temperatures, the main factors limiting the stability of matrices are the formation of silicate melts and the interaction with gases (e.g. hydrogen gas (H_2) or water vapor (H_2O), see Section 3.2). In order to prevent the formation of coexisting silicate liquids, matrices with over-stoichiometric mullite bulk composition (Al_2O_3 content >73 wt. %) can be a solution, until the fabrication of better fibers (e.g. yttrium aluminum garnet, YAG) is developed.

Dense mullite matrix composites   The fabrication of dense mullite matrix composites is a challenging procedure. On the one hand, sintering temperatures should not exceed 1200 °C, in order to avoid fiber degradation (see Section 6.2.2). On the other hand, it is difficult to achieve matrix densification in this temperature range, because of the low sinter activity of mullite on the one hand and the rigid fiber network on the other hand. Common problems of composite processing thus include composite densification with associated cracking of matrix during drying, sintering and mullitization. Furthermore, possible interactions between fibers, interphases and matrices can influence the damage tolerance of composites and therefore must be controlled carefully. The route used for fabrication of dense mullite matrix composites depends on the type and architecture of fiber preforms, the chemistry of matrix precursors and the fiber coatings.

Slurry infiltration of fiber preforms   Micro- or nano-sized mullite or mullite precursor powders, suspended in water or organic solutions are used for the slurry infiltration technique. Because of the limited thermal stability of commercial, polycrystalline oxide fibers (< 1200 °C, see Section 6.2.2) "normal" mullite powders, which

Table 7.2.1 Oxide fiber/mullite matrix composites: Fabrication and main characteristics.

Producer	Fabrication technique		Fibers	Matrix	Interface	Status
	Prematrix material	Shaping and joining of components				
COI Ceramics (U.S.A.)	Slurry and polymer	Laminating of fiber fabrics Winding of fiber bundles	Nextel 720 Nextel 610	Aluminosilicate (porous)	None	Commercial
Daimler Chrysler (Dornier, Germany)	Polymer	Laminating of fiber fabrics Winding of fiber bundles	Nextel 720 Nextel 610	Aluminosilicate	Fugitive (Gap)	Commercial (not currently available)
DLR (Germany)	Slurry	Winding of fiber bundles	Nextel 720 Nextel 610	Aluminosilicate (porous)	None	Precommercial (pilot process)
WPAB (U.S.A.)	Polymer	Laminating of fiber fabrics	Nextel 720	Aluminosilicate (porous)	None	Laboratory scale
UCSB (U.S.A.)	Slurry and polymer (with fillers)	Laminating of fiber fabrics	Nextel 720	Aluminosilicate/ zirconia (porous)	None	Laboratory scale
Pritzkow (Germany)	Slurry	Laminating of fiber fabrics	Nextel 440 Nivity	Aluminosilicate	None	Commercial (low grade)

DLR: German Aerospace Center, Köln
WPAB: Wright Patterson Airforce Base, Dayton
UCSB: University of California, Santa Barbara

sinter above 1600 °C, are not appropriate. Mullite pre-matrix systems, consisting of sol-gel-derived pre-calcined precursors, or of α-alumina or better γ-alumina particles, coated by thin non-crystalline silica layers, are more promising. The latter have the advantage of being easy to densify prior to mullitization by transient viscous sintering (TVS) (see Section 4.2.6). Slurries based on reaction-bonded mullite (RBM) systems can also be used: Silicon metal plus alumina systems doped with ceria (CeO_2) and yttria (Y_2O_3) sintering aids and with mullite seeds yield good results. Mechnich et al. (1998, 1999) were able to process mullite matrices below 1350 °C with this modified reaction bonding method. The advantage of reaction bonding for composite fabrication is evident: Oxidation of the silicon metal, which is accompanied by volume expansion, partially compensates for the sintering-induced shrinkage of green samples (the so-called near-net-shape process, Fig. 7.2.6, see also Section 4.2.4). Generally reaction bonded mullite (RBM) systems based on silicon are more favorable than those using aluminum because of their higher oxidation and melting temperatures. A special form of reaction bonding, although not realized for mullite yet, is infiltration of fiber preforms by molten aluminum silicon alloys (the so-called DIMOX process, Newkirk et al. 1987).

As described above, for composite fabrication fiber bundles or single fiber preforms and fiber fabrics are passed through slurries and stabilized by drying. The

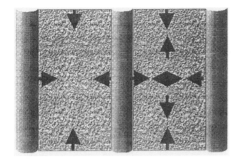

Fig. 7.2.6 Schematic drawing of reaction-sintered (below) and reaction-bonded (above) mullite matrix composites. Note the high and low shrinkage of the reaction-sintered and reaction-bonded materials, respectively. The latter is due to a near-net-shape process (after Saruhan, 2003).

fiber fabrics or preforms are then stacked to form green components. The fiber bundles after infiltration by the slurry are wound on mandrels unidirectionally or two dimensionally. The green tubes achieved via the winding process are cut, rolled out and stacked. The green bodies obtained from the different process routes are hot-pressed, typically between 1200 and 1400 °C (Fig. 7.2.3). Multiple infiltration steps may be required to achieve high matrix densities.

Nextel 720 aluminosilicate fiber-reinforced dense mullite matrix composites with double-layer fiber/matrix interphases have been produced by infiltration of fiber fabrics with a mullite precursor slurry and hot pressing in argon at 1300 °C (Saruhan et al. 2001a,b). Multilayer interphases consisting of zirconia/carbon and alumina/carbon are deposited on the fibers prior to the processing by chemical vapor deposition (CVD). Short-term heat treatments at 1200 °C produce gaps between fibers and matrix by burning out the carbon layers. These gaps act as weak interphases (designated as "fugitive coatings", see Section 7.2.5).

Colloid precursor infiltration This fabrication route starts with infiltration of fiber bundles or fabrics with homogeneously dispersed nano-sized colloidal particles followed by gelling the solutions by changing the acidity (pH) or by addition of water. A problem associated with this technique is the very high matrix shrinkage upon pyrolysis and sintering, producing highly porous composites (porosity >50%) containing multiple cracks. Therefore, repeated re-infiltration and hot pressing is indispensable for the consolidation of dense composites.

An alternative composite fabrication technique starting from colloidal precursors is electrophoretic deposition (EPD). This has the advantage that the infiltration rate can be precisely controlled. Westby et al. (1999) and Kooner et al. (2000) produced Nextel 720 aluminosilicate fiber-reinforced mullite matrix composites by means of electrophoretic infiltration of fiber fabrics using mullite-seeded colloidal precursors with subsequent sintering at 1400 °C. Kaya et al. (2002) processed neodymium phosphate (NdPO$_4$)-coated Nextel 720 aluminosilicate fiber fabrics. They obtained composites with densities higher than 85% after sintering at 1200 °C, and achieved attractive strength and favorable thermocyclic behavior of the materials (see Section 7.2.2). According to Kaya et al. the neodymium phosphate interphase, which is weakly bonded to fibers and matrices, and the presence of nanopores in the mullite matrix, are responsible for the good thermo-mechanical properties of composites. However, electrophoretic infiltration produces relatively thick matrix layers between the fiber fabrics, giving rise to low shear strengths.

Commercial polymer infiltrated oxide fiber-reinforced mullite matrix composites have been presented by DaimlerChrysler (see, for example, Peters et al. 2000). The process relies on infiltration and subsequent pyrolysis of carbon coated Nextel 610 alumina fiber prepregs (fugitive coating, see Section 7.2.5) by filament winding, resin transfer molding (RTM) or wet lamination and densification in an autoclave. In a further step the composite is annealed in air in order to reduce internal stresses and to remove the carbon coating of the fibers. The final composite consists of a matrix of mullite plus silicon-oxygen-carbon glass, arranged in cross-ply laminates with 0°/90° fiber orientations.

Porous mullite matrix composites Continuous fiber-reinforced porous mullite matrix composites without specific fiber/matrix interphases have been developed since the late 1990s. The idea behind this concept is that the matrix is weak enough to not allow stress transfer into the fibers. The behavior of these materials can best be compared to that of wood. Because of their good thermo-mechanical properties, and the simple and inexpensive manufacture, porous mullite matrix composites are attractive with a potential for many technical applications (see Section 7.2.6), since costly fiber coatings and matrix densification techniques such as hot pressing in specific atmospheres can be avoided. Major research and development work on porous mullite matrix composites has been conducted at the University of California (USA), COI Ceramics (USA), and the German Aerospace Center (DLR, Germany). Other research and development works have been performed at DaimlerChrysler (Germany) and at Pritzkow Spezialkeramik (Germany), (see Table 7.2.1).

Porous aluminosilicate fiber compacts without matrix can be considered as the simplest version of porous oxide fiber composites. Kanka et al. (2005) in an unpublished study give results on ceramic tiles produced by hot pressing Nextel 720 aluminosilicate fiber fabrics between 1250 and 1350 °C at 20 MPa. They found that a correlation exists between the hot pressing temperature and the degree of sintering, due to gradually increasing fiber/fiber contact areas (Fig. 7.2.7). Strongly sintered materials display higher strength but reduced toughness. The as-sintered tiles have suitable high temperature stability and good thermal shock resistance. The high damage tolerance of these compacts is explained by local crack bending

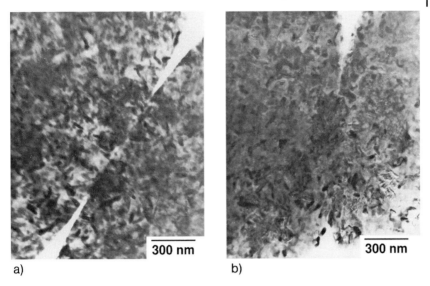

a)   b)

Fig. 7.2.7 Transmission electron micrographs of matrix-free
Nextel 720 aluminosilicate fiber fabric compacts. Samples are
produced by hot pressing at 20 MPa, at temperatures of (a)
1250 °C and (b) 1350 °C. Note the increased fiber contact area
at 1350 °C (Courtesy B. Kanka).

and splitting in the "pores" of the material. Obviously the fiber/fiber contact areas
are small enough to prevent brittleness but are sufficiently large to transfer load.

As mentioned above continuous oxide fiber-reinforced composites with porous
mullite matrices but without any fiber coatings have gained considerable intoest in
recent years. Levi et al. (1999) describe matrices consisting of filler particulates
which during sintering neither react nor shrink. Filler particulates are bonded
together by ceramic binders. While the distribution and packing density of the
filler control the porosity of the matrix, the type and amount of binders are crucial
for the load transfer, creep resistance and interlaminar shear strength of the composites. The lack of fiber/matrix interphases causes fiber/matrix interactions,
which can be controlled by increasing or reducing the fiber/matrix contact areas.
The processing opens many ways to modify the microstructural evolution and the
associated mechanical stability of the composites (Fig. 7.2.8). Levi's composites
consist of mullite or α-alumina filler particulates and of silica glass, α-alumina or
phosphate binders. Their composites display high strength at low and moderate
temperatures. However, softening of the silica glass binders below 1000 °C limits
the creep resistance of the material. As a solution to the problem Levi and coworkers suggest recrystallization of the glass binders, e. g. by admixing sintering and
crystallization aids such as ceria (CeO_2) or yttria (Y_2O_3).

According to Lange et al. (2001) matrix precursor powders can be packed into
fiber fabrics by slurry dipping or slurry injection, vacuum and pressure filtration,
and by vibro impregnation. All these techniques use particle suspensions in slur-

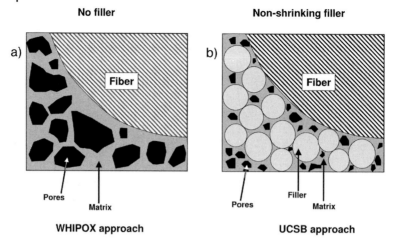

Fig. 7.2.8 Microstructural concepts of oxide fiber-reinforced porous mullite matrix composites. (a) Highly porous mullite matrix produced by sintering-induced shrinkage of the matrix precursor (WHIPOX technique, WHIPOX = Wound highly porous oxide ceramic matrix composite produced at the German Aerospace Center, DLR); (b) Porous mullite matrix produced by sintering-induced shrinkage of the binder, but not of the (inert) matrix filler (UCSB approach, UCSB = University of California at Santa Barbara).

ries. Of special importance for the achievement of homogeneously distributed particles in the slurry without formation of particle agglomerates are their repulsive potentials. Lange et al. introduced chemical coatings to achieve these repulsive potentials by absorbing ions or organic molecules at the particles' surfaces. The solid content of the final composite depends on the size, shape and distribution of particles in the slurry. Further factors are the solid-to-liquid ratios and the fiber diameters.

Slurry dipping or slurry injection of fabrics   This simple process to produce oxide fiber-reinforced mullite matrix composites has been used since the 1970s. Infiltration of matrices or matrix precursors is typically carried out by dipping the fiber prepregs into water-based slurries.

Vacuum and pressure infiltration of fabrics   Levi et al. (1998, 1999), on the basis of previous work by Tu et al. (1996) and Lange et al. (1995) produced mullite matrix composites reinforced by Nextel 720 aluminosilicate and Nextel 610 alumina fiber fabrics. The process involves a lay up of fiber fabrics followed by vacuum-assisted infiltration by a slurry consisting of 80 wt. % mullite and 20 wt. % α-alumina. The infiltrated fiber panels are then impregnated in several cycles with an aluminum hydroxide chloride ($Al_2(OH)_5Cl$) solution, followed by pyrolysis at 900 °C and sintering at 1200 °C (Fig. 7.2.9). Microstructural inspection shows homogeneous in-

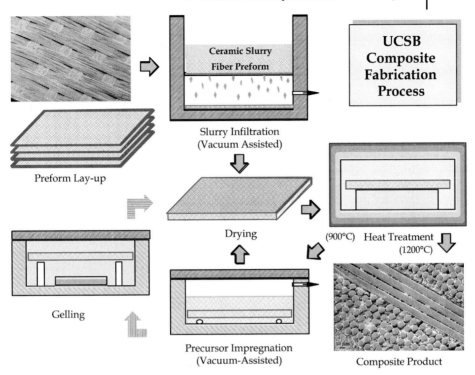

Fig. 7.2.9 Schematic flow sheet showing the fabrication route of oxide fiber-reinforced porous mullite matrix composites by vacuum-assisted pressure infiltration of fabrics according to the UCSB approach (UCSB = University of California Santa Barbara) (see Fig. 7.2.8. Courtesy C. G. Levi).

filtration (Fig. 7.2.10). Carelli et al. (2002) employed the same technique and identified a more or less regular distribution of shrinkage-induced matrix microcracks. It has been shown that in the as-produced state of the composite cracks stop at fiber fabrics oriented perpendicular to the crack planes, while after long term aging between 1000 and 1200 °C cracks traverse adjacent fabrics and penetrate into neighboring matrix areas. Crack opening is observed simultaneously. The effects are thought to be caused by densification of the matrix. Holmquist et al. (2001) infiltrated stacked Nextel 610 alumina fiber and Nextel 720 aluminosilicate fiber fabrics in vacuum with fine mullite and silica powders having a ratio of 3 : 1. After sintering at 1000 °C the composites' matrices consist of a mullite network, with individual mullite grains being bonded by silica bridges (see Fig. 7.2.8b), producing an overall porosity of composites of more than 40%. Strength enhancement of the as-sintered material is achieved by secondary infiltration with aluminum phosphate hydroxide ($Al(PO_4)(OH)_2$) solutions. The material exhibits good in-plane flexural strength at room temperature (see Section 7.2.2.2). However, high-temperature strength and creep resistance of these composites should be low,

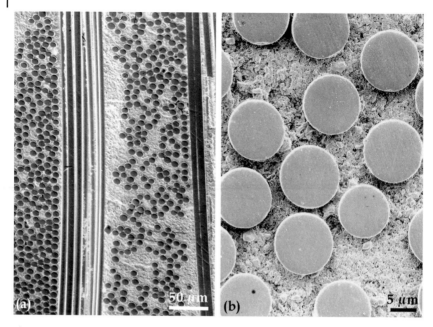

Fig. 7.2.10 Scanning electron micrographs of an oxide fiber-reinforced porous mullite plus α-alumina matrix composite. Composite fabrication has been performed via the vacuum-assisted pressure infiltration technique according to the UCSB approach (UCSB = University of California, Santa Barbara) (see Fig. 7.2.8. Courtesy C.G. Levi).

owing to the presence of silica glass bridges connecting the mullite particles in the matrix (see above).

Vibro infiltration of fabrics   This technique was introduced by Haslam et al. (2000). It uses specific interparticle pair potentials, by which powder compacts (e.g. 70 wt. % cubic zirconia and 30 wt. % mullite) are consolidated by pressure filtration. Thin ceramic sheets are formed by tape casting, vibrating and subsequent freezing. The frozen ceramic sheets are stacked between fiber fabrics. After thawing, the sheets are re-fluidized by vibrating, thus infiltrating the fiber fabrics. Bonding between particles is effected by an evaporation and condensation process of zirconia in a hydrochloric acid (HCl) atmosphere, without shrinkage during sintering. Sintering gives rise to grain coarsening, which makes the material stable against densification. More recently, Holmquist et al. (2003) have described the fabrication of Nextel 610 alumina fiber-reinforced porous mullite-alumina matrix composites by pressure infiltration of fiber fabrics. They achieve a high solid volume fraction and green bodies that can be fluidized by vibration. After drying, these bodies are strengthened by repeated infiltration with an alumina precursor and pyrolyzed at 900 °C. The final sintering step is at 1200 °C, and serves to develop α-alumina bridges between the mullite matrix network. Holmquist et al.

a)

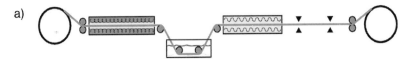

b)

| As recived fibers | Desizing | Water based slurry infiltration | Pre-drying | Fiber stress control | Computer controlled winding module |

Fig. 7.2.11 Processing of oxide fiber-reinforced highly porous mullite matrix composites using fiber bundle infiltration and subsequent winding according to the WHIPOX process (WHIPOX = Wound highly porous oxide ceramic matrix composite, see Fig. 7.2.8). (a) Schematic flowchart; (b) Winding equipment.

produced tubes with this technique by wrapping the infiltrated green prepregs around a mandrel. Pressurizing of the tubes causes delamination at hoop stresses above 50 MPa, and delamination frequently initiates within the porous matrix.

Holmquist and Lange (2003) fabricated oxide/oxide composites on the basis of Nextel 610 alumina and Nextel 720 aluminosilicate fiber cloths. Vibration-assisted injection of stacked cloth prepregs with mullite-alumina slurries in combination with a simple vacuum bag technique yields composites with homogeneous microstructures. The final composites display suitable room-temperature strengths (see Section 7.2.2.2).

Slurry infiltration of bundles – The WHIPOX process The German Aerospace Center (DLR) uses winding and subsequent sintering of slurry-infiltrated fiber bundles for the fabrication of oxide fiber-reinforced porous mullite matrix composites (designated as WHIPOX: Wound highly porous oxide ceramic matrix composites). In a first step the sizing of the Nextel 720 aluminosilicate and Nextel 610 alumina fibers is burned-off in a tube furnace. Fiber bundles are then spread apart mechanically and infiltrated with a water-based matrix slurry. Slurry particles with Al_2O_3 contents ranging from 68 wt. % ("SiO_2-rich") to > 95 wt. % ("Al_2O_3-rich") have been used. The infiltrated fiber rovings are then passed through a furnace for matrix stabilization. In a last step the fiber rovings are wound on plastic mandrels (diame-

Winding   **Forming and Joining**   **Machining**

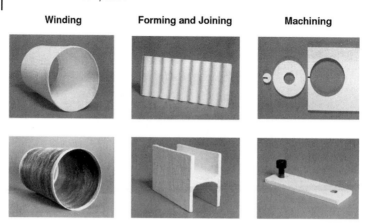

Fig. 7.2.12 Oxide fiber-reinforced porous mullite matrix components and structures fabricated according to the WHIPOX process (see Fig. 7.2.8).

ter: 20–200 mm). Both unidirectional (1D) and twodimensional (2D) composites with fiber bundle/fiber bundle orientations up to 90° can be realized (Fig. 7.2.11).

The green tubes obtained by this process have a leather-like consistency, and can be cut, stacked and joined to tapes and tiles and to more complex forms (see Fig. 7.2.3). Once in their final shape the WHIPOX components and structures are pressure-less sintered in air at 1300 °C. The low sintering temperature is important to prevent fiber damage during composite processing. After sintering, the components can further be machined e.g. by drilling or water jet cutting, according to the specific requirements (Figs. 7.2.12). The easy technical fabrication, and the fact that the composites need no fiber coatings, makes WHIPOX a low-cost product compared with other advanced ceramic matrix composites (see e.g. Kanka and Schneider, 2000, Schneider et al. 2000, Kanka et al. 2001a, 2001b, Schmücker and Schneider, 2004).

Although the WHIPOX process starts from mullite precursors, and not from inert fillers and reactive binders, there is some similarity in development of the mullite-based SiO_2-rich WHIPOX matrices to that seen in the work of Levi et al. (1998). After crystallization mullite stays unreacted, with a coexisting silica-rich glass acting as a "binder". Again a crucial point is the control of the (high) porosity during processing. Typical matrix porosities of WHIPOX range between 40 % and 80 %. While in Levi's material the porosity is lower and is determined by packing density of inert particles and the kind and amount of binder, the very high matrix porosity of WHIPOX is a consequence of the sintering-induced matrix shrinkage within the rigid fiber lattice. Thus systematic experimental work has been performed to control matrix shrinkage and mullitization of WHIPOX. Diphasic mullite precursors (type II, Schneider et al. 1993, see Section 1.4.2) consisting of transient alumina (γ-alumina) and amorphous silica have been used as starting materials. Owing to the presence of vitreous silica below 1200 °C, the sinterability of these precursors is high.

Because of their high specific surface the starting mullite precursors are pre-calcined in order to prevent extreme shrinkage and thus to obtain homogeneous matrices. If higher matrix densities are required, re-infiltration of the as-sintered composites with matrix precursors is a suitable way. The re-infiltration behavior of the slurry into the fiber bundles depends on the shape and diameter of pores and pore channels in the matrix, and on the concentration of the introduced slurry or solution. WHIPOX composites re-infiltrated ten times with aluminum chloride (AlCl$_3$) show reduced porosities from the original 66 % to 50 %. However, the pore distribution of these materials is heterogeneous with a lowered porosity near the surface, but no significant change in the interior of samples (Schneider et al. 2000, She et al. 2002). She et al. (2000) suggested that the re-infiltration occurs in two steps: A rapid intrusion of the aluminum chloride solution into the porous com-posites by capillary forces, and a slower migration of entrapped air bubbles through the solution to the surface of samples.

Macro-, meso- and microstructure   The controlling factors of the macrostructure of WHIPOX components are the fiber architecture (type of fiber bundles, fiber con-tent, fiber bundle density and fiber bundle angles), pre-matrix characteristics (com-position and calcination temperatures of precursors, solid content and viscosity of slurries) and the fabrication process (infiltration technique, winding angle and velocity, pre-drying of infiltrated bundles, etc.).

Ideally the mesostructural distribution of oxide fibers in the mullite matrix is homogeneous. In reality, however, fiber and matrix distributions are heteroge-neous. Schmücker et al. (2003) developed a method that describes the mesos-tructural fiber and matrix distributions of WHIPOX composites quantitatively and three-dimensionally. The procedure makes use of the optical transmittance and non-transmittance of fibers and matrices, respectively. It turns out that WHIPOX has a laminate character with matrix- and fiber-rich areas (Fig. 7.2.13).

The surfaces of WHIPOX components and structures, are disturbed process-inherently and by (partial) squeezing of the viscous matrix out of the composite's bulk to the surface during winding (Figs. 7.2.14a,b). These rough surfaces are susceptible to erosion and chemical attack, as they contain large numbers of flaws, which act as crack nuclei. Smooth environmental barrier coatings (EBCs), depos-ited on WHIPOX substrates by sol-gel, gas phase and thermal spray techniques, can help to improve the surface erosion and corrosion of materials.

The microstructural development of WHIPOX is very sensitive to (pre-)matrix characteristics and fabrication parameters. Again "SiO$_2$-rich" and "Al$_2$O$_3$-rich" ma-trices may be distinguished. Both compositions form threedimensional highly po-rous networks between individual fibers with relatively few fiber/matrix contacts (Fig. 7.2.15): SiO$_2$-rich matrices are built of a framework of mullite crystals with interstitial SiO$_2$-rich glass, while in Al$_2$O$_3$-rich matrices no glass phase occurs, and the major constituent is acicular α-alumina. Schneider et al. (2000) studied the influence of matrix bulk composition, pre-calcination temperature and sintering atmosphere on the microstructural evolution of oxide fiber-reinforced mullite ma-trix WHIPOX composites. They found that SiO$_2$-rich matrices exhibit strong sin-

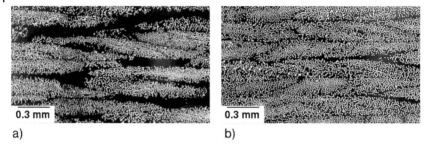

Fig. 7.2.13 Optical micrographs showing the mesostructure of Nextel 610 alumina fiber-reinforced porous mullite matrix composites processed according to the WHIPOX process (see Fig. 7.2.8). (a) Thick rovings (3000 DEN) with a heterogeneous laminate-type microstructure of the material; (b) Thin rovings (1500 DEN) showing a more homogeneous microstructure of the material. Fiber bundles are cut perpendicular to the imageplane with light points representing individual fibers. Dark areas correspond to the matrix. (Courtesy B. Kanka).

tering, which is triggered by viscous flow of the coexisting glass phase. The Al_2O_3-rich composition contains no glass, and thus displays lower sintering activity.

Schmücker et al. (2000) studied the temperature-induced interactions between Nextel 720 aluminosilicate fibers and mullite-rich matrices. Reactions between

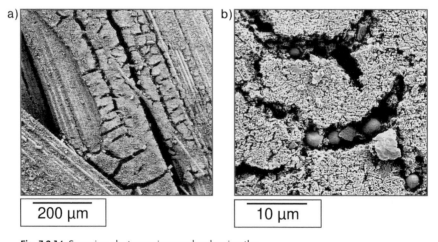

Fig. 7.2.14 Scanning electron micrographs showing the mesostructure of the surface of a Nextel 720 aluminosilicate fiber-reinforced porous mullite matrix composite processed according to the WHIPOX process (see Fig. 7.2.8) (a) Low magnification: It shows the fiber bundles with part of the matrix being squeezed out to the surface. Large drying- and sintering-induced cracks occur; (b) High magnification: It shows a detail of the squeezed-out matrix (Courtesy B. Kanka).

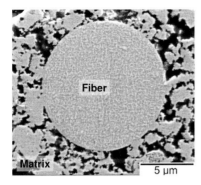

Fig. 7.2.15 Scanning electron micrograph of a Nextel 720 aluminosilicate fiber-reinforced porous mullite matrix composite processed according to the WHIPOX approach (see Fig. 7.2.8). Note the very high matrix porosity and the selective contacts between fibers and matrix (Courtesy B. Kanka).

matrix and fibers after short annealing (several hours) start above 1500 °C, while after long-term annealing (several 100 hours and more) the reaction is initiated at 1300 °C. In this process the silica of the glass phase of the matrix, coexisting with mullite in the as-sintered state, is mobilized and migrates to the fibers, where it reacts with α-alumina at the fiber peripheries according to the equation:

$$3\,Al_2O_3 + 2\,SiO_2 \rightarrow 3Al_2O_3 \cdot 2SiO_2 \qquad\qquad (1)$$
$$\text{α-alumina} \qquad\qquad \text{mullite}$$

The reaction continues until the entire "free" silica of the matrix is consumed, which is the case after annealing at about 1600 °C (Fig. 7.2.16). Schmücker et al. believe that growth of ab initio existing mullite crystals in the fibers' peripheries is driven by inter-diffusion of aluminum and silicon ions. The driving force is probably the Al_2O_3 concentration gradient in mullite, ranging from about 72 wt. % in contact with the SiO_2-rich glass in the matrix to about 74 wt. % in contact with the α-alumina in the fibers. The transport of the viscous silica phase to the fibers may take place by capillary forces. During these reactions the fiber-matrix contact areas gradually increase (Fig. 7.2.16), which implies a loss of damage tolerance of the WHIPOX composites.

7.2.2
Mechanical Properties

Oxide/oxide composites display suitable thermo-mechanical characteristics although they are inferior to those of advanced non-oxide composites, such as silicon carbide or silicon nitride matrices reinforced by carbon, silicon carbide or silicon boron nitrogen carbide fibers (designated as C_F/SiC, C_F/Si_3N_4, SiC_F/SiC, SiC_F/Si_3N_4, $SiBNC_F/SiC$). The big advantage of oxide/oxide composites is their intrinsic oxidation stability up to high temperatures, which makes these materials suitable for many long-term applications at moderate mechanical loads (see Section 7.2.6).

The technical application of structural oxide fiber-reinforced mullite matrix ceramics requires detailed knowledge of their thermo-mechanical stability. Depend-

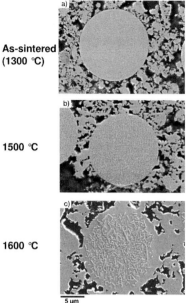

As-sintered (1300 °C)

1500 °C

1600 °C

5 µm

Fig. 7.2.16 Scanning electron micrographs of the temperature-dependent microstructural evolution of a Nextel 720 aluminosilicate fiber-reinforced porous SiO_2-rich mullite matrix composite processed according to the WHIPOX technique (see Fig. 7.2.8). (a) As-sintered (1300 °C); (b) 1500 °C, note the slight grain coarsening in the fiber and the increase of the fiber/matrix contact areas; (c) 1600 °C, note the strong grain coarsening in the fiber (acicular crystals correspond to α-alumina) and the large increase of the fiber/matrix contact areas. The α-alumina at the fiber's periphery is completely consumed by reaction with silica to mullite (Courtesy M. Schmücker).

ing upon the type of material, different failure mechanisms apply. Composites with dense matrices are characterized by a network of matrix cracks: Failure is controlled by fiber properties and the interaction between fibers and matrices. Fiber debonding and pullout, crack deflection, fiber cracks and friction between fibers and matrix contribute to energy dissipation, which is required in order to achieve non-brittle failure behavior. These failure mechanisms have been described in detail for non-oxide materials such as C_F/SiC and SiC_F/SiC composites. Similar principles can be applied to dense mullite matrix composites. Dense mullite matrices reinforced by continuous fibers of carbon (C_F/mullite), silicon carbide (SiC_F/mullite), alumina ($Al_2O_{3,F}$/mullite), and aluminosilicate (mullite plus α-alumina with dominance of mullite, i.e. Mu_F/mullite) exhibit strong fiber/matrix

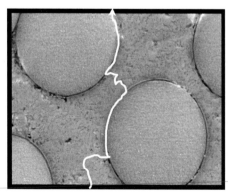

Fig. 7.2.17 Suggested crack path in an oxide fiber-reinforced dense mullite matrix composite with zirconia/gap (fugitive) double fiber/matrix interphase. The trace of the possible crack propagation is shown in white. Note the occurrence of crack deflection at the fiber/matrix interphases.

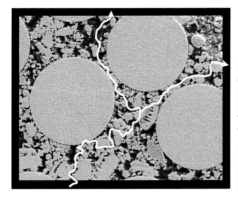

Fig. 7.2.18 Suggested crack paths in an oxide fiber-reinforced porous mullite matrix composite without fiber/matrix interphase, processed according to the WHIPOX technique (see also Fig. 7.2.8). The traces of possible crack propagations are shown in white. Note that the cracks run within the pores cutting weak particle-particle and particle-fiber necks ("sintering contacts").

bonding. As a consequence fiber/matrix interphases have to be introduced, which allow damage tolerant failure behavior of the composite (Fig. 7.2.17).

An alternative approach to achieve damage tolerance of composites starts from matrices with homogeneously distributed pores, but without applying any fiber/matrix interphase. Although, in this case fiber/matrix bonding is strong, there is no negative impact on the damage tolerance, since points of contacts between fibers and matrix occur only selectively. In these composites a continuous crack front does not develop, and crack extension takes place by continuous breaking of sintering necks between matrix particles instead (Fig. 7.2.18). This has the consequence that during extension of the crack in the porous matrix, the fibers will normally not see the crack front. Fiber fracture in this case is mainly initiated at flaws at the fiber's surface. Fiber pull-out in porous mullite matrix composites follows mechanisms different from those in dense mullite composites. No clean matrix holes are produced by fiber debonding from the matrix as is characteristic for dense fiber/interphase/matrix systems. Instead, owing to the strong fiber/matrix bonding, matrix particles are always attached at the fibers after fiber pull-out, and some of the matrix between the fibers is lost by frictional effects. In these composites the porous mullite matrix obviously has two functions: firstly that of "normal" matrix/fiber load transfer (matrix function), and secondly that of energy dissipation at the fiber/matrix interface (interphase function).

7.2.2.1 Non-oxide Fiber-reinforced Composites

Early studies on continuous carbon and silicon carbide fiberreinforced mullite matrix composites (C_F/mullite and SiC_F/mullite) were performed by Iwata et al. (1989) and Asaumi et al. (1991), respectively. The bending strength of these materials at room temperature can be as high as 600 MPa in C_F/mullite, and about 850 MPa in SiC_F/mullite composites (for comparison, monolithic mullite has a maximum strength of about 400 MPa). The high strength is retained up to 1200 °C in inert gas atmosphere. The enhancement of fracture toughness of the C_F/mullite and SiC_F/mullite composites is even more striking. Whereas the toughness of monolithic mullite is maximally 2.5 MPa m$^{0.5}$, values up to 18 MPa m$^{0.5}$ in C_F/

mullite and up to 34 MPa m$^{0.5}$ in SiC$_F$/mullite have been reported. Silicon carbide fiber-based mullite matrix composites also display favorable creep resistance and high thermal conductivity. The excellent mechanical properties of C$_F$/mullite and SiC$_F$/mullite materials are achieved by the high strength of carbon and silicon carbide fibers, and appropriate graphite easy-cleavage interphases between fibers and matrices. However, the mechanical properties of C$_F$/mullite composites deteriorate drastically if the composites are heat-treated in air, owing to degradation of carbon fibers and fiber/matrix interphases. SiC$_F$/mullite composites are not long-term stable at high temperature in air either. One reason is the oxidation-induced formation of silica layers at the silicon carbide fiber surfaces, which causes strong bonding between fibers and matrices. Another reason is spallation of the silica surface layers of the silicon carbide fibers and, as a consequence, further oxidation of the fibers.

The strong anisotropic character of unidirectionally (1D) aligned silicon carbide fiber composites can be reduced by using two- (2D) and three-dimensionally (3D) woven fabrics. Hirata et al. (1996) prepared laminated composites of 10 vol.% Si–Ti–C–O fiber woven fabrics with mullite matrices by infiltration or with the doctor blade method. At room temperature these composites show fracture toughness up to 4.7 MPa m$^{0.5}$ and moderate bending strength of about 270 MPa, which, however, decrease drastically at higher fiber contents. Hirata et al. demonstrated that weak graphite interphases form by an in-situ reaction at the contact between fibers and matrix. These graphite interphases enhance toughness, through fiber debonding and/or pull-out (see also Section 7.2.5). Strength values increase along with the process temperatures (maximum value about 500 MPa). However, the high fiber/matrix bonding in heat-treated composites comes at the expense of the toughness because of interphase degradation (Hirata et al. 2000, 2001).

Deng (1999) measured the compressive creep of 40 vol.% SiC$_F$/mullite composite between 1100 and 1200 °C. Compressive stresses (up to 55 MPa in air) were applied along (0°) and perpendicular (90°) to the unidirectionally arranged fibers. Both situations yield reduced creep with respect to monolithic mullite, although the strain rate of the (0°) composite was four to six times lower than that of the

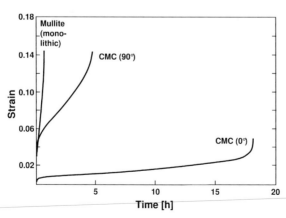

Fig. 7.2.19 Compressive creep curves of a 40 vol.% SiC$_F$/mullite composite (σ = 45 MPa at 1200 °C) with unidirectional fiber orientation. Compressive loads are parallel and perpendicular to the fiber direction in samples CMC (0°) and CMC (90°), respectively. For comparison the creep curve of monolithic mullite is given for the same load but at 1100 °C (after Deng, 1999).

(90°) composite (see Fig. 7.2.19). This is reasonable, since compressive creep of the (0°) material is essentially controlled by the silicon carbide fibers, while in the (90°) composite it is controlled by the mullite matrix. Deng observed a significant creep recovery behavior, which has been explained by relaxation of the residual stress between silicon carbide fibers and mullite matrix on load removal.

In summary, C_F/mullite and SiC_F/mullite composites display high strength and excellent fracture toughness. The big drawback of these composites, however, is their rapid degradation at high temperatures in air, owing to oxidation of the fibers and degradation of the fiber/matrix interphases.

7.2.2.2 Oxide Fiber-reinforced Composites

Since SiC_F/mullite and C_F/mullite composites rapidly degrade at high temperature in oxidizing environments, it has been desirable to develop composites that are stable against oxidation. Oxide fiber-reinforced mullite matrix composites can provide a solution to this problem. Again, a key issue is to properly engineer the interface between the fibers and the mullite matrix, so as to cause energy dissipation by crack deflection and fiber debonding and pull-out. One suitable way is to introduce a weak interphase between fibers and matrix (see also Section 7.2.5). Another approach is the use of a porous mullite matrix, without specific fiber/matrix interphases. In the latter case the porous matrix takes over the function of the interphase by crack energy dissipation (see also Section 7.2.1.2).

Composites with dense matrices The main objectives of dense oxide fiber-reinforced mullite matrix composites are high mechanical strength and toughness and maximized damage tolerance up to high temperatures. A key issue of this approach is the tailoring of the fiber/matrix interface (see Section 7.2.5). Nextel 720 aluminosilicate fiber-reinforced mullite matrix composites with double interphases consisting of carbon (fugitive component) and of dense alumina or zirconia coatings (protective component) have been statically and cyclically tested up to 1300 °C (Saruhan et al. 2001a,b, see Section 7.2.1.2). The stress-strain behavior of the composites, as-processed and after heat treatments at 1200 and 1300 °C, is given in Fig. 7.2.20 together with calculated Young's moduli. This shows that the dense oxide layer behind the fugitive carbon fiber coating plays an important role for stabilization of the interfacial gap and load transfer between matrix and fibers: The flexural strength of the composite with a reference carbon monolayer fiber coating is 135 MPa while composites with alumina-carbon and zirconia-carbon double fiber coatings display values of 180 MPa and 230 MPa, respectively. Prolonged sintering increases the contact areas between fibers and matrix, producing higher strength values and Young's moduli. However, simultaneously, the damage tolerance of the composites, as represented by the area under the stress-strain curve, decreases (Fig. 7.2.20). Saruhan et al. mentioned that composites are more stable under cyclic than under static conditions, which has been explained by the breakage of fiber/matrix contacts, which in turn leads to better maintenance of the interfacial gap.

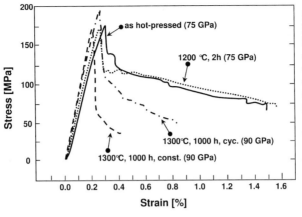

Fig. 7.2.20 Stress-strain behavior of a three-point bent Nextel 720 aluminosilicate fiber-reinforced mullite matrix composite with alumina/gap (fugitive) double fiber/matrix interphase. Numbers in parentheses correspond to Young's moduli (const. = static conditions, cyc. = thermocyclic conditions with 1000 cycles, after Saruhan et al. 2001a).

The tensile strength of polymer-infiltrated Nextel 610 alumina fiber composites with mullite plus silicon oxygen carbon matrices and fugitive fiber/matrix interphases has been determined by Peters et al. (2000; see also Section 7.2.1.2): At room temperature a value of about 175 MPa for the 0°/90° cross-ply laminates is given. The value decreases linearly but moderately to about 145 MPa at 800 °C and 40 MPa at 1200 °C. The Young's modulus of the 0°/90° laminate is about 100 GPa, and it is suggested that the matrix does not contribute much to the damage tolerance. Peters et al. measured very high intra-laminar shear strengths (about 35 MPa) at room temperature. This has been explained by the minor fiber pull-out effects in these materials.

Kaya et al. (2002) processed Nextel 720 aluminosilicate fiber-reinforced mullite matrix composites (see Section 7.2.1.2) to high densities (>85 % of the theoretical value). Their material is damage tolerant up to 1300 °C and exhibits flexural strengths of about 230 MPa, with no significant change after cycling the composite 300 times between room temperature and 1150 °C. Kaya et al. believe that the damage tolerance of this composite is due to the dense neodymium phosphate ($NdPO_4$) interphase between fibers and mullite matrix, which causes a weak bonding of fibers in the matrix. Additionally, homogeneously distributed nanopores in the matrix may contribute to the favorable mechanical behavior.

Composites with porous matrices   Damage tolerance in oxide/oxide ceramic matrix composites can be realized not only by weak fiber/matrix interphases but also by high matrix porosities. The failure of the latter materials is due to the lowered Young's modulus of the porous matrix which reduces the probability that cracks damage fibers. Additionally, matrix pores can act as sites of microcracking and disintegration, both effects being accompanied by substantial energy dissipation. These failure mechanisms also reduce the stress concentrations caused by fiber breaks (see Heathcote et al. 1999). Thus in porous composites a crack does not develop a continuous front, and matrix failure occurs by sequential breakage at intragranular contacts. This means that the fibers are "isolated" from the cracks, because the matrix is not strong enough to support the crack. Microcracking pro-

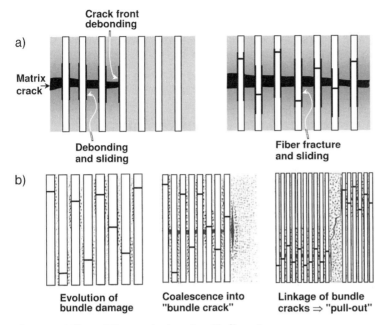

Fig. 7.2.21 Different failure mechanisms in oxide fiber-reinforced mullite matrix composites. (a) Debonding and sliding based on weak fiber/matrix interfaces; (b) Crack deflection and splitting in porous matrices (after Levi et al. 1999).

ceeds until the matrix is completely disintegrated, at which point fiber failure and finally failure of the composite take place. In dense fiber-reinforced matrix ceramic composites with specific fiber/matrix interfaces, the fiber/matrix bonding is weak, and fiber pull-out leaves a net hole in the matrix. Porous fiber-reinforced composites display different fiber pull-out mechanisms: Because of the strong bonding between matrix particles and the pulled-out fibers, coarse-grained fiber surfaces are revealed, with adherent matrix particles released by the powderized matrix adjacent to the pull-out holes (for references see, for example, Holmquist and Lange, 2003). A sequence of processes is taken into account: Matrix microcracking → local fiber failure → evolution of bundle damage → coalescence into "bundle cracks" → linkage of bundle cracks, i.e. "fiber pull-out" (see Fig. 7.2.21 and Levi et al. 1999).

Composites based on Nextel 610 alumina and Nextel 720 aluminosilicate fiber fabrics and mullite plus silica or aluminum phosphate matrices (Holmquist et al. 2001; see also Section 7.2.1.2) display strengths up to about 255 MPa and 145 MPa and non-brittle fracture behavior. Layered composites consisting of Nextel 720 aluminosilicate fiber-reinforced 70 wt. % zirconia plus 30 wt. % mullite matrices have been produced by Haslam et al. (2000) using the vibro infiltrating technique. With fibers oriented in-plane they yield bending strengths of 165 MPa, and moderate notch insensitivity in 0°/90° fiber orientations. The composites display shear fail-

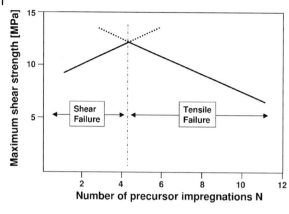

Fig. 7.2.22 Variation in shear stress of Nextel 720 aluminosilicate fiber fabric-reinforced porous α-alumina plus mullite matrix composites (45° off the fiber directions). Data were collected on the basis of short-beam shear tests (after Mattoni et al. 2001).

ure in the matrix in between the planes of the fiber fabrics, with shear strength in the range of about 10 MPa.

Mattoni et al. (2001) investigated the effect of porosity on Nextel 720 aluminosilicate fiber fabric-reinforced porous oxide matrix composites. The matrices consisted of 80 wt. % mullite plus 20 wt. % α-alumina with a porosity of about 30 %. The stabilized green panels were impregnated up to 12 times with an aluminum hydroxide chloride solution. With the number of impregnations the porosity of composites was gradually reduced and damage tolerance was lost. Reduction of the matrix porosity obviously leads to higher stress concentrations around broken fibers, which, on the other hand, increases the probability of failure of adjacent fibers. The interlaminar shear strengths of different impregnated specimens are shown in Fig. 7.2.22: For less than five impregnations, corresponding to more then 90 % of the initial porosity of the material, the load response is characterized by an initial linear curve and a load maximum. In these materials, displaying shear strength up to about 12 MPa, failure is caused by delamination between fiber plies. The interlaminar shear behavior of composites with matrices having less than 90 % porosity of the original matrix, corresponding to more than four impregnations, is different: They show brittle failure and can display shear strengths as low as 7 MPa (Fig. 7.2.22). Mattoni et al. suggest that in that case fracture occurs in a nearly pure tensile mode, with only a very small contribution from delamination.

Carelli et al. (2002) studied the effects of long-term thermal aging (up to 1000 h between 1000 and 1200 °C) on the mechanical properties of oxide/oxide composites, comparable to the studies by Mattoni et al. (2001). According to Carelli et al. a key issue is a stable matrix porosity, which is necessary for maintaining the damage tolerance of the material. Actually, in composites with 0°/90° fiber orientation, tensile strengths of about 140 MPa for both the as-processed and the 1200 °C (1000-h) aged samples have been observed, while the Young's moduli only increase slightly from about 60 to 70 GPa. On the other hand, there is significant enhancement of tensile strengths at 45°/45° fiber orientations from about 25 MPa (as-processed) to about 50 MPa (1200 °C for 1000 h, Fig. 7.2.23). Simultaneously the

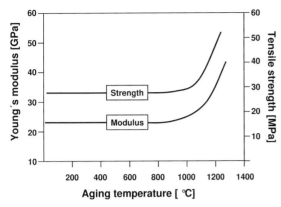

Fig. 7.2.23 Effect of thermal aging on tensile strength and elastic modulus of Nextel 720 aluminosilicate fiber fabric-reinforced porous α-alumina plus mullite matrix composites (45° off fiber directions, after Carelli et al. 2002).

elastic modulus of the composites increases from about 20 GPa to 40 GPa, respectively (Fig. 7.2.23). The large differences between the tensile properties of the 0°/90° and 45°/45° composites are explained by different failure mechanisms: The properties of the 45°/45° composites are dominated by matrix damage and interply delamination. This means that changes in the matrix porosity and phase content have a big influence on the composite's properties. In 0°/90° composites the mechanical behavior is fiber fracture dominated, while the matrix plays a less important role. Carelli et al. believe that the maximum matrix strength exists at a relatively low degree of matrix sintering. The retention of the fiber dominated properties after long-term aging at 1200 °C further stresses the potential of these materials for technical high-temperature applications, provided that a matrix with a stable pore structure is combined with a stable fiber.

Zawada et al. (2003) describe the high-temperature mechanical behavior of Nextel 610 alumina fiber fabric-reinforced porous aluminosilicate matrix composites. The stress-strain response of these composites is nearly linear to failure. In specimens with 0°/90° fiber orientation the tensile strength is about 200 MPa at room temperature and about 170 MPa at 1100 °C. Fiber fracture is believed to be the dominant damage mode. The loaded composites display intense matrix fragmentation, which causes a non-linear stress-strain behavior by fiber tow rotation during their withdrawal. Zawada et al. describe a slight in-plane strength increase from about 27 MPa at room temperature to about 32 MPa at 1100 °C in specimens with 45°/45° fiber orientation. Simultaneously the interlaminar shear strength increases more significantly from about 12 MPa to about 20 MPa. Both changes have been explained by an initial matrix sintering. Zawada et al. report the 1000 °C fatigue limit at 105 cycles to be 160 MPa and that the tensile strength of the run-out specimen is not affected by the fatigue loading. Furthermore a relatively poor rupture resistance of the composite has been described.

The temperature stability of porous mullite/α-alumina mixtures (60 to 100 wt. % mullite) for use in oxide fiber-reinforced oxide matrix composites has been investigated by Fujita et al. (2004). Fujita et al. found that both the Young's modulus and the toughness of composites increase linearly with the α-alumina content of the

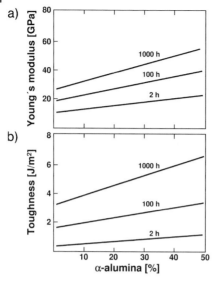

Fig. 7.2.24 Effect of α-alumina content and time on (a) Young's modulus, and (b) Toughness of porous mullite matrices (after Fujita et al. 2004).

matrix and the aging time, (Fig. 7.2.24a,b). They define a crack deflection parameter, which is proportional to the ratios of toughness and Young's modulus of fibers and matrices. This crack deflection parameter allows an extrapolation of experimental values to longer aging times. According to Fujita et al., the critical time, at which the damage tolerance of matrices is lost, is about 4000 h for a matrix with 60 wt. % mullite, and about 70 000 h for a matrix with 100 wt. % mullite. Fujita et al. conclude that for the design of porous matrices shear, interlaminar strength and compressive strength have to be considered. These values increase as the matrix is strengthened with the increase of the α-alumina/mullite ratio. At the same time, however, the crack deflection behavior deteriorates.

Holmquist and Lange (2003) evaluated the mechanical properties of Nextel 610 alumina and 720 aluminosilicate fiber fabric-reinforced porous aluminosilicate matrix composites. The in-plane fracture behavior of these materials is non-brittle, with room-temperature strengths above 280 MPa (Nextel 610 alumina fiber fabrics) and 170 MPa (Nextel 720 aluminosilicate fiber fabrics), respectively. However, the materials exhibit only moderate notch sensitivity (70 to 90 % of the un-notched material). The interlaminar shear strength is dominated by the matrix and changes from about 8 MPa to 12 MPa, while the matrix porosity decreases from about 43 to 37 vol. %. The porous microstructures of these materials are stable up to 1200 °C.

Strength and notch sensitivity of composites consisting of seven layers of Nextel 720 aluminosilicate fiber fabrics and porous matrices of mullite and α-alumina have been examined by Mattoni and Zok (2004). Mattini and Zok show that tensile and shear strengths increase if precursor-derived alumina is added to the matrix. However, these advantages are offset by deteriotation of the notch sensitivity. The authors introduced a finite element simulation model in order to rationalize the degree of notch sensitivity. The model accounts for interactions between notch tip

a)

b)

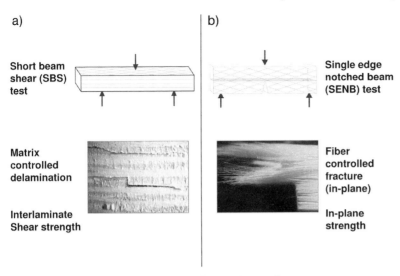

Short beam
shear (SBS)
test

Single edge
notched beam
(SENB) test

Matrix
controlled
delamination

Fiber
controlled
fracture
(in-plane)

Interlaminate
Shear strength

In-plane
strength

Fig. 7.2.25 Effect of fiber orientation on the failure behavior of
Nextel 720 aluminosilicate fiber-reinforced porous mullite
matrix composites processed according to the WHIPOX
approach. (a) Failure is matrix-controlled; (b) Failure is fiber-
controlled.

tensile and shear bands and agrees reasonably well with the experimental results.

A number of mechanical characterizations has been performed on highly porous Nextel 610 alumina or Nextel 720 aluminosilicate fiber-reinforced mullite
matrix composites (WHIPOX, see Section 7.2.1.2) developed at the German Aerospace Center. WHIPOX materials display laminate-type structures with a sequence
of fiber-rich and matrix-rich regions (see Fig. 7.2.13), and thus, like other oxide/
oxide composites, are sensitive against shear. This has been attributed to the weak
porous matrix, especially if macropores are present. Matrix failure occurs under
tension if loads are not in the fiber direction. If the tension is effective in fiber
direction, damage behavior is fiber dominated and fiber breaks reach deep into the
composites (Fig. 7.2.25). Tensile and interlaminar shear strengths of WHIPOX
Nextel 720 aluminosilicate fiber-reinforced mullite matrix composites are given in
Table 7.2.2. The values are similar to those of the other oxide fiber/mullite matrix
materials. A load-deflection curve of a WHIPOX composite with unidirectional
(1D) fiber orientation are given in Fig. 2.2.26. It shows bending strengths up to
280 MPa , due to the fiber dominance of the process. Beyond the linear branch of
the curves a stepwise strength decrease occurs, signalizing favorable damage tolerant behavior. Surprisingly, strength and damage tolerance in samples annealed up
to 1500 °C do not show the expected strong decrease. The reason for this favorable
behavior is not yet fully understood.

The cyclic fatigue of Nextel 720 aluminosilicate fiber-reinforced porous mullite
matrix WHIPOX composites has been studied by Göring et al. (2003). Göring and

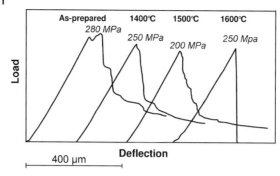

Fig. 7.2.26 Load-deflection curves of Nextel 720 alumino-silicate fiber-reinforced porous mullite matrix composites processed according to the WHIPOX approach. Curves are given for the as-prepared material, and after aging it at 1400, 1500 and 1600 °C (2 h each, after Schmücker et al. 2000).

coworkers measured the Young's moduli after different load cycles, and determined the "yield shear strength, τ_{el}" (i.e. the strength at the elastic limit) of the composite. In the elastic region of the material the applied stress τ_{ac} is below the yield shear strength τ_{el} consequently mechanical cycling in the elastic regime causes no degradation of the composite. However, if the actual loads are higher than the yield strength a continuous loss in stiffness is observed caused by the increasing number of defects in the composite, such as fiber cracks and local fiber/matrix debonding.

It is interesting to compare the crack resistance of WHIPOX mullite matrix composites with those of other fiber-reinforced ceramic matrix composites. Kuntz (1996) defined a "Formal strain intensity factor", which gives an estimate of the crack resistance of a composite. It turns out that WHIPOX displays a lower crack resistance than chemical vapor infiltration-produced SiC$_F$/SiC composites, whereas the WHIPOX values are well in the range of C$_F$/SiC composites and of other oxide/oxide materials (Fig. 7.2.27).

Table 7.2.2 Composition and properties of oxide fiber-reinforced porous mullite matrix composites produced on pilote or commercial scale.

Producer	Composition	Type	Room temperature tensile strength [MPa]	Interlaminar shear strength [MPa]	Temperature limitation
DLR-WHIPOX (Germany)	Nextel 720 fiber/mullite matrix	Porous matrix	130	> 10	> 1300 °C, 900 h
COI Ceramics (U.S.A.)	Nextel 610 fiber/mullite matrix	Porous matrix	365	15	< 1000 °C, 100 h
	Nextel 720 fiber/mullite matrix		220	14	≈ 1000 °C, 100 h
DaimlerChrysler (Dornier, Germany)	Nextel 610 fiber/mullite matrix	Dense matrix	177	36	< 1000 °C
	Nextel 720 fiber/mullite matrix		160	18	< 1100 °C

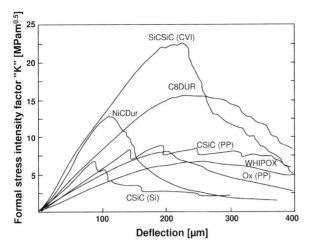

Fig. 7.2.27 Deflection behavior of a WHIPOX Nextel 720 aluminosilicate fiber-reinforced porous mullite matrix composite processed according to the WHIPOX approach. The results are shown in comparison to other continuous fiber-reinforced ceramic matrix composites. The "Formal strain intensity factor" on the ordinate gives the lower estimate of the achievable crack resistance. Data were derived from single edge notch beam-derived measurements (after Kuntz, 1996).

Oxide fiber-reinforced porous mullite matrix composites have also been produced by DaimlerChrysler (Dornier, Germany) and COI Ceramics (USA; see also Table 7.2.1). These composites display high room temperature strengths and interlaminar shear values (Table 7.2.2). Typical stress-strain curves of the COI porous mullite matrix composites reinforced by Nextel 610 alumina fibers, Nextel 720 aluminosilicate fibers, and the boron-rich Nextel 312 aluminosilicate fibers are given in Fig. 7.2.28. The application of these composites is limited to temperatures below 1100 °C, and it is suggested that their matrices contain relatively high

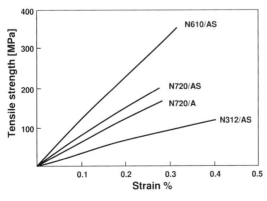

Fig. 7.2.28 Stress – strain behavior of oxide fiber-reinforced porous oxide matrix composites from COI Ceramics (U.S.A.). N 610 = Nextel 610 alumina fibers, N 720 = Nextel 720 aluminosilicate fibers, N312 = Nextel 312 boron aluminosilicate fibers, A = Alumina matrix, AS = Aluminosilicate matrix (Courtesy COI Ceramics, Inc.).

amounts of glass phase, which strengthens the materials at low temperatures but weakens them by viscous flow at elevated temperatures.

Parthasarathy et al. (2004) attempted to reduce the stresses generated from temperature gradients in service in laminated aluminosilicate matrix composites reinforced by Nextel 610 alumina and Nextel 720 aluminosilicate fiber fabrics. Two-dimensional laminate models are used to predict stress and strain in these composites. Parthasarathy et al. mention that the interlaminar shear stresses need to be controlled by gradually blending the two fiber types by means of hybrid fabrics or through cross-stitching.

7.2.3
Thermal Properties

Reliable data on the thermal properties of fiber-reinforced mullite matrix composites are important for many technical applications (see Section 7.2.6). The little information available in the literature mostly relate to the behavior of porous WHIPOX composites.

7.2.3.1 **Thermal Expansion**
Some unpublished thermal expansion data exist for Nextel 720 aluminosilicate fiber-reinforced porous mullite matrix WHIPOX composites. Expansion measurements have been performed in the fiber direction and perpendicular to it. There is no significant expansion up to about 200 °C, which corresponds to the typical behavior of mullite (see Section 2.2.2). Above this temperature the expansion up to 1200 °C is near to linearity with mean expansion coefficients parallel and perpendicular to the fiber direction of about 5.5×10^{-6} °C^{-1} in, and 5.0×10^{-6} °C^{-1}, re-

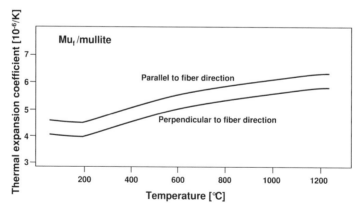

Fig. 7.2.29 Thermal expansions of Nextel 720 aluminosilicate fiber-reinforced porous mullite matrix composites processed according to the WHIPOX approach, in and perpendicular to the fiber direction. Note that there is no expansion between room temperature and about 200 °C (see Section 2.2.2).

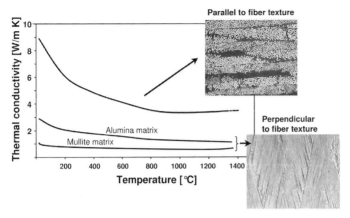

Fig. 7.2.30 Thermal conductivity of Nextel 610 alumina fiber-reinforced porous mullite matrix composites processed according to the WHIPOX approach, in and perpendicular to the fiber direction. Relationships for a Nextel 610 alumina fiber-reinforced porous alumina matrix composite are given for comparison (Courtesy M. Schmücker).

spectively (Fig. 7.2.29). The higher expansion in fiber direction has been explained by the dense character and the α-alumina content of the Nextel 720 aluminosilicate fibers.

7.2.3.2 **Thermal Conductivity**

The thermal conductivity of Nextel 610 α-alumina fiber-reinforced mullite matrix WHIPOX composites has been determined by Schmücker and Schneider (2004). The thermal conductivity of the composite perpendicular to the fiber orientation is about 1 W m^{-1} K^{-1}, while in the fiber direction a value above 8 W m^{-1} K^{-1} has been achieved at room temperature. The anisotropy of values is due to the higher conductivity of the dense α-alumina fibers in comparison to the porous mullite matrix (Fig. 7.2.30). With increasing temperature the thermal conductivity of the WHIPOX material decreases to values of about 3.3 W m^{-1} K^{-1} parallel and 0.7 W m^{-1} K^{-1} perpendicular to the fiber direction. At higher temperatures up to 1400 °C both thermal conductivities slightly increase again, possibly because the increasing influence of thermal conductivity caused by radiation, or matrix densification.

7.2.4
Miscellaneous Properties

Data are available for the gas permeability of oxide fiber-reinforced porous mullite matrix WHIPOX composites. The permeability strongly depends on the processing parameters, the matrix composition and the microstructure of the composite.

Gas flow experiments on WHIPOX composites consisting of Nextel 610 alumina fibers and a mullite plus α-alumina matrix have been performed by Schmücker and Schneider (2004), (flow rate: 10 to 100 $cm^{-2} h^{-1}$; gas pressure: 250 mbar). The gas flow is essentially controlled by the occurrence of mesopores, which typically occur in regions of matrix agglomeration. Such matrix agglomerations are more pronounced in composites produced from thicker 3000-DEN fiber rovings than from thinner more flexible 1500-DEN rovings. Furthermore Schmücker and Schneider suggest a higher gas flow in composites with two dimensional (2D) fiber orientation than in composites with unidirectional (1D) fiber orientation. These findings are rationalized in terms of fiber cross-over points, which typically contain agglomerations of porous matrix. The porous mullite matrix WHIPOX composites can be made airtight by sealing the surface with external coatings. First experiments on alumina plasmaspray- and sol-gel-coated WHIPOX composites show that these coatings do not only affect the air permeability but also improve the erosion resistance of the composites where as coating deposited by vapor phase techniques (CVD and PVD) do.

7.2.5
The Role of the Fiber/Matrix Interphase

Fiber/matrix interphases are required in dense continuous fiber-reinforced ceramic matrix composites to control the bonding of the fibers into the matrix, while porous matrix composites without fiber/matrix interphases follow other failure mechanisms. The specific relationships for dense and porous oxide fiber/mullite matrix composites are discussed in the introductory part of Section 7.2. Micromechanical parameters such as interfacial shear stress (τ) and debonding energy (G_i), and the long-term thermal stability of the interphase are critical issues for the fabrication and use of dense composites. The interphase should isolate the fibers from the matrix in order to protect them against local stress fields. It must be strong enough to transfer load from the matrix to the fibers, but, on the other hand, it should be weak enough to allow energy dissipating mechanisms such as crack deflection, fiber/matrix debonding and fiber pull-out. If these conditions are not fulfilled, dense fiber/matrix composites show brittle failure like monoliths. Thus, the essential property of the fiber/matrix interphase in dense composites is to debond in the presence of transverse matrix microcracks.

The discussion of stability and properties of dense mullite matrix composites has to take into account that fibers, matrices and interphases form a microstructural system that is not in thermodynamic equilibrium at high temperature, with the tendency to reduce the high surface area of fiber coatings and to react with each other. These undesirable effects have to be prevented during the fabrication and application of composites.

The deposition of interphases on the fibers is performed prior to composite fabrication. It is a complicated process, since continuous and homogeneous, very thin coatings (normally <1 μm), often with specific crystallographic orientations of the interphase, are required. Chemical vapor deposition (CVD), sputtering, physical vapor deposition (PVD) and chemical dipping techniques have been applied for fiber coatings (for references see, for example, Saruhan, 2003).

Interphase materials used for oxide fiber/mullite matrix systems typically fall in one of the following categories:

– Easy-cleavage materials with cleavage planes oriented parallel to the fiber surfaces
– Low toughness materials with weak phase boundaries
– Porous or "fugitive" materials leading to a thin gap between fiber and matrix.

7.2.5.1 Easy-cleavage Interphases

Easily cleaving carbon (graphite, turbostratic carbon) was identified as an excellent interphase for ceramic matrix composites in non-oxidizing environments (e.g. Brennan and Prewo, 1982). However, these materials suffer from instability in oxidizing atmosphere. Turbostratic boron nitride (BN) employed instead of carbon in mullite matrix composites displays only slightly better oxidation stability (Fig. 7.2.31a; see also, for example, Naslain et al. 1991). Some improvement can be achieved by silicon carbide protection of the boron nitride coatings in the form of SiC/BN double layers (Fig. 7.2.31b). When fired in an oxidizing atmosphere these SiC/BN double layers form thin silica scales at the contact of the SiC layer towards the mullite matrix. Since oxygen diffusion through silica is slow, further oxidation of the coating is retarded. This concept works up to 1200 °C; at higher temperatures, however, the double layer system also rapidly decomposes (Schmücker et al. 1997).

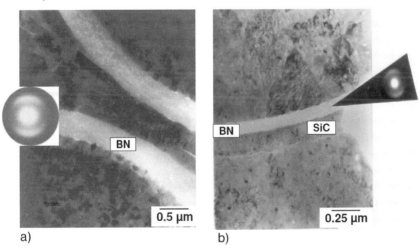

a)   b)

Fig. 7.2.31 Transmission electron micrographs of (a) Boron nitride fiber/matrix interphase and (b) Boron nitride and silicon carbide double fiber/matrix interphase in Nextel 720 aluminosilicate fiber-reinforced dense mullite matrix composites (inserts: Diffraction pattern of the BN layers. Courtesy M. Schmücker).

To overcome the oxidation problem of carbon and boron nitride fiber coatings, easily cleavable oxides, such as phyllosilicates, hexaaluminates and magnetoplumbites, have been considered as interphases for mullite matrix composites. While typical sheet silicates are prone to decomposition below 1000 °C, fluorophlogopite is stable above 1200 °C in dry atmospheres (Kerans et al. 2002). However, fluorophlogopite does react with mullite and α-alumina and, therefore is not suitable for mullite matrix systems. The basal plane cleavage of hexaaluminates such as hibonite ($CaAl_{12}O_{19}$) or lanthanum hexaaluminate ($LaAl_{11}O_{18}$) is inferior to that of micas, graphite or boron nitride (Kerans et al. 2002). Nonetheless, it has been demonstrated by Cinibulk and Hay (1996) that cracks propagate within hibonite coatings by transgranular cleavage. Only little fiber pull-out is found in composites with such fiber/matrix interphases, probably because of to the high roughness of the debonding surfaces. The application of hexaaluminate coatings in dense mullite matrix composites is further restricted by its limited stability in the presence of mullite.

7.2.5.2 Low-toughness Interphases

Fiber coatings such as tin oxide (SnO_2), monazite ($LaPO_4$), scheelite ($CaWO_4$) and other tungstates, niobates and vanadates may be used as interphases in oxide fiber/mullite matrix composites, although they are only stable together with mullite over a limited temperature range. These oxides usually have low cohesion with both fibers and matrices, caused by the high cation-induced polarization of oxygen bonds (see, for example, Morgan and Marshall 1995). The reinforcing mechanism is crack deflection, which is attributed either to weak bonding between coatings and fibers (Morgan and Marshall, 1995) or to an intrinsic weakness of interphases (Hay, 2000). However, sliding of such coated fibers in the matrix is low as evidenced by push-out experiments (Kerans et al. 2002).

7.2.5.3 Porous or Gap-producing Interphases

The low fracture energy of porous coatings facilitates enhanced debonding. Porous fiber coatings consisting e.g. of zirconia (ZrO_2) are deposited on the fibers by chemical vapor deposition (CVD), using excess carbon which is burned off subsequently (Fig. 7.2.32a). A crucial point of this approach is grain coarsening and pore agglomeration at elevated temperatures (Fig. 7.2.32b). Moreover, the roughness of the fracture surfaces within these porous interphases tends to be high, which works against fiber pull-out (e.g. Chawla, 2003). Nevertheless, porous fiber coatings have been considered for service as suitable low toughness interphases in mullite matrix composites. At this point it is interesting to note that a special form of fiber/matrix interphase exists in porous matrix composites, where the matrix itself takes over the function of a porous weak interphase (see introductory paragraphs of Section 7.2).

Fugitive coatings, producing a gap between fibers and matrix, are a simple but effective concept for ceramic matrix composites with damage tolerant fracture be-

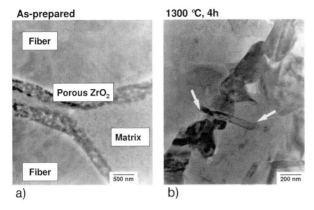

As-prepared   **1300 °C, 4h**

Fiber

Porous ZrO₂

Matrix

Fiber   500 nm

a)   b)

Fig. 7.2.32 Transmission electron micrographs of fiber/matrix interphases in porous zirconia (ZrO)₂-coated Nextel 720 aluminosilicate fiber-reinforced mullite matrix composites.
(a) As-processed; (b) Annealed at 1300 °C (4 h). Temperature-induced grain growth and pore agglomeration allow newly formed α-alumina crystals to penetrate coatings and fibers (arrows; Courtesy M. Schmücker).

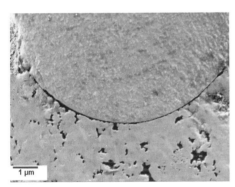

1 µm

Fig. 7.2.33 Scanning electron micrograph of the microstructure of a Nextel 720 aluminosilicate fiber-reinforced mullite matrix composite with a gap (fugitive) fiber/matrix interphase (Courtesy M. Schmücker).

havior (Fig. 7.2.33, see also Section 7.2.2.2). The typical fugitive coating material is carbon, which is burned-off during sintering of the composite. The evolving gap size must be in the range of the fiber surface roughness. If the gap is too small, friction due to roughness-induced clamping is high and fiber sliding is limited. On the other hand, if the gap is too big, there is lack of load transfer between fibers and matrices, and hence the strength of the material is poor (Kerans et al. 2002). Although the high-temperature stability of the gap is seen as a serious problem, Nubian et al. (2000) demonstrated that the favorable deformation behavior of oxide fiber-reinforced mullite matrix composites can be retained even after long-term heat treatments up to 1300 °C (see Section 7.2.2.2).

Table 7.2.3 Potential aerospace applications of oxide fiber-reinforced mullite matrix composites.

Application	Requirements
Combustor tiles for gas turbine engines	High temperature long-term stability in water-vapor rich gases, low surface erosion
Heat shields for re-entry space vehicles	High temperature short-term stability in gas plasmas, high emissivity of surface, little catalytic recombination, low surface erosion
Radome structures for missiles	High temperature short-term stability, no gas permeability, transmittance for infrared and radar radiation, low surface erosion

7.2.6
Applications

The technical and commercial importance of oxide fiber-reinforced mullite matrix composites significantly has increased in recent years. This development has been promoted because metals, and also non-oxide ceramic materials, such as C_F/SiC, SiC_F/SiC composites, display limited stability at high temperatures in air. Thus oxide fiber/oxide matrix composites have become candidate materials for many spacecraft, aero-engine (Table 7.2.3) and other industrial applications (see, for example, Newman and Schäfer, 2001). Most potential applications of oxide/oxide composites relate to materials with porous matrices, owing to their easy and low-cost fabrication process.

7.2.6.1 Spacecraft Applications
Partial replacement of silicon carbide fiber/silicon carbide matrix (SiC_F/SiC) composite heat shields of space re-entry vehicles by oxide fiber/mullite matrix composite structures has been considered (Fig. 7.2.34). A main advantage of oxide fiber/mullite matrix materials besides oxidation stability is the low-cost fabrication, especially in the case of porous mullite matrix composites. Because of the short duration of the thermal load during entry to the atmosphere (for re-entry space vehicles the total exposure time of the heat shield at maximum temperature can be taken as less than 20 h) oxide fiber/mullite matrix materials are considered to be usable up to peak temperatures of 1500 °C.

7.2.6.2 Aircraft Engine and Powerplant Applications
Nitrogen oxide (NO_x) emission has been identified as a main source for "acid rain" and the degradation of the earth's ozone layer over the poles. Therefore, aircraft engine and powerplant industries have been under increasing public pressure to

**Thermal protection systems (TPS)
for moderate temperature impact**

**Segmented engine
exhaust heat shields**

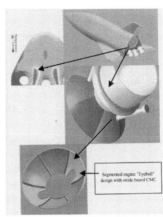

Fig. 7.2.34 Potential applications of oxide fiber-reinforced
porous mullite matrix composites for heat shields of
re-entry space vehicles. Oxide fiber-reinforced porous mullite
matrix composites may be suitable for temperature exposures
up to 1500 °C.

reduce the NO_x emissions of their systems. A possible way to achieve this aim is to
reduce the air film cooling of combustor liners and turbine airfoils, which has the
effect of a temperature increase in the combustor and at the combustor outlet
(Parks et al. 1991, Bannister et al. 1995). The nickel- and cobalt-based superalloys
used to date are operating at their upper temperature limit, while silicon carbide
fiber-reinforced silicon carbide matrix (SiC_F/SiC) composites, which have been
considered to replace these metals, suffer from a low stability in air, especially in
the presence of water vapor. It is anticipated that significant temperature increases
in aircraft and stationary gas turbine engines can be realized by implementation of
oxide fiber/mullite matrix composite liners, which have superior stability under
combustor environments (Fig. 7.2.35), especially if they are protected by thermal
and environmentally stable external coatings (environmental barrier coatings,
EBCs). In a final state a reduction in fuel consumption of about 20% and in the
NO_x emission of as much as 80% is envisaged (H. Schäfer, DaimlerChrysler, Dornier, pers. comm.). Intense research and development work, however, is still necessary in order to achieve this demanding aim (Fig. 7.2.36).

In principle the requirements for ceramic liners in aircraft and powerplant gas
turbine engines are comparable, although operation temperatures and time standards for stationary gas turbines are higher (>1400 °C for >50000 h) than for aircraft engines (>1300 °C for >20000 h). On the other hand extreme reliabilities have
to be achieved for aircraft components and structures, and high strength/weight
ratios of materials are important for economic flight transportation. Nevertheless
replacements of metal diffuser rings and exhaust cones by oxide fiber-reinforced
mullite matrix structures in the aircraft engine have also been envisaged.

a) **Aircraft turbine engine**   b) **Stationary gas turbine engine**

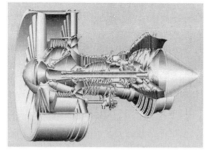   

General Electric GP7000 for Airbus A3XX   **Siemens Vx4.3A**

Fig. 7.2.35 Potential applications of oxide fiber-reinforced porous mullite matrix composites for thermal protection systems in combustors of (a) Aircraft and (b) Power generation gas turbine engines.

a)   b)

Fig. 7.2.36 Model combustor used for testing oxide fiber-reinforced porous mullite matrix composites processed according to the WHIPOX approach, under near-application conditions. (a) Overview, (b) Detail of the DLR equipment.

7.2.6.3 Other Industrial Applications

Although oxide fiber-reinforced mullite matrix composites have been developed for aerospace applications they also have potential for other technical areas that require high temperature stability in severe environments and high damage tolerance. One such field is that of chemical reactors, where, depending on the specific process, temperature stability and resistance against chemical attack are necessary. Other potential applications of oxide fiber-reinforced mullite matrix composites are efficient catalytic convertors, kiln funitures, hot gas filters, heat exchangers, and surface and pore burners (Fig. 7.2.37). Kitaoka et al. (2005) prepared con-

Pore burner   **Surface burner**

Fig. 7.2.37 Pore and Surface burners made of oxide fiber-reinforced mullite matrix composites processed according to the
WHIPOX technique (DLR components).

tinuous alumina fiber-reinforced mullite matrix composites for high temperature
gas filters with good damage tolerance properties and high corrosion resistance to
fly ash. They achieved excellent separation efficiency and low pressure drop in the
separation of fly ash. The low pressure can be realized by mullite whiskers formed
on the surface of the filters. This configuration is thought to be a good candidate
for hot gas dust-trap filters.

7.3
Platelet- and Particle-reinforced Mullite Matrix Composites
H. Schneider and K. Okada

7.3.1
Basic Principles

The main aim of platelet and particle reinforcement of mullite matrices is improvement of the thermo-mechanical behavior, although other properties can be of
interest as well. Different strengthening mechanisms by addition of platelets and
particles have been considered, which will be described in the following.

7.3.1.1 Transformation Toughening
This process refers to the strengthening of mullite matrices by addition of zirconia. The different strengthening mechanisms involved with zirconia transformation (monoclinic → tetragonal → cubic) are shown in Fig. 7.3.1:

- Stress-induced toughening, caused by volume change during phase transformation of zirconia from tetragonal to monoclinic.
- Microcrack toughening, caused by thermal expansion mismatch of zirconia and
  mullite.
- Grain boundary strengthening, caused by incorporating metastable zirconia into
  mullite.

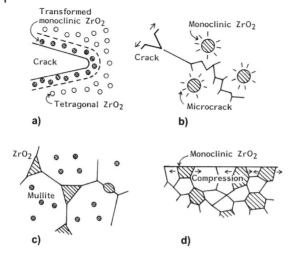

Fig. 7.3.1 Mechanisms of toughening of mullite matrices by zirconia (ZrO₂) particle addition. (a) Stress-induced toughening, (b) Microcrack toughening, (c) Zirconia incorporation toughening, (d) Surface compression toughening.

- Surface compression toughening caused by the occurrence of monoclinic zirconia on the surface of mullite specimens.

7.3.1.2 Crack-deflection Toughening

This toughening process is generally valid for particle-reinforced composites. Strengthening of the mullite matrix is achieved by crack deflection or bowing at the platelet/mullite and particle/mullite interfaces. Crack deflection needs weak platelet/mullite or particle/mullite interphases.

7.3.1.3 Toughening by Modulus Load Transfer

This toughening is effected by stress transfer from the crack tip to the bulk of the mullite matrix. This process requires that the elastic modulus of the reinforcing component is higher than that of the matrix (so-called "modulus load transfer", see Nischik et al. 1991). While crack deflection needs a weak platelet/matrix interface, the modulus load transfer mechanism requires a strong interface. Thus, depending on the interface characteristics, either of the mechanisms may work. The most frequently used reinforcing elements are silicon carbide particles and platelets, while other systems, such as α-alumina particles, are less important.

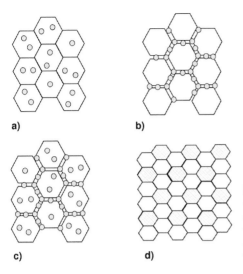

Fig. 7.3.2 Schematic view of mullite matrix reinforcement by granularly dispersed nanoparticles. (a) Intra-granular-type, (b) Inter-granular-type, (c) Intra- and inter-granular-type, (d) Nanograin type (after Niihara et al. 1989).

7.3.1.4 Nanoparticle Toughening

In this case reinforcement of mullite matrices is effected by inter- and intra-granularly dispersed nanoparticles (Fig. 7.3.2; see Niihara et al. 1989). Inter-granular (grain boundary type) strengthening makes use of "classical" mechanisms such as crack deflection at the particle/mullite interface and of stress intensity reduction at the tip of the crack (Fig. 7.3.1). In the case of intra-granularly distributed nanoparticles, expansion mismatches between reinforcing particles and mullite matrix can produce subgrain boundaries in the matrix, by which the strength of the composite is enhanced. Nanoparticles must have high elastic moduli, little temperature-induced grain growth, and low thermal expansion. Silicon carbide nanoparticles are most suitable for mullite toughening.

7.3.2
Zirconia Particle-reinforced Composites

A large number of papers has been published on zirconia particle-reinforced mullite matrix composites (designated as $ZrO_{2,P}$/mullite). In this review we focus on basic works only. For more information the reader is referred to Schneider et al. (1994).

7.3.2.1 Fabrication Routes

Different processing routes of $ZrO_{2,P}$/mullite composites have been reported with the following categories:
- Sintering of mullite and zirconia (ZrO_2)
- Reaction sintering of zirconia (ZrO_2) and mullite precursors, or of zirconia (ZrO_2), alumina (Al_2O_3) and silica (SiO_2)

– Reaction sintering of zircon ($ZrSiO_4$) and alumina (Al_2O_3)
– Reaction bonding of aluminum metal (Al), alumina (Al_2O_3) and zircon ($ZrSiO_4$)
– Crystallization of rapidly quenched melts of the ZrO_2–Al_2O_3–SiO_2 system
– Miscellaneous fabrication methods

Sintering of mullite and zirconia (ZrO_2) Prochazka et al. (1983) prepared zirconia particle-reinforced mullite matrix composites ($ZrO_{2,P}$/mullite) by sintering mixtures of fused-mullite and zirconia (ZrO_2) powders at 1610 °C. The addition of zirconia is effective in enhancing densification and retarding grain growth of mullite. It also reduces the development of a glassy phase at the grain boundaries, so that nearly glass-free microstructures are obtained. Similar techniques to fabricate zirconia-toughened mullite matrix composites were applied by Deportu and Henney (1984), Yuan et al. (1986a,b) and Nishiyama et al. (1991). Mizuno et al. (1988) prepared $ZrO_{2,P}$/mullite composites by the sol-gel method using mullite and zirconia sols as starting materials. Nearly full density is reached by firing at only 1550 °C. The mullite particles of this material are both fine equiaxed and elongated, with grain sizes ranging between about 0.3 and 0.9 μm with most of the zirconia grains occurring intra-granularly in mullite. Yoon and Chen (1990) produced dense $ZrO_{2,P}$/mullite ceramics from zirconia and mullite, derived from co-precipitation of tetraethyloxysilane (TEOS) and aluminum nitrate nonahydrate solutions, and firing the green compacts at 1400 °C. The average grain sizes of mullite and zirconia particles in these materials are very small (0.2 μm), probably because of the low firing temperature. Composites with 10 to 50 vol.% zirconia display superplastic deformation, in which the mullite grains behave as hard inclusions during deformation. Koyama et al. (1994) prepared nearly fully dense $ZrO_{2,P}$/mullite composites from commercial mullite and unstabilized zirconia powders at 1520 °C. At elevated temperature mullite releases silica as a liquid phase, which enhances elongated grain growth of mullite.

Some papers have also been published on mullite matrix composites toughened with yttria (Y_2O_3) partially stabilized zirconia (Y-TZP). Hirano et al. (1990) started from mullite particles coated with partially (4 vol.%) yttria stabilized zirconia (4Y-TZP) consolidated by a colloidal filtration method. The size distribution of pores in these compacts are narrow and small, and full density is achieved at 1600 °C. The fracture toughness of mullite matrix composites with 10 vol.% 4Y-TZP/mullite is slightly higher than that obtained by conventional processing. Composites of 3Y-TZP/mullite were prepared by Ishitsuka et al. (1987). They describe a rod-like crystal growth of mullite, which has been explained as being due to liquid-phase sintering out of a Y_2O_3–Al_2O_3–SiO_2 melt.

Rincon and Romero (2000) published data on mullite composites produced by sintering of mullite plus zirconia without and with addition of 1 wt. % CaO sintering aid, and of mullite reinforced with alumina plus zirconia. Transmission electron microscopy of these materials shows inter-granular zirconia grains often being twinned at the contact to mullite and mullite crystals with high stress contrast lines (Fig. 7.3.3).

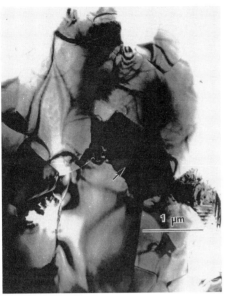

Fig. 7.3.3 Transmission electron micrograph of a $ZrO_{2,P}/Al_2O_{3,P}$/mullite matrix composite, prepared by sintering mullite, alumina and zirconia at 1570 °C (16 h). Note the inter-granular occurrence of zirconia (dark, arrow) between alumina and mullite (light). The mullite crystals depict high-stress field lines (Courtesy J. M. Rincon).

Imose et al. (1998a) followed a low temperature route to produce mullite composites reinforced with 2 vol.% partially yttria stabilized zirconia (2Y-TZP) and obtained nearly dense materials by sintering at 1450 °C in air. They used commercial zirconia and highly reactive mullite, synthesized by the hydrazine method (Imose et al. 1998b). The grain size of mullite decreases with the zirconia content of the composite, while that of zirconia increases. Strength and fracture toughness of these composites are relatively high. Miao et al. (1998) produced $ZrO_{2,P}$/mullite composites at 1550 °C from silicon carbide, aluminum metal alumina and zirconia by reaction bonding. The bulk of the as-sintered samples consists of mullite, tetragonal zirconia, residual alumina and silicon carbide. Simultaneously a porous outer oxidized layer of mullite, large zirconia grains and a small amount of zircon ($ZrSiO_4$) is formed. Indentation tests indicate compressive stress in the outer layer, which is associated with the occurrence of monoclinic zirconia and the difference in thermal expansion coefficients.

Reaction sintering of zirconia (ZrO_2) and mullite precursors or of zirconia (ZrO_2), alumina (Al_2O_3) and silica (SiO_2) Most work on these synthesis started from alkoxides or the sol-gel process. Moya and Osendi (1984) reported the use of zirconia and of a "pre-mullite" phase, prepared by thermal and chemical treatment of kaolinite. The 15 vol.% $ZrO_{2,P}$/mullite composite densifies at about 1570 °C. Interestingly, about 0.5 wt. % of ZrO_2 is metastably incorporated into the mullite structure (see Section 1.3.2). The enhancement of the mechanical properties in these specimens is explained by grain boundary strengthening, due to the presence of zirconia. Preparation of $ZrO_{2,P}$/mullite composites by the alkoxide process has been reported by several workers (Yuan et al. 1986a,b, Ruh et al. 1988, Suzuki and Saito, 1990, Rahaman and Jeng, 1990, Colomban and Mazerolles, 1991). Suzuki and Saito (1990) prepared a precursor solution with mullite composition by mixing pre-hydrolyzed tetraethyloxysilane (TEOS), aluminum isopropoxide (Al(O^{i}Pr)$_3$) and zirconium isopropoxide (Zr(O^{i}Pr)$_4$). They believe that specimens with uniform microstructures can be achieved if fine zirconia particles are dispersed in the mullite matrix. However, densification of compacts is poor and bending strength is low. Crystallization of xerogels prepared from alkoxides, and a similar process with salts instead of alkoxides, have been examined by Low and McPherson (1989a,b), Suzuki and Saito (1990) and Colomban and Mazerolles (1991). Fairly large amounts of zirconium are considered to be incorporated in mullite by these processes (see Section 1.3.2). Shiga et al. (1991) used the sol-gel method to prepare ZrO$_{2,P}$/mullite composites at about 1630 °C. Starting materials are boehmite sol, colloidal silica and zirconium hydrate sol. Nearly full density is obtained for composites containing 20 vol.% zirconia, while higher zirconia contents again produce lower densities. This has been explained by microcracking because of the large thermal expansion differences between mullite and zirconia. Kubota and Takagi (1986) used a similar method but with aluminum nitrate instead of a boehmite sol.

Boch and Chartier (1991) prepared ZrO$_{2,P}$/mullite composites by tape casting. The starting fine-grained powders of α-alumina, quartz and zirconia sintered to nearly full density below 1600 °C. Similar effects of ZrO$_2$ addition as reported by Prochazka et al. (1983) have been recognized. Rundgren et al. (1990) used very small zirconia particles (1–20 nm) and salts for the sol-gel method to prepare yttria stabilized ZrO$_{2,P}$/mullite composites, and the microstructure of their sintered bodies is controlled by the particle size of the starting materials. Addition of yttria produces a liquid phase during sintering and therefore the mullite crystals develop an acicular shape with zirconia grains occurring intra-granularly. Koyama et al. (1994) showed that the favorable sinterability of ZrO$_{2,P}$/mullite, produced by reaction sintering of alumina, silica glass and zirconia at 1520 °C, is due to viscous-flow mechanisms, and that equiaxed mullite grains and intra-granularly dispersed zirconia crystals are formed.

Yaroshenko and Wilkinson (2001) made an interesting approach to the preparation of ZrO$_{2,P}$/mullite composites. They use silica-coated alumina microcomposite powders admixed with yttria stabilized zirconia (Fig. 7.3.4). Reaction sintering proceeds in two steps. In the first stage at about 1200 °C the composites densify by transient viscous sintering (TVS, see Section 4.2.6). An important point at that stage is to prevent crystallization of the amorphous silica layers around alumina, which is achieved by rapid heating. The second step is mullitization. Mullite seeds accelerate the process, moving it towards lower temperature. The reaction sequence without seeds develop a microstructure with thick SiO$_2$-rich glassy grain-boundary films, which is explained by the transient formation of zircon and Al$_2$O$_3$-rich mullite. Seeding prevents the formation of these unfavorable transient reaction products (Fig. 7.3.5). Yaroshenko and Wilkinson (2001) suggest that the crystallization of zirconia inclusions and of mullite grains is a thermally activated process. The increase of the zirconia particle size goes along with a decrease of the tetragonal fraction of zirconia.

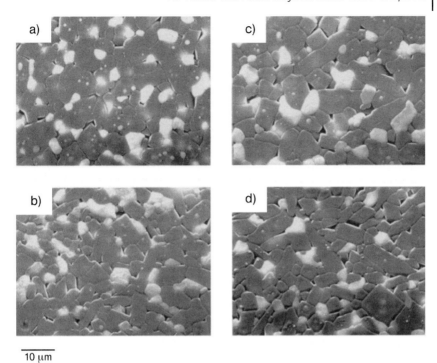

10 µm

Fig. 7.3.4 Scanning electron micrographs of 15 vol.% $ZrO_{2,P}/$
mullite composites (mullite: dark, ZrO_2: light). The composites were produced by transient viscous sintering (TVS) of
silica-coated alumina particles and subsequent mullitization.
(a) 1550 °C, 0.5% mullite seed content; (b) 1550 °C, 1% mullite seed content; (c) 1580 °C, 0.5% mullite seed content; (d)
1580 °C, 1% mullite seed content (Courtesy D.S. Wilkinson).

Reaction sintering of zircon ($ZrSiO_4$) and alumina (Al_2O_3) The effectiveness of zirconia addition in enhancing the mechanical properties of mullite-based ceramics
was first demonstrated by Claussen and Jahn (1980). They prepared composites by
reaction sintering of zircon and alumina. A two-step process has been considered.
In the first step, at about 1450 °C, zircon and alumina particles sinter to a maximal
relative density of 95%. During the second step, at about 1575 °C, the reaction of
zircon and alumina to mullite and zirconia takes place with the sintered specimens having about 98% relative density. The schematic overall reaction is
(Fig. 7.3.6; Wallace et al. 1984):

$$ZrSiO_4 + 3\ Al_2O_3 \rightarrow 3Al_2O_3 \cdot 2SiO_2 + ZrO_2 \qquad (2)$$
zircon   α-alumina mullite   zirconia

The moderately high bending strength and fracture toughness of these composites
is associated with microcrack toughening. Boch et al. (1990a,b) examined various

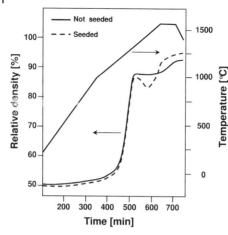

Fig. 7.3.5 Kinetics of densification of a $ZrO_{2,P}$/mullite composite containing 15 vol.% partially stabilized zirconia. The composite has been produced by transient viscous sintering (TVS) of silica-coated alumina particles (1000 °C) and subsequent mullitization (1200 °C, see Fig. 7.3.4). The densification curves of unseeded and 2 vol.% seeded materials are shown (after Yarashenko and Wilkinson, 2001).

processing parameters during reaction sintering of zircon and alumina, and found that sinterability increased if the starting materials had small particle sizes and the mullitization was sluggish. They also found that the size and shape of zirconia grains change depending on the Al_2O_3/SiO_2 ratio of specimens, which in turn has a direct influence on the mechanical properties of $ZrO_{2,P}$/mullite composites. Materials with 68 mol% Al_2O_3 yield the best results.

The influence of additives on reaction sintering and the mechanical properties of $ZrO_{2,P}$/mullite composites has been studied frequently. The effect of titania (TiO_2) addition was examined by Rincon et al. (1986), Leriche (1990) and Descamps et al. (1991). The microstructures of these composites consist of juxta-

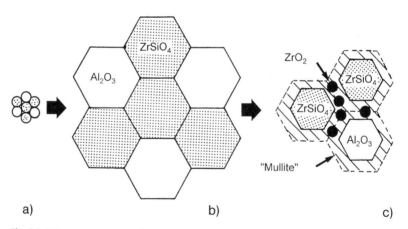

Fig. 7.3.6 Reaction sintering of zircon and alumina to $ZrO_{2,P}$/ mullite matrix composites. (a) Starting powders; (b) Sintering and grain growth of α-alumina (Al_2O_3) and zircon ($ZrSiO_4$); (c) Nucleation and growth of mullite and zirconia (ZrO_2) (after Wallace et al. 1984).

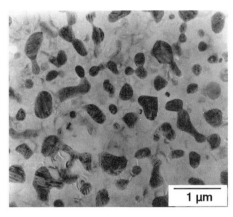

Fig. 7.3.7 Transmission electron micrograph of a zircon plus alumina-derived $ZrO_{2,P}$/mullite matrix composite heated to 1635 °C and then quenched to room temperature. Dark granis are zirconia, which are embedded in a mullite matrix. (Courtesy K. Okada).

posed equiaxed mullite without a glassy phase (Leriche 1990). Added titania (TiO_2) is incorporated in both mullite and zirconia (Rincon et al. 1986) and/or forms tieillite (Al_2TiO_5; Descamps et al. 1991). On the other hand, addition of magnesia, (Leriche, 1990, Descamps et al. 1991, Orange et al. 1985) or calcia, (Moya, 1990) develops microstructures with cross-linked mullite grains and an inter-granular glassy phase, indicating liquid phase sintering. The different effects of titania and magnesia or calcia addition is considered to be caused by the difference in the eutectic temperature in the respective metal oxide-Al_2O_3–SiO_2 ternary system. Wu and Lin (1991) examined the effect of ceria (CeO_2) addition and found that it promotes mullitization and increases the fraction of tetragonal zirconia. The fracture toughness of these composites is apparently higher than that of the other $ZrO_{2,P}$/mullite composites.

Koyama et al. (1992) prepared $ZrO_{2,P}$/mullite composites and achieved nearly full density on firing below 1500 °C. The composites consist of small rounded zirconia particles, dispersed inter- and intra-granularly in the matrix. The composites display favorable fracture toughness. Koyama et al. (1994, 1996) also prepared $ZrO_{2,P}$/mullite from alumina and zircon as starting powders. They observe transient formation of Al_2O_3-rich mullite (up to about 65 mol%), and mention that up to about 1 wt % ZrO_2 is incorporated into the mullite structure (see Section 1.3.2). These $ZrO_{2,P}$/mullite composites are composed of irregularly shaped mullite and intra- and inter-granularly dispersed zirconia (Fig. 7.3.7).

Bhattacharjee et al. (2000) prepared zirconia-toughened mullite ceramics from zircon plus alumina starting powder mixtures using a specific thermal plasma reaction. Bhattacharjee et al. found that the grain size of zirconia in Al_2O_3-rich admixtures is significantly smaller than in SiO_2-rich admixtures. Since the thermal plasma technique produces melts of the starting materials in a very short time, no problems occur with densification of materials. Das et al. (1998) and Das and Banerjee (2000) produced $ZrO_{2,P}$/mullite composites from zircon flour and calcined alumina without or with rear earth oxide additives (4.5 mol% Y_2O_3 or Dy_2O_3) by reaction sintering between 1400 and 1650 °C. Sintering aids enhance the process by forming a transient liquid phase and by stabilizing the tetragonal zirconia. Das and Banerjee obtained the best mechanical properties for the 5 mol% yttria (Y_2O_3)- and 2.5 mol% dysprosia (Dy_2O_3)-doped materials. Das and Banerjee mention that hardness, bending strength and fracture toughness of the materials are slightly increased by these additions. Ebadzadeh and Ghasemi (2000) produced fully reacted $ZrO_{2,P}$/mullite composites by reaction sintering of aluminum nitrate and zircon powders at 1600 °C. The as-produced zirconia particles are fine and homogeneously distributed inter- and intra-granularly.

Rocha-Rangel et al. (2000) prepared $ZrO_{2,P}$/mullite composites by spark plasma sintering of a mixture of zircon, aluminum and alumina at 1420 °C. Nearly full reaction and high density (>98%) are obtained after isothermal treatment at 1550 °C. The composites of Rocha-Rangel et al. display relatively high fracture toughness but only moderate elastic moduli. The mullitization-induced expansion of samples is high (about 10%), which causes microcracking. Rocha-Rangel et al. (2000) also prove that addition of titania enhances reactions by transient liquid formation with simultaneous decrease of the processing temperature (to 1460 °C for full mullitization). The toughness of these systems increases only slightly, owing to the low content of tetragonal zirconia. Khor et al. (2003) performed studies on spark plasma-sintered $ZrO_{2,P}$/mullite systems. They started from highly reactive plasma-spheroidized zircon and alumina powders. Mullitization following decomposition of zircon was initiated above 1000 °C and dominated the process at 1200 °C. The as-sintered material at 1300 °C exhibits high toughness. The combination of spheroidization of starting powders and spark plasma sintering is clearly an effective way to produce $ZrO_{2,P}$/mullite composites. Koyama et al. (1994) obtained irregularly shaped mullite and intra- and inter-granularly distributed zirconia particles after heat-treatment at 1520 °C, with the Al_2O_3 content of mullite still being markedly high. There is also evidence for some zirconium incorporation into these mullites (see Section 1.3.2).

Reaction bonding of aluminum metal, alumina and zircon   Lathabai et al. (1996) produced $ZrO_{2,P}$/composites by reaction bonding of aluminum metal, alumina and zircon. The aluminum metal first oxidizes to γ-alumina which transforms to α-alumina above 1100 °C. Zircon dissociation starts at about 1400 °C. The $ZrO_{2,P}$/mullite composites display high densities (about 95% of the theoretical value) and consist of mullite, tetragonal and monoclinic zirconia, and residual α-alumina. The $ZrO_{2,P}$/mullite composites show a relatively high strength, and approximately 80% of the room-temperature strength is retained up to 1200 °C. However, there is only little improvement of toughness of these composites in comparison to pure mullite.

Crystallization of rapidly quenched melts of the system ZrO_2–Al_2O_3–SiO_2   Yoshimura et al. (1990) prepared $ZrO_{2,P}$/mullite composites by rapid melt quenching. For that purpose a mixture of alumina, silica and zirconia powders is melted in a xenon arc imaging furnace, and droplets of the melt are rapidly quenched by a steel twin roller. The primarily obtained amorphous material with less than 10 wt. % ZrO_2

shows crystallization of mullite only, whereas material with 20 wt. % ZrO_2 produces both mullite and tetragonal zirconia at about 970 °C. The latter mullite is Al_2O_3-rich and contains significant amounts of ZrO_2 (see Section 1.3.2), probably due to the low crystallization temperature (see Section 2.5.1). The porosity of the composites can be diminished by hot pressing. Pilate and Cambier (1989, 1990) injected mixtures of zircon and alumina, or alumina, silica and zirconia powders into a nitrogen plasma. The molten particles were rapidly quenched in water, and $ZrO_{2,P}$/mullite composites have been obtained by reaction sintering from the transient amorphous materials.

Miscellaneous fabrication methods Marple and Green (1991) proposed an infiltration technique to prepare $ZrO_{2,P}$/mullite composites. Porous alumina bodies are infiltrated by tetraethyloxysilane (TEOS) and subsequently fired in order to obtain mullite. The same technique was previously applied by Yoshida et al. (1989). They used mullite porous bodies with 20 to 40 % open porosity, which were infiltrated with zirconium oxychloride ($ZrOCl_2$) solutions and fired at 1620 °C. Lin and Chen (2001a) cyclically infiltrated porous mullite ceramics and zirconia-mullite and SiC-mullite preforms with alcoholic solutions of tetraethyloxysilane (TEOS) and aluminum nitrate nonahydrate. A small reduction of the sintering shrinkage is achieved with this technique, but the density is little affected. The as-sintered ceramics exhibit two types of mullite: Relatively large crystals grown from the preforms and smaller crystals grown from infiltrants. Zirconia occurs inter-granularly at mullite grain boundaries and grain junctions. The effect of cyclic infiltration of zirconia preforms by mullite precursors has been investigated by Lin and Chen (2001b). Heat treatment of the infiltrated zirconia specimens achieves nearly full density. The newly formed mullite crystals are heterogeneously distributed and have largely elongated shapes. In spite of the microstructural evolution, the toughness of these materials is not significantly improved.

7.3.2.2 Mechanical Properties

The fracture toughness of zirconia particle-reinforced mullite tends to increase from about 2 to 3 MPa m$^{0.5}$ in monolithic mullite to about 3 to 5 MPa m$^{0.5}$ (Figs. 7.1.1 and 7.1.2). These toughness increases are rather weak, although higher values have also been reported: Wu and Lin (1991), Imose et al. (1998a,b), Das et al. (1998) and Das and Banerjee (2000) provide fracture toughnesses between 5 and 6 MPa m$^{0.5}$, while Khor et al. (2003) published values above 10 MPa m$^{0.5}$. The strength improvement of these materials is low, typically from about 200 to 450 MPa in monolithic mullite to about 200 to 550 MPa in mullite matrix composites, even at zirconia contents of 50 vol.%. In contrast to these earlier published values Imose et al. (1998a,b) measured strengths up to 780 MPa in systems produced from partially yttria stabilized zirconia and mullite.

Different toughening mechanisms have been proposed to be effective in $ZrO_{2,P}$ / mullite composites (Ruh et al. 1986, see also Schneider et al. 1994):

- Stress-induced toughening, due to the tetragonal to monoclinic phase transformation of zirconia
- Microcrack toughening, due to the thermal expansion mismatch of zirconia particles and mullite matrix
- Grain-boundary strengthening, due to metastable zirconia incorporation into the mullite matrix
- Surface-compression toughening, due to the presence of monoclinic zirconia at the surface of samples
- Strengthening, due to crack bowing at zirconia grains
- Toughening, due to crack deflection at zirconia grains, strengthened by thermal compression

Claussen and Jahn (1980) believe that stress-induced toughening is the major mechanism for $ZrO_{2,P}$/mullite composites. If so, the mechanical properties of these composites should improve with an increase in the zirconia fraction. Results compatible with this expectation have been reported by Shiga and Ismail (1990), Rundgren et al. (1990) and Nishiyama et al. (1991). By contrast, Kubota and Takagi (1986) reported the opposite trend, i.e. strength decreasing with increasing tetragonal zirconia fraction, while no clear relationship between mechanical properties and tetragonal zirconia fraction was reported by Yuan et al. (1986a) and Ruh et al. (1988).

No improvement of the mechanical properties is achieved in $ZrO_{2,P}$/mullite composites with respect to monolithic mullite at very high temperature, but the high strength level is maintained up to 1300 °C (Fig. 7.1.3; Ismail et al. 1991). The bending strength of $ZrO_{2,P}$/mullite materials rapidly decreases at high temperatures if a glassy phase is present at the grain boundaries of the mullite matrix. This is often the case in materials produced by reaction sintering of alumina and zircon. As reaction sintering is a slow process, the transiently produced silica glass can persist, enveloping individual mullite grains, and this may have a negative effect on the mechanical properties (Koyama et al. 1996).

Hamidouche et al. (1996) measured the temperature-dependent elastic moduli of $ZrO_{2,P}$/mullite composites. They mention that the elastic modulus of $ZrO_{2,P}$/mullite materials increases, whereas that of monolithic mullite decreases, with temperature. Kong et al. (1998) examined wear rates of $ZrO_{2,P}$/mullite composites in water and oil, in comparison to monolithic mullite. As expected, the wear resistance increases with the degree of zirconia reinforcement. Also the hardness of $ZrO_{2,P}$/mullite produced by reaction sintering from zircon plus alumina with addition of yttria and dysprosia has been increased considerably compared to the values of the undoped composites (10 GPa instead of 5.5 GPa, Das et al. 1998, Das and Banerjee, 2000).

The static fatigue behavior of a 10-vol.% $ZrO_{2,P}$/mullite composite was studied by Ashizuka et al. (1998) at 1200 °C. The fatigue parameter, which correlates with the fatigue resistance of the material, increases with the Al_2O_3 content of the mullite matrix in a similar way as in monolithic mullite ceramics. An explanation may be the lack of a glassy grain-boundary phase in Al_2O_3-rich composites. On the other hand, the fracture toughness increases with the decrease of the Al_2O_3 content of the matrix, due to the blunting of crack tips in the grain-boundary glassy phase.

The creep behavior of $ZrO_{2,P}$/mullite composites was investigated by Tkalcec et al. (1998), Descamps et al. (1991) and Ashizuka et al. (1991). However, a comparison of data turns out to be difficult, since various chemical compositions and different processing methods have been applied. Furthermore, the composites of Tkalcec et al. were possibly contaminated by the grinding media (magnesia-stabilized zirconia balls) during milling of the starting materials. Anyway, 10 vol.% $ZrO_{2,P}$/mullite composites display higher creep rates than monolithic mullite ceramics, and viscous glass phase grain-boundary sliding has been considered at temperatures above 1300 °C. Descamps et al. (1991), investigating the creep of titania- and magnesia-doped $ZrO_{2,P}$/mullite systems (see Section 7.3.2), suggested two different creep mechanisms: At high stresses and/or low temperatures, the major creep mechanism is microcracking, which has been derived from stress exponents ($n \approx 4$), activation energies of creep ($Q = 520$–680 kJ mol^{-1}) and from microscopic observation. At low stresses and/or high temperatures, creep more probably occurs by deformation, following a solution-precipitation reaction which accommodates grain boundary sliding ($n \approx 2$, $Q = 820$–1280 kJ mol^{-1}). Descamps et al. (1991) describe relatively low creep rates in magnesia-doped samples, which has been attributed to the microstructure with interlinked elongated mullite grains. Ashizuka et al. (1991) used mullite powders with addition of 10 vol.% zirconia, and determined higher creep rates of the composites than of monolithic mullite. The effects are attributed to stress relaxation at grain triple points by creep deformation of zirconia grains.

In summary, in spite of the innovative approach of toughening mullite matrices by zirconia phase transformations and the many attempts made, there is no real breakthrough in their mechanical behavior. New ideas and technologies are required to achieve higher strength and toughness, in order that this technique becomes competitive to continuous fiber-reinforced mullite matrix composites.

7.3.2.3 Thermal Properties

The presence of zirconia reduces the heat capacity (c_p) of $ZrO_{2,P}$/mullite composites compared to that of monolithic mullite (Fig. 7.3.8, Russell et al. 1996). The same is true for the thermal conductivity of $ZrO_{2,P}$/mullite (Russell et al. 1996). The thermal expansion of zirconia ($\approx 11 \times 10^{-6}$ °C^{-1}) is more than twice as high than that of mullite ($\approx 5.0 \times 10^{-6}$ °C^{-1}) and, as a consequence, the addition of zirconia increases the value for mullite matrix composites (see Table 7.3.1).

7.3.3
Silicon Carbide Platelet- or Particle-reinforced Composites

An alternative to zirconia toughening is the reinforcement of mullite matrices by silicon carbide platelets and particles on a micro- or nanometer scale (designated SiC$_P$/mullite).

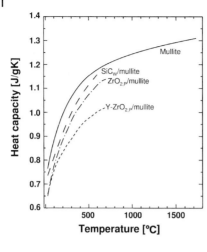

Fig. 7.3.8 Heat capacities of particle- and whisker-reinforced mullite matrix composites. SiC_W/mullite = silicon carbide whisker-reinforced mullite matrix composite; $ZrO_{2,P}$/mullite = zirconia particle-reinforced mullite matrix composite; Y-$ZrO_{2,P}$/mullite = yttria-stabilized zirconia particle-reinforced mullite matrix composite (data from Russell et al. 1996). Calculated data for monolithic mullite are given for comparison (Barin and Knacke, 1973).

Table 7.3.1 Linear thermal expansions of particle-reinforced mullite matrix composites.

Sample	Thermal expansion coefficient ($\times 10^{-6}\,°C^{-1}$)	Temperature range (°C)
Mullite (Reference)	5.5	50–1000
Mullite (Reference)	6.0	100–1000
Mullite/Al_2TiO_5 (50 vol.%)	5.3	25–1000
Mullite/Cordierite (40 vol.%)	5.0	50–1000
Mullite/Cordierite (74 vol.%)	3.6	50–1000
Mullite/Y (3 mol %)-ZrO_2 (30 vol.%)	8.7	100–1000
Mullite/Y_2O_3 (1.5 wt.%)-ZrO_2 (35.4 wt.%)	7.2	25–1000
Mullite/Y_2O_3 (1.5 wt.%)-ZrO_2 (35.4 wt.%)/SiC (30 wt.%)	6.2	25–1000

7.3.3.1 Fabrication Routes

Sakai et al. (1991) and Ando et al. (2001) produced SiC_p/mullite composites with nearly full density by hot pressing between 1500 and 1800 °C and pressures of 30 MPa. The starting powders used by Sakai et al. contained 30 vol.% silicon carbide platelets. Nischik et al. (1991) applied a colloidal processing route for production of SiC_P/mullite matrix composites and obtained a reduction in the amount of flaws together with a substantial improvement of the mechanical properties. Sacks et al. (1991) prepared SiC_P/mullite matrix composites with high density by transient viscous sintering (TVS) (see Section 4.2.6) at temperatures as low as 1300 °C.

A most important factor in improving the toughness of SiC_P/mullite composites is the interfacial bonding strength between silicon carbide particles and mullite matrix. Rezaie et al. (1999) prepared a 20 vol.% SiC_p/mullite matrix composite by hot pressing (1500 and 1600 °C, 5 and 15 MPa) and obtained a toughness increase of about 30% with respect to monophase mullite. Rezaie et al. believe that the reason for the toughening increase is crack deflection around silicon carbide platelets rather than crack bridging. Another successful fabrication route for SiC_P/mullite matrix composites is infiltration of porous silicon carbide preforms by aluminosilicate melts as proposed by Tian and Shobu (2003a,b): Dense composites are obtained in closed boron nitride (BN) crucibles above 1830 °C, which is necessary to prevent decomposition of the material into volatile silicon monoxide (SiO). The same papers also report infiltration of silicon carbide powder compacts with a melt of near mullite composition at 1900 °C. The 50 vol.% SiC_p/mullite composite exhibits very good oxidation resistance in air at 1500 °C (Fig. 7.3.9a,b).

Lin and Tsang (2003) prepared SiC_P/mullite composites from carbon black, colloidal silica, and alumina plus amorphous silica powder mixtures in a three-stage process: Formation of silicon carbide (1500 °C), removal of excess carbon (600 °C), and reaction sintering to mullite and silicon carbide (1500 °C). The density of these composites is low and drastically decreases with the amount of silicon carbide particles. Lin and Tsang believe that mullite formation is a main reason for the low densification, although oxidation of silicon carbide to silica with subsequent decomposition to volatile silicon monoxide (SiO) can also play an important role.

Lin et al. (1999, 2000) studied the oxidation behavior of SiC_p/mullite, SiC_P/$ZrO_{2,P}$/mullite and SiC_w/$ZrO_{2,P}$/mullite systems. They identified two different oxidation modes of silicon carbide (either particles or whiskers). The first mode, observed between 1000 and 1350 °C, involves strong oxidation of silicon carbide at the outer rims of specimens. The second mode, frequently observed in SiC_W/$ZrO_{2,P}$/mullite composites, partially oxidizes silicon carbide, but deeper into the composites. Obviously in the latter case oxygen diffusion is more effective in the matrix than through the silica layers formed by oxidation around silicon carbide particles and whiskers. These silica coatings, at least at low and moderate temperatures, act as diffusion barriers, thus retarding the ongoing oxidation process.

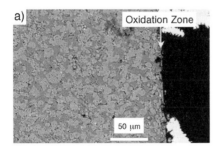 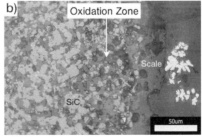

Fig. 7.3.9 Scanning electron micrographs of (a) 50 vol.% silicon carbide particle-reinforced mullite matrix composite (SiC_P/mullite), in comparison to (b) 50 vol.% silicon carbide particle-reinforced alumina matrix composite (SiC_P/alumina). Both materials are oxidized at 1500 °C in air. Note that the oxidation zone in SiC_P/alumina is much thicker than in SiC_P/mullite (Courtesy K. Shobu).

Park et al. (1998) processed silicon carbide and mullite in an argon plasma. They observed reactions between silicon carbide and mullite according to the equation:

$$3Al_2O_3 \cdot 2SiO_2(s) + SiC(s) \rightarrow \alpha\text{-}Al_2O_3(s) + 3\,SiO(g) + CO(g) \qquad (3)$$

mullite α-alumina

s = solid, g = gaseous

Since the gaseous reaction products are continuously removed from the system, there is a driving force for an ongoing decomposition of mullite, in spite of the fact that the reaction is endothermic. Therefore, plasma sintering is not suitable for the fabrication of SiC_P/mullite composites.

In recent years, reinforcement of mullite matrices with silicon carbide nanoparticles has become of interest (designated as SiC_{NP}/mullite). Takada et al. (1991) compacted mixtures of kaolinite plus alumina and of kaolinite plus silicon carbide nanoparticles at 1700 °C in nitrogen atmosphere. Warrier and Anilkumar (2001) produced dense SiC_{NP}/mullite composites by coating 5 vol.% silicon carbide nanoparticles (about 200 nm in diameter and 50–80 nm in thickness) with a seeded mullite precursor sol and annealed the compacts between 1450 and 1550 °C. Soraru et al. (2000) developed a route to process composites based on the "active filler-controlled pyrolysis (AFCOP)" method. Starting materials are alumina powders and silicon alkoxides ($MeSi(OEt)_3$). Above 900 °C the alumina-filled siloxane is transformed into a material consisting of alumina particles dispersed in a non-crystalline silicon oxycarbide (SiOC) matrix. This material has the advantage of densifying easily by viscous flow at about 1300 °C. Above 1350 °C the silicon oxycarbide glass decomposes into nano-sized silicon carbide particles and silica. Finally, mullitization occurs by reaction of alumina and silica above 1500 °C. In the as-prepared SiC_{NP}/mullite composite, silicon carbide nanoparticles (5 to 10 nm in size) are dispersed as intra-granular precipitates in individual mullite grains. Larger inter-granular grains (up to 10 nm in size) also occur at triple grain junctions (Fig. 7.3.10).

Spark plasma sintering was adopted for the preparation of SiC_{NP}/mullite composites by Gao et al. (2002). The authors dispersed 5 and 10 vol.% silicon carbide nanoparticles (about 70 nm in size) in a mullite precursor consisting of tetraethyloxysilane (TEOS) and aluminum chloride. Surpressed matrix grain coarsening and enhanced mass transport caused by electric discharge allowed nearly fully dense specimens to be achieved at temperatures as low as 1500 °C. Gao et al. (2002) found clean mullite-mullite and silicon carbide nanoparticle-mullite grain boundaries in their systems (Fig. 7.3.11). They ascribe this observation to the difference between the "mobility" of the glass phase and the reinforcing silicon carbide nanoparticles: The high mobility of the low viscosity glassy phase enables it to move with the sintering front advancing from the surface towards the interior of specimens. On the other hand, the low mobility of silicon carbide nanoparticles causes them to be left behind the reaction front and thus become entrapped in the mullite grains, producing an intra-granular-type SiC_{NP}/mullite composite. These

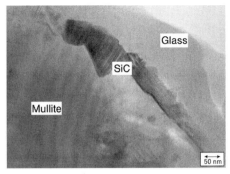

Fig. 7.3.10 Transmission electron micrograph of a silicon carbide nanoparticle-reinforced mullite matrix composite (SiC$_{NP}$/mullite) produced by pyrolysis at 1500 °C (1 h) of alumina particle-filled polymethylsiloxane gels. The picture shows mullite with a nanosized silicon carbide particle coexisting with a glassy phase (Courtesy H.-J. Kleebe).

a)
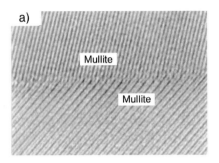

b)
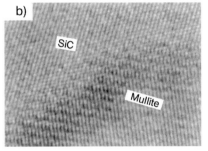

Fig. 7.3.11 High-resolution transmission electron micrographs of spark plasma-sintered silicon carbide nanoparticle-reinforced mullite matrix composite (SiC$_{NP}$/mullite). (a) Mullite/mullite grain boundary; (b) Mullite/silicon carbide particle grain boundary. Note that both grain boundaries are clean, without any interphase (Courtesy L. Gao).

nanocomposites display superior strength in comparison to monolithic mullite, but there is no significant improvement of toughness (see Section 4.3.1).

7.3.3.2 Mechanical Properties

Awaji et al. (2000) propose three major toughening mechanisms for particulate-dispersed ceramic composites:
– Frontal process-zone toughening (intrinsic type, example: Nanocomposites)
– Crack-bridging toughening (extrinsic type, example: Conventional composites)
– Micro- and macroscopic crack-deflection (example: Laminate composites with combination of tough and weak, e.g. dense and porous layers)

Intensive research has been performed on mullite matrix composites with nanosized silicon carbide particles, because of their high strength and creep resistance (Niihara, 1991). The concept of strengthening of nanocomposites is schematically depicted in Fig. 7.3.2. Studies have been carried out on mullite matrix composites with silicon carbide nanoparticles (Takada et al. 1991, Gao et al. 2002) and silicon

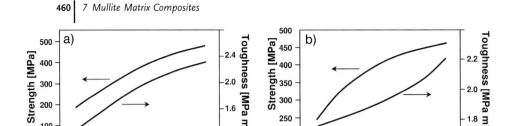

Fig. 7.3.12 Dependence of the mechanical properties of spark plasma-sintered silicon carbide nanoparticle-reinforced mullite matrix composite (SiC$_{NP}$/mullite) from the sintering temperature. (a) 5 vol.% SiC$_{NP}$/mullite; (b) 10 vol.% SiC$_{NP}$/mullite (after Gao et al. 2002).

carbide nanoparticles plus zirconia particles (Jin et al. 2002). Takada et al. (1991) give a bending strength of 500 MPa for a SiC$_{NP}$/mullite composite with 10 vol% SiC$_{NP}$ nanoparticles (Fig. 7.3.12, Gao et al. 2002), which is higher than the value for monolithic mullite (maximally 400 MPa). The fracture toughness of this material with 2.3 MPa m$^{0.5}$ is in the range of monolithic mullite ceramics. The bending strength of an 8 vol.% SiC$_{NP}$/mullite composite increases from about 400 MPa to about 540 MPa compared to monolithic mullite, while the fracture toughness increases from about 2.8 to 4.2 MPa m$^{0.5}$ in a 12 vol.% SiC$_{NP}$/mullite composite.

The bending strength of the nanocomposites can be further improved by addition of micrometer-sized silicon carbide particles. The studies have shown that nano-sized silicon carbide particle-reinforced mullite composites require a lower volume fraction of reinforcing compounds than microparticle-reinforced materials to achieve similar enhancement of the mechanical properties. This is explained by the higher number of individual particles (higher surface areas) in the latter case.

The crack-healing behavior of SiC$_P$/mullite composites has been examined by Tian and Shobu (2003b), Lin and Tsang (2003) and by Lin et al. (1999). Crack healing takes place by glass wetting of cracks. Crack healing effectively proceeds at 1300 °C in SiC$_p$/mullite composites, and is related to the softening of the glassy phase. Increases in strength from 400 to 580 MPa in a 15-vol.% silicon particle-dispersed composite (Lin and Tsang, 2003) and from 340 to 520 MPa in a 20-vol.% silicon carbide-dispersed composite (Lin et al. 1999) have been determined. Crack healing of SiC$_P$/mullite composites has also been investigated by Ando et al. (2001). A complete healing of cracks with diameters up to 200 μm at 1300 °C has been found. The crack-healed specimens exhibit higher strength than the as-cracked samples, owing to closure of cracks by oxidation-induced formation of viscous silica glass.

7.3.4
Miscellaneous Oxide Particle-reinforced Composites

A number of studies dealt with oxide particle-reinforced mullite composites (designated as oxide$_P$/mullite), which have improved mechanical properties or lower thermal expansion coefficients than monolithic mullite. This section draws on a selection of publications. Khan et al. (1998, 2000) studied the toughening of a 50-vol.% alumina particle-reinforced mullite matrix composite (designated as $Al_2O_{3,P}$/mullite) by addition of coarse-grained (>15 μm) and fine-grained (<2.5 μm) alumina agglomerates, with the diameter of agglomerates being about 50 μm in both cases. These $Al_2O_{3,P}$/mullite composites display improved toughness, which indicates high resistance of the material against crack propagation. Khan et al. (1998, 2000) derive the favorable mechanical properties of their $Al_2O_{3,P}$/mullite composites from the presence of grain-grain contacts ("unbroken grain bridges"). Ledbetter et al. (2001) measured the elastic moduli of alumina platelet-reinforced mullite ceramics using the dynamic method. Their composites were hot-pressed between 1500 and 1600 °C at 34.5 MPa. Instead of the expected stiffening the authors describe a softening of the composites with respect to monolithic mullite, with an elastic modulus slightly lower than that of monolithic mullite (220 instead of 230 GPa). Ledbetter et al. (2001) believe that this is due to grain-boundary gliding along the glassy phases enveloping individual mullite grains. A possible way to obtain improved $Al_2O_{3,P}$/mullite composites, therefore, is the use of matrices with Al_2O_3 bulk compositions higher than 72 wt. %, which are glass-free.

Functionally graded ceramics (FGCs) containing layers gradually changing from monophasic mullite to 20 wt. % alumina and 80 wt. % mullite were manufactured by Bartholome et al. (1998) by means of sequential slip casting and subsequent sintering at 1650 °C. The graded material exhibits lower strength than the monolithic sample of equal composition (graded material: about 180 MPa, monolithic mullite: about 230 MPa), but enhanced fatigue resistance. Tensile stresses at the surface of samples are responsible for the lower strengths, while compressive stresses within the bulk hinder crack propagation and thus improve the fatigue life of the material under cyclic conditions.

Specific catalytic convertors, microelectronic substrates or kiln furnitures require thermal expansion values lower than that of mullite. Such components can be designed by admixing mullite to refractory low-expansion oxide phases such as tieillite (Al_2TiO_5), cordierite ($Mg_2Al_4Si_5O_{18}$) or silica glass (SiO_2). Since these phases suffer from low strength, the addition of mullite does improve their mechanical properties. Huang et al. (1998, 2000) reaction-sintered mullite/tieillite composites (designated as Mu/tieillite) at 1600 °C. They worked with excess silica in the starting materials, thus enhancing densification by formation of a transient liquid phase at sintering temperature. Melendez-Martinez et al. (2001) measured the compressive deformation of Mu/tieillite composites. They used material reaction-sintered from alumina, titania and 10 wt. % mullite at 1460 °C with equiaxed, large tieillite grains and small mullite crystals, occurring at tieillite grain junctions.

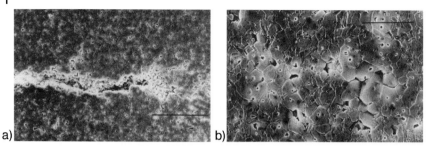

a) b)

Fig. 7.3.13 Scanning electron micrographs of a mullite-reinforced tieillite composite (Mu/tieillite) after compression (1400 °C, strain rate 4×10^{-4} sec^{-1}). (a) Single crack propagation; (b) Coalescence of cavities in stress direction (large grains are tieillite), (Courtesy J. J. Melendez-Martinez).

Melendez-Martinez et al. (2001) describe brittle and ductile deformation regimes in these composites (Fig. 7.3.13). The brittle deformation occurs at low strains and is characterized by inter-granular fracture. Typically crack growth by coalescence of cavities, nucleated at processing flaws, is observed. The ductile deformation regime is associated with high strain and diffusion-controlled grain boundary sliding. Singh and Low (2002) carried out depth profile analyses on phase composition and orientation of grains in an tieillite/mullite/alumina composite. They infiltrated porous alumina preforms with silica and titania sols and sintered the samples at 1600 °C. Owing to the incomplete infiltration of preforms a distinct gradation in phase abundance occurs. Near to the surface of samples, mullite and tieillite occur in a zone about 2-µm thick, while alumina stays unreacted in the interior of specimens. Singh and Low (2002) found a preferred orientation of mullite in the composites, which depends on the depth from the surface.

Ebadzadeh and Lee (1998) presented data on mullite-reinforced cordierite materials (designated as Mu/cordierite), where mullite is prepared from mixed starting powders and from sol-gel precursors. As expected, the strength increased with the mullite content from 120 MPa in composites with mullite/cordierite ratios of 10/90 to 150 MPa in 90/10 materials. Simultaneously the dielectric constant ε of the material changes from 5 to 7. Camerucci et al. (2001) used cordierite and mullite powders to prepare 30 wt. % Mu/cordierite composites. This material displays an elastic constant near 140 GPa, a hardness of 9 GPa, and a toughness of 2 MPa m$^{0.5}$. The studies indicate that mullite addition improves the strength but not the toughness of the composites. Mullite matrix composites with tieillite (thermal expansion: 1.8×10^{-6} °C^{-1}) and cordierite (thermal expansion: 2.2×10^{-6} °C^{-1}) have frequently been used to adjust the thermal expansion of electronic chip substrates to that of silicon (thermal expansion: 3.5×10^{-6} °C^{-1}; for comparison the thermal expansion of mullite is about 5.5×10^{-6} °C^{-1}, see, for example, Camerucci et al. 2001a,b). The same effect can be achieved by admixing mullite and silicate glass (thermal expansion e.g.: 1.21×10^{-6} °C^{-1}; see Ishihara 1999).

Very low thermal expansions have been measured for composites based on spodumene (LiAlSi$_2$O$_6$) /mullite and β-sialon (Si$_{6-x}$Al$_x$O$_x$N$_{8-x}$)/mullite (Yamuma et al. 2001, Yamagishi et al. 1990, Kobayashi et al. 1990). Composites with low thermal expansions have also been produced by admixing mullite and aluminosilicate and borosilicate glasses (e.g. Aoki et al. 1987). Data on other systems, such as titania-coated boron carbide (B$_4$C)/mullite (Zhao et al. 2003), microporous zeolite-coated mullite honeycombs (Komarneni et al. 1998) and magnetite/mullite (Morales et al. 2000), have also been published. The magnetite/mullite composite is interesting because it shows a high coercive coefficient of 500 Oe, which is maintained up to 1400 °C.

7.4
Metal-reinforced Mullite Matrix Composites
H. Schneider

Metal-reinforced mullite matrix composites (designated as Me/mullite, metals: Essentially aluminum, molybdenum) are new and innovative materials that have technical relevance as metal filters, catalyst supports, radiant burners and sensors. Metal-reinforced mullite matrix composites exhibit excellent crack-growth resistance effected by specific bridgings of the ductile metal grains, by which the stress intensity at the crack tip is reduced and crack growth is hindered. A serious drawback of Me/mullite materials are the low melting points of metals (normally below 1000 °C) and their poor resistance to oxidation. Another limiting factor can be the chemical incombatibility of metals and mullite. On the other hand, the incombatibility of aluminum melts with mullite has been successfully used to produce homogeneous aluminum-silicon alloy-reinforced alumina composites (designated as (Al,Si)$_{All}$/alumina).

7.4.1
Aluminum Metal-reinforced Composites

Composite fabrication has been carried out with the "reactive metal penetration technique" (e.g. Loehman et al. 1996). The reactive aluminum melt penetration of mullite substrates yields homogeneous aluminum/mullite and aluminum/alumina composites (designated as Al$_{Me}$/mullite and Al$_{Me}$/alumina, respectively). Processes are feasible, because the aluminum melt wets and penetrates the surface of porous and dense mullite preforms. Depending on the processing temperature either Al$_{Me}$/mullite (<900 °C) or Al$_{Me}$/alumina composites (>900 °C) are being formed. The overall reaction is:

$$(8 + x) \text{ Al} + 3 \text{ Al}_2\text{O}_3 \cdot 2\text{SiO}_2 \rightarrow 13 \text{ Al}_2\text{O}_3 + 6 \text{ Si} + x \text{ Al} \qquad (4)$$
$$\text{mullite} \qquad \qquad \alpha\text{-alumina}$$

x = Amount of aluminum in excess of that required to completely decompose mullite to alumina plus silicon.

As mullite decomposition proceeds, silicon diffuses away from the reaction front and out of the body, resulting in a material consisting mainly of aluminum and alumina (Fahrenholtz et al. 1998, Bandyopadhyay, 1999). Fahrenholtz et al. presented an overview of processes taking place during aluminum melt penetration of mullite performs. The authors immersed dense mullite preforms in a bath of molten aluminum at temperatures between 900 and 1300 °C in argon. Microstructures are shown in Fig. 7.4.1. Up to 1100 °C the reaction layer formed at the contact between aluminum melt and mullite grows with the duration of the experiment. However, the reaction slows down above 1150 °C, and finally stops at above about 1300 °C. This has been explained by the build-up of a silicon layer at the reaction front (Fahrenholtz et al. 1998). Obviously, at high temperature mullite decomposition and production of free silicon are much faster than the transport of silicon away from the reaction front, while the contrary is the case at lower temperatures. Transmission electron microscopy studies yield further evidence for this mechanism: At 900 °C, the reaction front is free of silicon, because its transport away from the reaction front is faster than decomposition of mullite (Fig. 7.4.1a). At 1100 °C, some free silicon occurs at the reaction front, although mullite decomposition is still going on (Fig. 7.4.1b). Finally, at 1300 °C large amounts of silicon precipitate appear, and the reaction zone becomes very thin (Fig. 7.4.1c).

Saiz et al. (1999) studied the reaction of mullite substrates with molten aluminum. In dense mullite substrates, the reaction between aluminum melt and mullite produces an $(Al,Si)_{All}$/mullite composite. The rate-controlling factor is the chemical reaction between aluminum and mullite, with maximum reaction rates between 1000 and 1200 °C. In the case of porous mullite preforms, a critical temperature has to be reached before pore infiltration starts. Saiz et al. (1999) suggest that the microstructure of the final $(Al,Si)_{All}$/alumina product can be tailored by the characteristics of the mullite substrate. Bandyopadhyay (1999) and Soundararajan et al. (2001) used the so-called "Fused deposition modeling (FDM)" technique to achieve completely infiltrated honeycomb mullite structures having three-dimensionally interconnected pores (Fig. 7.4.2). The pores are infiltrated by dipping the mold into an aluminum bath at 900 °C, and, after cooling, a $(Al,Si)_{All}$/mullite composite is obtained. Heat treatment of Al_{Me}/mullite composites above 1100 °C yields Al_{Me}/alumina composites (see above). Owing to the honeycomb structure, aluminum and silicon metal alloy and mullite or alumina form three-dimensional (3D) composites. The fabrication of Al_{Me}/mullite and $(Al,Si)_{All}$/alumina materials with the FDM technique allows control of the shape and microstructure of specimens. It is a good tool for the design of graded microstructures, in which one end is rich in aluminum and the other in mullite or alumina.

Moya et al. (1999) carried out sessile drop experiments at 800 °C on mullite with molten aluminum containing magnesium or magnesium plus copper. Mullite is completely attacked by these melts, which penetrate into the material along grain boundaries. The addition of magnesium or magnesium plus copper to the aluminum melt lowers the decomposition temperature of monophase mullite but not of $ZrO_{2,P}$/mullite composites. Moya et al. (1999) explain this observation by the relatively densely packed network of zirconia particles, with interparticle spaces as low

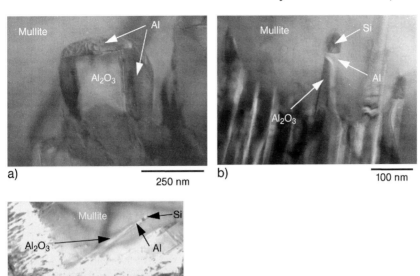

Fig. 7.4.1 Transmission electron micrographs of an aluminum metal-reinforced alumina matrix composite (Al$_{Me}$/alumina) produced by reactive penetration of aluminum melts into mullite. Reaction steps are shown at (a) 900 °C; (b) 1100 °C; (c) 1300 °C (Courtesy W. G. Fahrenholtz).

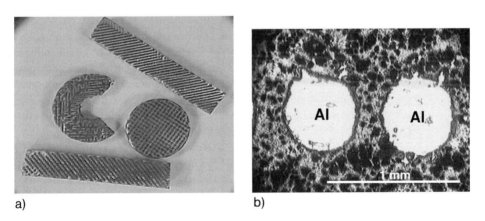

Fig. 7.4.2 Functionally designed aluminum metal-reinforced mullite matrix composites (Al$_{Me}$/mullite) processed via the "Fused deposition modeling (FDM)" process (Bandyopadhyay 1999). (a) Low magnification picture showing different Al$_{Me}$/mullite components; (b) High magnification picture showing infiltration of the pores of mullite by aluminum metal (Courtesy A. Bandyopadhyay).

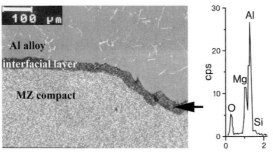

Fig. 7.4.3 Scanning electron micrograph of a zirconia particle-reinforced mullite composite ($ZrO_{2,P}$/mullite = MZ) after long-term attack at 800 °C by an aluminum alloy (Si 7.1, Mg 0.6, Ti 0.1, Fe 0.1 and traces of other metals; see Moya et al. 1999, Courtesy J. S. Moya).

as 1.5 μm, and low wetting of the melts. Both facts prevent penetration of the melt into the bulk of the mullite substrate. Moreover, a thin and dense spinel ($MgAl_2O_4$) layer forms, which is stable in the melt and hinders further melt impregnation (Fig. 7.4.3).

Reinforcement of mullite nanoparticles (67 nm) by nanocrystalline aluminum (27 nm) was described by Zhang et al. (2002). Samples are fabricated by hot isostatic pressing at 300 to 400 °C and pressures up to 1.5 GPa. These Al_{Me}/mullite composites need further improvement, since mullite particles tend to agglomerate, producing voids in the material. Zhang et al. showed that the aluminum, silicon and oxygen contents change linearly over the phase boundary of aluminum metal and mullite, and interpreted this result in terms of a diffusion-controlled process.

7.4.2
Molybdenum Metal-reinforced Composites

Molybdenum/mullite composites (designated as Mo_{Me}/mullite) have become of interest in recent years because of their actual and potential use for technical high-temperature electronic substrates, thermal barrier coatings and electrical conductor or insulator elements. A big advantage of molybdenum/mullite composites is the small difference of thermal expansion coefficients of molybdenum ($\alpha_{Mo} \approx 5.75 \times 10^{-6}$ °C^{-1}) and mullite ($\alpha_{Mu} \approx 5.5 \times 10^{-6}$ °C^{-1}), by which residual thermal stresses on rapid heating or cooling of the composites are avoided (Diaz et al. 2000, Bartholome et al. 2002). A further advantage is the low reactivity between molybdenum and mullite under non-oxidizing conditions at high temperatures. Bartholome et al. (1999) prepared Mo_{Me}/mullite composites starting from molybdenum and mullite powders, which are sintered at 1650 °C in vacuum and under reducing atmosphere. Depending on the molybdenum grain size and reaction atmosphere strengths and toughnesses up to about 530 MPa and 7 MPa m$^{0.5}$ have been achieved. During sintering in vacuum, a solid state de-wetting of molybdenum oxide (MoO_2) from the surface of the Mo particles takes place. Under reducing reaction conditions and in the absence of molydenum oxide the strength and toughness of the composites increase by a factor of 2 to 4 compared to monophase mullite.

Diaz et al. (2000) prepared 32 vol.% Mo_{Me}/mullite composites by wet techniques using water or alcohol. Water-based suspensions show pseudo-plastic behavior without any segregation, producing composites free of agglomerates at 1650 °C. By contrast, alcoholic suspensions processed in the same way display segregation and agglomeration. Significant improvement is achieved in the homogeneity of composites by controlling the deflocculant contents, solid loadings and suspension media. "Pulse electric current sintering (PECS)" is another technique to produce dense Mo_{Me}/mullite composites (Sivakumar et al. 2002). Favorable mechanical properties are obtained in molybdenum-poor composites owing to the fine-grained microstructure, producing only few critical flaws. On the other hand, the high strength of these composites is correlated to the network of interconnected ductile molybdenum particles. Pinning of molybdenum and mullite grains may reinforce effects. Frontal process zone toughening is the reason for the good pseudo-plasticity of the materials: A damage zone is created in front of the crack tip, by which the crack energy is partially consumed. Further effects are crack bridging and molybdenum particle pull-out. Similar results have been reported by Jayaseelan et al. (2002).

Tomsia et al. (1998) fabricated nearly dense graded Mo_{Me}/mullite composites at 1650 °C. The composites with 35 wt. % molybdenum show higher toughness than monophase mullite. These materials are electrical insulators if they contain less than 20 wt. % molybdenum, but change to electrical conductors in composites with higher molybdenum contents, owing to the occurrence of molybdenum-molybdenum contacts. Similar composites are described by Sivakumar et al. (2003).

References

Ando, K., Furusawa, K., Chu, M. C., Hanagata, T., Tuji, K. and Sato, S. (2001). Crack-healing behavior under stress of mullite/silicon carbide ceramics and the resultant fatigue strength. J. Am. Ceram. Soc. **84**, 2073–2078.

Aoki, S., Imanaka, Y., Kamehara, N. and Niwa, K. (1987). Crystallization of low firing glass-ceramics composite. In: Abstracts of the Annual Meeting of the Ceramic Society of Japan, Paper No. 3H08. Ceramic Society of Japan, Tokyo.

Asaumi, J., Enomoto, N., Naito, K., Yoshida, H. and Yamagishi, C. (1991). Mechanical properties of mullite-SiC fibre composite. In: Abstracts of the Annual Meeting of the Ceramic Society of Japan, Paper No. 3E25. Ceramic Society of Japan, Tokyo.

Ashizuka, M., Honda, T. and Kubota, Y. (1991). Creep in mullite ceramics containing zirconia. J. Ceram. Soc. Jpn. **99**, 357–360.

Ashizuka, M., Nakamura, S. and Kubota, Y. (1998). Fatigue behavior and fracture toughness of mullite containing ZrO_2 at 1200 °C. J. Ceram. Soc. Jpn. **106**, 460–464.

Awaji, H., Choi, S.-M., Ebisudani, T. and Jayaseelan, D. D. (2000). Toughening mechanisms of structural ceramics. J. Ceram. Soc. Jpn. **108**, 611–613.

Bandyopadhyay, A. (1999). Functionally designed 3-3 mullite-aluminum composites. J. Adv. Engin. Mater. **1**, 199–201.

Bannister, R. L., Ceruvu, N. S., Little, D. A. and McQuiggan, G. (1995). Development requirements for an advanced gas turbine system. Trans. ASME **117**, 724–733.

Barin, J. (1989). Thermochemical Data of Pure Substances. VCH, Weinheim, p. 61.

Barin, J. and Knacke, O. (1973). Thermochemical Properties of Inorganic Substances. Springer, Berlin, p. 41.

Bartholome, J. F., Moya, J. S., Requena, J., Llorca, J. and Anglada, M. (1998). Fatigue crack growth behavior in mullite/alumina functionally graded ceramics. J. Am. Ceram. Soc. **81**, 1502–1508.

Bartholome, J.F., Diaz, M., Requena, J., Moya, J.S. and Tomsia, A.P. (1999). Mullite/molybdenum ceramic-metal composites. Acta Mater. **47**, 3891–3899.

Bartholome, J.F., Diaz, M. and Moya, J.S. (2002). Influence of metal particle size on the crack growth resistance in mullite-molybdenum composites. J. Am. Ceram. Soc. **85**, 2278–2284.

Bhattacharjee, S., Singh, S.K. and Gagali, R.K. (2000). Preparation of zirconia toughened mullite by thermal plasma. Mater. Lett. **43**, 77–80.

Boccaccini, A.R. and Panton, C.B. (1995). Processing ceramic-matrix composites using electrophoretic deposition. J. Opt. Microsc. **47**, 34–37.

Boccaccini, A.R., Trusty, P.A., Taplin, D.M.R. and Panton, C.B. (1996). Colloidal processing of a mullite matrix material suitable for infiltrating woven fibre preforms using electrophoretic deposition. J. Europ. Ceram. Soc. **16**, 1319–1327.

Boccaccini, A.R., MacLaren, I., Lewis, M.H. and Panton, C.B. (1997). Electrophoretic deposition infiltration of 2-D woven SiC fibre mats with mixed sols of mullite composition. J. Europ. Ceram. Soc. **17**, 1545–1550.

Boccaccini, A.R., Kaya, C. and Chawla, K.K. (2001). Use of electrophoretic deposition in the processing of fibre reinforced ceramic and glass matrix composites: A review. Composites **A 32**, 997–1006.

Boch, P. and Chartier, T. (1991). Tape casting and properties of mullite and zirconia-mullite ceramics. J. Am. Ceram. Soc. **74**, 2448–2452.

Boch, P., Chartier, T. and Giry, J.P. (1990a). Zirconia-toughened mullite/The role of zircon dissociation. Ceramic Trans. **6**, 473–494.

Boch, P. Chartier, T. and Rodrigo, P.D.D. (1990b). High-purity mullite ceramics by reaction-sintering. Ceramic Trans. **6**, 353–374.

Brennan, J.J. and Prewo, K.M. (1982). Silicon-carbide-fiber-reinforced glass-ceramic composites exhibiting high strength and toughness. J. Mater. Sci. **17**, 2371–2383.

Camerucci, M.A., Urretavizcaya, G. and Cava-

lieri, A.L. (2001a). Mechanical behavior of cordierite and cordierite-mullite materials evaluated by indentation techniques. J. Europ. Ceram. Soc. **21**, 1195–1204.

Camerucci, M.A., Urretavizcaya, G., Castro, M.S. and Cavalieri, A.L. (2001b). Electrical properties and thermal expansion of cordierite and cordierite-mullite materials. J. Europ. Ceram. Soc. **21**, 2917–2923.

Carelli, E.A.V., Fujita, H., Yang, J.Y. and Zok, F.W. (2002). Effects of thermal aging on the mechanical properties of a porous-matrix ceramic composite. J. Am. Ceram. Soc. **85**, 595–602.

Chawla, K.K. (2003). Ceramic Matrix Composites. 2nd edition. Kluwer Academic Publishers, Boston, pp. 291–354.

Cinibulk, M.K. and Hay, R.S. (1996). Textured magneto plumbite fiber-matrix interphase derived from sol-gel fiber coatings. J. Am. Ceram. Soc. **79**, 1233–1246.

Claussen, N. and Jahn, J.W. (1980). Mechanical properties of sintered, in situ-reacted mullite-zirconia composites. J. Am. Ceram. Soc. **63**, 228–229.

Claussen, N. and Petzow, G. (1986). Whisker-reinforced oxide ceramics. J. Phys. **47**, 693–702.

Colomban, P. and Mazerolles, L. (1991). Nanocomposites in mullite-ZrO_2 and mullite-TiO_2 systems synthesized through alkoxide hydrolysis gel routes: Microstructure and fractography. J. Mater. Sci. **26**, 3503–3510.

Cooke T.F. (1991). Inorganic fibres – A literature review. J. Am. Ceram. Soc. **74**, 2959–2978.

Das, K. and Banerjee, G. (2000). Mechanical properties and microstructures of reaction sintered mullite-zirconia composites in the presence of an additive. J. Europ. Ceram. Soc. **20**, 153–157.

Das, K., Mukherjee, B. and Banerjee, G. (1998). Effect of yttria on mechanical and microstructural properties of reaction sintered mullite-zirconia composites. J. Europ. Ceram. Soc. **18**, 1771–1777.

Deng, Z.-Y. (1999). Effect of different fiber orientations on compressive creep behavior of SiC fiber-reinforced mullite matrix composites. J. Europ. Ceram. Soc. **19**, 2133–2144.

Deportu, G. and Henney, J.W. (1984). The microstructure and mechanical properties of mullite-zirconia composites. Trans. Brit. Ceram. Soc. **83**, 69–72.

Descamps, P., Sakaguchi, S., Poorteman, M. and Cambier, F. (1991). High-temperature characterization of reaction-sintered mullite-zirconia composites. J. Am. Ceram. Soc. **74**, 2476–2481.

Diaz, M., Bartholome, J. F., Requena, J. and Moya, J. S. (2000). Wet processing of mullite/molybdenum composites. J. Europ. Ceram. Soc. **20**, 1907–1914.

Ebadzadeh, T. and Ghasemi, E. (2000). Influence of starting materials on the reaction sintering of mullite-ZrO$_2$ composites. J. Mater. Sci. Eng. **A 283**, 289–297.

Ebadzadeh, T. and Lee, W. E. (1998). Processing-microstructure-property relations in mullite-cordierite composites. J. Europ. Ceram. Soc. **18**, 837–848.

Faber, K. T. and Evans, A. G. (1983). Crack deflection process – II. Experiment. Acta Metal. **31**, 577–584.

Fahrenholtz, W. G., Ewsuk, K. G., Loehman, R. E. and Lu, P. (1998). Kinetics of ceramic-metal composite formation by reactive metal penetration. J. Am. Ceram. Soc. **81**, 2533–2541.

Fujita, H., Jefferson, G., McMeeking, R. M. and Zok, F. W. (2004). Mullite/alumina mixtures for use as porous matrices in oxide fiber composites. J. Am. Ceram. Soc. **87**, 261–267.

Gao, L., Jin, X., Kawaoka, H., Sekino, T. and Niihara, K. (2002). Microstructure and mechanical properties of SiC-mullite nanocomposite prepared by spark plasma sintering. J. Mater. Sci. Eng. **A 334**, 262–266.

Gebhardt, R. A. and Ruh, R. (2001). Volume fraction and whisker orientation dependence of the electrical properties of SiC-whisker-reinforced mullite composites. J. Am. Ceram. Soc. **84**, 2328–2334.

Göring, J., Hackemann, S. and Schneider, H. (2003). Oxid/Oxid-Verbundwerkstoffe. Herstellung, Eigenschaften und Anwendungen. In: Krenkel, W. (ed.), Keramische Verbundwerkstoffe. Wiley-VCH, pp. 123–148.

Hamidouche, M., Bouaouadja, N., Osmani, H., Torrecillas, R. and Fantozzi, G. (1996). Thermomechanical behavior of mullite-zirconia composite. J. Europ. Ceram. Soc. **16**, 441–445.

Haslam, J. J., Berroth, K. E. and Lange, F. F. (2000). Processing and properties of an all-

oxide composite with porous matrix. J. Europ. Ceram. Soc. **20**, 607–618.

Hay, R. S. (2000). Monazite and scheelite deformation mechanisms. Ceram. Eng. Sci. Proc. **21**, 203–218.

Heathcote, J. A., Gong, X.-Y., Yang, J. Y., Ramamurty, U. and Zok, F. W. (1999). In-plane mechanical properties of an all-oxide ceramic composite. J. Am. Ceram. Soc. **82**, 2721–2730.

Hirano, S., Hayashi, T. and Kato, C. (1990). Preparation and evaluation of mullite/zirconia ceramics by colloidal filtration method. Funtai Funmatsuyakin **37**, 371–375.

Hirata, Y., Matsuda, M., Takeshima, K., Yamashita, R., Shibuya, M., Schmücker, M. and Schneider, H. (1996). Processing and mechanical properties of laminated composites of mullite/woven fabrics of Si–Ti–C–O fibers. J. Europ. Ceram. Soc. **16**, 315–320.

Hirata, Y., Matsura, T. and Hayata, K. (2000). Infiltration and pyrolysis of polytitanocarbosilane in a Si–Ti–C–O fabric/mullite porous composite. J. Am. Ceram. Soc. **83**, 1044–1048.

Hirata, Y., Yamashita, R. and Shibuya, M. (2001). Microstructures and crack extension in laminated Si–Ti–C–O fabrics/mullite matrix composites. J. Mater. Chem. Phys. **69**, 252–260.

Holmquist, M. G. and Lange, F. F. (2003). Processing and properties of a porous oxide matrix composite reinforced with continuous oxide fibers. J. Am. Ceram. Soc. **86**, 1733–1740.

Holmquist, M. G., Kristoffersson, A., Lundberg, R. and Adlerborn, J. (1997). Hot isostatically pressed silicon carbide fibre/mullite composites. In: Key Engineering Materials, **127–131**, pp. 239–246.

Holmquist, M. G., Hoffer, L., Kristoffersson, A. and Lundberg, R. (2001). Aluminium phosphate bonded oxide fibre reinforced porous mullite based matrix composites. In: Krenkel, W., Naslain, R. and Schneider, H. (eds.) High Temperature Ceramic Matrix Composites. Wiley-VCH, Weinheim, pp. 627–632.

Holmquist, M. G., Radsick, T. C., Sudre, O. H. and Lange, F. F. (2003). Fabrication and testing of all-oxide CFCC tubes. Composites **A 34**, 1–8.

Huang, Y. X., Senos, A. M. R. and Baptista, J. L.

(1998). Effect of excess SiO_2 on the reaction sintering of aluminium titanate-25 vol% mullite composites. Ceramics Intern. **24**, 223–228.

Huang, Y.X., Senos, A.M.R. and Baptista, J.L. (2000). Thermal and mechanical properties of aluminum titanate-mullite composites. J. Mater. Res. **15**, 357–363.

Imose, M., Ohta, A., Takano, Y., Yoshinaka, M., Hirota, K. and Yamaguchi, O. (1998a). Low-temperature sintering of mullite/ yttria-doped zirconia composites in the mullite-rich region. J. Am. Ceram. Soc. **81**, 1052–1062.

Imose, M., Takano, Y., Yoshinaka, M., Hirota, K. and Yamaguchi, O. (1998b). Novel synthesis of mullite powder with high surface aera. J. Am. Ceram. Soc **81**, 1537–1540.

Ishihara, S. (1999). Ph. D. Thesis. Tokyo Institute of Technology, Tokyo.

Ishitsuka, M., Sato, T., Endo, T. and Shimada, M. (1987). Sintering and mechanical properties of yttria-doped tetragonal ZrO_2 polycrystal/mullite composites. J. Am. Ceram. Soc. **70**, C 342–C 346.

Ismail, M.G.M.U., Shiga, H., Katayama, K., Nakai, Z., Akiba, T. and Somiya, S. (1991). Microstructure and mechanical properties of ZrO_2 toughened mullite synthesized by sol-gel method. In: Carlsson, R., Johansson and Kahlman, L. (eds.) Proc. 4th Intern. Symp.: Ceramic Materials and Components for Engines. Elsevier Applied Science, London, pp. 381–388.

Iwata, M., Oshima, K., Isoda, T., Arai, M., Nakano, K. and Kamiya, A. (1989). Mechanical properties and micro-structures of fibre reinforced silicon nitride and mullite composites. In: Abstracts of the Annual Meeting of the Ceramic Society of Japan, Paper No. 2E36. Ceramic Society of Japan, Tokyo.

Jayaseelan Doni, D., Rani, A., Nishikawa, T., Awaji, H. and Ohji, T. (2002). Sintering and microstructure of mullite-Mo composites. J. Europ. Ceram. Soc. **22**, 1131–1117.

Jin, X.H., Gao, L., Gui, L.H. and Gui, J.K. (2002). Microstructure and mechanical properties of SiC/zirconia-toughened mullite nanocomposites prepared from mixtures of mullite gel, 2Y-TZP, and SiC nanopowders. J. Mater. Res. **17**, 1024–1029.

Kamiaka, H., Yamagishi, C. and Asaumi, J. (1990). Mechanical properties and micro-

structure of mullite-SiC–ZrO_2 particulate composite. Ceramic Trans. **6**, 509–517.

Kanka, B.J. and Schneider, H. (2000). Aluminosilicate fiber/mullite matrix composites with favorable high-temperature properties. J. Europ. Ceram. Soc. **20**, 619–623.

Kanka, B.J., Schmücker, M. and Schneider, H. (2005). Nextel 720 fiber fabric compacts, a simple oxide/oxide composite: Processing and mechanical behavior. Unpublished results.

Kanka, B.J., Schmücker, M., Luxem, W. and Schneider, H. (2001a). Processing and microstructure of WHIPOX™. In: Krenkel, W., Naslain, R. and Schneider, H. (eds.) High Temperature Ceramic Matrix Composites. Wiley-VCH, pp. 610–615.

Kanka, B.J., Göring, J., Schmücker, M. and Schneider, H. (2001b). Processing, microstructure and properties of Nextel™ 610, 650 and 720 fiber/porous mullite matrix composites. Ceram. Eng. Sci. Proc. **22**, 703–710.

Kaya, C., Butler, E.G., Selcuk, A., Boccaccini, A.R. and Lewis, M.H. (2002). Mullite (Nextel™ 720) fibre-reinforced mullite matrix composites exhibiting favorable thermomechanical properties. J. Europ. Ceram. Soc. **22**, 2333–2342.

Kerans, R.J., Hay, R.S., Parthasarathy, T.A. and Cinibulk, M.K. (2002). Interface design for oxidation-resistant ceramic composites. J. Am. Ceram. Soc. **85**, 2599–2632.

Khan, A., Chan, H.M., Harmer, M.P. and Cook, R.F. (1998). Toughness curve behavior of an alumina-mullite composite. J. Am. Ceram. Soc. **81**, 2613–2623.

Khan, A., Chan, H.M., Harmer, M.P. and Cook, R.F. (2000). Toughening of an alumina-mullite composite by unbroken bridging elements. J. Am. Ceram. Soc. **83**, 833–840.

Khor, K.A., Yu, L.G., Li, Y., Dang, Z.L. and Munir, Z.A. (2003). Spark plasma sintering of ZrO_2-mullite composites from plasma spheroidized zircon/alumina powders. J. Mater. Sci. Engin. A **339**, 286–296.

Kitaoka, S., Kawashima, N., Komatsubara, Y., Yamaguchi, A. and Suzuki, H. (2005). Improved filtration performance of continuous alumina-fiber-reinforced mullite composites for hot-gas-cleaning. J. Am. Ceram. Soc. **88**, 45–50.

Kobayashi, H., Ishibashi, N., Akiba, T. and

Mitamura, T. (1990). Preparation and thermal expansion of mullite β-spodumene composite. J. Ceram. Soc. Jpn. **98**, 1023–1028.

Kojima, Y., Usuki, A., Kawasumi, M., Okada, A., Fukushima, Y., Kuranchi, T. and Kamigaito, O. (1993). Mechanical properties of nylon 6-clay hybrid. J. Mater. Res. **8**, 1185–1189.

Komarneni, S., Katsuki, H., Furuta, S. (1998). Novel honey comb structure: A microporous ZSM-5 and macroporous mullite composite. J. Mater. Chem. **8**, 2327–2329.

Kong, Y., Yang, Z., Zhang, G. and Tuan, Q. (1998). Friction and wear characteristics of mullite ZTM and TZP ceramics. Wear **218**, 159–166.

Kooner, S., Westby, W. S., Watson, C.M.A. and Farries, P.M. (2000). Processing of Nextel™ 720/mullite composition composite using electrophoretic deposition. J. Europ. Ceram. Soc. **20**, 631–638.

Koyama, T., Hayashi, S., Okada, K. and Otsuka, N. (1992). Mechanical properties of mullite/ZrO_2 composites prepared by various methods. In: Abstracts of the Annual Meeting of the Ceramic Society of Japan, Paper No. 3H06. Ceramic Society of Japan, Tokyo.

Koyama, T., Hayashi, S., Yasumori, A. and Okada, K. (1994). Preparation and characterization of mullite-zirconia composites from various starting materials. J. Europ. Ceram. Soc. **14**, 295–302.

Koyama, T., Hayashi, S., Yasumori, A., Okada, K., Schmücker, M. and Schneider, H. (1996). Microstructure and mechanical properties of mullite/zirconia composites prepared from alumina and zircon under various firing conditions. J. Europ. Ceram. Soc. **16**, 231–238.

Kubota, Y. and Takagi, H. (1986). Preparation and mechanical properties of mullites and mullite-zirconia composites. In: Howlett, S. P. and Taylor, D. (eds.) Special Ceramics, Vol. 8. Institute of Ceramics, Staffs, pp. 179–188.

Kumazawa, T., Ohta, S., Tabata, H. and Kanzaki, S. (1989). Mechanical properties of mullite-SiC whisker composites. J. Ceram. Soc. Jpn. **97**, 895–902.

Kumazawa, T., Ohta, S., Kanzaki, S. and Nakagawa, Z. (1990). Mechanical properties of mullite-cristobalite composite. In: Abstracts of the Annual Meeting of the Ceramic Society of Japan, Paper No. 1B15. Ceramic Society of Japan, Tokyo.

Kuntz, M. (1996). Rißwiderstand keramischer Faserverbundwerkstoffe. Shaker, Aachen, pp. 1–128.

Lange, F.F., Tu, W.C. and Evans, A.G. (1995). Processing of damage-tolerant, oxidation-resistant ceramic matrix composites by a precursor infiltration and pyrolysis method. J. Mater. Sci. Eng. **A 195**, 145–150.

Lange, F.F., Radsick, T.C. and Holmquist, M.G. (2001). Oxide/oxide composites: Control of microstructure and properties. In: Krenkel, W., Naslain, R. and Schneider, H. (eds.) High Temperature Ceramic Matrix Composites. Wiley-VCH, pp. 587–599.

Lathabai, S., Hay, D.G., Wagner, F. and Claussen, N. (1996). Reaction-bonded mullite/zirconia composite. J. Am. Ceram. Soc. **79**, 248–256.

Ledbetter, H., Kim, S., Dunn, M., Xu, Z., Crudele, S. and Kriven, W. (2001). Elastic constants of mullite containing alumina platelets. J. Europ. Ceram. Soc. **21**, 2569–2576.

Leriche, A. (1990). Mechanical properties and microstructures of mullite-zirconia composites. Ceramic Trans. **6**, 541–552.

Levi, C.G., Yang, J.Y., Dagleish, B.J., Zok, F.W. and Evans, A.G. (1998). Processing and performance of an all-oxide ceramic composite. J. Am. Ceram. Soc. **81**, 2077–2086.

Levi, C.G., Zok, F.W., Yang, J.-Y., Mattoni, M. and Lofvander, J.P.A. (1999). Microstructural design of stable porous matrices for all-oxide ceramic composites. Z. Metallk. **90**, 1037-1047.

Lewinsohn, C.A., Hellmann, J.R., Messing, G.L. and Amateau, M.F. (1994). Fabrication of silicon carbide whisker-reinforced mullite tubes via tape casting and hot isostatic pressing. J. Mater. Syn. Proc. **2**, 407–414.

Lin, C.-C., Zangvil, A. and Ruh, R. (1999). Modes of oxidation in SiC-reinforced mullite/ZrO_2 composites: Oxidation vs depth behavior. Acta Mater. **47**, 1977–1986.

Lin, C.-C., Zangvil, A. and Ruh, R. (2000). Phase evolution in silicon carbide-whisker-reinforced mullite/zirconia composite during long-term oxidation at 1000 to 1350 °C. J. Am. Ceram. Soc. **83**, 1797–1803.

Lin, Y.-J. and Chen, Y.-C. (2001a). Fabrication of mullite composites by cyclic infiltration and reaction sintering. J. Mater. Sci. Engin. **A 298**, 179–186.

Lin, Y.-J. and Chen, Y.-C. (2001b). Cyclic infiltration of porous zirconia preforms with a liquid solution of mullite precursor. J. Am. Ceram. Soc. **84**, 71–78.

Lin, Y.-J. and Tsang, C.-P. (2003). Fabrication of mullite/SiC and mullite/zirconia/SiC composites by 'dual' in-situ reaction synthese. Mater. Sci. Eng. **A 344**, 168–174.

Loehman, R. E., Ewsuk, K. and Tomsia, A. P. (1996). Synthesis of Al_2O_3–Al composites by reactive melt penetration. J. Am. Ceram. Soc. **97**, 27–32.

Low, I. M. and McPherson, R. (1989a). Crystallisation of gel-derived mullite-zirconia composites. J. Mater. Sci. **24**, 951–958.

Low, I. M. and McPherson, R. (1989b). Crystallisation of gels in the SiO_2–Al_2O_3–ZrO_2 system. J. Mater. Sci. **24**, 1648–1652.

Marple, B. R. and Green, D. J. (1991). Mullite/alumina particulate composites by infiltration processing: III, Mechanical properties. J. Am. Ceram. Soc. **74**, 2453–2459.

Mattoni, M. A., Yang, J. Y., Levi, C. G. and Zok, F. W. (2001). Effects of matrix porosity on the mechanical properties of a porous matrix, all-oxide ceramic composite. J. Am. Ceram. Soc. **84**, 2594–2602.

Mattoni, M. A. and Zok, F. W. (2004). Notch sensitivity of ceramic composites with rising fracture resistance. J. Am. Ceram. Soc. **87**, 914–922.

Mechnich, P. Schneider, H., Schmücker, M. and Saruhan, B. (1998). Accelerated reaction bonding of mullite (RBM). J. Am. Ceram. Soc. **81**, 1931–1937.

Mechnich, P., Schmücker, M. and Schneider, H. (1999). Reaction sequence and microstructural development of CeO_2-doped reaction-bonded mullite. J. Am. Ceram. Soc. **82**, 2517–2522.

Melendez-Martinez, J. J., Jimenez-Melendo, M., Dominguez-Rodriguez, A. and Wötting, G. (2001). High temperature mechanical behavior of aluminium titanite-mullite composites. J. Europ. Ceram. Soc. **21**, 63–70.

Meng, J., Cai, S., Yang, Z., Yuan, Q. and Chen, Y. (1998). Microstructure and mechanical properties of mullite ceramics

containing rodlike particles. J. Europ. Ceram. Soc. **18**, 1107–1114.

Miao, X., Scheppokat, S., Claussen, N. and Swain, M. V. (1998). Characterisation of an oxidation layer on reaction bonded mullite/zirconia composites by indentation. J. Europ. Ceram. Soc. **18**, 653–659.

Mizuno, M., Shiraishi, M. and Saito, H. (1988). Preparation and properties of mullite-ZrO_2 composites. In: Abstracts of the Annual Meeting of the Ceramic Society of Japan, Paper No. 3F29. Ceramic Society of Japan, Tokyo.

Morales, M. P., Gonzales-Carreno, T., Ocana, M., Alonso-Sanudo, M. and Serna, C. J. (2000). Magnetic iron oxide/mullite nanocomposite stable up to 1400 °C. J. Solid State Chem. **155**, 458–462.

Morgan, P. E. D. and Marshall, D. B. (1995). Ceramic composites of monazite and alumina. J. Am. Ceram. Soc. **78**, 1553–1563.

Moya, J. S. and Osendi, M. I. (1984). Microstructure and mechanical properties of mullite/ZrO_2 composites. J. Mater. Sci. **19**, 2909–2914.

Moya, J. S., Steier, H. P. and Requena, J. (1999). Interfacial reactions in aluminum alloys/mullite-zirconia composites. Composites **A 30**, 439–444.

Moya, S. (1990). Reaction sintered mullite-zirconia and mullite-zirconia-SiC ceramics. Ceramic Trans. **6**, 495–507.

Naslain, R., Dugne, O., Guette, A., Sevely, J., Robin-Brosse, C., Rocher, J. P. and Cotteret, J. (1991). Boron nitride interphase in ceramic-matrix composites. J. Am. Ceram. Soc. **74**, 2482–2488.

Newkirk, M. S., Lesher, H. D., White, D. R., Kennedy, C. R., Urquhart, A. W. and Claar, T. D. (1987). Preparation of Lanxide ceramic matrix composites: Matrix formation by the directed oxidation of molten metals. Ceram. Eng. Sci. Proc. **8**, 879–885.

Newman, B. and Schäfer, W. (2001). Processing and properties of oxide/oxide composites for industrial applications. In: Krenkel, W., Naslain, R. and Schneider, H. (eds.) High Temperature Ceramic Matrix Composites. Wiley-VCH, pp. 600–609.

Niihara, K., Hirano, T., Nakahara, A., Ojima, K., Izaki, K. and Kawakami, T. (1989). High-temperature performance of Si_3N_4–SiC composites from fine, amorphous Si–

C–N powder. In: Doyama, M., Somiya, S. and Chang. R.P.H. (eds.) Proceedings of MRS International Meeting on Advanced materials 5, Structural Ceramics Fracture Mechanics. Materials Research Society, Pittsburgh, pp. 197–112.

Niihara, K. (1991). New design concept of structural ceramics – ceramics nanocomposites. J. Ceram. Soc. Jpn. **99**, 974–982.

Nischik, C., Seibold, M.M., Travitzky, N.A. and Claussen, N. (1991). Effect of processing on mechanical properties of platelet-reinforced mullite composites. J. Am. Ceram. Soc. **74**, 2464–2468.

Nishiyama, A., Sasaki, T., Sasaki, H., Hamano, K. and Okada, S. (1991). Influence of ZrO_2 grain size on the mechanical properties of mullite/ZrO_2 composites. In: Abstracts of the Annual Meeting of the Ceramic Society of Japan, Paper No. 3E24. Ceramic Society of Japan, Tokyo.

Nixon, R.D., Chevacharvenkul, S., Davis, R.F. and Tiegs, T.N. (1990). Creep of hot-pressed SiC-whisker-reinforced mullite. Ceramic Trans. **6**, 579–603.

Nubian, K., Saruhan, B., Schmücker, M., Schneider, H. and Wahl, G. (2000). Chemical vapor deposition of ZrO_2 and C/ZrO_2 on mullite fibers for interfaces in mullite/mullite fibre-reinforced composites. J. Europ. Ceram. Soc. **20**, 537–543.

Okada, K. and Otsuka, N. (1991). Synthesis of mullite whiskers and their application in composites. J. Am. Ceram. Soc. **74**, 2414–2418.

Orange, G., Fantozzi, G., Cambier, F., Leblud, C., Anseau, M.R. and Leriche, A. (1985). High temperature mechanical properties of reaction-sintered mullite/zirconia and mullite/alumina/zirconia composites. J. Mater. Sci. **20**, 2533–2540.

Park, Y., McNallam, M.J. and Butt, D.P. (1998). Endothermic reactions between mullite and silicon carbide in an argon plasma environment. J. Am. Ceram. Soc. **81**, 233–236.

Parks, W.P., Ramey, R.R., Rawlins, D.C., Price, J.R. and Van Roode, M. (1991). Potential applications of structural ceramic composites in gas turbines. J. Eng. Gas Turbines Power **113**, 628–634.

Parthasarathy, T.A., Keller, K.A., Mah, T.-I., Kerans, R.J. and Butner, S. (2004). Reduction of thermal-gradient-induced stresses in composites using mixed fibers. J. Am. Ceram. Soc. **87**, 617–625.

Peters, P.W.M., Daniels, B., Clemens, F. and Vogel, W.D. (2000). Mechanical characterization of mullite-based ceramic matrix composites at test temperatures up to 1200 °C. J. Europ. Ceram. Soc. **20**, 531–535.

Pilate, P. and Cambier, F. (1989). Preparation par hypertrempe de poudres mixtes destinées a la synthèse de composites a dispersion de particules. Silicates Ind. **5–6**, 89–94.

Pilate, P. and Cambier, F. (1990). Ultra rapid quenching: A new route to produce high temperature ceramic composites. In: Davidge, R.W. and Thompson, D.P. (eds.) Brit. Ceramic Proc. **45**, Fabrication Technology. Institute of Ceramics, Staffs, pp. 71–78.

Prochazka, S., Wallace, J.S. and Claussen, N. (1983). Microstructures of sintered mullite-zirconia composites. J. Am. Ceram. Soc. **66**, C-125–C-127.

Rahaman, M.N. and Jeng., D.-Y. (1990). Sintering of mullite and mullite matrix composites. Ceramic Trans. **7**, 753–766.

Rezaie, H.R., Rainforth, W.M. and Lee, W.E. (1999). Fabrication and mechanical properties of SiC platelet reinforced mullite matrix composites. J. Europ. Ceram. Soc. **19**, 1777–1787.

Rincon, J.M. and Romero, M. (2000). Characterization of mullite/ZrO_2 toughness ceramic materials by medium voltage analytical electron microscopy. J. Mater. Charact. **45**, 117–123.

Rincon, J.M., Moya, J.S. and Demelo, M.F. (1986). Microstructural study of toughened ZrO_2/mullite ceramic composites obtained by reaction sintering with TiO_2 additions. Trans. Brit. Ceram. Soc. **85**, 201–206.

Rocha-Rangel, E., De la Torre, S.D. and Balmori-Ramirez, H. (2000). Production of ZrO_2-reinforced mullite ceramics by spark plasma sintering and isothermal reaction. Ceramic Trans. **115**, 285–292.

Ruh, R., Mazdiyasni, K.S. and Mendiratta, M.G. (1986). Mechanical and microstructural characterisation of mullite and mullite-SiC-whisker and ZrO_2-toughened-mullite-SiC-whisker composites. J. Am. Ceram. Soc. **71**, 503–512.

Rundgren, K., Elfving, P., Tabata, H. and Kanzaki, S. (1990). Microstructures and mechanical properties of mullite-zirconia composites made from inorganic sols and salts. Ceramic Trans. **6**, 553–566.

Russell, L. M., Donaldson, K. Y., Hasselman, D.P.H., Ruh, R. and Adams, J. W. (1996). Thermal diffusivity/conductivity and specific heat of mullite-zirconia-silicon carbide whisker composites. J. Am. Ceram. Soc. **79**, 2767–2770.

Sacks, M. D., Bozkurt, N. and Scheiffele, G. W. (1991). Fabrication of mullite and mullite-matrix composites by transient viscous sintering of composite powders. J. Am. Ceram. Soc. **74**, 2828–2437.

Saiz, E., Foppiano, S., Moberly Chan, W. J. and Tomsia, A. P. (1999). Synthesis and processing of ceramic-metal composites by reactive metal penetration. Composites **A 30**, 399–403.

Sakai, H., Matsuhiro, K. and Furuse, Y. (1991). Mechanical properties of SiC platelet reinforced ceramic composites. Ceramic Trans. **19**, 765–771.

Saruhan, B. (2003). Oxide-Based Fiber-Reinforced Ceramic-Matrix Composites-Principles and Materials. Kluwer, Boston-Dordrecht-London, pp. 1–199.

Saruhan, B., Schmücker, M., Bartsch, M., Schneider, H., Nubian, K. and Wahl, G. (2001a). Effect of interphase characteristics on long-term durability of oxide-based fiber-reinforced composites. In: Chawla, K. K., Mortensen, A. and Månson, J.-A. E. (eds.) Processing of Fibers and Composites. Composites **A 32**, 1095–1103.

Saruhan, B., Bartsch, M., Schmücker, M., Schneider, H., Nubian, K. and Wahl, G. (2001b). Correlation of high-temperature properties and interphase characteristics in oxide/oxide fiber-reinforced composites. Intern. J. Mater. Prod. Techn. **16**, 259–268.

Schmücker, M. and Schneider, H. (2004). WHIPOX all oxide/oxide ceramic matrix composites. In: Bansal, N. P. (ed.) Handbook of glass and ceramics. Kluwer, Dordrecht, pp. 423–436.

Schmücker, M., Schneider, H., Chawla, K. K., Xu, R. and Ha, J. S. (1997). Thermal degradation behavior of BN fiber coatings. J. Am. Ceram. Soc. **80**, 2136–2140.

Schmücker, M., Kanka, B. J. and Schneider, H. (2000). Temperature-induced fibre/matrix interactions in porous alumino silicate

ceramic matrix composites. J. Europ. Ceram. Soc. **20**, 2491–2497.

Schmücker, M., Grafmüller, A. and Schneider, H. (2003). Mesostructure of WHIPOX all oxide CMCs. Composites **A 34**, 613–622.

Schneider, H., Saruhan, B., Voll, D. Merwin, L. and Sebald, A. (1993). Mullite precursor phases. J. Europ. Ceram. Soc. **11**, 87–94.

Schneider, H., Okada, K., Pask, J. A. (1994). Mullite and Mullite Ceramics. Wiley, Chichester, pp. 1–251.

Schneider, H., Schmücker, M., Göring, J., Kanka, B., She, S. and Mechnich, P. (2000). Porous alumino silicate fiber/mullite matrix composites: Fabrication and properties. In: Bansal, N. P. and Singh, J. P. (eds.) Innovative Processing and Synthesis of Ceramics, Glasses and Composites IV. Am. Ceram. Soc., 415–434.

She, J. H., Mechnich, P., Schneider, H., Kanka, B. and Schmücker, M. (2000). Infiltration behaviors of porous mullite/mullite preforms in aluminum-chloride solutions. J. Mater. Sci. Lett. **19**, 1887–1891.

She, J. H., Mechnich, P., Schneider, H., Schmücker, M. and Kanka, B. J. (2002). Effect of cyclic infiltrations on microstructure and mechanical behavior of porous mullite/mullite composites. J. Mater. Sci. Eng. **A 325**, 19–24.

Shiga, H. and Ismail, M.G.M.U. (1990). Synthesis of MgO doped mullite/ZrO_2 composite powders by sol-gel method and its sintering. In: Abstracts of the 3rd Autumn Symposium on Ceramics, Paper No. 6-1A02. Ceramic Society of Japan, Tokyo.

Shiga, H., Ismail, M.G.M.U. and Katayama, K. (1991). Sintering of ZrO_2 toughened mullite ceramics and its microstructure. J. Ceram. Soc. Jpn. **99**, 798–802.

Singh, M. and Low, I. M. (2002). Depth-profiling of phase composition and preferred orientation in a graded alumina/mullite/aluminium-titanate hybrid using X-ray and synchrotron radiation diffraction. Mater. Res. Bull. **37**, 1279–1291.

Sivakumar, R., Jayaseelan Doni, D., Nishikawa, T., Honda, S. and Awaji, H. (2002). Mullite-molybdenum composites fabricated by pulse electric current sintering technique. J. Europ. Ceram. Soc. **22**, 761–768.

Sivakumar, R., Nishikawa, T., Honda, S., Awaji, H. and Gnanam, F. D. (2003). Processing of mullite-molybdenum graded hollow cylinders by centrifugal melting technique. J. Europ. Ceram. Soc. **23**, 765–772.

Soraru, G. D., Kleebe, H.-J., Ceccato, R. and Pederiva, L. (2000). Development of mullite-SiC nanocomposites of filled polymethylsiloxane gels. J. Europ. Ceram. Soc. **20**, 2509–2517.

Soundararajan, R., Kuhn, G., Atisivan, R., Bose, S. and Bandyopadhyay, A. (2001). Processing of mullite-aluminum composites. J. Am. Ceram. Soc. **84**, 509–513.

Suzuki, H. and Saito, H. (1988). Preparation of precursor powders of cordierite-mullite composites from metal alkoxides and its sintering. J. Ceram. Soc. Jpn. **96**, 659–665.

Suzuki, H. and Saito, H. (1990). Preparation and sintering of fine composite precursors of mullite-zirconia by chemical copolymerisation of metal alkoxides. J. Mater. Sci. **25**, 2253–2258.

Takada, H., Nakahira, A., Niihara, K., Ohnishi, H. and Kawanami, T. (1991). High temperature properties of mullite/SiC nanocomposites. In: Abstracts of the 29th Symposium on the Basic Science of Ceramics, Paper No. 1B21. Ceramic Society of Japan, Tokyo.

Tian, J. and Shobu, K. (2003a). Fabrication of silicon carbide-mullite composite by melt infiltration. J. Am. Ceram. Soc. **86**, 39–42.

Tian, J. and Shobu, K. (2003b). Improvement of the oxidation resistance of oxide-matrix silicon carbide-particulate composites by mullite infiltration. J. Am. Ceram. Soc. **86**, 1806–1808.

Tkalcec, E., Nass, R., Krajewski, T., Rein, R. and Schmidt, H. (1998). Microstructure and mechanical properties of slip cast sol-gel derived mullite ceramics. J. Europ. Ceram. Soc. **18**, 1089–1099.

Tian, J. and Shobu, K. (2003a). Fabrication of silicon carbide-mullite composites by melt infiltration. J. Am. Ceram. Soc. **86**, 39–42.

Tian, J. and Shobu, K. (2003b). Improvement of the oxidation resistance of oxide-matrix silicon carbide-particulate composites by mullite infiltration. J. Am. Ceram. Soc. **86**, 1806–1808.

Tomsia, A. P., Saiz, E., Ishibashi, H., Diaz, M., Requena, J. and Moya, J. S. (1998). Powder processing of mullite/Mo functionally graded materials. J. Europ. Ceram. Soc. **18**, 1365–1371.

Tu, W.-C., Lange, F. F. and Evans, A. G. (1996). Concept for a damage-tolerant ceramic composite with strong interfaces. J. Am. Ceram. Soc. **79**, 417–424.

Wallace, J. S., Petzow, G. and Claussen, N. (1984). Microstructure and property development of in situ-reacted mullite-ZrO_2 composites. Adv. Ceram. **12**, 436–442.

Warrier, K.G.K. and Anilkumar, G. M. (2001). Densification of mullite-SiC nanocomposite sol-gel precursors by pressureless sintering. J. Mater. Chem. Phys. **67**, 263–266.

Wei, G. C. and Becher, P. F. (1985). Development of SiC-whisker-reinforced ceramics. Amer. Ceram. Soc. Bull. **64**, 298–304.

Westby, W. S., Kooner, S., Farries, P. M., Boother, P. and Shatwell, R. A. (1999). Processing of Nextel 720/mullite composition composite using electrophoretic deposition. J. Mater. Sci. **34**, 5021–5031.

Wu, J.-M. and Lin, C.-M. (1991). Effect of CeO_2 on reaction-sintered mullite-ZrO_2 ceramics. J. Mater. Sci. **26**, 4631–4636.

Wu, M., Messing, G. L. and Amatean, M. F. (1991). Laminate processing and properties of oriented SiC whisker-reinforced composites. Ceramic Trans. **19**, 665–675.

Wu, M. and Messing, G. L. (1994). Fabrication of oriented SiC-whisker-reinforced mullite matrix composites by tape casting. J. Am. Ceram. Soc. **77**, 2586–2592.

Yamade, Y., Kawaguchi, Y., Takeda, N., Kishi, T. (1995). Interfacial behavior of mullite/SiC continuous fiber composite. J. Am. Ceram. Soc. **78**, 3209–3216.

Yamagishi, C., Yoshida, H., Kamiaka, H., Asaumi, J. and Miyata, N. (1990). Fabrication of mullite-sialon composites. Funtai Funmatsuyakin **37**, 362–266.

Yamuma, A., Devanarayanan, S. and Lalithambik, M. (2001). Mullite-β-spodumene composites from aluminosilicates. J. Am. Ceram. Soc. **84**, 1703–1709.

Yaroshenko, V. and Wilkinson, D. S. (2001). Sintering and microstructure modification of mullite/zirconia composites derived from silica-coated alumina powders. J. Am. Ceram. Soc. **84**, 850–858.

Yoon, C. K. and Chen, I.-W. (1990). Superplastic flow of mullite-zirconia composites. Ceramic Trans. **6**, 567–577.

Yoshida, H., Asaumi, J., Miyata, N. and Yamagishi, C. (1989). Surface modification and mechanical properties of mullite by partial infiltration of zirconia. In: Meeting Abstracts of the 1st International Ceramic Science and Technology Congress, Paper No. 16-SVIP-89C. American Ceramic Society, Westerville.

Yoshimura, M., Hanaue, Y. and Somiya, S. (1990). Non-stoichiometric mullites from Al_2O_3–SiO_2–ZrO_2 amorphous materials by rapid quenching. Ceramic Trans. **6**, 449–456.

Yuan, Q.-M., Tan, J.-Q. and Jin, Z.-G. (1986a). Preparation and properties of zirconia toughened mullite ceramics. J. Am. Ceram. Soc. **69**, 265–267.

Yuan, Q.-M., Tan, J.-Q., Shen, J.-Y., Zhu, S.-H. and Yand, Z.-F. (1986b). Processing and microstructure of mullite-zirconia composites prepared from sol-gel powders. J. Am. Ceram. Soc. **69**, 268–269.

Zawada, L. P., Hay, R. S., Lee, S. S. and Staehler, J. (2003). Characterization and high-temperature mechanical behavior of an oxide/oxide composite. J. Am. Ceram. Soc. **86**, 981–990.

Zhang, H., Maljkovic, N. and Mitchell, B. S. (2002). Structure and interfacial properties of nanocrystalline aluminum/mullite composites. J. Mater. Sci. Engin. **A 326**, 317–323.

Zhao, H., Hiragushi, K. and Mizota, Y. (2003). Densification and mechanical properties of mullite/TiO_2-coated B_4C composites. J. Europ. Ceram. Soc. **23**, 1485–1490.

Zhien, L., Jianjun, Y., Jiehua, W. (1995). Mechanical properties of SiC_w/mullite composites doped with LAS glass-ceramics. J. Mater. Sci. Lett. **14**, 190–191.

Index

a

acrylamide, monomers 271
activation energy 103f, 108ff, 126
– of creep 320
adamite 34f
adelite 40
aerosol decomposition technique 281
aircraft engine 440f
^{27}Al MAS NMR spectroscopy 200
– solid-state 268
– spectrum 194
– triple-quantum spectra 199
^{27}Al NMR spectroscopy 97, 116, 121, 123, 269
– solid-state 268
albite 389
alkali aluminates 4, 29, 33
– mullite-type 33
alkali gallate 4, 33
Al_2O_3-rich mullite 356
Al_2O_3-SiO_2 glasses, tetrahedral tricluster 124
Al_2O_3-SiO_2 phase diagram 227ff
– simulation 236
Altex fibers 384
alumina 448
– decomposition 182
α-alumina 31, 233, 252
– crystals 234f
– liquidus 231, 233, 237
– nucleation 234
– particles 232
– porous layers 240
– skeleton 369
ι-alumina 28
alumina microcomposite, silica-coated 448
alumina particle reinforcement 461
alumino siloxane 272
aluminosilicate
– homogeneous 232
– non-crystalline 94

aluminosilicate fiber compacts
– porous 412
aluminosilicate fibers 381, 392
aluminosilicate gels 100f
aluminosilicate glass 238
aluminosilicate glass fibers
– activation energy 103f
– nucleation and growth 104
– viscosity 103
aluminosilicate phases 242
aluminosilicate powder 232
– heterocoagulation 276
aluminum, fivefold coordinated 101
aluminum borate 76
– mullite-structured 203
aluminum carboxylates 383
aluminum dibutoxide ethylacetoacetate 272
aluminum fluoride 252, 292
aluminum hydroxide 252f, 256, 269, 274, 277, 281, 299, 308
aluminum hydroxide chloride 415
aluminum isopropoxide 381
aluminum metal 447
aluminum metal reinforcement 463ff
– reactive metal penetration 463
aluminum oxygen 178
aluminum phosphate hydroxide 415
aluminum sec-butoxide 382
aluminum-silicon disorder 151
aluminum tris-isopropoxide, synthesis 264
andalusite 2, 4, 7, 19, 33f, 46, 101, 151f, 174ff, 180, 252
– decomposition processes 179
– mullite transformation 176f
anorthite 391
antiferromagnetism 37
apuanite 7, 14f, 18
– group 14
aristotype 1ff, 9, 12
arsenate 33

arsen-descloizite 41
atomic coordination 122f
atomic diffusion 156
Auger electron spectroscopy (AES) 76
austinite 41
average structure 57, 60

b
ball milling 95, 148
Bärnighausen tree 5, 7, 9, 12, 37
barrier coatings 369
bauxite 252
bending strength 257, 298, 308ff.
$Bi_2M_4O_9$ group
– edge-shared MO_6 octahedra 14
– MO_4 tetrahedra 14
– MO_5 square-pyramidal polyhedra 14
binder 418
blunting of crack tips 455
B_2O_3 109
Boltzmann-Matano interface 231
bond lengths 149
bond strength 149
boralsilite 7, 33, 42ff
– group 42
boron aluminates 4, 33, 37ff, 43, 46, 181
– $Al_{6-x}B_xO_9$ 43
Brillouin scattering 146
brittle failure 428, 436
bulk modulus 312
burner rigs 367

c
calciovolborthite 41
calcium oxide 254, 261
calorimetry 151
catalyst supports 398
catalytic convertors 442
ćechite 38
ceramic belts 301
ceramic liners in aircraft 441
ceramic matrix composites
– graphite 437
– interphase 437
– porous matrices 426
– turbostratic carbon 437
ceria (CeO_2) 164, 410
chains 178
charge-transfer (CT) reactions 167
chemical dipping 436
chemical resistance 324
chemical vapor deposition (CVD) 283, 290, 336, 350
-- coatings 350ff

– metalorganic (MOCVD) 351
chemical-mullite 32, 256
chromium clusters 77
chromium-doped mullite
– Rietveld refinement 81
chromium-doped-mullites 77
cleavable oxides
– hexaaluminates 438
– magnetoplumbites 438
– phyllosilicates 438
coatings 368
– ceramic 337
– environmental barrier 349
– flame-sprayed 367, 372
– fugitive 411, 438
– functionally graded 357
– multi-layered environmental barrier 371
– oxidation barriers 364
– plasma-sprayed 369
– plasma-sprayed thermal barrier 372
– protective 337, 373
– thermal shock resistance 357
cobalt-doped mullite 89
compressive stress 143
congruent melting 227ff
conichalcite 41
continous fiber formation 381ff
continous fiber reinforcement
– fabric processing 405
– fiber compacts without matrix 412f
– matrix fillers 413
– mechanical properties 421ff
– non-oxide composites 423ff
– oxide/oxide composites 425ff
– slurry infiltration of bundles 417ff
– slurry injection 414
– winding technique 404
contrast simulation (HREM) 49ff
copper arsenate 35f
corrosion resistance 327, 443
crack bridging 457
crack deflection 422, 436, 457
crack resistance 432
creep 141, 317f
– resistance 423
cristobalite 252, 366
critical thermal shock 315
crystal field spectra, unpolarized 78
crystal growth techniques
– Czochralski method 259
– solution-sol-gel process 259
crystal structures
– mullite-type 1ff
crystalline mullite 355

d

damage tolerance 403
decomposition 179ff, 240, 242
deformation 312
– mechanism 143
delamination 417, 428
descloizite 38, 40f
dielectric constant 323
dielectric loss tangent 324
differential thermal analysis (DTA) 263
– kaolinite 170
– metakaolin 170
diffractometry in situ single-crystal 191
diffuse scattering 49, 54ff
diffusion couples 231
diphasic gels 276f, 282, 300ff
diphasic mullite 420
diphasic precursor 94
diphasic sols 277
diphasic xerogels 299
– mullite crystalline seeds 300
– mullite nanocomposites 300
dissolution 179
doctor blade method 424
double angle spinning (DOR) 191, 200
duftite 41
dynamic stress response 149

e

easy-cleavage interphases 437
èechite 38
effect of reaction atmosphere 256
effect of structural defects 256
elastic constant 312
elastic modulus 142ff, 149, 312
electrical conductivity 165ff
electrical conductors 467
electrical insulators 467
electrical properties
– dielectric constant 323
electrical resistivity 169
electron diffraction 48f
– pattern 30
– studies 171
electron micrograph 179
electron paramagnetic resonance (EPR)
   spectroscopy 74, 189, 192, 206ff
– Cr-doped mullites 207
– Co-substituted mullite 209
– Fe-doped mullite 206
– Ti-doped mullite 209
– V-doped mullite 209
electronic chip substrates 463
electronic packaging 332

electronic substrates 466
electrophoretic impregnation 405, 411
enthalpy 149, 151
entropy 151
environmental barrier coatings (EBCs) 243
epidote 3,4
epitactical nucleation 179, 187
eptactical growth, mullite on muscovite 172
equiaxed grain structures 287
equilibrium phase diagram 229
ethylene glycol 269, 281
ethylenediaminetetraacetic acid (EDTA) 271
europium-doped mullites 93
eutectic melting 86
eutectic temperature 291
eveite 34f
exsolution 179
– of aluminum from mullite 184
extended X-ray absorption fine structure
   (EXAFS) 80

f

fatigue parameter 316
ferromagnetism 37
fiber coatings 436
fiber cracks 422
fiber debonding 422f
fiber fabric infiltration routes
– electrophoretic deposition 404
– polymer impregnation pyrolysis 404
fiber failure 390
fiber fracture 429
fiber pull-out 427, 436
fiber/matrix bonding 427
fiber/matrix debonding 436
fiber/matrix interfaces 425, 427
fiber/matrix interphase
– easy cleavage 437ff
– fugitive 438f
– low toughness 438
– porous 438
fiber-matrix contact 421
fiber-reinforced composites, non-oxide 423
fibers, creep behavior 382
filters, hot gas dust-trap 443
fireclay 327
fivefold coordination 19, 30, 33, 35, 39, 42f
fluorine salts 242f
foreign cation incorporation 70
– boron 91
– chromium 77
– cobalt 89
– europium 93
– gallium 91

– iron 83
– manganese 81
– molybdenum 93
– other 91
– sodium 91
– tin 92
– titanium 74
– vanadium 76
– zirconium 92
formation of mullite
– from andalusite 172ff
– from kaolinite 167ff
– from muscovite 170ff
– from sillimanite 175ff
– from X-sialon 178
Fourier-transform infrared (FTIR)
  spectroscopy 207, 209
fracture behavior in-plane 430
fracture toughness 400, 447
friction coefficient 316
fused-mullite 31f, 251, 260ff, 289ff, 327f

g
gallate 33
gas permeability 437
gas-phase-deposited Al_2O_3-rich mullite, the
  structure 357
gas-phase-deposited coatings
– alumina 352
– ceramic substrates 352
– chemical (CVD coatings) 350
– growth kinetics 352
– microstructure of 352
– mullite 352
– silicon carbide 352
– silicon nitride 352
– thermodynamics 350
gas-phase-deposited mullite
– high-temperature phase transformations
  359
– hot-corrosion 364
– oxidation 364
– recession protection 364
gas-transport reactions, whisker formation
  378
germanium mullite 32f, 40
– Al_2O_3-rich 40
Gibbs energy 150
glass forming ability 95f
glass phase 262, 383
glasy film, intergranular 290
gottlobite 41
grain boundary mass transport 286, 290, 312,
  314, 317ff

grain boundary phases 146
grain coarsening 438
grain growth 162ff
– inhibitor 164
– intergranular inclusions 164
grandidierite 4, 7, 19f, 33, 46, 206
– group 19

h
Hall-Petch correlation 381
Hall-Petch equation 309f
heat capacity c_p 144, 147, 151f, 154f, 186f,
  189
– high-temperature 189
heat exchangers 329, 442
heat of fusion 151
heat regenerator 331
high creep resistance 377
high thermal conductivity 423
high-alumina mullite 160
high-purity mullite 256
high-resolution electron microscopy (HREM)
  47ff
– diffuse scattering 49
– oxygen-vacancy distribution 48
– videographic real-structure simulations 54
high-resolution transmission electron micrograph (HRTEM) 302
high-temperature behavior 238
high-temperature engineering materials 331
high-temperature gas filters 336
hollow mullite microspheres 277
homogeneous precipitation method 274
hot gas filters 442
hot isostatic pressing (HIP) 290, 301
hot pressing 457
hot-corrosion barrier 367
hydralsite 282
hydrolysis 266
– incomplete 279
hydrothermal technique 72
hydrothermally produced mullite 281

i
illite 172
inclination angle ẅ 1ff
incongruent melting 227ff
infiltration
– colloid precursor 411
– metal melt 404
– sol-gel 404
– slurry 404, 417ff
– vacuum or pressure 414f
– vapor 403

– vibro 416
interstitial Cr^{3+} incorporation 77
interphases
– low-toughness 438
– porous or gap-producing 438
inter-vacancy correlation vectors 59, 62ff
intragranular contacts 428
ionic conductivity 159
– oxygen vacancy hopping 169
IR spectroscopy 2f, 16f, 33, 189, 192, 207, 210ff
– mullite formation 208
– spectrum 217
iron oxide 244, 254, 255

j
Jahn-Teller distortion 40, 42, 82, 90
Jahn-Teller effect 42

k
kanonaite 34
kaolinite 172, 251
– formation 169
– NMR spectroscopic studies 200
– thermal decomposition 171
kiln furnishings 442
Knudsen equation 309
kyanite 35, 151, 180, 252
– decomposition 179

l
laser flash technique 157
laser heating 391
layered composites 427
libethenite 34f
linear thermal expansion 322
liquid-phase sintering 296f, 302, 311
liquid-state-derived mullite 259
load-deflection curve 431
low temperature eutectics 229
low thermal expansions 463
low-expansion oxide phases
– cordierite 462
– silica glass 462
– tieillite 462

m
magic angle spinning (MAS) 191, 197, 204
magnesium oxide 254, 261
magnetic structure 37
manganese oxide 261
manganese-doped mullites 81
– ground state 82
– high-spin state 82

manganese-substituted mullites 90
MAS NMR spectroscopy 120
– ^{27}Al 269, 276
– materials for heat exchangers 331
– ^{29}Si 268, 276
matrix composites 397ff
– dense mullite 409
– dense oxide 409ff, 425f
– hot pressing 407
– non-oxide fiber-reinforced 407
– oxide fiber-reinforced 409
– silicon carbide fibers 407
matrix porosity 430
mechanical properties 141ff, 399, 421, 460
– bending strength 310
– corrosion resistance 350
– creep resistance 307f, 317, 327, 350
– elastic modulus 307, 312
– fatigue behavior 307, 316
– fracture toughness 307f, 310, 316f
– hardness 307, 314f, 334
– high-temperature strength 350
– mechanical strength 303, 307ff, 310, 332, 336
– microhardness 314ff
– mullite ceramics 307ff
– thermal shock resistance 307, 315
– toughness 141, 307ff, 350
– wear resistance 307, 316
melting behavior 236
melting point 236
metal-reinforced composites
– aluminum 463
– molybdenum 466
metastable crystallization 233
metastable immiscibility 101f
– aluminum silicate glasses 102
– immiscibility gap 103
– miscibility gap 103
– in SiO_2-Al_2O_3 101
metastable liquid immiscibility 283
microcomposites 306
microcrack toughening 450
microcracking 438, 455
microhardness 146f, 314
– amorphization 148
– temperature dependence 147ff
microwave furnace 417
mineralizer
– effects 254
– potassium carbonate 255
miscellaneous oxide particle-reinforced
    composites 461
miscellaneous particle reinforcement 461ff

miscellaneous properties 435
miscibility gap 28
molecular dynamics (MD) simulations 191
molten aluminum silicon alloys 410
molten sodium salts 241
molybdenum metal reinforcement 466
molybdenum/mullite composites 466
monolithic mullite 400, 456, 459
mosaic-type mullite 386
Mössbauer spectroscopy 190, 192, 203, 210ff
– ^{57}Fe 205
– Fe-containing mullites 213
– ferric iron in mullite 217
– of iron depend mullite 84
– of iron-rich mullite 84
– of 3:2 mullite 217
mottramite 38
mozartite 7, 40ff
– group 40
mullite
– 2/1-type 231
– aluminum-rich 187
– chromium-doped 168
– congruent melting 230, 231
– incongruent melting 230f, 234
– liquidus 232
– orthorhombic 29, 271
– oxygen vacancies 191
– peritectic temperature 232
– in porcelain bodies 377
– pseudo-tetragonal 28, 265, 268, 270ff, 277, 281, 286
– real structure 46
– silica-rich 187
– structure variants 60
– tetragonal 28
– vanadium-rich 77
3/2-mullite, solid-solution range 235
mullite beads 336
mullite ceramic fabrication techniques 288
– reaction bonding of mullite 288
– reaction sintering of Al_2O_3– and SiO_2-containing reactants 288
– reaction sintering of chemically produced mullite precursors 288
– sintering of mullite powder compacts 288
– transient viscous sintering of composite powders 288, 306
mullite ceramics
– applications 328
– damage tolerance 397
– dense 274
– hot pressed 311
– infrared-transparent 299

– optically translucent 167
– polycrystalline 146
– porous 328, 336
– processing 286
mullite coatings 283
– chemically vapor-deposited (CVD) 365
– deposition of 372
– miscellaneous techniques 372
– plasma-sprayed 367
mullite composites
– by tape casting 399
– continuous fiber-reinforced 403
– hot pressing 399
– silicon nitride 402
– whisker-reinforced 399, 401
mullite crystallization 104, 124, 453
– heterogeneous 125
– homogeneous 125
– kinetics 108
– mechanisms 108
– type I single-phase aluminosilicate glasses 125
mullite crystals
– growth 104, 128
– with high aspect ratios 303
– prismatic 291
mullite decomposition 237
mullite fiber mats 277
mullite fibers 377
– application of 393
– continuous melt-grown 391
– polycrystalline 144
– single-crystal 391
mullite formation
– diffusion-controlled 111
– interface-controlled 111
– kinetics 108
mullite glasses 96
– aluminium coordination 120
– single phase 94
mullite growth 109
mullite matrices
– porous 413
– nano-pores 412
mullite matrix composites 298, 307, 327f, 397ff
– continuous fiber-reinforced 398
– dense oxide fiber-reinforced 425
– fiber-reinforced 397
– metal-reinforced 398, 463
– platelet- and particle-reinforced 443
– porous 412, 433, 440
– whisker-reinforced 402
– zirconia-toughened 446

mullite matrix materials, whisker-reinforced
338
mullite needles 282
mullite nucleation 108f
– epitactical 111
– topotactical 111
mullite powders
– microwave heating 271
– sintering atmosphere 287
– sintering mechanisms 286
– ultrafine 270
mullite precursors 94, 117
– aluminum coordination 120
– aluminum sec-butylate 96, 108
– aluminum-isopropylate 96
– condensation 98
– dehydration 98
– differences 117
– microcomposites 105f
– nanocomposite 106
– preparation 282, 299
– produced by coprecipitation 301
– similarities 117
– single phase 94
– tetrahedral triclusters 120
– of zirconia 448
mullite properties
– creep 142, 162
– creep resistance 142, 162
– strength 143
– toughness 143
mullite refractories 240, 262
mullite seeds 449
mullite single crystals 230
– compressive strength 143
– elastic constants 144
mullite solid solution 20ff
mullite synthesis
– fused-mullite 251
– hydrothermal process 251
– sinter-mullite 251
– solid-state processes 251
– solution-sol-gel (SSG) processes 251
– vapor-state process 251
mullite-type family of crystal structures 1
mullite whiskers 240, 377
mullite-based refractory 328
mullite-glass composite 332f
mullite-mullite transformations
– compositional 180ff
– phase 183
– structural 186ff
mullite-type gels 93
mullite-type glasses 93

mullite-type structure family 1
mullitization 171
– activation energies 109
– diffusion distance 126
– hydrothermal treatment 270, 282
– microcomposites 109
– process 127
multicomponent systems 246
multiple quantum (MQMAS) NMR 191, 193,
196, 198, 201
multi-slice approximation 49
muscovite 172, 251
– dehydroxylation 172

n
Nabarro-Herring equation 318
nanocomposites 461f
nanocrystalline 355
nanocrystalline layer 356
nanoparticles 445
needle-like mullite 274
neodymium phosphate 412, 426
neutron diffraction data 24
Nextel 610 alumina 431
Nextel 720 aluminosilicate fibers 163, 386,
390, 411, 420
Nextel fibers 385ff
nickel-austinite 41
Nivity fibers 385
NMR spectroscopy 24, 28, 39, 46, 198ff
– ^{27}Al 191f, 195f, 198ff, 269
– ^{27}Al triple-quantum MAS NMR 195
– aluminosilicate mullite 194
– amorphous aluminosilicate gels 200
– ^{11}B 201f, 204f
– ^{71}Ga 199f
– gallium and germanium mullites studies
201
– ^{25}Mg 203ff
– ^{23}Na 191ff
– nutation studies 198
– ^{17}O 191, 195, 199f, 203
– octahedral and tetrahedral resonances
198
– powders 269f
– ^{29}Si 24, 191, 194, 198f, 202ff, 269
– T* tricluster sites 198
– ^{89}Y 203
non oxide ceramics 283, 307, 311
nucleation 103f, 108f, 111, 114, 120, 124,
128
– barrier 187
– period 294
nucleophilic reaction 266f

o

^{18}O diffusivity measurements 373
^{17}O MAS NMR spectrum 197
olivenite 7, 35f
– group 35
ominelite 20
optical materials 333f
optical properties 164f
– mullite single crystals 324
ordering scheme of vacancies 54, 68f
oxide fiber/mullite matrix composites,
    thermal protection systems 372
oxide fiber-reinforced composites 413, 425
oxide/oxide composites 393
oxygen conductivity 17f
oxygen diffusion 156ff
oxygen ion conductivity 336
oxygen ion conductor 336f
oxygen ion implantation 375
oxygen vacancies 17, 24f, 29ff, 33, 40, 46
– distribution 68f

p

packing density 418
particle suspensions 414
particle-reinforced composites 456
peritectic composition 229
peritectic reaction 233
perovskite, derivative structures of 9
phase diagram, Al_2O_3-SiO_2 229
phase equilibria 229
phase separation, nanometer-scale 187
phase-equilibrium diagrams
– alkaline earth oxide-Al_2O_3-SiO_2 246
– alkaline oxide-Al_2O_3-SiO_2 245
– iron oxide-Al_2O_3-SiO_2 245
– MnO-Al_2O_3-SiO_2 247
– ternary X-Al_2O_3-SiO_2 245
– TiO_2-Al_2O_3-SiO_2 247
phosphates 33
physical vapor deposition (PVD) 95, 117ff,
    350, 372, 436
plasma spraying 96, 373
– conventional 368
plastic deformation 150
platelet and particle reinforcement 441ff
P_2O_5 109
Poisson's ratio 312f, 315
polycrystalline fibers 393
polycrystalline mullite, creep 143
polymerization 266ff, 270, 281, 285, 306
pore agglomeration 438
porous composites 426
porous materials, filters 336

porous microstructures 430
porous mullite 324, 335ff
porous networks 420
porous oxide matrices 412, 426ff
porous weak interphase 438
potassium oxide 254
powder particle agglomerations 287
powerplant applications 440f
precursor layer, nanocrystalline 360f
precursors
– chemically homogeneous 279
- dehydration 98
pressure filtration 414
primary mullite 286
prismatic mullite 287, 291, 302f, 305
properties
– electrical 168
– infrared 167
– miscellaneous 167
– optical 167
protective coatings
– chemical vapor deposition (CVD) 350
– deposition from an aqueous slurry 350
– physical vapor deposition (PVD) 350
– plasma and flame spraying 350
– self oxidation 350
pseudo-tetragonal metric 28, 43
pulse electric current sintering (PECS) 467
pulsed-laser deposition (PLD) 373
pyrobelonite 38
pyrophyllite 251
– thermal decomposition 172

r

radial distribution function 100, 122f
radiant burners 398
Raman spectroscopy 207, 209
rapid solidification 96
reaction bonding 447
– of ceramics 288, 298
– technique 298
reaction couples 232
reaction sintering 448ff
– alumina 297
– alumina and silica 253, 288, 291
– silica 297
reaction-bonded mullite (RBM) systems 410
reactive metal penetration technique 463
reactive wetting 165
real structure 46ff
– mullite 143
reducing conditions 240
reducing environments 236
refractory bricks 258

refractory materials 327
resonance ultrasonic spectroscopy 312

s

satellite transition (SATRAS) NMR 192ff, 199
scanning electron micrographs
– cross-section 186
– glass/mullite system 186
schafarzikite 7, 13ff, 18
– group 13
seeding 449
sensors 398
shear modulus 312
shear strength 428
short-range order 68ff, 121, 125
^{29}Si NMR spectroscopy 24, 99, 107, 115,
  117f, 196, 269
– solid-state 268
β-sialon 463
signal transmision delay time 332
silica 448
– devitrification 361
silica glass binders 413
silica sources
– α-alumina 253
– alumina sources 253
– boehmite 253
– cristobalite 253
– diaspore 253
– fumed silica 253, 281
– fused silica 253
– gibbsite 253
– quartz 253
– silica glass 253
– silicic acid 253
silicon alkoxide 264, 266, 269ff, 275, 279,
  281, 285, 383
silicon aluminum ester 272
silicon carbide 283, 298, 307f, 311, 313ff,
  325f, 328, 334, 445, 447
– oxidation 366
silicon carbide matrix composites (SiC$_F$/SiC),
  thermal cycling conditions 368
silicon carbide nanoparticles 458
silicon carbide platelet reinforcement
– fabrication 456ff
– mechanical properties 459f
silicon diffusion 159ff
silicon nitride 283, 298, 307f, 311, 314ff, 317,
  323, 325f
silicon oxycarbide (SiOC), non-crystalline
  458
silicon tetrafluoride (SiF$_4$) 244
silicon tetrakis-isopropoxide 264

sillimanite 4, 7, 9, 21, 24f, 28ff, 35ff, 76, 141,
  148, 151f, 174, 180, 229, 252
– decomposition 182
– growth 179
– transformation to 3/2 mullite 178
sillimanite group 35
– compounds 37
simulation techniques 56
single phase 252, 269, 282
– gels 282, 382f
– precursor 94
single-crystal mullite, elastic properties 144
single-crystal X-ray diffraction (XRD), high-
  pressure 146f
sintering
– characteristics 286
– composite powders 288, 306
– liquid-phase 451
– liquid-state 305
– of mullite and zirconia 446
– powder compacts 288
– solid-state 286, 289, 302, 305
– spark plasma 452
– transient viscous 306
sinter-mullite 31, 251, 256ff, 289, 327f
small-angle X-ray scattering (SAXS) 97, 120
smectite 172
sodium aluminate, mullite-structured 202
sodium oxide 254, 261
sol precursors, hydro-thermally treated 276
sol-gel precursors 462
sol-gel process 95f
sol-gel route synthesis
– chromium 72
– iron 72
– vanadium 72
sol-gel-derived mullites 251, 262, 291
solid-state nuclear magnetic resonance
  (NMR) 190f
– double angle spinning (DOR) 191
– dynamic angle spinning (DAS) 191
– multiple quantum (MQ) techniques 191
solid-state reaction
– α-alumina particles 295, 306
– silica grains 295
sol-plus-sol process 275
solution/precipitation 112
solution-plus-sol process 272
solution-plus-solution process 263
– in situ hydrolysis 264
– monophasic gels 264
solution-sol-gel (SSG) approach, water-free
  269
solution-sol-gel (SSG) process 251, 262, 269

– sol plus sol 262, 275
– solution plus sol 262, 272
– solution plus solution 262f
solution-sol-gel-derived mullite 262
spacecraft applications 440
spinodal demixing 94
spin-off 442f
splat cooling 96
spodumene 463
spray pyrolysis 95f
spray pyrolysis approach
– atomizers 278
– ultrasonicators 278
spray pyrolyzed mullite 279, 291
sputtering 436
stability of mullite 229
stable crystallization 233
standardization method 4ff, 12f, 37
starch 271
static fatigue behavior 455
staurolite 252
steady-state creep rate 317
stiffness 432
strenght 141
stress-strain curves 435
stress-strain rate 320
structural materials 331
structural transformation 188
structure refinement of mullite 190
STRUCTURE TIDY 5
structure variants 60ff
structures rutile-typ 9
sub-critical crack growth 390
subsolidus liquid 292
superalloys 441
surface and pore burners 442

t
tangeite 41
tape casting 416
tensile strengths 391f
ternary phase diagram of $AlCl_3$-$SiCl_4$-CO_2-H_2
   353
ternary systems 243ff
tetracluster 29
tetraethyloxysilane (TEOS) 263, 383
– acid hydrolyzed 276
tetragonal zirconia 447, 452
tetragonal-like mullite 286
tetrahedral double chains 149
tetrahedral triclusters 120f, 124f, 187
tetramethoxysilane (TMOS) 264
thermal barrier coatings 468
thermal conductivity 155ff, 322, 435

thermal decomposition reactions
– montmorillonite 201
– muscovite mica 201
– pyrophyllite 201
thermal diffusivity 157, 322
thermal expansion 152ff, 434, 448
– behavior 154
– coefficients 154, 315
thermal properties 434
– enthalpy 151
– entropy 151
– Gibbs energy 151
– thermal conductivity 322
– thermal expansion 322
– thermochemical data 151
– whisker-reinforced mullite matrix
   composites 402
thermal properties of mullite ceramics 149ff
– enthalpy 321
– enthalpy of formation reaction 321
– entropy 321
– Gibbs energy 321
– Gibbs energy of formation 321
– heat capacity 321
– thermal conductivity 321f
– thermal expansion 321f
titanium oxide 109, 254, 262
topaz 252, 380
– thermal decomposition 380
topological material 335
topotactical transformation 175
toughening
– crack-deflection 444
– by modulus load transfer 444
– nanoparticle 445
– transformation 443ff
toughening mechanisms
– grain-boundary strengthening 454
– microcrack toughening 454
– strengthening due to crack bowing 454
– stress-induced toughening 454
– surface-compression toughening 454
– toughening due to crack deflection 454
transient viscous sintering (TVS) 105, 399
transition alumina 106f
transition metal, stability relations 72
transition metal incorporation 70, 89
– aluminum substitution 74
– chromium substitution 77
– cobalt substitution 89
– general remarks 89
– iron substitution 83
– manganese substitution 81
– titanium substitution 74

– vanadium substitution 76
transitional mullite 286
translucent mullite 335
transmission electron microscopy (TEM)
    150, 160, 171, 282, 306
– cross-sectional 360
– high-resolution 179
transparent gels 267
tribological materials 335
tricluster 24
triethanolamine 269, 281
trippkeite 14
turbine engines 367
turbostratic boron nitride (BN) 437
type I glasses
– co-precipitation 95
– melt-derived 96
– preparation 95
– sol-gel method 95
– spray hydrolysis 95
type I precursors 94f, 103, 117, 119, 265,
    285
– aluminum sec-butylate 97
– chemically derived 95
– co-precipitation 95
– dried gel stage 97
– gelation 97
– hydrolysis 97
– hydrolysis-polymerization 95
– preparation 95
– single-phase 118
– sol-gel method 95
– spray hydrolysis 95
– tetraethyloxysilane 95, 97
– wet (solution) stage 97
type II precursors 94, 105f, 276, 285
– diphasic 105, 118
– synthesis 105
type III precursors 114f, 117
– aluminum sec-butylate 115
– synthesis 115

v
vanadium-doped mullites 76
vapor-state-derived mullite 283
versiliaite 7, 18f
– group 17
vibro impregnation 414
videographic method 56ff
– fourier transformation 58
– supercell 58

videographic reconstructions, two-dimen-
    sional 62
videographic simulations, three-dimensional
    64
viscosity of the melt 392
viscous flow 304, 420, 433
viscous silica phase 421
volatile compounds 239
vuagnatite 40ff

w
water-gas shift reaction 352
werdingite 3f, 7, 33, 43ff
– group 46
wetting behavior 162ff
WHIPOX 417ff
whisker formation
– from gas-transport reactions 378
– from melts 377
– process 378
whisker reinforcement 398ff
– fabrication 399
– mechanical properties 399ff
– miscellaneous properties 403
– thermal properties 402

x
X-phase sialon, silicon aluminum oxynitride
    compounds 204
X-ray diffraction (XRD) 24, 54ff, 76, 263
– single-crystal 175
– situ high-temperature 157
X-ray single-crystal diffraction 47
X-sialon 4, 180
– formation from 180

y
Young's modulus 312ff, 431
yttria 410

z
zeolites 6
zirconia 447f
– cyclic infiltration 453
– teragonal 453
– yttria stabilized 449
zirconia particle-reinforced composites 446
zirconia particle reinforcement
– fabrication 445ff
– mechanical properties 453ff
– thermal properties 455